Explorations in the History of Machines and Mechanisms

HISTORY OF MECHANISM AND MACHINE SCIENCE

Volume 15

Series Editor

MARCO CECCARELLI

Aims and Scope of the Series

This book series aims to establish a well defined forum for Monographs and Proceedings on the History of Mechanism and Machine Science (MMS). The series publishes works that give an overview of the historical developments, from the earliest times up to and including the recent past, of MMS in all its technical aspects.

This technical approach is an essential characteristic of the series. By discussing technical details and formulations and even reformulating those in terms of modern formalisms the possibility is created not only to track the historical technical developments but also to use past experiences in technical teaching and research today. In order to do so, the emphasis must be on technical aspects rather than a purely historical focus, although the latter has its place too.

Furthermore, the series will consider the republication of out-of-print older works with English translation and comments.

The book series is intended to collect technical views on historical developments of the broad field of MMS in a unique frame that can be seen in its totality as an Encyclopaedia of the History of MMS but with the additional purpose of archiving and teaching the History of MMS. Therefore the book series is intended not only for researchers of the History of Engineering but also for professionals and students who are interested in obtaining a clear perspective of the past for their future technical works. The books will be written in general by engineers but not only for engineers.

Prospective authors and editors can contact the series editor, Professor M. Ceccarelli, about future publications within the series at:

LARM: Laboratory of Robotics and Mechatronics
DiMSAT – University of Cassino
Via Di Biasio 43, 03043 Cassino (Fr)
Italy
E-mail: ceccarelli@unicas.it

For other titles published in this series, go to
www.springer.com/series/7481

Teun Koetsier and Marco Ceccarelli (Eds.)

Explorations in the History of Machines and Mechanisms

Proceedings of HMM2012

Springer

Editors
Teun Koetsier
Department of Mathematics
Faculty of Science
VU University
Amsterdam
The Netherlands

Marco Ceccarelli
University of Cassino
Laboratory of Robotics and Mechatronics
LARM
Cassino
Italy

ISSN 1875-3442
e-ISSN 1875-3426
ISBN 978-94-007-4131-7
e-ISBN 978-94-007-4132-4
DOI 10.1007/978-94-007-4132-4
Springer Dordrecht Heidelberg New York London

Library of Congress Control Number: 2012931925

Printed on acid-free paper

Springer is part of Springer Science+Business Media (www.springer.com)

Preface

The organization of an international symposium on the History of Machines and Mechanisms (HMM) every four years is the main activity of the Permanent Commission (PC) for the History of Mechanism and Machine Science of IFToMM, the International Federation for the Promotion of Mechanism and Machine Science. The first two symposia, HMM2000 and HMM2004 were held at the University of Cassino in Cassino, Italy in the years 2000 and 2004. The third symposium, HMM2008, was held at the National Cheng Kung University in Tainan, Taiwan in 2008. The present volume contains the proceedings of HMM2012, the 4th International Symposium on the History of Machines and Mechanisms that was held at VU University in Amsterdam, The Netherlands, from May 7 until May 11, 2012.

The mission of IFToMM is to promote research and development in the field of machines and mechanisms by theoretical and experimental methods, along with their practical applications. The aim of the international symposia on HMM is to maintain an international forum for the exploration of the history of machines and mechanism. Although the emphasis is on the history of technical systems and their applications, the scope of the symposia is wide. Relevant topics are also the history of theories and design methods, biographies, the history of the institutions involved, the relations with other disciplines, the history of engineering education and the social and cultural aspects of machines.

History is not only full of exciting and entertaining stories. Historical investigations put our own present day activities in a wider perspective. They help us define who we are. Moreover history remains a source of ideas.

This book is meant for researchers, graduate students, engineers and all others with an interest in the history of machines and mechanisms. We believe that it can inspire and motivate them.

After the review process 40 papers by authors representing 20 different countries were accepted for publication in the proceedings of HMM2012. One glance at the table of contents is enough to see that we succeeded in bringing together an interesting group of people with a stimulating variation in subjects. We are very satisfied with this result and we thank the authors for their valuable contributions and for the efforts in submitting the final versions of the papers in time.

We would like to express our sincere gratitude to the members of the scientific committee: Gerard Alberts (The Netherlands), Hanfried Kerle (Germany), Alexander Golovin (Russia) , Carlos Lopez-Cajùn (Mexico), Jammi S. Rao (India), Lu Zhen (P. R. China), Hong Sen Yan (Taiwan), Thomas Chondros (Greece; Chair of IFToMM's PC for History), Baichun Zhang (P. R. China). Moreover, we would also like to thank the colleagues who helped us in the review process: Jean Pierre Merlet, I-Ming Chen, Emilio Bautista, Juan Ignacio Cuadrado, Teresa Zielinska, Klaus Mauersberger,

Agamenon Oliveira, Erwin Lovasz, Ton Klein Breteler, Burkhard Corves, Stephanos Paipetis, Just Herder, Vera Chinenova, John Gal, Pier Gabriele Molari, Junichi Takeno, Juan A. Carretero, Manfred Husty.

We also thank the sponsors of the symposium: IFToMM, the European Society for the History of Science, the Institute for History and Social Aspects of Science of VU University and the Department of Mathematics of the Faculty of Sciences of VU University. Moreover we are very grateful for the support we received from many friends and colleagues at the Faculty of Sciences: Saskia van Es, Cees van Gent, Dick Hoogendoorn, Hubertus Irth, Ger Koole, Frans van Lunteren, Ronald Meester, Andre Ran, Dirkje Schinkelshoek, Maryke Titawano. Without their support we would not have been able to organize HMM2012.

May 2012
Amsterdam

Teun Koetsier
Marco Ceccarelli

Introduction

Teun Koetsier

Erewhon

In 1859 a young Englishman, Samuel Butler (1835-1902), immigrated to New Zealand. New Zealand was not an industrialized country, but England was. Shortly after 1800 Britain had become the first industrialized nation in the world. Machine based industry had replaced agriculture as the most important economic activity. The Industrial Revolution had begun in the textile industry, but soon moved to other areas of industry. Steam engines had been introduced. The iron industry grew and with it the coal industry. By 1850 most of England's areas were connected by railroads and in coastal and river transportation steamships had taken over. Butler had witnessed it. In New Zealand in 1863 he wrote a letter with the title "Darwin among the machines". He wrote:

> *"Day by day, however, the machines are gaining ground upon us; day by day we are becoming more subservient to them; more men are daily bound down as slaves to tend them, more men are daily devoting the energies of their whole lives to the development of mechanical life. The upshot is simply a question of time, but that the time will come when the machines will hold the real supremacy over the world and its inhabitants is what no person of a truly philosophic mind can for a moment question. Our opinion is that war to the death should be instantly proclaimed against them. Every machine of every sort should be destroyed by the well-wisher of his species."* (See [7])

Butler's memories of Britain together with his impressions of New Zealand stimulated his fantasy. In 1874 he anonymously published the novel called *Erewhon*. The word 'Erewhon' is an anagram of 'nowhere'. Erewhon is a country, somewhere hidden behind the mountains in New Zealand, where the inhabitants used to use machines but not anymore. The machines have all been destroyed together with all books on mechanics and all engineer's workshops in a war, the war between the machinists and the anti-machinists. The anti-machinists got the victory. After his arrival in Erewhon the protagonist of the novel learns about all this in the City of the Colleges of Unreason. Some machines survived the war, because of the intervention of the professors of inconsistency and evasion. These machines are studied in Erewhon in the way we study long forgotten religious practices.

In this fascinating novel machines are compared to a new life form that is evolving rapidly and threatens to become superior to man. It is argued that there are no good reasons to suppose that they cannot have consciousness. They have started to do things that are very similar to eating and talking. Moreover, they can reproduce. Of course, their reproduction is different from the way mankind reproduces itself. However, in nature there are many kinds of reproduction. The author of the *Book of the Machines* foresaw that the machines would enslave man in the distant future. In Erewhon this had not happened, but a war had been necessary to prevent it.

Butler's book nicely illustrates how the industrial revolutions were perceived in the 19th century. They were seen as revolutions in which machines became omnipresent in society. *Erewhon* is the first novel in which this technological development is taken as a central theme. The book is satirical, but also critical of the ubiquity of machines. Many similar stories would follow. In 1909 Edward Morgan Forster published a story in *The Oxford and Cambridge Review* called "The Machine Stops". In 1920 the Russian novelist Yevgeny Zamyatin wrote the novel *We*. In 1921 Karel Čapek wrote the play *R. U. R. (Rossum's Universal Robots)*. In 1926 Fritz Lang made the film *Metropolis*. Aldous Huxley's *Brave New World* (1932) and George Orwell's *1984* from 1949 were written in a similar vein. For an extensive list see [6]. Although many of such novels and movies reflect a romantic longing for a distant past that is usually idealized and often does not have much in common with the real past, they are great works of art that draw our attention to negative aspects and possible dangers of technological progress.

The Escape from the Malthusian Trap

Of course, there is an entirely different way to look at the industrial revolutions. Instead of emphasizing possible downsides of the proliferation of machines one can also stress their positive effects. The *First Industrial Revolution* took place in the 18th century. In a sense it was the most striking development in the entire history of mankind. Economists sometimes say: The industrial revolutions made it possible for mankind to escape from the *Malthusian trap*. They mean this. Before the Industrial Revolution technology progressed – many examples of useful inventions can be given easily – but the standard of living of the average individual did not really change. Increases in the production of food were always followed by growth of the population leaving the average income per capita unchanged. Mankind was trapped. Yet the industrial revolutions changed it all. Mechanization in combination with the use of fossil fuels made such an increase of production possible that growth of the population could no longer neutralize it. Mankind had escaped from the trap.

Moreover, the First Industrial Revolution, dominated by textile machines, steam engines, iron, coal and the introduction of the factory system, was only the beginning. In the middle of the 19th century it was followed by a new wave of innovations: the *Second Industrial Revolution*. In this revolution railroads, steamships, steel, rubber and farm machines played the central role. The application of the steam engine to transportation had an enormous impact. Steam ships had been limited to coastal shipping because of the impossibility to bunker enough coal. The invention of much more efficient compound engines changed all this. Moreover in the iron industry finally one succeeded in producing steel cheaply. Some other industries emerged in addition to the railroads. The telegraph was invented by Samuel Morse and others. The electric dynamo made it possible to produce electricity. The transformer made long-distance transportation of electricity possible. Electricity could be used in industry. Charles Goodyear invented the process of vulcanization, which made rubber less sensitive to changes in temperature. It no longer became brittle in cold weather and sticky in hot weather. The petroleum industry got started and the chemical industry grew. Agriculture changed, because of new machines and fertilizers. In the industrializing nations production per capita rocketed.

With the First Industrial Revolution mankind seems to have entered an entirely new phase in its history. In the industrialized countries recurring waves of technological innovations pushed up production higher and higher. At the beginning of the 20th century the *Third Industrial Revolution* took place. It involved automobiles, airplanes, electricity, radio, petroleum, movies. The economy and daily life were transformed by the automobile and by the use of electricity. This is the phase in which science began to play an important role in technology. The impact on the way wars were fought was also dramatic. The cavalry was replaced by tanks and aircraft began to play a role. The automobile industry and the aircraft industry received an immense stimulus. In 1938 the United States 3.600 airplanes were built. This number had become 96.000 in 1944 ([5], p. 199). The next wave came in the second half of the 20th century when the *Fourth Industrial Revolution* took place: television, computers, plastics, the Internet and many of us wonder what the next industrial revolution will be like. Although Japan played an important role during the Fourth Industrial Revolution all four of them took place in the Western World. It is probable that in the *Fifth Industrial Revolution* Asian nations will dominate.

The escape from the Malthusian trap has considerably improved the lives of most people in the industrialized nations. And although as for the distribution of wealth worldwide there is much to be desired, technological progress is basically a good thing. This is fortunate because it happens to be an inevitable development. It is something we cannot escape from. It is the human way and if technology creates problems, we can only hope for good government and the development of more technology to solve them.

Machines, the Science of Machines and IFTOMM

In everyday language a machine is an artifact that can be used to assist us in the execution of a specific task. This is a very broad definition. It covers simple tools like a hammer or a can opener, but also an airplane and a computer are machines in this sense. The definition may even be too broad. Is a frying pan a machine, are knitting needles machines? According to the Dictionary [1] of IFToMM , the *International Federation for the Promotion of Mechanism and Machine Science*, a machine is a technical system "built to transform energy or matter or to transfer and transform movement and force in accordance with certain control information" (Translated from the German: "zum Umformen von Energie oder Stoff oder zum Weiterleiten und Umformen von Bewegung und Kraft gemäß zugeführten Steuerinformationen"). IFToMM's definition refers to traditional machines, the machines that dominated during the first two industrial revolutions and continued to play a central role during the revolutions that followed. In IFToMM mechanisms are "constrained system of bodies designed to convert motions of, and forces on, one or several bodies into motions of, and forces on, the remaining bodies". Often, mechanisms are parts of machines. It is not unusual that in a big machine many different mechanisms can be distinguished.

Although attempts to understand and design machines and mechanisms on a scientific basis go back to antiquity, scientific mechanical engineering as an established discipline of great practical importance must be associated with the Third Industrial Revolution. Scientific mechanical engineering was born when mechanical engineers discovered the value of a scientific approach. This started on the continent in Europe with sophisticated and often graphical methods that were used in the design of mechanisms. It took some time before these methods spread to the US. Once they had established themselves there, after World War II they were replaced with the introduction of electronic computers in the 1960s by analytical methods. This led to extremely fertile research programs in mechanism and machine science.

One might have the impression that only during the first and the second industrial revolutions machines played a central role. That impression is wrong. Many machines have become less visible. They have become smaller and less noisy. They have also become more autonomous, needing less people to control them. Moreover, we got very much used to them. Yet, in industry, in construction works, in transportation, in agriculture, in the army, in hospitals and at home, machines are everywhere. Although one does not immediately associate the new technologies in the second half of the 20th century with them, machines and mechanisms continued to play a central role in our interaction with nature. Machines were and are one of the pillars of modern scientific technology and they are, moreover, from an intellectual and engineering point of view challenging. So it is not strange that the Fourth Industrial Revolution was accompanied by considerable interest in the theory of machines and mechanisms. The birth of IFToMM as part of the Fourth Industrial Revolution wasn't accidental.

In 1969 engineers from behind the Iron Curtain, the Western World and some crucial non-aligned countries, got together and founded IFToMM. This is remarkable. IFToMM was founded in the middle of the Cold War. In the 1970s, I had the pleasure of meeting a vice president of a major American aircraft industry. This gentleman by the name of Barton Evans told me he had worked for the CIA in the 1950s and 1960s contacting American scientists who were going to be or had been in touch with colleagues from communist countries. He proudly showed me his copy of *Who is Who in CIA*, a booklet prepared in 1966 and published in 1969 by the Department of Disinformation in Czechoslovakia in order to make life for the CIA harder ([2] and [3]). The name of my CIA agent was in the booklet. Although Evans had not been involved he was sure that the foundation of IFToMM had been surrounded by considerable CIA and KGB interest and briefing and debriefing from all sides. Right from the start IFToMM was in every respect an international organization. The USSR, Bulgaria, the German Democratic Republic, Hungary, Poland and Rumania represented Eastern Europe. The West was represented by the USA, Australia, the German Federal Republic, Italy and the United Kingdom. India and Yugoslavia were the non-aligned countries. On the other hand, big parts of the World were not represented. Right now IFToMM has 48 member organizations. The Ibero-American community is represented and so is Asia. The organization has grown but, inevitably, it has also changed its focus. In the 1960s, for example, classical kinematics of mechanisms remained a core discipline. In the past 50 years fast computing, sophisticated software and new applications have changed the theory of machines and mechanisms. Mechanical engineering has always been a multidisciplinary activity, but the number of disciplines involved has only grown. Kinematics, dynamics and gearing are classical subjects but the computer has changed them. We now have, for example, computational kinematics, multibody-dynamics and tribology. Really new subjects are robotics, mechatronics and micromachines.

IFTOMM'S Permanent Commission for History

The history of Mechanism and Machine Science is part of the history of science and technology. The word *technology* was coined by the German Johann Beckmann (1739-1811). He used it for a description and classification of all the existing crafts and methods of manufacture. The 1971 edition of *Webster's Third International Dictionary* says that technology is "The science of the application of knowledge to practical purposes". Definitions and distinctions are useful. However, it is difficult to draw a sharp border line between practical problems and non-practical problems. Basically a practical problem is a problem that requires some action outside of the study or the laboratory. In this context it makes sense to distinguish *knowledge-how* from *knowledge-that. Knowledge-how* is related to functionality; it concerns what should be done to reach some goal. We know how to get somewhere, how to do something, sometimes without even knowing why the method works. *That is knowledge-how.* Technology is or concerns always *knowledge-how. Knowledge-that* is related to truth; we know that something is the case, nothing more, nothing less. It may be completely useless. Pure science is *knowledge-that.* It is the multidisciplinary character of Mechanism and Machine Science in combination with the fact that it encompasses both knowledge-that and knowledge-how, which has led to a situation in which the history of machines and mechanisms and the scientific theories related to them is not often studied in its own right. Obviously historians of science are interested in the history of machines, but usually only in so far as it concerns mathematics, physics or one of the other sciences. On the other hand, historians of technology tend to concentrate on what machines can do and discuss their economic, social and cultural impact. Although such work is very valuable, the focus is usually not on machines and mechanisms in its own right. That is where IFToMM's Permanent Commission (PC) for History is filling a gap. The commission was established in 1973 because of the strong support from the first IFToMM President, the impressive Ivan I. Artobolevskii from Russia. In the PC for History almost all nations participating in IFToMM are represented. In the course of time the PC has been chaired by Jack Phillips (Australia), Elisabeth Filemon (Hungary), Teun Koetsier (The Netherlands), Marco Ceccarelli (Italy), Hong Sen-Yan (Taiwan) and Hanfried Kerle (Germany). The present chair is Thomas Chondros (Greece).

Originally the PC primarily dealt with the institutional history of IFToMM but in the course of time it broadened its scope. Now, the activities of the commission cover all aspects of the history of machines and mechanisms and the theories dealing with them. I think this is a very good thing. Given the central role of machines and mechanisms in the development of technology their history indeed deserves special attention.

This Volume

The 38 chapters of this volume nicely illustrate the activities of IFToMM's PC for History. Most of the authors are engineers and not professional historians. Modern professional historians of science and technology put great emphasis upon understanding the actual historical development, while many of the authors in this volume deal with what is sometimes called "heritage": they look at the past as modern engineers. What happened in the past from an engineering point of view? is for them the central question. That in itself leads to very interesting results. It leads, for example, to the possibility of a modern analysis of an ancient machine or the application of modern design methodology to answer the question what kind of machine could have been built in the past.

It is interesting to view the contributions in this book against the background of the big history of mankind. In his extremely well written *Why the West Rules – For Now* [4] Stanford historian Ian Morris describes the history of mankind as a race between the East and the West. After the last Ice Age the so-called "Hilly Flanks" in an arc shaped area covering parts of Iran, Turkey, Syria and Israel was in an extremely favorable position. By 10.000 BC the possibility of the domestication of plants and animals had been discovered there. The cereals that would develop into wheat and barley grew there abundantly and the climate was suitable for agriculture. The West, defined broadly by Morris, took the lead. The East followed two thousand years later. In both the East and the West social development took place along similar lines. In the course of time big agricultural empires developed. A major step forward was the transition from kingdoms that functioned on the basis of relations between the members of the elite to so-called high-end states in which taxes were raised and professional bureaucrats and soldiers represented the state. In this respect the Roman Empire in the West can be compare to the empire of Qin Chi Huang Ti, the emperor whose grave in Xi'an is protected by the well known army of terracotta warriors. Rome started as a city in Italy and became a superpower in the 2nd century BC. Qin was a small Chinese state, but it conquered its last enemy in 221 BC. At the time the West went ahead as for social development. However, in the last centuries of its existence the Roman Empire entered a period of decline which created an opportunity for the East. Things were different in the East. In the 6th century the East took over the West. When Marco Polo visited China in the 13th century everything about China amazed him: in many respects he faced a superior civilization. This supremacy would last until the 18th century when in the West the First Industrial Revolution started. The West bounced back and in the 19th century it took over most of the world.

Yet, as we know, in the 20th century things started changing. The West still rules. For how long, however? Morris argues that in order to understand social development in the past and in the future geography is a crucial factor. The domestication of plants can only be discovered if the geographical circumstances are favorable. Morris explains the fact that the Industrial Revolution took place in the West and not in the East also in terms of geography: the West was favorably positioned in order to discover America, which had in many different ways an enormous impact, Together with other geographic factors like the availability of coal in England this led to the First Industrial Revolution.

The present volume on the history of machines and mechanisms consists of six sections. In Section I the emphasis is on the role of institutions. Section II is devoted to primarily biographical contributions. The Sections III, IV and V are on mechanical systems. Section VI is devoted to papers on concepts and theories. If we consider the book against the background of Morris' big history, most of the 38 chapters in this book are understandably clearly directly related to developments in the West. They concern the industrialized nations in the 19th and 20th centuries. There are, for example, contributions on the history of MMS in German, Russia, Spain, France Italy, Romania, Mexico, Serbia and England. Yet several contributions refer to the wider context. Kuo Hung-Hsiao and Hong Sen-Yan apply a modern design methodology to an ancient Chinese cross bow. Section II starts with a paper by Nam Moon-Hyon that nicely illustrates the technological potential of the East in the 15th century. He discusses the Korean Chief Royal Engineer Jang Yeong-sil, who designed a complex water clock.

Several chapters in the book are devoted to Greek and Roman culture. In section III Finlay McCourt studies mechanisms of movement in Heron of Alexandria's work. In this same section Junich Takeno and Yoshihiko Takeno attempt to unravel the mystery of the defense chain that once protected Constantinople. Section IV starts with a paper by Michael Wright on the Antikythera mechanism, an unbelievably complex planetarium from the second century BC. In the last section Giuseppe Boscarino compares the epistemological status of concepts in Aristotle and Archimedes. Such papers demonstrate unique aspects of Western culture. The East never developed anything comparable to Greek mathematics and astronomy. This is a significant point. Could the first industrial revolution have taken place in the East? Morris argues that in the 11th and 12th centuries such a revolution was brewing in Kaifeng, at the time a very big and prosperous city. Momentous changes took place there, in the textile industry, in coal and in iron works. Yet, the development didn't continue. In the 15th century huge Chinese Treasure Fleets explored the Indian Ocean, but they did not cross the Pacific; geographically China was positioned less favorably with respect to the Americas than Europe. At certain moments China consciously rejected Western culture. In section I Michela Cigola describes the role of the Jesuits in the spreading of mechanical knowledge from Europe to China in the 16th and 17th centuries. That is the period in which the East was still superior. In some respects Jesuit astronomical knowledge was clearly superior to Chinese astronomical knowledge, but the Chinese emperor felt he did not need Western science.

One of the crucial elements in Western culture that did not have its Chinese counterpart is the clockwork model of nature. Western philosophers like Descartes realized that it was not necessary to consider nature as a living organism. One could view it as a machine and understand its functioning from this extremely fertile point of view. Several chapters in this book are on the Western tradition of automata and clocks. The clockwork model of nature was a crucial element in the Scientific Revolution in the West and is clearly linked to the development of mathematics during and after the Renaissance. Western culture had at the time one great asset that the East did not have: the Greek scientific heritage. The question why the Industrial Revolution did take place in the West and not in China is not a simple one. Geographic factors were certainly important – Morris is right there - but it seems highly probable that essential elements in the Western cultural tradition, that go back to the Greeks and include a theoretic interest in machines, also played a role. At this point the work of Margaret Jacob must be mentioned. See [8]. She did show how the Scientific Revolution, which was influenced decisively by Greek culture, had in its turn a cultural impact that in a subtle way contributed to the First Industrial Revolution.

References

[1] The IFToMM Dictionary, http://www.iftomm.3me.tudelft.nl/
[2] Bittman, L.: The Deception Game. Ballantine Books, New York (1981)
[3] Julius, M.: Who is Who in CIA, Berlin (1969)
[4] Morris, I.: Why the West Rules ~ For Now. Profile Books (2010)
[5] Nolan, P., Lenski, G.: Human Societies, An introduction to macrosociology. Paradigm Publishers (2006)
[6] Daniel Chandler's website, http://www.aber.ac.uk/media/Documents/SF/chrono.html
[7] Butler, S.: Darwin among the machines, Digitally available at http://www.nzetc.org/tm/scholarly/tei-ButFir-t1-g1-t1-g1-t4-body.html
[8] Jacob, M.C.: The Cultural Meaning of the Scientific Revolution. Alfred Knopf, sold to McGraw-Hill, New York (1988)

Contents

Section I

History of Institutions

A Social Network Model for Innovation in Early Aviation History

Francis C. Moon

Sibley School of Mechanical and Aerospace Engineering
Cornell University, Ithaca, NY
fcm3@cornell.edu

Abstract. Using data from many historical studies of early aviation, we present evidence that early advances in aviation were the result of an evolving social network with more than eighty nodes and over three hundred links. The node-link distribution is shown to be similar to a scale-free, power law behavior, characteristic of modern social networks such as the World-Wide-Web. This study suggests that technical innovation and invention in the late nineteenth and early twentieth centuries flourished through complex social networks commonly assumed to be unique to the twenty-first century.

1 Introduction

Historians of science and technology have long made a narrative case for the evolution model for scientific and technical innovation. As early as 1929, the historian Abbott Usher wrote; "The more serious recent literature (c.1929) on the theory of invention recognizes the necessity of presenting invention as a cumulative synthesis of a relatively large number of individual items." More recently the aviation historian Anderson (1993) has presented narrative evidence that there were many developments in the science and technology of flying machines before the Wright brothers advances (1899-1904) and it would have been only a matter of one or two years before someone else would have created a controllable flying machine. Our goal in this paper has been to assess the role of social networks in invention and innovation in technology as well seek similarities between late 19th century historical networks and modern day social networks. This example is also a test case for exploiting the techniques of complex network theory for relatively small network node sets. (Barabasi 2003, Newman 2003, Watts 2003).

In this study we have culled a number of classic histories of aviation beginning with Chanute (1893), Gibbs-Smith (1954), Crouch (1981) and Anderson (2004), and have created an influence matrix establishing links between approximately eighty nodes spanning the period 1810, the time of George Cayley's writings on flying machines to 1910 the year of the great international aviation meet in Belmont Park in New York City.

For the reader not familiar with the history of aviation, there were a dozen major players in the evolution of flight technology, ten of whom are listed here;

T. Koetsier, M. Ceccarelli (Eds.): Explorations in the His. of Machines & Mech., HMMS 15, pp. 3–19.
springerlink.com

George Cayley: {1809-1850} Early writings on the topology of aircraft;
Gottlieb Daimler; {1875-1900} Developed Otto's four-cycle engine into a lightweight engine for flight.
Otto Lilienthal; {1898-1896} Experimented with monoplane and biplane gliders;
Samuel Langley: {1898-1903} Measured lift and drag and built powered models;
Octave Chanute; {1898-1906} Published review of the world's research as well as built classic biplane glider;
Orville and Wilbur Wright; {1899-1910} Built and flew the first controllable human carrying aircraft;
Alberto Santos-Dumont; {1900-1906} Built flyable aircraft in Europe;
Gabriel Voisin; {1903-1910} Designed and built aircraft for many aviators;
Louis Bleriot; {1905-1910} First to fly the English Channel;
Glenn Curtiss: {1904-1910] Developed lightweight engines and fast planes as well as built the first seaplanes.

These principals were linked by a historical network of more than seventy other individuals and institutions.

2 A Social Network of Early Aviation

'H*istorical innovation networks'* consist of a set of persons [inventors, scientists, artisans], organizations [e.g. universities, Royal Societies] and social events such as international expositions, each of which is called a *node*. This collection of nodes has inter-nodal connections called *links* (or edges). Links between person-nodes are created by friendship, letter exchange, writings, books, and financial interest, to mention a few. Links between person-nodes and organizations involve membership, professorial appointments, education of students etc. Connections between event nodes and person-nodes are created by attendance or by proceedings etc.

Each link is either directional or bidirectional. For example links between nodes in different generations are unidirectional. Thus Lilienthal could influence the Wright brothers but they did not influence Lilienthal because he died in 1896 three years before the brothers became interested in flight (1899). In social network theory, nodes can have attributes called *fitness,* which measures the importance of the node or link. For example the Wright brothers would be a high fitness node as would be Lilienthal. Unlike social networks such as the World Wide Web, or *Facebook*, a history network spans generations linking person-nodes living in one generation with dead person-nodes in an earlier generation. In a history network, nodes can be connected across generations. Historical Innovation/Invention Networks [HIN] also evolve as new nodes are added and new links are created between new players. In HIN, there can be *sub-networks* that can focus on a specific component; e.g. in aviation one can focus on lightweight engine evolution, or the wing design as well as the underlying aerodynamic theory development.

The difference between a pure narrative history and a history network theory, is that in the latter, the historian is obliged to determine if there are links between all of the nodes in the network. Thus if a new node (player) is mentioned, it's links to all the other nodes must be established. The analysis in this paper uses an influence matrix (Figure1), a network graph (Figure 2) and an event-node timeline graph (Figure 3b).

An overall picture of early aviation network is shown in Figure 2. For print legibility, only about 60% of the nodes are shown and about 50% of the links. Even at this level one can see a significant complexity in this network. Not shown here is the directedness of the links nor the fitness or weighted value of the nodes and links. Lists of person nodes, association nodes and university nodes are given in Appendices I, II, III respectively.

Node	Cayley	Lilienthal	Chanute	Wrights	Bleriot	#Links
Cayley	X	1	1	0	0	2
Lilienthal	0	X	1	1	1	3
Chanute	0	1	X	1	1	3
Wrights	0	0	1	X	1	2
Bleriot	0	0	0	1	X	1

Fig. 1. Sub-matrix of Influence Links; Rows are the source nodes and columns are the dependent nodes. The "link" column sums the number of influence links for a given source node.

The data in Figures 2 was obtained from an influence matrix M[NS, NT] where NS are the source nodes of the influence and NT are the terminus nodes of the directed influence links. Thus Cayley is an NS link for Chanute, while Chanute is a source node for the Wright brothers. The matrix M is 80 x 80 nodes and the entries are not weighted and contain only zeroes and ones. [See Figure 1.] In general node-pairs that were contemporary were assigned both outgoing and incoming links; it was assumed that each influenced the other. Earlier generation nodes connecting later nodes were assumed to have only outgoing links. A more detailed HIN would include weighted, as well as directed links. One could also scale the importance of the nodes with respect to the advance of the technology, however we did not attempt to assign a node scale in this study. An example of a sub-influence matrix is shown in figure 1.

The top of the graph in Figure 2 dates from 1810 with George Cayley as a node. The lower middle section around 1890 has Otto Lilienthal and Octave Chanute as major nodes and the lower part of the graph dating from 1900 to 1910 has nodes for Orville and Wilbur Wright in the US and Gabriel Voisin and Louis Bleriot from France. The reader can find a list of short biographies of most of these nodes in Longyard (1994).

Also shown in this graph are a few institutional nodes such as the Aeronautical Society of Great Britain [later the Royal Aeronautical Society] founded in 1866, the Technische Hochschule Berlin, where Lilienthal attended, Cornell University in upstate NY where a number of secondary but essential nodes where associated, and the Smithsonian Institution, where Langley was Director and from whom the Wright brothers obtained their early information. [See e.g. Mackersey, 2003.]

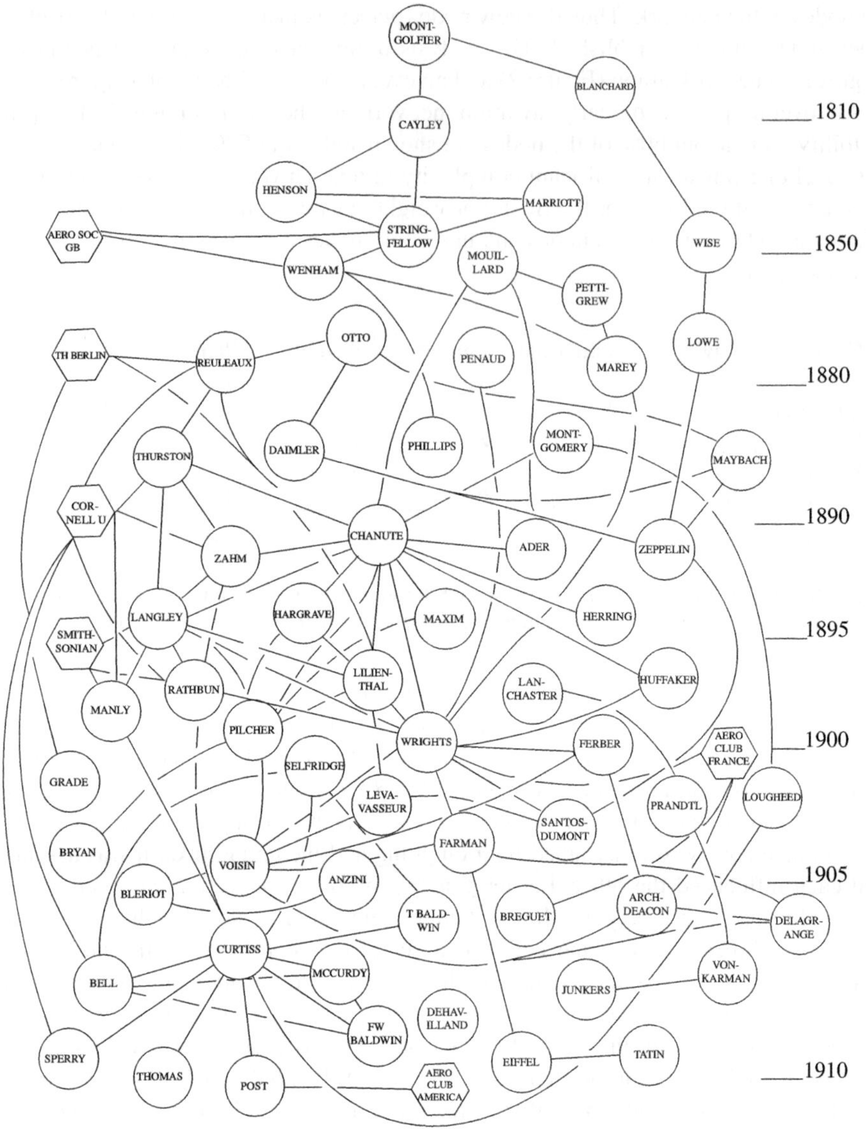

Fig. 2. Partial aviation network of innovator nodes and selected links, including four institutional nodes, for the period 1810-1910. Directional nature of the links is not shown. Most ballooning innovation nodes are not included.

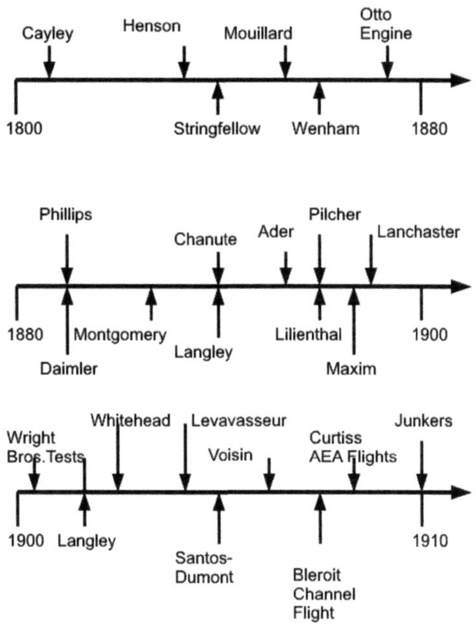

Fig. 3. a. Selected aviation events timeline, 1800-1910, showing increasing density with time.

This network includes only a few ballooning nodes. A 'lighter-than-air' machine network would show similar complexity to that for 'heavier-than-air' machine development. However a number of ballooning nodes are included in the network of Figure 2. Cayley, it was written, was inspired by the ballooning exploits of the Montgolfier Brothers. Also, the American balloonist T. Lowe mentored Zeppelin. Zeppelin later employed a Daimler engine in his dirigible that led to lightweight engines for 'heavier-than-air' machines. Similarly, the American balloonist, Thomas Baldwin used a Curtiss motorcycle engine in his dirigible that brought Curtiss to the attention of Alexander Bell.

Another measure of the progress of heavier-than-air development is a plot of the growth of event nodes with time, shown in Figures 3a,b below. The addition of new nodes shows the beginnings of exponential growth several years before the seminal year of 1903 in which the Wright brothers demonstrated a controllable flying machine.

One commentator on the dual slope result of Figure 3b, suggested that the low slope growth reflects the "passion years" of aviation and the steeper slope the emergence of the "profit years" through air meet prizes and aircraft sales.

One can also associate with many of the nodes the creation of actual flying machines. A plot of the growth of engine-powered flying machines in the period 1900-1915 is shown in Figure 4 using data from Jane (1917) *All the World's Flying Machines.* This illustrates the fact that, although the Wrights did not fly publicly until 1908 in Paris, the world's knowledge base in aviation science and technology was ripe by the turn of the century enabling the creation of dozens of aircraft in a half dozen nations.

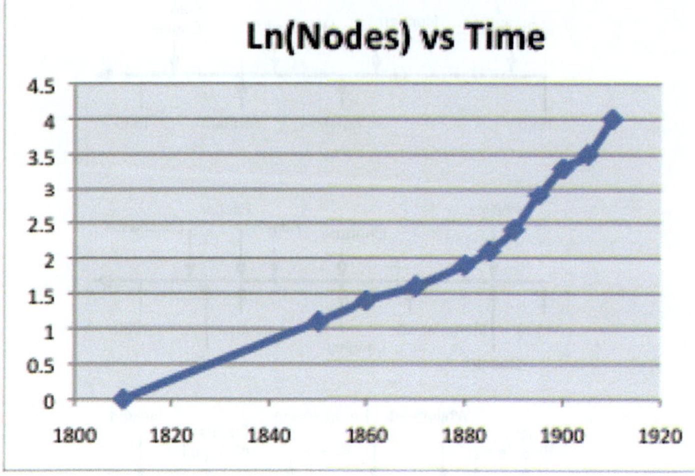

Fig. 3. b. The growth of aviation innovator nodes vs. time showing periods of different exponential growth.

Modern network analysts usually try to measure the degree of connectedness by plotting the distribution of nodes and links. [See e.g. Watts 2003, Barabasi 2003] Using a data base of 66 nodes and over 300 links, the distribution of number of nodes N(k) with 'k' links for our early aviation network is shown in Figure 5;

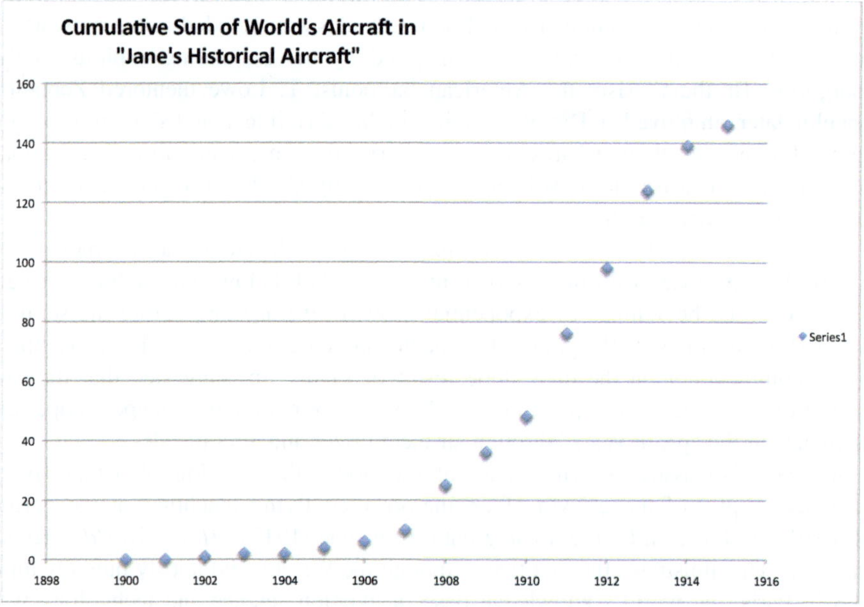

Fig. 4. Growth of flyable human carrying aircraft in the post Wright brothers era. From Jane (1917).

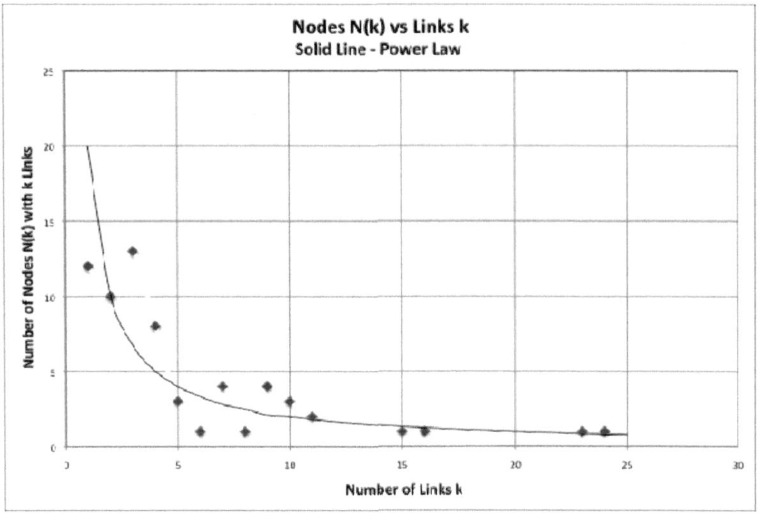

Fig. 5. Distribution of number of nodes N(k) with 'k' links for the aviation network in Figure 1. Solid Line; Power law fit N(k) = 20/k, for the data in Figure 5. Number of nodes = 66 and number of links = 320.

3 Discussion of Results

The nature of modern networks such as airports in the US or the World-Wide Web, has been characterized as scale-free networks [See e.g. Watts (2003), Barabasi (2003)] in which a small number of nodes have the majority of links and the majority of nodes have a small number of links. This behavior seems to be similar to the early aviation network of Figure 2. The two largest nodes are the Wright brothers and Octave Chanute with 24 and 23 links respectively. The Wrights influence is understandable. However, Chanute's role as one of the most important nodes in the aviation social network is largely ignored in popular histories except that of serious aviation historians. [At this writing the author could not find one biography written on Chanute] Chanute was a retired civil engineer who communicated with almost all the principals in early aviation, including hundreds of letters between him and the Wright brothers (MacFarland, 1953). Chanute also wrote definitive essays on flying machines in 1893, a decade before the Wright's successful flights.

A sub-network centered around Octave Chanute is shown in Figure 6. Here we include more detail but we omit links to other nodes outside Chanute's influence.

The Chanute sub-network in Figure 6 shows much greater detail but does not contain links outside Chanute's influence circle. The graph has a more 'star'-like shape. When this graph is sewn together with other sub-networks, such as the Wrights or the French aviators, the overall network has the complex qualities illustrated in Figure 2.

The two next largest nodes are Otto Lilienthal (1899) [who died in 1896 in a glider test] and Samuel Langley (1911) with 16 and 15 influence links respectively. Both innovators had published articles and books on flying machines that served to extend their influence. Although Langley's 1903 machine did not fly, he had performed careful experiments in the early 1890's and had succeeded in creating large unmanned engine-powered flying machines in the late 1890's that gave encouragement to the growing number of inventors and would-be aviators.

Further down the list in Appendix I are Lawrence Hargrave and Robert Thurston with 10 influence links. Although Hargrave worked in Australia, he belonged to a local scientific society and published more than a dozen papers on flying which he sent to others in Europe and North America including Robert Thurston of Cornell.

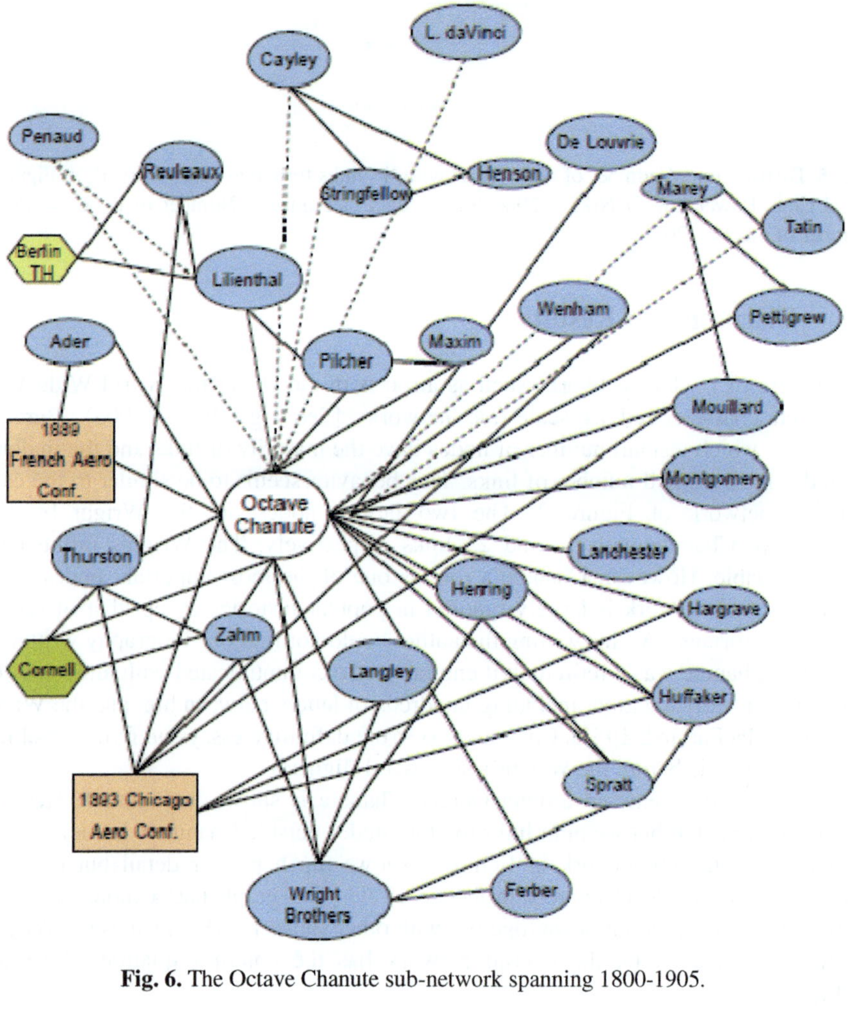

Fig. 6. The Octave Chanute sub-network spanning 1800-1905.

The Thurston node is an example of an enabling node – one that is not a principal contributor [he presented a paper on "materials of aeronautic engineering" at Chanute's 1893 Chicago Conference] but one that linked important players together. Thurston was an expert on the steam engine and thermodynamics and taught at the US Naval Academy and Stevens Tech before coming to Cornell in 1885. Thurston invited Octave Chanute to Cornell in 1890 to lecture on aerial navigation when Chanute was worried that he would be ridiculed for speculation about flying machines. Alfred Zahm (1911) was Thurston's student who attended these lectures and graduated from Cornell in 1892 and convinced Chanute to organize an international conference on aerial navigation in 1893 at the Columbian Exposition in Chicago. Around 1900, Thurston was asked by his friend Langley to recommend a Cornell engineer for the scale-up project of Langley's flying models and Charles Manly took the job. Thurston was a friend of Alexander Bell who was also a Cornell lecturer. Bell later organized the Aerial Experiment Association in 1907 out of which came Glenn Curtiss's 'June Bug' machine. (See e.g. Curtiss and Post, 1911, Hatch 1942) Thurston also had links to the Berlin nodes of Reuleaux and the engine nodes in Germany. Finally William Durand was a younger colleague of Thurston at Cornell who would later go to Stanford and start propeller research and head up the new National Advisory Committee on Aeronautics in 1914.

Association, Print Journals

In addition to individual human nodes, there were institutional nodes such as Cornell University and the Smithsonian Institution. Cornell was the site of Chanute's 1890 lectures on 'Aerial Navigation' that were subsequently published in Cornell's Sibley Journal of Engineering. Cornell's Charles Manly was the engineer and pilot for the Langley machine. Also Albert Francis Zahn, a Cornell grad of 1892, was instrumental in helping Chanute organize an international conference on aviation at the Columbian Exposition in Chicago in 1893. Elmer Sperry, who developed a gyro controller for later aircraft, also attended Cornell around this time. Lionel Marks another Cornell grad taught at MIT and wrote a textbook on aircraft engines.

There were other institutional nodes such as the Aeronautical Society of Great Britain founded in 1866 or the Aero Club de France and many of the Ivy Colleges had glider 'aero' clubs in the eastern US before WWI, including Cornell University. [NY Times, June 1, 1913] See Appendix II for a list of early aeronautical societies, as well as Figure 7. Some of these aeronautical associations were focused on ballooning as well as heavier that air flight. However as the list in Appendix II shows, there were hundreds of members in these aero clubs who helped spread the word that human flight was not only possible but was close to reality at the turn of the 20th century.

Other nodes, not shown but equally important, were journals such as *Scientific American*, the *American Engineer and Railroad Journal*, and *Means Aeronautical Journal* to mention a few in the United States and *L'Aerophile* in France. Another transient node class were international meetings such as one held in France in 1889 to which Chanute attended and the 1893 conference in Chicago that Zahm and Chanute

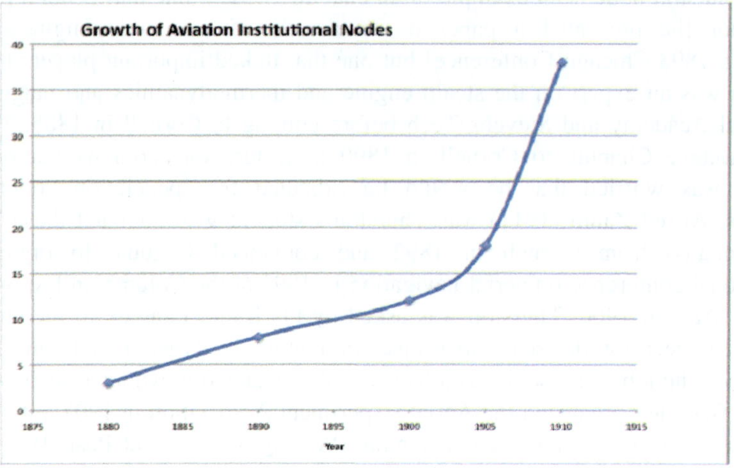

Fig. 7. Growth of Aviation Related Institutional Nodes [See e.g. Moebedeck 1906]

organized. [Third International Conference on Aerial Navigation] Chanute's linguistic ability in French and English also helped spread the news of US progress to Europe as well as interpret research from such obscure nodes as Louis-Pierre Mouillard in French controlled Cairo, Egypt.

The Gas Engine Sub-Network

Often forgotten in the history of flight is the development of the gasoline engine, which would certainly be labeled an *enabling technology* for heavier than air machines. A sub-network for gas-engine development would show similar complexity as in Figures 2, 6. Non-steam engines have origins in the work of the French Lenoir engine and the German Otto-Langen gas engine, both shown at the Paris Exposition in 1867. [Langen was a friend of Reuleaux in Berlin, who had connections to Cornell.] Otto set up a US subsidiary *The Otto Gas Engine Works* in Philadelphia in 1878 after exhibiting the engine at the 1876 Expo in Philadelphia. Daimler worked for Otto before setting up his own business with a subsidiary business in New York City. Later Daimler made a lightweight engine for Zeppelin.

By the time of the Wright brothers powered flight, the gas engine and the gasoline engine had entered the knowledge base of workshops all over the industrial world, making it difficult to trace the influence path from Otto to the Wrights. In *Scientific American* of 1900, one can find advertisements and articles about gasoline engines from Otto, Daimler, Charter, Webster, Olin and Olds, all in the United States to mention a few. In Buffalo there was the E.R. Thomas Co. making engines, automobiles and motorcycles. Curtiss bought engine castings from Thomas. The development of the internal combustion engine had reached beyond the 'tipping point' or avalanche stage of technology making it difficult to create an accurate influence network.

The Role of Aeronautical Science Nodes

In his seminal work *A History of Aerodynamics,* Anderson describes the lag between mathematically based fluid mechanics theory and practical advances in 'heavier-than-air" machines up to early 20th century. Theorists such as Joukowski, Kutta, Prandtl, developed a mathematical theory of lift, for example, after dozens of machines were already in the air. The history of science however often recognizes mathematical codification of theory rather than empirical science. For example, Langley, Lilienthal, Lanchester developed theories of flight in the 1890's. Langley and Lilienthal performed extensive experiments with wind tunnel facilities and published data on lift and drag before the Wright brothers' experiments. These men were not amateurs. Langley had training at Harvard in Astronomy, Lilienthal had training at TH Berlin under Franz Reuleaux, and Lanchester studied at the Royal College of Science in London. Yet books in fluid mechanics do not often refer to the Lilienthal theory of flight or the Lanchester vortex theory of wing-lift [Anderson is the exception]. The Wrights of course made extensive scientific experiments on lift, but they had the hindsight and examples of Langley and Lilienthal to follow. The aviation network also makes clear the connection between aeronautical research in the 19th C. and the biomechanics studies of bird flight such as Marey, Pettigrew and Mouillard. Mathematical theory took a back seat in the 19th century vis a vis aeronautics, even though giants such as Navier, Stokes, Reynolds, and Helmholtz were mathematically codifying fluid mechanics. Helmholtz is even quoted as pronouncing that heavier than air flying machines were not possible.

Other Historical Invention/Innovation Networks

Finally we suggest that similar historical invention networks were probably at work in the development of other technologies such as wireless and radio communication in the early 20th century. An event growth curve for early radio shown in Figure 8, shows a similar shape to that of early aviation in Figure 3b.

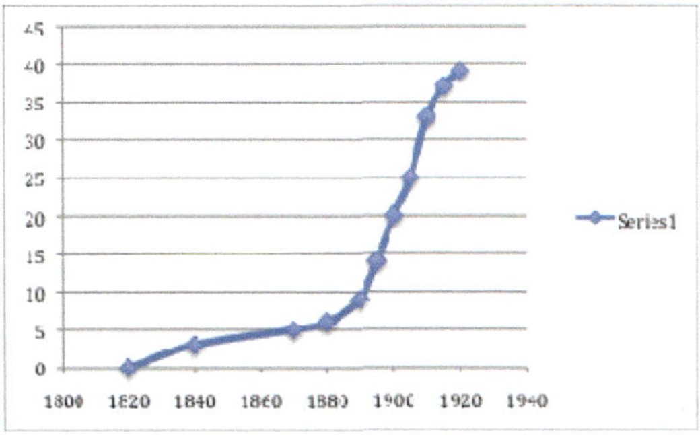

Fig. 8. Growth of wireless and early radio nodes, beginning with the work of Faraday and Maxwell and ending with the work of Marconi, Deforest and Armstrong.

The data in Figures 3b, 4, 7, 8, suggest that one might be able to find an innovation node growth model of the form $X(n+1) = F[X(n)]$, where X is the number of innovation nodes and 'n' is the year or unit time. Similar models, such as the "logistic equation" have been used to capture population dynamics in biology.

The Author has recently published a study of the development of kinematic mechanisms in the history of machines that shows a continuous network of artisans, artists and engineers from the age of Leonardo da Vinci to the late 19th century to the network surrounding Franz Reuleaux at the Berlin Technische Hochschule. [Moon 2007].

Often when historians write a history focusing on one inventor, the network described can be characterized as a 'star' network, with a central node and radiating branches to other nodes as in the Chanute sub-network in Figure 6. In order to create a network such as the one in Figure 2, one needs to read the histories of many inventors and connect the separate star networks with common links. Sometimes this is done in surveys of a technology as in the work of Gibbs-Smith (1954) or Hallion (2003). However the minor nodes and their links are not always easy to uncover. The links from Ernest Archdeacon, Chairman of the Aero Club de France for example, were important in connecting Chanute, Voisin, Beriot and many other French aviators. Also Alexander Bell was a link to a half dozen innovators including Glenn Curtiss. Another set of minor but important nodes are siblings and family members. The Wrights were not the only sibling pair, as shown in Appendix IV, suggesting that innovation is sometimes a family affair.

The aviation historian Anne Millbrooke (2006) suggested to the author that the methodology used in this paper is similar to *prosopography,* used in historical research. According to one definition, prosopography is the investigation of the common background characteristics of a group of person nodes in history by means of a collective study of their lives. This technique of analyzing history apparently has its roots in the late nineteenth century.

In this study there are many other minor and major nodes and links that are not included here but are listed in Chanute's book. However, further study will only add more nodes and links to the already complex patterns of Figures 2, 6. It can only serve to confirm the truth that human innovation in science and technology is the result of collective action and communication, rather than the product of lone 'geniuses' working in isolation. Inventors do not flourish on the *'shoulders of giants'* but are part of a vast creative network of other talented humans.

4 Summary

In his seminal history of the steam engine, Robert H. Thurston, (1878) the first President of the American Society of Mechanical Engineers, wrote; "*Great Inventions are never, and great discoveries are seldom, the work of any one mind.* Every Great invention is really the aggregation of minor inventions or the final step of a progression. It is not a creation, but a growth- as truly as is the trees of the forest." A century later, the historian George Basalla (1988) echoed a similar thesis that technology evolves through contributions of many human and institutional nodes. A more recent variant of this theme may be found in Bekun (2010).

In this short paper we have presented evidence for a social network theory of invention in early aviation primarily based on historical books and documents from the period 1890-1910. This model makes explicit what earlier historians have posited on the creation of heavier than air machines; namely that the seeds for human flight were laid in the mid to late 19th century and grew out of the contributions of many artisans, engineers, fluid scientists, industrialists as well as educational and professional institutions. Such models challenge the so-called "genius inventor" theory of innovation. The network model makes explicit the contributions of earlier studies of bird flight as well as the critical development of the internal combustion engines. This model also highlights the importance of communication and transportation technologies in the globalization of the effort to create heavier than air flying machines through publications and international meetings. The statistics of the node-links in the early aviation network appear to have anticipated those of modern 21st century networks. Finally there is preliminary evidence that other technologies such as wireless and early radio followed similar innovation patterns of exponential network growth.

Acknowledgments. The ideas for this paper developed in part from a history of science social network conference in Syracusa Italy, on the "Genius of Archimedes", recently published by Springer in 2010. The author wishes to thank Prof. Teun Koetsier of the Vrije Universiteit, Amsterdam for our lively discussion as well as for his comments on an earlier version of this paper. The author also acknowledges the help of Cornell University Rare and Manuscript Division, Elaine Engst, University Archivist, and Steve Strogatz of Cornell for references on modern social network theory. The author also thanks aviation historian Anne Millbrooke for suggestions and David Ollis of North Carolina State University for his editing comments.

Selected References

[1] Abbot, C.G.: Great Inventions, Smithsonian Scientific Series, vol. 12, pp. 225–239. The Airplane (1932)

[2] Anderson Jr., J.D.: Inventing Flight: The Wright Brothers and Their Predecessors. John Hopkins Press, Baltimore (2004)

[3] Anderson Jr., J.D.: A History of Aeronautics and it Impact on Flying Machines. Cambridge University Press, Cambridge (1997)

[4] Barabasi, A.-L.: Linked. Plume-Penguin Publ., Cambridge (2003)

[5] Basalla, G.: The Evolution of Technology (1988)

[6] Berkun, S.: The Myth of Innovation. O'Reilly Media, CA (2010)

[7] Cayley, G.: On Aerial Navigation. Nicholson's Journal November 1809, February 1810, March 1810 (1809-1810)

[8] Chanute, O.: Progress in Flying Machines. Reprinted by Dover Publ. (1997) (1896)

[9] Crouch, T.: A Dream of Wings. Norton, NY (1981)

[10] Curtiss, G.H., Post, A.: The Curtiss Aviation Book. Frederick A. Stokes Co., NY (1912)

[11] Gibbs–Smith, C.H.: A History of Flying. Frederick A. Praeger, NY (1954)

[12] Graham-White, C., Harper, H.: The Aeroplane, Past, Present and Future. J.B. Lippincott Co., Philadelphia (1911)
[13] Hallion, R.P.: Taking Flight. Oxford Press, NY (2003)
[14] Hatch, A.: Glenn Curtiss: Pioneer of Aviation. The Lyon's Press, Guilford (1942)
[15] Herring-Curtiss Archive. Cornell University
[16] Jane, F.: Jane's Historical Aircraft, Facsimile of 1917 Printing (1917)
[17] Lanchester, F.W.: Aerodynamics. A. Constable & Co., London (1907)
[18] Langley, S.: In: Manly, C. (ed.) Langley Memoir on Mechanical Flight. Smithsonian Institution, Wash., DC (1911)
[19] Longyard, W.H.: Who's Who in Aviation History: 500 Biographies. Presidio Press, CA (1994)
[20] Lougheed, V.: Vehicles of the Air. The Reilly and Britton Co., Chicago (1909)
[21] Lilienthal, O.: Bird Flight as the Basis of Aviation, Reprinted 2001, in translation from the French, by American Aeronautical Archives (1889)
[22] MacFarland, M.W.: The Papers of Wilbur and Orville Wright: Including the Chanute – Wright Letters and Other Papers of Octave Chanute, vol. 1,2. McGraw-Hill Book Co., NY (1953)
[23] Mackersey, I.: The Wright Brothers. Oxford Press, NY (2003)
[24] Magoun, F.A., Hodgins, E.: A History of Aircraft. Whittlesey House, McGraw-Hill Book Co., NY (1931)
[25] Maxim, H.: Artificial and Natural Flight. Whittaker & Co., London (1908)
[26] Means, J.: The Aeronautical Annual. W. B. Clarke Publ., Boston (1896)
[27] Milbank Jr., J.: The First Century of Flight in America. Princeton University Press, Princeton (1943)
[28] Millbrooke, A.: Aviation History. Jeppesen Press, Englewood Colo (2006)
[29] Moebedeck: Pocket Book of Aeronautics (1906)
[30] Moon, F.C.: The Machines of Leonardo da Vinci and Franz Reuleaux. Springer, Heidelberg (2007)
[31] Mouillard, L.: L'Empire de l'air [The Empire of the Air] (1881)
[32] Newman, M.E.J.: The structure and function of complex networks. SIAM Review (2003)
[33] Sibley J. Engineering (1993-1903),The Crank (1886-1892). Sibley College, Cornell University
[34] Thurston, R.H.: A History of the Growth of the Steam Engine. D. Appleton and Co., NY (1878)
[35] Von Karman, T.: Aerodynamics: Selected Topics in the Light of Their Historical Development. Cornell University Press, Ithaca (1954)
[36] Watts, D.J., Strogatz, S.H.: Collective dynamics of small world networks. Nature 393, 440–442 (1998)
[37] Watts, D.J.: Six Degrees. Norton, NY (2003)
[38] Zahm, A.F.: Aerial Navigation. D. Appleton and Co., NY (1911)

Appendix I: Human Nodes 1800-1910

Clemant Ader [1841-1925]; Alessandro Anzini [1877-1956]; Ernest Archdeacon [1863-1957]; William Avery; Thomas Baldwin [1860-1923]; Bedell; Alexander Bell [1847-1922]; Karl Benz [1844-1929]; Louis Bleriot [1872-1936]; Louis Breguet [1880-1955]; George Bryan [1864-1928]; George Cayley [1773-1857]; Octave Chanute [1832-1910]; Glenn Curtiss [1878-1930]; Gottlieb Daimler; Geoffrey deHavilland [1882-1965]; Leon Delagrange [1873-1910]; William Durand; Jacob Ellehammer [1871-1946]; Henry Farman [1874-1958]; Ferdinand Ferber [1862-1909]; Hans Grade [1879-1946]; Claude Grahame-White [1879-1959]; Lawrence Hargrave [1850-1915]; Helmholtz; William Henson [1812-1888]; Edward Huffaker [1856-1936]; Augustus Herring [1867-1927];J. Frederick Jane [1865-1916]; Nikolai Joukowski [1847-1921]; Hugo Junkers [18559-1935]; Martin Kutta [1867-1944]; Frank Lahm [1877-1963]; Frederick Lanchester [1868-1946]; Eugen Langen [1833-1895]; Samuel Langley [1834-1906]; Lenoir; Leonardo da Vinci [1452-1519]; Leon Levavasseur [1863-1922]; Otto Lilienthal [1848-1896]; Thaddeus Lowe [1832-1913]; Charles Manly [1876-1927]; Etienne Marey [1830-1904]; Marks; Marriott; Martin; Hiram Maxim [1840-1916]; Wilhelm Maybach [1846-1929];James Means; Joseph & Jacques de Montgolfier [1740-1810, 1745-1799]; John Montgomery [1858-1911]; Louis P. Mouillard [1834-1897]; Alexander Mozhaiski [1825-1890]; Nikolas Otto [1832-1891]; Alphonse Penaud [1850-1880]; James Pettigrew [1832-1908]; Horatio Phillips [1845-1926]; Percy Pilcher [1867-1899]; Ludwig Prandtl [1875-1953]; Franz Reuleaux [1829-1905]; Ferdinand Redtenbacher; Charles Rolls [1877-1910]; Alberto Santos-Dumont [1873-1932]; Thomas Selfridge [1882-1908]; Elmer Sperry [1860-1930]; Spratt; John Stringfellow [1799-1883]; Victor Tatin [1843-1913]; Edwin Thomas; Robert Thurston [1839-1903]; Gabriel Voisin [1880-1973]; Theodore von Karman [1881-1963]; Orville & Wilbur Wright [1871-1948, 1867-1912]; Francis Wenham [1824-1908]; Gustav Whitehead [1874-1927]; Albert Zahm [1862-1954]; Ferdinand von Zeppelin [1838-1917];

Source Nodes vs Unidirectional Links

Wright Bros.	k = 24
Octave Chanute	k = 23
Samuel Langley	k = 15
Otto Lilienthal	k = 16
Glenn Curtiss	k = 11
George Cayley	k = 10
Franz Reuleaux	k = 10
Robert Thurston	k = 10
Alexander Bell	k = 9
Lawrence Hargrave	k = 9
John Montgomery	k = 9
Percy Pilcher	k = 9
John Stringfellow	k = 8

Appendix II: Aeronautical Societies Nodes

1866 The Aeronautical Society of Great Britain [120 members in 1906]
1872 Societe Francais de Navigation Aerienne [103 members in 1906]
1880 Russian Aeronautical Society
1881 Berliner Verein Fur Luftschiffahrt [807 members in 1906]
1887 Flugtechnischer Verein in Wein [90 members in 1906]
1889 Munchener Verein fur Luftschiffahrt [383 members in 1906]
1886 The International Commission for Scientific Aeronautics [Paris]
1897 Aeronautical Club de France [500 members in 1906]
1898 Aero-Club, Societe d'encouragement a la locomotion aerienne
1900 The Permanent International Aeronautical Committee [33 elected members; included O. Chanute]
1900 Svenska Aeronautiska Sallskapet [80 members in 1906]
1901 Aero-Club de Belgique [300 members in 1906]
1901 Schweizerischer Aero-Club [140 members in 1906]
1902 Deutscher Luftschiffahrt-Verband
1902 The Aero-Club of the United Kingdom- A society for the encouragement of aerial locomotion [700 members in 1906]
1904 Societa Aeronautica Italiana [182 members in 1906]
1905 Federation Aeronautique Internationale [Paris]
1905 The Aero-Club of America [275 members in 1906]
1905 El Real Aero-Club de Espana [105 members in 1906]
1910 Intercollegiate Soaring Clubs [USA]

Source: Moedebeck, (1906) *The Pocket Book of Aeronautics* pp 436-444.

Appendix III

University Nodes in the Early Aviation Network

Technische Hochschule Berlin [Lilienthal, Junkers, Spade]
Cornell University [Chanute Lectures, Zahm, Manly, Sperry, Grumman]
Santa Clara College CA [Montgomery]
Royal College of Science (Now Imperial College) [Lanchester]
Notre Dame University [Zahm]
Catholic University of America [Zahm wind tunnel]
Gottingen University [Prandtl wind tunnel]
Allegheny College [Langley]
MIT [Lionel Mark, Aircraft Engines]
Stanford University [William Durand, propeller lab]
Columbia University [first aeronautical engineering degree]
US Army Schools of Military Aeronautics [Cornell, Princeton, Ohio State, Texas]
Intercollegiate Soaring Association [Cornell, Yale, Princeton, Columbia, c. 1910]

US Naval Academy[Langley, Thurston]
Moscow University [Joukowski]
Munich University [Kutta]

Appendix IV: Sibling Nodes

Breguet, Louis and Jacques
Farman, Marurice A. and Henry
Lanchester, Frederick W. and George
Lilienthal, Otto and Gustavus
Lougheed, Alan and Malcolm
Moisant, John, Alfred, and Matilde
De Montgolfier, Joseph-Michel and Jacques-Etienne
Percy and Ella Pilcher
Stinson, Majorie, Katherine and Eddie
Voisin, Gabriel and Charles
Wright, Orville and Wilbur

Source: Longyard, W.H. (1994) *Who's Who in Aviation History: 500 Biographies.*

Franz Reuleaux and the Rhenish Roots of Mechanism Theory

Reinhard Braune[1] and Burkhard Corves[2]

[1] Leibniz University Hannover, Germany
[2] RWTH Aachen University, Germany

Abstract. In 2009, the Department of Mechanism Theory and Dynamics of Machines of Aachen University (RWTH) celebrated the 50th anniversary of its foundation. Among the few higher education institutions still active in mechanisms theory in Germany, it is one of the older establishments, but compared with Dresden, for example, it is still relatively young. In a historical retrospective, this article examines whether, through the personal ties of FRANZ REULEAUX to his Rhenish homeland, there are even older relationships between his Berlin sphere of activity and the still young Aachen University. The result is very likely doubly interesting not only for the young, present day members of the department but also for foreigners to RWTH Aachen University because it sheds an interesting light on an important development phase of mechanism theory and also on the other hand explains the development of RWTH Aachen University during the early years of scientific consideration of mechanical engineering.

Fig. 1. Franz Reulaux (1829-1905)
Newspaper print on 70th birthday

1 The Aachen Industrial Region at the Beginning of the 19th Century

At the beginning of the 19th century, Aachen and its surroundings formed what was probably the most highly developed early industrial region in the German speaking world. As in the industrial mother country England, here too a modern industry, with its main characteristical use of technology and factory operation, developed starting from the mechanisation of the existing numerous, still water-powered cloth mills. In

T. Koetsier, M. Ceccarelli (Eds.): Explorations in the His. of Machines & Mech., HMMS 15, pp. 21–38.
springerlink.com

addition there was mining and metalworking in an economic region, at the time still unified, which is today called Euregio and extends over parts of Belgium, Germany and the Netherlands, between Jülich, Düren, Stolberg, Monschau, Verviers and Eupen reaching as far as Liège and Maastricht. For example, already in 1794, one of the first (still English) steam engines was started up on German soil in Eschweiler as pump drive for pit dewatering in a pumphouse that still exists today. From this came the name of "Eschweiler-Pump" for the district in which FRANZ REULEAUX was born in 1829.

During the Napoleonic occupation until 1814, Rhineland had a modern, liberal administrative policy with freedom of trade, which remained even under the subsequent Prussian administration of the "Rhine Province" and wasn't introduced for the whole of Prussia until 1845. This facilitated the creation of confident, middle class entrepreneurship in the West of Germany, while in the East the old titled rural junkers still called the tune. In addition to this, there was the "locational advantage" of longstanding economic and highly personal relationships with the Walloon region which, after England, quickly developed as the first industrial model region on the continent and became nothing short of *"a gateway to progressive technologies"* in the Aachen area. The Prussian state government promoted this development in many ways, e.g. by giving up-and-coming companies modern machines modelled on the English example, however, with the provison that they make them accessible to other companies to examine and assess [1]. The formation of the subsequent RWTH Aachen University in 1870 in Aachen was also one of these regional "structural development" acts.

The special economic situation of the Aachen area at that time, with its east-west technology transfer typical of the region, was also reflected in the REULEAUX family in Eschweiler. The family's ancestors had emigrated from the area around Lüttich in the 18th century. As so-called "Kunstmeister" – that is to say, specialists for the operation of pumping installations in mines – in Eschweiler they were highly welcome specialists for the coal mining at great depths. The REULEAUX family name can also be traced back to this vocational origin, in which the stems of the French words "relever = lift" and "eaux = waters" are found. According to T. Becker *"... from this it was possible to construct the (substantive) "Relève-eaux = the water lifter", analogous to the actually existing word "Relève-rail = the rail-lifting jack"* [2].

The entry into industrialisation and entrepreneurship was made by the father of FRANZ REULEAUX, who in 1818/19 founded the company "Englerth, Reuleaux & Dobbs" with a German and an English co-owner. The result was one of the first engineering works in Rhineland, in which one of the first steam engines in Germany was built, and which still exists today in Eschweiler, as "ESW-Röhrenwerke GmbH".

2 REULEAUX' Life in His Time

Franz REULEAUX was born into this sphere of successful factory owners during the time of early industrialisation on September 30th, 1829. He was therefore born with a certain relationship to practical application of technology, but not necessarily towards a path of a more theoretically minded technical scientist with worldwide charisma difficult to comprehend today. Before things quietened down towards the end of his extremely rich life, characterised by diverse commitment – he died almost at the age of 76 on 20/8/1905 in Berlin – in general terms he was "the" personality of his century in the young field of engineering sciences. Apart from his main fields, which we would now call "theory of engineering design" and "mechanism theory", for decades he was deemed by all to be a renowned and recognised "institution" in all fields of engineering.

This is exemplarily substantiated in a sentence found by the authors purely by chance in the inaugural speech of Prof. WÜLLNER as Rector of the RWTH Aachen in 1883, in which he discusses the enormous advances made in the practical application of the then still young electrical engineering and without further reference to whom he was talking about, that "... *REULEAUX in a speech says that in engineering we are at the boundary between two ages, the age of steam is beginning to become the age of electricity*" [3]. WÜLLNER was therefore apparently able to assume that all those attending to this academic event in Aachen knew who he was quoting and that REULEAUX in Berlin was naturally a weighty "principal witness" for such a general statement. Only a charismatic personality and outstanding intelligence, as well as extreme hard working effort and definitely a certain self-assertiveness in the then still very conceited, often decidedly "peacock-like" behaviour of most university lecturers amongst each other, could have led to such recognition in a university career.

Having moved from Eschweiler to Koblenz on the confluence of the rivers Mosel and Rhine – due to the early death of his father – REULEAUX spent only 9 years at school (graduating from a "higher vocational school") and then had 3-years of practical training as a "machine builder“ in an engineering works and iron foundry, before in 1846, still only 17 years old, he started 4 years as a " trainee", again in Eschweiler, in the former company of his father, which was now run by an uncle, where he initially worked in the design department and then spent most of his time outside the factory premises on the company's installation sites.

Then, from 1850, at the age of 21 the scientific life of FRANZ REULEAUX began, starting at one of the very top establishments, namely in Karlsruhe as a student of FERDINAND REDTENBACHER (*1809,+1863), who having worked in Vienna and Zurich, at the time taught at the "polytechnic school" in Karlsruhe now Karlsruhe Institute of Technology. Founded in 1825, as one of the first technical universities in the German speaking worlds, it was based on the model of the highly renowned "École Polytechnique" in Paris and REDTENBACHER, later called the founder of scientific mechanical engineering [4], must have been an impressive role model for the young REULEAUX and many other, later famous engineers, who passed through the Karlsruhe school.

This institution and the 8 others, all founded roughly at the same time or still in the 19th century, the so-called polytechnic schools, which later became the classic German technical universities (Aachen, Berlin, Braunschweig, Darmstadt, Dresden, Hannover, Munich, Stuttgart, i.e. today's so-called "TU9") soon detached themselves from the French model and taught in their very own teaching style. While in England, the origin and heartland of industry, for many years mechanical engineers still mainly opted for practical experience in machine development and in France mechanical engineering tended to be considered a "nice" example field by the mathematicians and physicists in the universities, REDTENBACHER and after him REULEAUX and others tried to achieve a symbiosis of natural sciences and practical engineering in the form of an independent scientific discipline, an approach which in today's education at technical universities seems so natural, that it is difficult to imagine anything else.

After just about 2 years in Karlsruhe, REULEAUX had already learned all that he could learn about mechanical engineering at universities of the time, but he supplemented his academic training with a further 2 years in Berlin and Bonn where he studied the humanities and natural sciences. Then, in 1854, at the age of 25 he was as it were "ready" and became the head of the design department of the machine factory Emil Bährens & Co in Cologne, which produced steam boilers but also steam engines.

Already in his first book "Construktionslehre für den Maschinenbau" (Design Theory for Mechanical Engineering) written together with his university friend C.L. MOLL and published in 1854 REULEAUX dealt with the relationship between technology and culture as BRAUN points out in [5]. In the last chapter of the work REULEAUX even made considerations about the "Maschinenbaustil" (Mechanical Engineering Style) in a last chapter headed "Beitrag zur Begründung einer Formenlehre für den Maschinenbau" (Contribution to the establishment of an accidence for engineering", where he distinguished between "basic form", "purpose form" and "relational form".

This book quickly found great recognition due to its clear layout and exemplary drawings, but also earned him the reproach that he had "unfairly" used quite a lot of the material from the lectures of his teacher REDTENBACHER. It also drew the attention of the then already famous physicist and thermodynamics specialist ZEUNER at the establishment preceding today's ETH Zurich to the young REULEAUX, and so in 1856, at the age of 27 we find him as a well-established professor of mechanical theory at this highly respected institution, both at that time and still today. As this establishment has never suffered the effects of war, it still has a lot of material and with a

Fig. 2. Franz Reuleaux in Zurich

few mouse clicks in its digital university archive, in the "Portrait of the month, August 2005" on the 100th anniversary of his death, it is possible to open the personal details, a photo and a sample of the writing of the very distinguished looking "young professor" [6], fig. 2.

In Zurich, on the one hand he wrote the first of his three truly great books, namely the "Constructeur", which following the first edition in 1861 became the worldwide standard work on machine elements for 30 years and was also published in 5 editions in English, French, Russian and Swedish. The 4th edition of 1899 available to the authors has around 1200 pages with just as many high-quality drawings printed as wood engravings. Even more important from the view of a mechanisms engineer, were the initial beginnings developed by REULEAUX in Zurich of what still affects today's mechanism theory, the so called "theory of constrained motion", parts of which he also presented in several lectures at the time.

This awakened the interest of other universities, and after a rejected call to go to Riga, in 1864, aged 35, REULEAUX went to Berlin to the then "königliche Gewerbeinstitut" (Royal Industrial Institute), from which the present-day Technical University (TU) Berlin later developed, and worked there for 32 years – many of which as the Rector – until he retired at 67. We can only speculate about what made the call to go to Berlin so interesting for REULEAUX, because in scientific terms, the Berlin establishment did not (yet) have a top reputation. He did recognise the guiding influence of Prussia within the German speaking world and apart from the university position; he probably saw that he would have far more wide-reaching opportunities in Berlin than in the political province. Last but not least, the apparently very generous benefits offered to him with his chair in Berlin probably attracted him too. Unfortunately, due to the effect of the wars of the last centuries, there are no longer any specific documents about this at the TU Berlin, although there are clear indications. For example, in an obituary following his death in 1905 we can read that his *"great life's works" "supported by the rich funds afforded them by the Royal Prussian State Administration were accomplished in Germany"*. The production, discussed later, of more than 800, some very complicated, kinematic models, made by him would otherwise not have been possible.

The most important result of the Berlin years, having a lasting effect up to this day, was without doubt the main scientific work of REULEAUX, his book on "Theoretische Kinematik" (Theoretical Kinematics), examined more closely later, which was published as a book in 1875 with around 600 pages and approximately 500 illustrations, based on material he had taught for many years in his lectures, fig. 3. The follow-up work, long announced but not published until 1900, "The practical relationships between Kinematics and Geometry and Mechanics" had approx. 800 pages and 700 illustrations, of which an estimated more than 200 showed in detail then known and practically implemented non-uniformly transmitting gear mechanisms in today's sense. Despite this breadth, the book failed to achieve the fundamental significance and emanation of the previous book.

Initially, in Berlin REULEAUX very quickly acquired a brilliant reputation as a university lecturer. According to many witnesses, he must have been a charismatic lecturer, on whose lips *"his students hung with ardent admiration"*. Impressive evidence of this, among other things, is provided by the wonderful handwritten lecture notes of students, later printed as a manuscript [7].

Apart from his work for his own university and others just being founded, amongst which was also the subsequent RWTH Aachen University and even in America, REULEAUX also worked extensively as an "industrial consultant" – with connections to major industrial personalities, such as, SIEMENS, MANNESMANN and LANGEN, who produced the OTTO (petrol) engine in Cologne-Deutz – and was also an active member of numerous state "commissions".

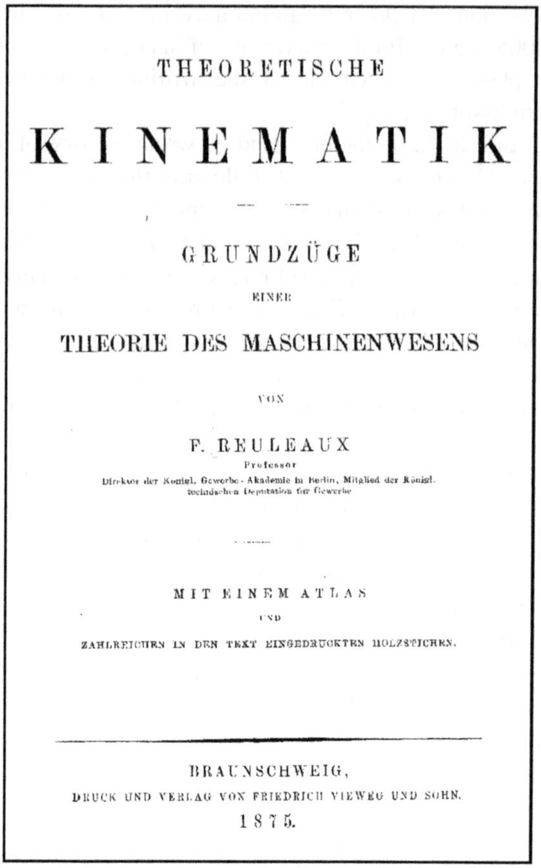

THEORETISCHE

KINEMATIK

GRUNDZÜGE

EINER

THEORIE DES MASCHINENWESENS

VON

F. REULEAUX

Professor

Director der Königl. Gewerbe-Akademie in Berlin, Mitglied der Königl. technischen Deputation für Gewerbe

MIT EINEM ATLAS

UND

ZAHLREICHEN IN DEN TEXT EINGEDRUCKTEN HOLZSTICHEN.

BRAUNSCHWEIG,

DRUCK UND VERLAG VON FRIEDRICH VIEWEG UND SOHN.

1875.

Fig. 3. "Theoretical Kinematics" Title page dated 1875

Only two of these commissions are mentioned here by way of example, namely his decisive participation in the drawing up of uniform modern patent legislation for the German Empire that had been created in the meantime and his representation of Germany as a judge and organizer of the German participation in 6 World Exhibitions between 1862 and 1881 in London, Paris, Vienna, Philadelphia, Sydney and Melbourne, to which he travelled, and that during a time in which the first telegraph cable had just been laid between Europe and North America and modern steam-driven seagoing ships still had sails as a safety backup. REULEAUX' public dismissal of German industrial products at the World Exhibition of 1876 in Philadelphia with the withering judgment that they were *"cheap and shoddy!"* [8] became generally famous and infamous, earning him harsh criticism and the accusation of having an unpatriotic attitude, but the effect was lasting and resulted in an actual turnaround of German industry switching from the production of simple mass goods to "high-tech for a high price", which has remained a recipe for success to this day. In [8] one can also find another sentence which presumably did not represent the general feeling of the German public in those days but nevertheless hinted at darker times to come: " *In fact (...) we are taken by an uneasy feeling when we walk through the Exhibition and look at our compartment with virtually marching battalions of Germanias, Borussias, Kaisers, Crown Princes, "Red Princes", Bismarcks, Moltkes, Roones, made from china, semi-porcelain, bronze, zinc, iron, clay and being painted, embroidered, knitted, printed, lithographed, woven, approaching us from every nook and cranny.*"

Despite all his activities described above, REULEAUX still found time to write articles about arts and crafts, philosophy and social politics (including – in the authors' opinion – fairly "romantic" thoughts about the then brewing "worker issue") [9] and to learn Arabic and Sanskrit and to write an apparently very entertaining travel book about India, which he visited on a return trip from Australia. In a eulogy on his 70th birthday in the Berlin "Ilustrierte Zeitung" it is written that he had translated the so-called Hiawatha Poem of the American poet LONGFELLOW (an "Indian Saga") into German in original metre, and the authors roughly estimate that this ballad had around 10000 lines! With all this, as present day professors of mechanism science, it would be good to know how many staff positions REULEAUX had in his department, but the authors were unable to find a source to retrieve such information.

Towards the end of his active life REULEAUX had to defend himself against sharp criticism and finally even hostility from his own university. This was partly because he could be quite an arrogant person. Evidence of this is provided by a – in fact probably not fully unjustified – review of the „Lehrbuch der Kinematik“ (Kinematics Test Book) published in 1886 by BURMESTER (which was also one of the great classics!), which REULEAUX brilliantly worded in 1889 on 20 closely printed pages, but also inserted with caustic sharpness in the foreword to the 4th edition of his „Constructeurs“. A particularly smug excerpt states, for example: "*Even the most careful author can let an error slip through and this earns reasonable allowance. But especially the error considered here has already caused so much material damage and has already so often been urgently censured that it takes unbelievable superficiality to make it again, namely when with such rattling armamentarium, the tasks are approached with all guns blazing, as in Burmester's book.* And further on: "*It is me who first identified the kinematic elements and chains; more than half of my book is dedicated to the unambiguous differentiation of link and element, and now Mr Burmester tells me, and this in terrible German, that I got the two terms mixed up. Mr Burmester quite simply proves himself be a dilettante in his subject.*" The fact that BURMESTER'S book also develops the basis of the entire subsequent theory of dimensional mechanism synthesis is not even acknowledged by REULEAUX in his "wholly damning criticism".

But the main reason for his late loss of charisma was that towards the end of the century he "was somehow no longer in keeping with the times", and the development of universities with its classical, deductive theory approach, for which only paper and pencil were really needed, was opposed by the newer trends of younger professors, who – of course quite rightly – saw that the future need would be for empirical research in laboratories on real machines. In particular, a certain ALOIS RIEDLER aged 38, who had already spent 8 years as a professor in Vienna and Aachen; as he became professor of mechanical theory in Berlin in 1888, seriously harassed REULEAUX and was one of the main reasons why in 1896 he retired from active university work earlier than he would have had to at that time, and apparently with a certain resignation or even bitterness. But he remained active during the final 9 years of his life as a "private scholar" with numerous publications, above all of course with the publication of his already mentioned "Practical Kinematics" in 1900. The final years until his death in 1905 due to a stroke were however overshadowed by worries about his financial situation, as he had invested a large part of his fortune in Mannesmann shares, which did not begin to yield a profit until later [10].

After the many honours and awards during his lifetime, in 1912, seven years after his death, "*friends and grateful students*" erected an impressive memorial to him in the park of his university in Berlin-Charlottenburg, fig. 4, which can still be seen today, however, unfortunately robbed of the two original larger than life-size bronze figures – in the taste of that time – posted next to the memorial stone. The memorial bears an inscription, understandable in view of the admiration of the writer, but seeming somewhat grandiloquent in today's parlance, fig. 4:

FRANZ REULEAUX

RESEARCHER AND TEACHER

IDENTIFIER OF THE

RELATIONSHIP OF

ENGINEERING WITH

SCIENCE AND LIFE

Fig. 4. The memorial to Franz Reuleaux at TU Berlin in its original form

The Association of German Engineers, the VDI, honoured REULEAUX by using the name "Reuleaux Society" for its mechanisms working group, which existed in the period before World War II, and with its series of texts, the „Reuleaux-Nachrichten" (Reuleaux News). In 1993, a great grandson of REULEAUX, Dr. med. HENNING REULEAUX, who still lives in Berlin, managed to persuade the City of Berlin to take the grave of REULEAUX into its care as a famous grave [10]. In 1929, REULEAUX' hometown of Eschweiler named him an honorary citizen, also named a street after him in honour of his 100th birthday, had a commemorative plaque put up on the house in which he was born, however, it was destroyed with the house during World War II.

3 REULEAUX' Relationships with RWTH Aachen University and Early Presence of Mechanism Theory in Aachen

After many years of swaying between Cologne and Aachen, in 1863, the Prussian government with the highest approval of the crown "*in that King Wilhelm I issued the submitted draft order with this signature*", decided to build Prussia's long-planned first "polytechnic university" in Aachen. The so-called "Königliche Gewerbeinstitut" (Royal Industrial Institute) founded in Berlin in 1821, in which REULEAUX worked, had not yet formally reached this rank by then.

From the mountains of arguments, which the respective representatives gave for and against Aachen over the years, only the following are mentioned by way of example: First, the two reasons still understandable today, that the Aachen party would be able to "*attract*" more "*funds*" from interested commercial and industrial circles, and that the numerous industrial companies that could be reached in the close vicinity in Germany, Belgium and the Netherlands, would enable cost-effective day trips without the need for overnight accommodation. By contrast, the following two points seem rather curious today, namely that such an establishment with highly trained scientists from other parts of Germany would perhaps dampen the city's keen Catholicism somewhat, which was unpopular with the Prussian government, and finally that the smaller number of wine bars compared to Cologne would be "*expedient for the enthusiasm and virtue of the pupils*". In 1870, the university was then officially opened as the last of the already mentioned present-day TU9 in a purpose built new building, today's historical main building. It was an uplifting moment for all present, even though the highest ranking majesties were unfortunately not present, as they were at the time "on the battlefield". (Author's comment: They were on the verge of defeating France and as a consequence, becoming "German Emperor".)

In fact, REULEAUX was also frequently involved during the planning and development phase: Initially he was approached – during the World Exhibition 1867 in Paris (Paris Exposition Universelle), where the gentlemen met – to ask whether he would like to accept the post of founding director. REULEAUX was apparently definitely interested, but this appointment did not happen, "*because REULEAUX' demands appeared so exaggerated to the Aachen personalities.*" This is also an indication of the good facilities apparently available to REULEAUX in Berlin, and he did not remain the only "*peacock*", whose appointment ultimately failed due to the stinginess of the people in Aachen. Despite this, REULEAUX continued to be involved in the planning and development of the RWTH, for example, he drew up a draft budget, and was a

Fig. 5. Adolf von Gizycki, Professor of "Kinematics Theory, (Reuleaux System)" at Aachen Technical University from 1870 to 1891

member of a commission advising on the structure and curriculum of the new university [11].

And there is another circumstance, which the authors came across during her searches for this article, which add a completely new facet to the history of mechanism theory at RWTH, and in which the authors cannot believe that REULEAUX was not involved.

It concerns a certain ADOLF VON GIZYCKI (*1834, +1891), fig. 5, who from the start in 1870 was appointed as only one of three other professors of still young mechanical engineering at RWTH and later, from 1880 - 1883 was its first elected rector. This Professor V. GYZICKI had previously been an assistant and lecturer at the Royal Industrial Institute in Berlin and in Aachen represented the fields of *"Descriptive Machine Theory"*, *"Theoretical Machine Theory"* and – would you believe it – "*Theory of Kinematics (Reuleaux System)*"! "Modern" mechanism theory was therefore at home in Aachen from the outset, since 1870 [11] and was therefore even introduced here somewhat earlier than, for example, in Dresden, where it was not introduced until 1874 by TRAJAN RITTERSHAUS (*1834, +1899, also a student of REULEAUX) with his appointment and move from Berlin to Dresden.

The authors suspects, probably rightly, that in the appointment of the lecturer V. GIZYCKI and his move from Berlin to Aachen "behind the scenes" influence of REULEAUX was also involved, say in the form that also apparently still occurs in similar cases today, in that after his own appointment as founding director "failed", he could well have said: " *....then I won't come in person, but I might have a capable young man..."*. The fact that this "*Theoretical Kinematics (Reuleaux System)*" course of Prof. V. GYZYCKI did not have an entirely insignificant status at the new university can be deduced from another very interesting clue:

13. Die kinematische Sammlung

enthält:

195 Demonstrationsmodelle, welche in den Werkstätten der Kgl. Gewerbe-Akademie zu Berlin angefertigt worden sind und zusammen 15200 Mark gekostet haben.

8 Wandtafeln.

12 Tafeln auf Pappe.

Fig. 6. Inventory of the "Kinematic Collection" of RWTH Aachen in 1880

In 1880, an "Industry and Art Exhibition" was held in Düsseldorf, in which RWTH Aachen University also took part. The exhibition catalogue produced by the university has survived [12] and, apart from the exhibition items, it also lists the inventories of the various "teaching collections" in Aachen. This list also includes a "*Kinematic Collection*" and the inventory is given as follows: *"195 demonstration models, produced in the workshops of the Royal Industrial Academy of Berlin, costing 15200 Mark in total."!* The authors can't help the following speculation: The good man REULEAUX gave his "young man" a decent start-up and quite coincidentally of course also acquired a respectable order volume for his own university workshop.

This sum, with a purchase power that would equate to around 150,000€ today, was certainly not pocket money for Aachen university, still in its early development phase. It can only be concluded that at that time in Aachen, the precursor of today's "mechanism theory" was considered to be a very important subject. Unfortunately, these – today very valuable in engineering history terms – so-called REULEAUX models, discussed later in this article, apparently disappeared from RWTH, and the authors were also unable to find any other direct witnesses to the "Theoretical Kinematics (Reuleaux System)" teaching of Prof. GIZYCKI in Aachen. However, there is a very good handwritten school exercise book of a student named EDW. NEUFANG from the winter semester 1876/77 on the lecture on "Descriptive Machine Theory" by Prof. GIZYCKI and in this there are also various mechanisms sections on a Reuleaux basis [13], Figure 7. The section about *"Transmission mechanisms with rotating motion"* begins in the same way as a present day beginner's course in mechanism theory could begin: *"The mechanism unusually applied is the four-bar linkage."*

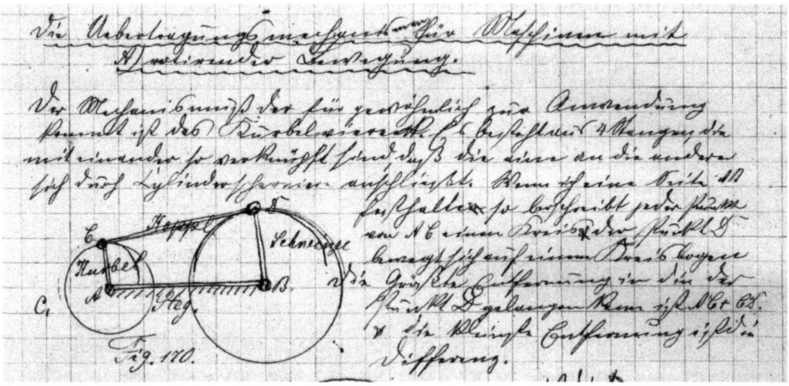

Fig. 7. Lecture notes of the student Edw. Neufang at RWTH Aachen University in the winter semester 1876/77

Prof. V. GZIYCKI apparently rendered outstanding services to teaching and as a rector to the further development and organization of the university and became extremely valued as a person among students and colleagues. Very warm words in an obituary following his early death by stroke at the age of 57 provide very eloquent evidence of this [14]. Apart from a travel report on an excursion with students to England and Scotland, however, the RWTH library only has one technical publication by him. After Prof. V. GYZYCKI, the subject of "Kinematics" appears to have gradually lost significance as an independent field in Aachen. In the chronicles on the 50 year existence of the university [11], "Kinematics" appears in the documented subjects for the last time in 1906 with a certain Prof. LANGER, but now only "included" in the subject "Machine Elements".

However, in 1909 there is a "mechanisms" dissertation by a certain Eugen Essich under Prof. Langer with the title "*About control mechanisms with rolling cams*" [15]. In the ten years following the right to confer doctorates from 1899, until 1909 only 11 doctorates in Aachen were in mechanical engineering, and that one of these related to "cam mechanisms" in today's language, convincingly shows that mechanism theory – taken from REULEAUX' environment and probably with his involvement by v. GIZYCKI from Berlin to Aachen – held no small significance during the early years of the subsequent RWTH and did not only start with RAUH very much later [16]. With the various references presented here, the authors indeed see today a very much closer and effective relationship between Franz Reuleaux and RWTH Aachen University than could have been expected at the start of the search just based on the spatial vicinity between REULEAUX' birthplace Eschweiler and Aachen.

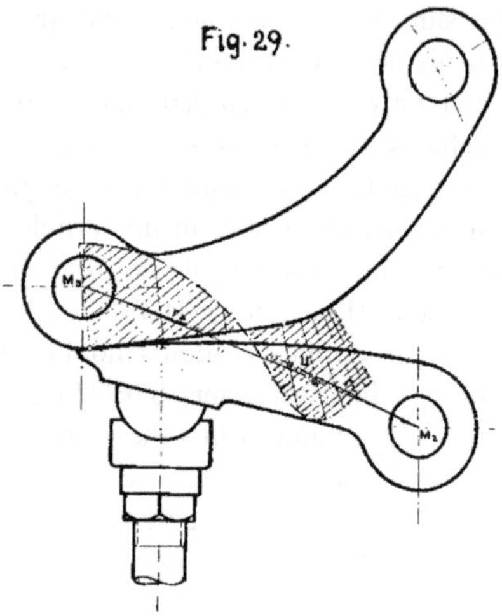

Fig. 8. Fig. 29 from the first mechanisms dissertation at RWTH Aachen 1909: Determination of the sliding motion at a rolling cam pair

4 "Father of Kinematics"?

Numerous subsequent biographies of REULEAUX contain decidedly ecuberant estimations of his complete work, roughly in the style of the already quoted inscription on his memorial in Berlin. Much of that which has been written until now is based on the writings of CARL WEIHE from 1925 and 1942, who was one of the last students of REULEAUX. For example, WEIHE – specifically referring to the "Kinematics" of REULEAUX – speaks of an "*overwhelming edifice of ideas, which reveals its creator as a master of engineering, as a culture researcher and as a philosopher*" and he closes with the summary: "*In theory of constrained motion he gave law to motion, this basis of all life and all engineering, and he brought order and naming into the diverse world of engineering forms*" [17]. This almost sounds like religious fervour and awakens associations with Christian creation theory. REULEAUX may well have liked to see himself in such a "God-father roll". At least this impression involuntary comes to the mind of today's reader, when they see in "Theoretical Kinematics" the subtitle "Fundamentals of a theory of mechanical engineering" and then in the foreword with the objective, repeatedly reiterated in more or less the same way later, of developing the "fundamentals of a *theory encompassing the whole of mechanical engineering*", a truly very comprehensive claim. From today's perspective – and even from the point of view at that time – the authors are unable to follow how Reuleaux could arrive at such an exaggerated categorization of the status of his own kinematic considerations. Even at that time, for example in a steam engine, apart from kinematics, a bit of thermodynamics was also involved in the transmission of motion and the controls!

Thus, some contemporaries of REULEAUX viewed his "kinematic life's work" far more modestly, through to damning criticism of it as being unrealistic and practically useless theorization [18]. The truth probably lies somewhere in the middle and as principal witness, whom the author can fully agree with from today's point of view, TRAJAN RITTERSHAUS [19] is quoted, who – as a former assistant of REULEAUX and himself only one year earlier appointed Professor of Kinematics in Dresden – wrote a 13-page review of "Theoretical Kinematics" immediately after it was published. First, on four pages he briefly describes the history of the development of kinematics to date and gives a concentrated summary of the REULEAUX core theses, before then meticulously pulling to pieces individual detail statements of REULEAUX on the remaining 9 pages. As an introduction to this review, however, he writes about the value of the book as a whole "... here I would first like to categorically acknowledge the far-reaching significance of the same, hoping that the sincerity of this assurance is not questioned as a result of the corrections contained in the following".

The truly new approach of REULEAUX became the basis of what today we call "Topological mechanism classification" and – to put it very briefly – is taught in the first hour of each beginner's course [20]:

"*Each mechanism – initially fully detached from its purpose and design – consists of links and joints, which connect the links with more or less constrained motion relative to each other. A joint is formed by two elements forming a (kinematic) pair. The forms of paired surfaces determine the respective possible relative motion typical for the joint. As long as in such a system of connected links it is not yet decided, in a specific use, which of the links is to be the fixed frame and into which joints the actuating movements are introduced, it is called a kinematic chain. Provided this is "constrained" – and unlike today, REULEAUX was solely interested in constrained chains – the possible relative movement of all links is defined in the (still free) kinematic chain (Figure 9). Depending on which link in the chain is made the fixed (stationary) frame, different motion characteristics can be derived from the same kinematic chain relative to the frame. REULEAUX calls the kinematic chain where the frame link has been already chosen a*

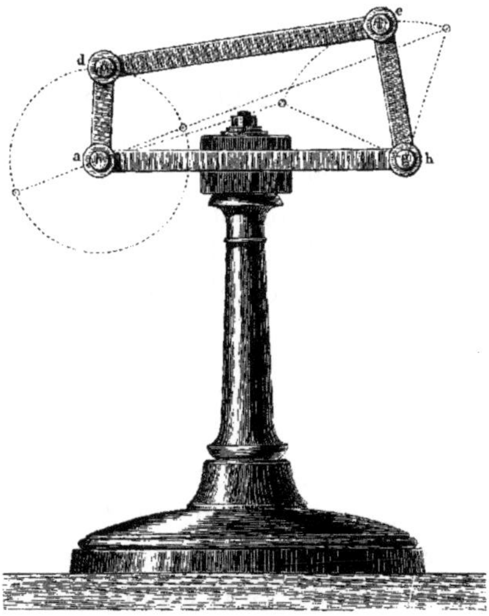

Fig. 9. Demonstration model of Reuleaux for a core theses of his “constraints theory”, the generation of different mechanisms from a single chain by selecting different links of the chain as a fixed frame while retaining all relative movements.

"mechanism". Today often it is also required to name the driven link before defining the system as a mechanism."

Far-reaching conclusions can be drawn from such an approach, for example, to systematically develop all possible structures of kinematic chains. REULEAUX also tackled this, but only in part and not in a logically consistent way. However, the basic approach of such a systematic classification of mechanisms is – like the renowned "Egg of COLUMBUS" (i.e. inspired discovery) – so simple and immediately plausible (and mostly somewhat boring for first-year students) that today it is difficult to fully appreciate the remarkable step REULEAUX took with his liberation from the previously standard, mostly completely improper and unsystematic approach of solely considering existing embodiments of mechanisms. But he also stopped with the status described in his first kinematics book in 1875. Although he protested vehemently – as shown above in the BURMESTER example – if he felt (rightly in this case) sloppily quoted and even wilfully interpreted, but he did not respond to important continuations of his successors, even those such as RITTERSHAUS who did not fail to acknowledge REULEAUX' contribution, although publication of the second kinematic book in 1900 – to a certain extent as a continuation of the first book for practical application – would have provided an ideal opportunity to do so. After all, GRASHOF and GRÜBLER, both in 1883, had already published their wonderfully simple solutions to the rotatability of the four-bar linkage or to determine the degrees of freedom of arbitrarily formed chains, which are today elementary components of every lesson on topological mechanism classification.

Summarising, it is hard to imagine a present day mechanism theory lecture that doesn't includes sections on topological mechanism classification. REULEAUX made the decisive start to this and it would be presumptuous to criticise him for the fact that he didn't create everything that is important to us today. On the other hand, many sophisticated, present day individual mechanism solutions, for example, the fascinating VALVETRONIC mechanical valve lift adjustment in BMW engines, would have been developed anyway, even without REULEAUX.

5 The REULEAUX Model Collection

A paper about REULEAUX should not ignore his famous model collection. As already mentioned above, during his 32 years in Berlin he had created a collection of more than 800 kinematic demonstration models made, very finely developed – mainly to his own designs – with a uniform style in the workshops of his university. Unlike the occasional previously existing collections, mostly very arbitrarily thrown together, the "Reuleaux Collection" – as it was subsequently officially named in his honour – captivated people because of its uniformity and above all because of its consistent didactic concept, as each model was especially designed for a very specific demonstration purpose. All mechanism engineering departments today maintain a comprehensive collection (naturally of newer, more contemporary) mechanism models, a fact that can certainly be attributed to the attraction of the Reuleaux Collection in Berlin. The importance the collection had in teaching is impressively

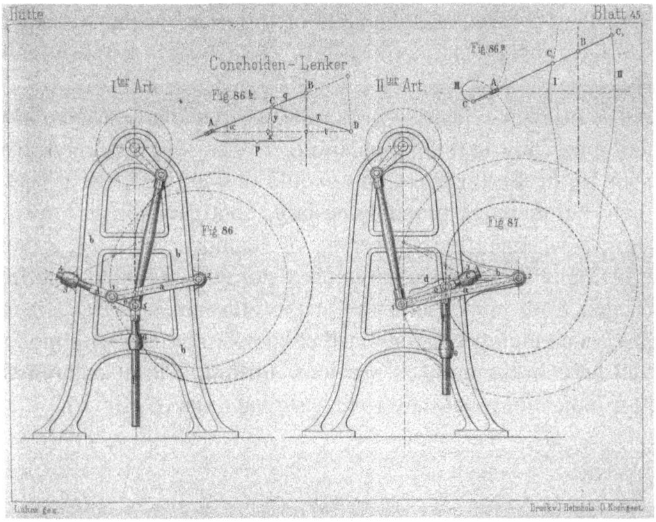

Fig. 10. Reuleaux models for straight line guidance from the "sketchbook" of the "Hütte"

shown in a 2-volume work with illustration and text part [21], self-published by a student society named "Hütte" in 1880, in which 5 students described more than 200 models in very good drawings and supplemented and explained these with numerous secondary drawings and described them in the accompanying text volume – "*according to the lectures given by the "Geheimen Regierungsrath" (Privy Councillor) Prof. Reuleaux*" – and then dedicated this work to "*their extraordinary honorary member ... in gratitude and esteemed admiration*".

Parts of this collection have also been shown outside the university in many exhibitions, e.g. regularly at the Leipzig fair and even at the world exhibitions in America. The collection became world famous and various universities expressed the wish of also setting up such collections for their teaching. This demand was covered by a firm founded in Berlin for precisely this purpose by GUSTAV VOIGT, who as a former student of REULEAUX was ideally prepared for the task. The author dares the supposition that indeed REULEAUX was also somehow involved in this business, as after all, he was the intellectual father of the collection, and a handwritten letter dated 1882, from REULEAUX to ANDREW D. WHITE, at the time the president of Cornell University in the State of New York, has survived, which is about the purchase of a large number of models.

In fact, 266 models in total were sold to Cornell University. At an average list price of approx. 150 Mark according to a surviving VOIGT catalogue of 1907 [22] that would be the equivalent to around 350,000€ with today's purchasing power. Of these models, around 220 are still in existence today and are accessible to the public – in real life and in an excellently designed Internet presentation [22, 23]. The current curator of this collection, Prof. FRANCIS C. MOON, among other things, has written a wonderful book with the title "The Machines of Leonardo Da Vinci and Franz Reuleaux" [24] and in it he writes that according to his research, apart from the models in his collection, 10 models from the VOIGT workshop exist at ETH Zurich, 113 in the University of Porto and a unnamed number exists in a university in Kyoto. However, older sources report that, for example, models were also delivered to Petersburg and Montreal. The Deutsche Museum in Munich also has an inventory of more than 50 Reuleaux models of different origin, of which approximately 20 are exhibited in glass cabinets. Most of these models also originally came from the VOIGT workshop.

The original collection in Berlin was so badly destroyed at the end of World War II that apparently nothing was saved. It was long assumed that the "original models" made in the Berlin University would possibly still exist here and there as individual items, but there are probably no large contiguous inventories. It was therefore a great surprise when, in the University of Hannover in 1994, around 20 Reuleaux models were found in a long since unused storage basement awaiting a clear-out, and were rescued from being scrapped at the eleventh hour by one of the authors, after an observant employee had noticed certain similarities with models he had coincidentally seen once in the author's previous institute. Closer examination of the models gave clear indications that they were "original models", fig. 11.

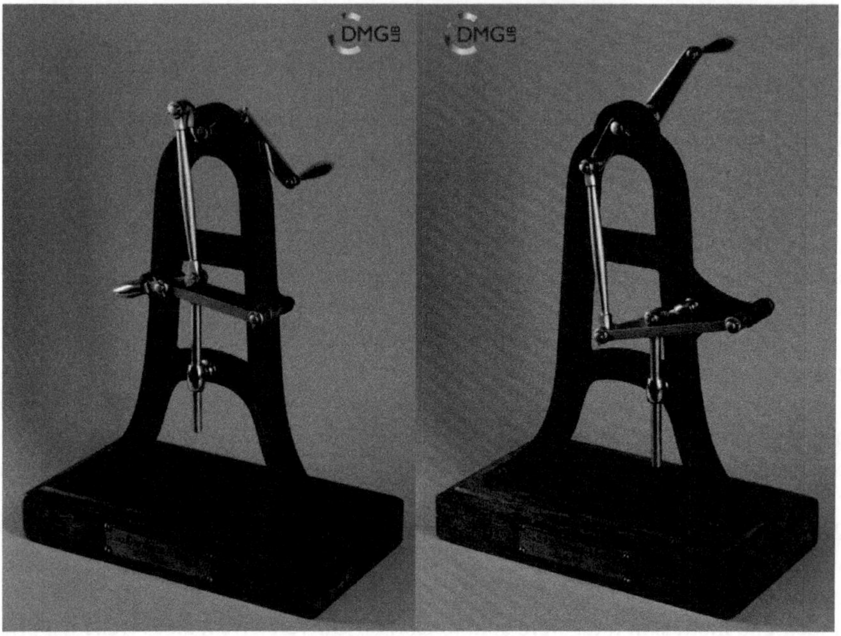

Fig. 11. Reuleaux models from the collection of the Institute of Mechanism Theory, Leibniz University, Hannover, Germany"

However, the company nameplate found on all Voigt models is missing, the models also had a brown base instead of the black base found on most Voigt models and finally, it was found that on the back of the brass plates with the model name, some scratched or partly painted over, was the lettering "KOENIGL. GEWERBE-INSTITUT" and some uninformative number. This suggests that the previously unused backs of some old signs, no longer needed for their previous purpose, were "recycled" in the Berlin industrial institute for new Reuleaux models, and that the models therefore also came from there. However, no information could be found about when and how the models could have landed in Hannover. Since they were found, a small selection of particularly fine models has been exhibited in the Institute of Mechanism Engineering of the Leibniz University, Hannover and since 2008 they have also been presented in the DMG-Lib on the internet [25].

Acknowledgements. This paper has been compiled by the authors educated as mechanical engineers specialized in the field of mechanism and machine theory without any special experience of performing historical searches and presenting historical events. Therefore the authors can only hope that they have not been taken in by any incorrect sources and have not made any serious errors or infringed the rights of others to the texts or illustrations used. Just in case, we ask for mild judgment by "learned" historians. The authors would like to take this opportunity to thank the staff of RWTH Aachen's archive and library, who assisted quickly and competently on repeated occasions, especially Dr. Graf and Dr. Rappmann, as well as Mr Simon Küpper of Eschweiler History Society and Toni Becker for his "interpretation of the name" in French and – last but not least – Dr. Henning Reuleaux for his personal comments on the life and work of his great great-grandfather.

Also acknowledgement is given to the German Research Foundation (DFG) which supported the DMG-Lib project and the European Union which supported the thinkMOTION project, both aimed at digitally preserving past and present knowledge in Mechanism and Machine Science.

References

[1] Ricking, K.: Mens agitat molem – 125 Jahre Geschichte der RWTH. HISTECH EV – Verein für regionale Technikgeschichte, `http://www.histech.org/00011_00000.html`
[2] Becker, Toni: Personal communication, Aachen (2009)
[3] RWTH Aachen University Library, Die Einführung des ersten Rektors am 15 November 1880, `http://darwin.bth.rwth-aachen.de/opus3/volltexte/2008/2558/`
[4] Wauer, J., Moon, F.C., Mauersberger, K.: Ferdinand Redtenbacher (1809-1863): Pioneer in Scientific Machine Engineering. Mechanism & Machine Theory 44, 1607–1626 (2009)
[5] Braun, H.-J.: Technik als "Kulturhebel" und "Kulturfaktor", Zum Verhältnis von Technik und Kultur bei Franz Reuleaux. In: Dietz, B., Fessner, M., Maier (Hrsg.), H. (eds.) Technische Intelligenz und "Kulturfaktor Technik": Kulturvorstellungen von Technikern und Ingenieuren zwischen Kaiserreich und früher Bundesrepublik Deutschland, Münster, New York, München, Waxmann, Berlin (1996)
[6] ETH Zürich – Library, Portrait des Monats (August 2005), `http://www.ethbib.ethz.ch/aktuell/galerie/reuleaux/`
[7] N.N.: Vorträge über Kinematik von Prof. F. Reuleaux, Berlin, Leibniz University Hannover – Library, Signature 4B4483 (1870)
[8] Reuleaux, F.: Letters from Philadelphia. Looked through by author and increased issue with additions, 1st edn. Braunschweig, Vieweg (1877)

[9] Reuleaux, F.: Die Maschine und die Arbeiterfrage. In: Soziale Zeitfragen, Heft 2. hg. v. Ernst Henriet Lehnsmann, Minden i.W (1895)
[10] Reuleaux, H.: Personal Communication, Berlin (1995/2009)
[11] Gast, P. (ed.): Die Technische Hochschule zu Aachen, – RWTH Aachen – Library, Signature Ka21=3 (1870-1920)
[12] RWTH Aachen University Library, Gewerbe- und Kunstausstellung in Düsseldorf – Ausstellungskatalog (1880), http://darwin.bth.rwth-aachen.de/opus/volltexte/2009/2792/
[13] Neufang, E.: Skript zur Vorlesung "Beschreibende Maschinenlehre" von Prof. v. Gizycki im WS 1876/77, RWTH Aachen – Signatur Ff 2117/HBRA
[14] Bibliothek der RWTH Aachen, Nachruf auf Prof. v. Gizycki in: Die Übergabe des Rectorates am ersten Juli (1892), http://darwin.bth.rwth-aachen.de/opus3/volltexte/2009/2802/
[15] Essich, E.: Über Steuerungsgetriebe mit Wälzhebeln, Dissertation bei Prof. Langer, RWTH Aachen, RWTH Aachen – Bibliothek, Signatur S2305 (1909)
[16] Corves, B.: 50+ - Getriebetechnik an der RWTH Aachen. In: 8. Kolloquium Getriebetechnik Aachen 2009, Verlagshaus Mainz, Aachen, pp. 209–234 (2009) ISBN: 3-86130-984-X
[17] Weihe, C.: Franz Reuleaux und die Grundlagen seiner Kinematik. In: Deutsches Museum München, Abhandl. und Berichte, 14. Jg., Heft 4 VDI-Verlag, Berlin (1942)
[18] Lüders, J.: Wieder Herrn Reuleaux! - Eine Würdigung der wissenschaftlichen Arbeiten und Methoden des Herrn Professor Reuleaux, Kiel, Homann, 1877 ETH Zürich – Bibliothek. In: Sammelband ETH-BIB [004583529]
[19] Rittershaus, T.: Zur heutigen Schule der Kinematik; zugleich kritische Besprechung von Reuleaux's "Theoretische Kinematik". In: Der Civilingenieur, Digitale Mechanismen- und Getriebebibliothek – DMG-Lib (1875), http://www.dmg-lib.org/dmglib/handler?docum=740009
[20] Braune, R.: Grundlagen der Getriebetechnik Vorlesungen an der Leibniz Universität Hannover (1986-2008)
[21] Althof: Skizzenbuch der Angewandten Kinematik, Selbstverlag Verein "Hütte", Berlin, Leibniz Universität Hannover – Bibliothek, Signatur C1036, DMG-Lib und =26009 (1880), http://www.dmg-lib.org/dmglib/handler?docum=25009
[22] KMODDL – Kinematic Models for Design Digital Library. Cornell University, Ithaca, http://kmoddl.library.cornell.edu/
[23] Moon, F.C.: Mechanical Engineering Heritage Collection, Reuleaux Collection of Kinematic Mechanisms at Cornell University 1882, American Society of Mechanical Engineers (2004)
[24] Moon, F.C.: The Machines of Leonardo Da Vinci and Franz Reuleaux. Springer, Dordrecht (2007)
[25] DMG-Lib: Modellsammlung des Instituts für Getriebetechnik. der Universität Hannover, http://www.dmg-lib.org/dmglib/main/portal.jsp?mainNaviState=browsen.browse&collId=26046

The Birth of the Moscow School of Applied Mechanics in the 2nd Half of the 19th Century and the Early 20th Century

Vera Chinenova[1], Alexander Golovin[2], and Irina Tyulina[1]

[1] Assistant Professor, Mathematics and Mechanics Department,
Moscow State University
chinenova_v@mail.ru

[2] Professor, Theory of Machines and Mechanisms Department,
Bauman Moscow State Technical University
aalgol@mail.ru

Abstract. One of the results of Russia's defeat during the Crimean war (1853-55) was increased interest for technical education and especially for applied mechanics (theories of mechanisms, machines etc.). A particular place in this historical process is occupied by the Moscow school of applied mechanics that was created by efforts of the professors of the Moscow University (MU) and the Imperial Technical School (ITS). This paper presents the contributions of MU professors (A. Yershov, N. Zhukovskii, F. Orlov, D. Zernov, N. Mertsalov) to the development of the Applied Mechanics training course at the ITU. Mertsalov's activity is shown up to the revolution of 1917. The evolution of the course is shown from encyclopedic compiled content to an original high grade theoretical course. The conformity between the courses and the engineering training requirements of that period is described.

1 Moscow University and Imperial Technical School: The Basic Timeline of the Development of the Applied Mechanics Course

1.1 Moscow University [1, 2, 8]

Moscow University was founded by Mikhail Lomonosov in 1755. According to the advanced constitution of the Russian universities adopted in 1863, the universities received a relative autonomy: their rectors and deans could be elected by the staff, the administration could make decisions concerning the number of the departments to be organized, and the universities began to receive higher financial support [1]. Due to these events, the department of Theoretical and Practical Mechanics appeared at the MU [2]. The name of the department was fixed according to the constitution of 1884. Professor N.D.Brashman insisted on formation of two separate departments (theoretical and practical), but he was not able to be in charge of the newly organized department of theoretical and practical mechanics because of his illness. It was

T. Koetsier, M. Ceccarelli (Eds.): Explorations in the His. of Machines & Mech., HMMS 15, pp. 39–53.
springerlink.com

F.A. Sludskii who became a head of the department in 1886. Apart from defending the PhD dissertation on astronomy, he presented and defended the dissertation for the degree of doctor of applied mathematics. Sludskii insisted on solving some specific actual problems of mechanics and had a great effect on the widening of the course of theoretical mechanics. Sludskii was the tutor of N.E.Zhukovskii.

1.2 The Imperial Technical School [3 – 7]

The Founding Hospital was transformed into Moscow Educational Industrial Institute (MEII) in 1830 and received a status of "Technical University". The course of practical mechanics appeared in 1839. After having graduated from the department of Mechanics and Mathematics in 1843, A.Yershov became a lecturer at the MEII. Owing to his efforts, MEII lecturers acquired the status of master's and higher degrees since 1850. The course of practical (applied) mechanics was appreciably extended since 1857. Yershov was the rector of MEII from 1859 to 1867 and the author of the first MEII constitution. Due to Yershov's efforts, the status of the MEII lecturer was made equal to the status of the MU lecturer according to the Table of Ranks. Yershov prepared a decree on transformation of MEII into the Imperial Moscow Technical School (ITS) but it was issued only in 1868, after Yershov's death.

The only models [5, Fig. 2.2] in the Bauman University collection of mechanisms remained from the MEII epoch are dated as far back as 1861 and 1862.

N.Zhukovskii appeared at the ITS in 1870. The applied mechanics was taught by F.Orlov at the ITS from 1872 to 1892. Zhukovskii founded the department of theoretical mechanics in 1872. The department of applied mechanics was lead by D.Zernov from 1892 to 1898 and by N.Mertsalov from 1898 to 1929. Mertsalov founded the department of thermodynamics at the ITS in 1898.

1.3 Russian Professors in the 2nd Half of the 19th Century [1, 2, 8]

The higher education in Russia in the first half of the 19th century was aimed not only at increasing the number of competent lecturers and specialists but also at extending the methods of scientific research among students. The modest estimates yield approximately 40 scientific investigations carried out in that period of time. Eight doctors, sixteen masters, and more than a hundred candidates of sciences [8] were prepared at the faculty of physics and mathematics between 1825 and 1854.

The 2nd half of the 19th century began for Russia with the defeat in the Crimean war of 1853-55. On the one hand, the heroic defense of Sevastopol was a subject of admiration even among enemies [9]; on the other hand, the technical backwardness and gigantic corruption were specific features of the Russian social life of that period [9, 10]. Alexander II of Russia carried out some reforms that stimulated the interest towards the natural and technical sciences. A special role in that process belonged to the MU and ITS. The rise of the scientific and technical thought was especially significant because the Russian science did not have prolonged historical traditions and the professors in Russia, who had appeared only at the end of the 18th century, were not organized as a specific social community until the middle of the 19th century.

According to the Table of Ranks, which appeared as a result of the constitution of 1804, an Ordinary Professor was equivalent to a V class official, that is, Civil Councilor (or General)[1]. However, the salary did not correspond to such a high rank: the rate of wages did not grow since the adoption of the constitution of Russian universities in 1863 until the end of the 19th century. The professors were forced to work in other educational institutions along with their lecturing activity at the university. According to the population census of 1902, there were 1.6 million people in Moscow and among them, 20 thousand intellectuals (teachers, scientists, writers, people of art, etc.) and approximately 200 professors (including 99 at the MU and 18 at the ITS [1]). As to the number of ITS students, there were 1251 students at the ITS in 1905 [4].

The interest in applied sciences grew anew in Russia in the beginning of the 20th century, which was due to the industrial development of the country. The increased social activity, however, indicative of the early 20th century was followed by the savage reaction of the regime. In 1911, the minister of public enlightenment, Kasso, fired the elected rector of the MU (Prof. A.A.Manuilov) and his pro-rectors. As a response to this repression, over 100 lecturers (including S.A.Chaplygin and N.I.Mertsalov) left the university; this led to decreasing its role as a scientific center [2; 10].

The education requirements for engineers were changing according to the demands of the times. Accordingly, the requirements for the levels of competence in natural sciences and their practical applications were increasing too. This tendency touched obviously the applied mechanics as a basic training discipline for engineers. Requirements for qualification of ITS lecturers increased in accordance with this challenge. It was natural for the ITS executives to address to the neighboring MU for help. Thus, the MU graduates (Yershov, Zhukovskii, Orlov, Zernov, and Mertsalov) became the lecturers and then professors at the ITS; as a rule, they did not disconnect from the *alma mater*. It is these scholars who laid the foundations of the Moscow School of Applied Mechanics.

1.4 A.YERSHOV (1818 - 1867) [5, 6, 8, 11, 12]

Yershov (Fig. 1) entered the Moscow University in 1839. In 1844, he defended the dissertation for the degree of master of pure and applied mathematics: "About water as an engine". Yershov began to work at the ITS in 1843.

New subjects were introduced into the curriculum at the university department of physics and mathematics in 1844: descriptive geometry and practical mechanics. A special course of practical mechanics was an initiative of N.D.Brashman, professor of mechanics at the Moscow University, who had the engineering educational background and understood that the theoretical and technical mechanics should be linked together in a teaching course. Brashman was a tutor of young scientists and made much effort to create a research direction in the area of applied mechanics in Russia. After having defended his master dissertation, Yershov began to give lectures on

---

[1] Table of Ranks: "Everybody who received any of the eight highest ranks in the Civil or Court authorities will be considered to belong, along with their descendants, to the highest and senior nobility, even if they were not of noble birth".

practical mechanics and descriptive geometry at the MU faculty of physics and mathematics since 1844. He finished his lecturing activity at the university in 1859.

Yershov developed the program of the course of applied mechanics. In 1849, the minister of public enlightenment Shirinskii-Shikhmatov asked P.L.Chebyshev to make a review of this course. Chebyshev analyzed thoroughly the course and wrote his remarks in the form of the report "On teaching the practical mechanics and descriptive geometry at the Moscow University" [11]. Chebyshev, who understood deeply the demands of the times and major requirements of the social practice directed towards the exact sciences, began to give similar lectures at the St. Petersburg University since 1849. Both the Yershov's program and the advanced Chebyshev's course emphasized the idea that *the mechanics must be tightly connected to the wide distribution of machines* [12].

Fig. 1. A. Yershov

Since 1853, Yershov was approved as "a person executing the position of an extraordinary professor of the Moscow University". In 1854, he created the first Russian textbook on applied mechanics: "Foundation of Kinematics, or Elementary Theory of Motion in General and Mechanism of Machines in Particular". This work was mostly of the encyclopedic character, although a due attention was also given to the practical questions concerning the design of mechanisms. For example, the theory of toothed gearing was described by Yershov excellently.

Yershov paid also great attention to drawing and producing the profile of cam mechanisms. He was among the pioneers to propose devices in studying and demonstrating the motions. A.N.Bogolyubov noted [13] that Yershov was the first to have used experimental methods in studying the motion because the results of Chebyshev's experiments were published later. Yershov explored the following devices: a cylinder driven by a spring or a weight to study "fast movements", a circle spinning uniformly around an axis, an infinite paper ribbon enveloping two cylinders one of which was a leading cylinder. All these devices were very elementary gadgets but they were the first facilities used in mechanical laboratories of higher technical schools for making the first experiments. Yershov considered also the graphical method for determining the ratio of angular velocities of mechanism parts.

1.5 N.E.Zhukovskii (1847–1921) [2, 13, 14, 16]

Nikolai Egorovich Zhukovskii (Fig. 2) entered the faculty of Physics and Mathematics of the Moscow University in 1864 and graduated from the university in 1868. While being a first-year student, he showed a great interest in mathematics and was recommended by the astronomer M.F.Khandrikov for the admittance into the society of amateurs of mathematical sciences created by the initiative of N.D.Brashman. This society evolved into the Moscow Mathematical Society subsequently.

Fig. 2. N. E. Zhukovskii

Zhukovskii began to give lectures on practical mechanics at the Moscow Practical Academy (specialized secondary school) since 1872 and simultaneously worked as a teacher of mathematics at the Moscow Technical School. Zhukovskii was constantly interested in popularization of the scientific knowledge. His "Elementary Course of Practical Mechanics" lectured at the Moscow Practical Academy of Commercial Sciences was an interesting example. This excellent course characterized by brevity and clarity of presentation was issued lithographically in 1901. It contains an extremely clear and simple description of various mechanisms, water-raising and air-blowing machines, water wheels and turbines, etc. [14]. In 1874, he was recommended by F.E.Orlov and approved as an associated professor at the Analytical Mechanics department of this school. Zhukovskii highly appreciated the friendship with Orlov who introduced him into the Polytechnic Society associated with the school and acquainted him with the Moscow engineers.

Since 1872, Zhukovskii worked as an associated professor at the ITS Analytical Mechanics department where he became an acting professor five years later. He defended the master dissertation entitled "Kinematics of the liquid body" in 1876 and doctoral dissertation "On the endurance of motion" (dealing with the orbital stability) in 1882. After the defense of his doctoral dissertation, Zhukovskii became a private docent of the Moscow University (where he gave a special course of lectures on hydromechanics) and then received the position of an acting professor at the mechanics department in July, 1886. By that time, Sludskii retired from service although continued to lecture some special courses; it was Zhukovskii who became the head of the department. A year later, he received the position of a full professor at the ITS also.

Zhukovskii was elected the Corresponding Member of the Academy of Sciences in 1894. In 1900, he had to go to the polls as a Real Member of the Academy. But he refused to participate in the elections because it became clear that it was necessary to move to St. Petersburg to be a Real Member. Zhukovskii did not want to part with his students and colleagues and break the connections with the Moscow University and the Technical School where the *Russian School of Machinery Mechanics* began to form. In 1905, Zhukovskii was elected the president of the Moscow Mathematical Society [2].

Zhukovskii's interest towards the problems of the theory of mechanisms was not occasional; it was a natural part of his creative scientific activity. He was not only a great theoretician but also an outstanding inventor. For example, he invented and built the models of centrifugal rotating steam generator, gyroplane, helicopter screw, and the project of an air thermal jet engine in 1903-05. N.E. Zhukovskii studied various engineering problems of applied mechanics, which were extremely topical for the practice of that time. The works of Zhukovskii are not only an appreciable contribution into the area of applied mechanics but also an effective method of the modern university teaching of mechanics and mathematics as well as way of studying: geometrical representation and simulation. Zhukovskii attached great importance to problem solving workshops. He carried out the training courses himself and made up a problem book that was repeatedly published.

In 1902, Zhukovskii built a small aerodynamic tube at the university with a square cross section of 75 x 75 cm^2. It was one of the first suction pipes ever built in the world. Zhurovskii began to carry out his first aerodynamic experiments using this tube. Since that year, he concentrated his efforts on solving the aircraft problems.

Later, Zhukovskii created the theoretical foundations of aircraft and used the obtained results for lecturing the courses at the Moscow Higher Technical School (MHTS). The scientific research in the area of aircraft was appreciably stimulated by the war of 1914. The aviation specialty at the MHTS faculty of mechanical engineering transformed gradually into the separate faculty, which subsequently grew into the Moscow Aviation Institute.

It can be said that he popularized the ideas of the machinery mechanics: he presented his reports in the Polytechnic Society, Society of Amateur Natural Philosophers, Anthropologists, and Ethnographers as well as in the Moscow Mathematical Society.

Zhukovskii was lecturing a special course on controlling the machines at the ITS for several years. The course consisted of two parts: Statics of the Controller and Dynamics of the Controller. The course was prepared using the notes for students and published in 1909. At the Moscow University, Zhukovskii gave lectures on the fundamental course of theoretical mechanics and the private docents V.M.Kovalenskii and N.I.Mertsalov taught additional training courses on applied mechanics. Zhukovskii took care of teaching the practical mechanics at the Moscow University. He invited the professors of applied mechanics from the Technical School to university for teaching as private docents; thus, N.I.Mertsalov worked at university from 1897 to 1920.

Zhukovskii lectured many special courses on topical problems of science and technology, for example, on centrifugal controllers (1906), the theory of controllers (1908-1909), outer ballistics (1916-1918), and other areas of mechanics. He came regularly to university on Mondays, Wednesdays, and Fridays until his latest years: between 9 am and 4 pm, Zhukovskii gave different advisory classes for students and visiting scientists and engineers, along with the major courses.

In the 1930s, four departments of mechanics were created at the Moscow University on the basis of the wide range of scientific problems considered by Zhukovskii: aeromechanics, hydromechanics, theory of elasticity, and applied mechanics (in 1941). Such eminent scientists as S.A.Chaplygin (a full professor of mechanics at the Moscow University since 1903), D.N.Goryachev, N.N.Bukhgol'ts, I.V.Stankevich, G.G.Appel'rot, and V.P.Goryachkin were among his disciples.

It was owing to Zhukovskii that the Moscow school of mechanics appeared. His studies on different problems of practical mechanics and hydraulics were decisive in choosing the modern research directions in the area of the theory of mechanisms, theory of machine controlling, and water-pipe and machine hydraulics. His name is connected with a number of new directions and scientific schools of mechanicians and engineers, who developed successfully the national science and technology. "He founded several schools rather than one school," said Chaplygin on Zhukovskii's grave, "He was the best combination of the science and technology; he was almost a university!"

So, Zhukovskii seems to have foreseen the most promising research directions.

1.6 F.E.Orlov (1843–1892) [2, 16]

Along with the professors N.Brashman and A.Yershov, F. Orlov (Fig. 3) deserves much credit for increasing the scientific and technological level in teaching the practical (applied) mechanics in the Russian higher school.

Fyodor Orlov entered the faculty of Physics and Mathematics of the Moscow University in 1859, where he gained the sympathy of Prof. N.Brashman. He graduated from the university in 1863 and remained at the department of pure mathematics to "prepare for the professorship". In 1869, Orlov defended the master dissertation "On the reciprocal differential equations".

Fig. 3. F. E. Orlov

The department of theoretical and practical mechanics that was organized at the university in 1866 was led by F.A. Sludskii. At that time, the cabinet of mechanical models was separated from the cabinet of physics and given under Orlov's supervision. Orlov lectured the course of practical mechanics both at the MU and ITS. The themes of candidate dissertations proposed by Orlov were of experimental character. This was due to the excellent condition of the cabinet of mechanical models and devices.

His teacher, A.Yu.Davydov, recommended Orlov for lecturing the course of practical mechanics. Later Orlov was sent abroad for a business trip. His goal was to get ideas about ways of practical mechanics teaching at the best European Polytechnic Schools. Orlov learnt the basic principles of engineering, found out how to design machines, and became a close Zeuner's disciple. The next year, Orlov arrived in Berlin where he met F.Reuleaux. Later Orlov used Reuleaux's conceptions and examples in his inaugural lecture at the Moscow University. Reuleaux's kinematic methods were also developed by Orlov.

Orlov returned to Moscow in the autumn of 1872 and began to lecture the practical mechanics at the Moscow University.

Orlov's scientific works were related to the theory of mechanisms and the theory of roulettes. From the very beginning of his activity as a lecturer at the university, Orlov endeavored to elaborate the process of teaching the practical mechanics.

His "Lectures on Applied Mechanics" formed the basis of the most complete manual on the theory of machines and mechanisms in Russia in the 19th century. The textbook contained almost all necessary information, which could be useful for the students specializing in all spheres of applied mechanics.

In 1874, owing to active Orlov's participation, Zhukovskii was assigned as Assistant Professor at the Analytical Mechanics department at the ITS. Orlov introduced him to many Moscow engineers and brought him in the Polytechnic Society organized on the basis of the ITS. Along with his activity at the university and ITS, Orlov gave lectures on practical mechanics at the Moscow Practical Academy of Commercial Sciences [2].

He enlarged appreciably the cabinet of practical mechanics at the university that had been in a neglected condition.

It is Orlov who can be considered as a main founder of the applied direction in the mechanics, which began to develop at the Moscow University since that time.

After having returned to Russia in 1872, Orlov began to give introductory lectures at the MU and ITS, where he was in charge of the applied mechanics department. His inaugural lecture "On the Machines" was published in the "Mathematical Collection" [15] that contained the most recent results on the engineering mechanics, in particular, Reuleaux's ideas concerning the kinematic pairs and chains. Since 1872, he taught the descriptive geometry, the theory of mechanisms, and hydraulics at the MU and ITS. These courses gained the applicative character due to his efforts.

Zhukovskii became acquainted with Orlov at the ITS while listening to his inaugural lecture. Recalling this lecture, Zhukovskii wrote: "His lecture was composed excellently but the young lecturer presented it timidly. There were not yet the calmness and certitude, which were formed through the next years of his teaching activity" [16].

It is necessary to say that Orlov began lecturing the practical mechanics at the Moscow University as an external teacher. The university authorities undertook active efforts to establish a new staff position for Orlov at the mechanics department. Such activity of the faculty and university clearly characterized the attitude towards teaching the practical mechanics.

In response to the faculty request, the position of Assistant Professor was allocated to Orlov who was elected to this position unanimously by the faculty Council on November 5, 1873 [16].

Orlov began to attract private docents for more qualified teaching some disciplines of applied mechanics. Thus, for example, the private docent V.A.Mikhel'son lectured the course "The Mathematical Theory of Elasticity of Solid Bodies" in 1887; the assistant A.S.Shestakov gave the lessons of technical drawing under Orlov's supervision; the private docent M.M.Cherepashinskii (who taught the building mechanics at the ITS) lectured the course "Building Mechanics and Graphic Statics" at the university (1881–1887). The course on applied mechanics lectured by F.E.Orlov in the 1873-74 academic year were first issued lithographically (written by hand and reproduced in collotype).

The course of practical mechanics was lectured at the Moscow University four hours a week [17]. It contained the theory of mechanisms (with exercises), hydraulics, theory of turbines, thermodynamics, theory of strength of materials, and descriptive geometry with elements of technical drawing. Totally, there were ten hours allocated to the classes in practical mechanics by 1888. The contemporary literature was recommended on all specified course sections. Zhukovskii mentioned the high quality of Orlov's teaching of practical mechanics: "Teaching the practical mechanics at the Moscow University was raised by F.E. to such a high level that exceeded any other Russian universities… It can be said that a half of all candidate works at the university were prepared on practical mechanics" [2]. The textbook covers almost all necessary information for the students specializing in any areas of usage of the applied mechanics. Orlov foresaw the industrial 'boom' in Russia in the future to come and mentioned repeatedly that the level of education must be improved and that the university must prepare the lecturers for other educational institutions.

"He spent much time publishing the lectures. He endeavored to make his lectures not only to be resistant to criticism but also accessible to students; the lectures had to be easily understood by the students and kept in their memory. He valued his students' work and time sometimes more than his own pains and often indicated that the erudition of a professor did not have any value when delivered to students with great difficulty or effort."[2]

Orlov obtained the additional financial support to organize and manage the cabinet; in particular, while discussing the project of a new faculty staff structure (preparing a new constitution), he proposed to include the following term: "It is necessary to have one assistant and one technician in the cabinet of practical mechanics". This proposal was implemented much later in a fragmentary form: the position of technician in practical mechanics was established in 1912 only [2].

Zhukovskii endeavored to widen the staff of private docents at the MU mechanics department. In 1886, the lecturer A.S.Shestakov, who had graduated from the ITS, was invited to the position of assistant. His duties included lecturing the descriptive geometry, teaching the technical drawing, and giving the special course "Hydraulic Machines". In 1892, Shestakov was promoted to the position of the guardian of the cabinet of mechanical models after Prof. Orlov's death. It was difficult to find an adequate substitution for Orlov, an excellent lecturer and master of engineering mechanics, but the private docent Zernov coped perfectly with the lecturing the practical mechanics at the Moscow University. He was giving this course from 1893 to 1898; simultaneously, he was lecturing the special course on the theory of mechanisms and machines, participating in the selection of problems for a workbook, and carrying out the practical training on mechanics.

After Shestakov's death in 1897, Zhukovskii proposed to hand over the position of the guardian of the cabinet of mechanical models to Mertsalov, who was also his student. Subsequently, Mertsalov became one of the greatest specialists on applied mechanics.

1.7 D.S.Zernov (1860-1922) [2, 17]

Dmitrii Stepanovich Zernov (Fig. 4) was an engineer and social activist. He was born in 1860. Zernov graduated from the Moscow University and St. Petersburg Technological Institute. He was an associated professor of the Kharkov Technological Institute. Later he became a professor of the department of applied mechanics at the Moscow Technical School and lectured the course of descriptive geometry and applied mechanics at the Moscow University at the same time. He was appointed director of

---

2 A.K.Eshliman, *Memories of the Deceased Fedor Evplovich Orlov*. Politechnic Society Associated to the Imperial Technical School. Moscow, 1892, pp. 18-22.

the Kharkov Technological Institute in 1898 and became the director and professor of the St. Petersburg Technological Institute in 1902. Zernov was also a professor at the Mining Institute and Nikolai Naval Academy, where he taught the applied mechanics. Zernov devoted himself to social activities in September 1904; he played a prominent role in the union of professors of higher educational institutions and was one of the initiators who suggested writing a memorandum of February 12, 1905 signed by 342 scientists. As a result of this action, Zernov was dismissed from the position of director of the institute. He was elected and approved a director of the St. Petersburg Technological Institute in September 1907 and kept holding that position until September 1912; in 1913, he was again dismissed, this time because of the superannuation. Zernov was a chairman of the Society of Technologists, the 2nd department of the Imperial Russian Technical Society, the compassionate benefit society to help the students in need, and a member of the city council of the St. Petersburg Municipal Duma. His major works were the following: "Test of the Kerosene Engines at the Moscow Exhibition", "Test of Strength of Materials", "Goals of Higher Technical Education". He gave a course of lectures on various sections of applied mechanics. After the October Revolution, a question was raised that Zernov should return to the university. In 1920, he was elected director of the St. Petersburg Technological Institute for the third time and was working in this position until his death.

Fig. 4. D.S. Zernov

The course created by D.S.Zernov was used at the Russian Higher Technical School for as long as 35 years; it was published for the first time in 1896. The course consisted of two parts: the theory of mechanisms and general theory of machines. By Zernov's definition, "the machine and the mechanism are the same things but considered from different points of view; the machine is considered as a unit on which the forces are exerted that produce the work, the mechanism is reckoned as a device consisting of individual parts that can have various movements relative to each other" [17].

In the author's preface, some data from the kinematic geometry were given: the concepts of planar motion, centers of rotation, the theory of poloids and envelopes, Savary's theorem, the concept of head swivel. Zernov explained the principles of kinematics according to Reuleaux; he takes Willis's scheme of classification but with several modifications introduced by Bour. "The imperfection of this classification appears clearly if we pay attention to the fact that, for example, the mechanisms derived from the same plane four-link chain are distributed into all three classes although they are related to each other by common origin from the same chain and, as we could see, should be considered together" [17].

This observation of D.S.Zernov was adopted by most Russian teachers of applied mechanics who used the classification of Willis while correcting it as needed. Zernov singled out three groups in his classification, which took into account the type of transmission (direct contact, intermediate hard segments, flexible bodies), and three classes (with the ratio of angular velocities being constant, variable, and constant and variable jointly).

Zernov considered thoroughly the mechanics of toothed gearing. He measured all sizes of the wheels in fractions of a step (the term "module" was not used).

Zernov began the second part of the course devoted to the general theory of machines with basic concepts of steam engines. In fact, three sections of the second part were related to the dynamics of machines: general conditions of machines, controlling the machine running, and the basics of the theory of parasitic resistances (sliding friction, rolling friction, and friction of flexible bodies).

That is what the level of knowledge in applied mechanics was like at the end of the 19th century. But before the decade was over, in 1909, the introductory course, the theory of mechanisms, the general theory of machines, and a number of special courses were lectured at the ITS complemented with graphics works. Special courses included those lectured by Zhukovskii who gave the lectures on the controlling of machine running and the theory of controllers. A.I Sidorov lectured the introductory course entitled "Elements of the Machine Science" that was a kind of encyclopedia on the mechanical engineering: it presented general concepts of the theory of mechanisms and machines, science of machines, usage of the mechanics of elastic bodies in calculating the machine parts, and general design principles of machinery and individual machines. This obligatory course was two hours a week in the first half-year period.

1.8 N.I.Mertsalov (1866-1948) [2, 18-20]

Nikolai Ivanovich Mertsalov (Fig. 5) was the famous Russian mathematician and engineer. His major works were devoted to the theory of machines and mechanisms (including works on the theory, calculation and design of toothed mechanisms), thermodynamics, theory of friction. He was one of the founders of the dynamics of machines and the theory of spatial mechanisms.

Mertsalov entered the Moscow University (mathematics department) in 1884. He sought to devote himself to science while being financially independent; this desire forced him to manage his small inheritance so that the money would be enough till the end of the learning as well as for the travel abroad (as he planned). Mertsalov assigned to himself some monthly amount of money. At that time, teaching of natural and exact sciences was at a very high level at the Moscow University, the curriculum of the Mathematics Faculty was very extensive and compacted; the students had to take examinations even in optional subjects. From the very beginning of his learning, Mertsalov used original sources and studied self-dependently the works of Lobachevskii, Chebyshev, Lagrange, and Euler.

Fig. 5. N. Mertsalov

The most demanding teachers and professors were impressed by his knowledge. Zhukovskii drew his attention to the student Mertsalov too. In turn, Mertsalov considered Zhukovskii not only as an outstanding scholar of mathematics and mechanics but also as a first-class engineer with the vast practical knowledge and experience who was able to implement the findings of the modern science into practice. Mertsalov seemed to be under the direct influence of his teacher when he decided to become a well-informed scientist and engineer to be able to contribute into the development of national science and technology.

Systematic work with the originals enriched the knowledge of Nikolai Ivanovich in such a way that he presented his graduate paper "On the curvature of curves according to Lagrange" at the end of his university course and received the degree of candidate of mathematical sciences. Mertsalov graduated from the Moscow University in 1888. Professor P.A.Nekrasov invited him to stay at the department but Mertsalov was eager to fulfill his vital plan and rejected this attractive and natural proposal. He left for Germany in autumn of the same year and worked as a machinist, serviceman or mechanic at various machine factories for three years. He acquired there the craft skills and qualifications that became subsequently a subject of his pride. Mertsalov used the acquired skills in manufacturing the models of mechanisms and in other works. The hard work at the factories did not prevent Mertsalov from studying science in his free time, widening and deepening his knowledge; he attended lectures on selected subjects at the Higher Technical School in Dresden at the same time.

While getting acquainted with the foreign engineering industry and the ways of teaching the technical sciences, Mertsalov became more and more interested in new problems of theoretical engineering appeared in the science of machines and mechanisms, namely, applied (or practical) mechanics.

In 1892, Mertsalov returned to Russia and began extensive preparations for the master's exams at the Moscow University and for the external exams at the Moscow Higher Technical School. He was deeply convinced that the people working in the area of technological development were able to fulfill their duties towards the Motherland in the most productive way.

Academician I.I.Artobolevskii described the state of the technical science of machines at the turn of the 19th and the 20th centuries as follows:

"Summing up the state of practical mechanics in Russia and, in particular, at the Moscow University and Higher Technical School by the end of the 19th century, we can say that the Russian practical mechanics was at the verge of great scientific discoveries. New original scientific schools began to form that gave rise to our modern... school on the theory of mechanisms and machines. The foundations of the teachings of these schools were laid using the outstanding geometrical methods of investigations developed by N.E.Zhukovskii and the analytical methods developed and perfected by P.L.Chebyshev. The synthesis of these two directions gave birth to a new Russian... school of the theory of mechanisms and machines. It is this moment in the development of the Russian school that is considered to be the beginning of Mertsalov's research and pedagogical activity" [18].

Despite his sudden serious illness, Mertsalov successfully passed the examinations at the ITS during 18 months and received the rank of mechanical engineer in 1894. Then he passed his master's examinations with flying colors at the university and in 1895, after having received the master's degree, was involved by Zhukovskii in lecturing the applied mechanics and was appointed a guardian of the cabinet of applied mechanics at the university. Mertsalov and Zhukovskii took a shine to each other; they were on friendly terms until Zhukovskii's death. The professor appreciated the audacious thoughts of his disciple and colleague, his desire to seek knowledge, the ability and readiness to solve assigned tasks in his own way. Under Zhukovskii's recommendation, Mertsalov was elected an associate professor at the department of Applied Mechanics at the Bauman University (ITS) at the end of 1897. In the same year, he occupied the position of the private docent at the university and began to teach the optional courses on the theory of mechanisms and machines and to lecture a number of special courses (on descriptive geometry, thermodynamics, etc.). I.I.Artobolevskii wrote: "*With arriving of N.I.Mertsalov at the Moscow University and Higher Technical School, the period of his largest creative activity started when he laid the foundations of the theory of mechanisms and machines*" [18].

Nikolai Ivanovich had to teach the course of applied mechanics at the Moscow Higher Technical School, which was leactured by Prof. Zernov before him. The contents of Zernov's course on applied mechanics reflected, to a certain degree, the level of knowledge on the subject at the end of the 19th century.

As always, Mertsalov decided to follow his own way in the teaching of applied mechanics. Artobolevskii wrote: "*There were two ways for Mertsalov to create the foundations of the theory of mechanisms and machines: a simpler way was to cultivate and develop the Reuleaux's ideas that were appreciably elaborated in the works of a number of scientists by that time. Mertsalov chose the second and more difficult way of independent research, pioneering new directions in the science of mechanisms. This was the path specified in the works of Chebyshev and Zhukovskii ... Mertsalov was put forth by Zhukovskii into the pleiad of eminent scientists and engineers who formed the glory of our ... technical science*" [18]. Mertsalov produced his course on strict statements and laws of theoretical mechanics. He extensively used the developed theory of kinematic geometry while analyzing and synthesizing the mechanisms; in particular, it was Mertsalov who first used the so-called "head-swivel" technique. Students were deeply impressed by Mertsalov's lectures characterized by the clarity, neatness, and beauty of presentation.

The section "*Theory of Mechanisms*" included the following topics: general kinematic theorems on the motion of an invariant system, motion parallel to a plane, link mechanisms, screw mechanisms, gears, cam mechanisms, Hooke's joint, and guidance mechanisms. The section "General Theory of Machines" contained the teaching on transfer of the work in machines and their efficiency, parasitic resistances, vibrations of machines and their parts, and description of the methods to measure the work in machines and control the machine running.

Mertsalov's lectures at the ITS in 1898-1900 formed the basis for his courses on thermodynamics (1898) and applied mechanics (1900/1901).

The course created by Mertsalov covered the kinematics of mechanisms and dynamics of machines and was a great event in the Russian technical literature. The name of Mertsalov became widely known due to his outstanding works on the theory of mechanisms and machines, which considered the laws of motion of mechanisms and methods of their design. According to Artobolevskii, the course of Mertsalov was a profound scientific treatise summarizing the theoretical problems of mechanisms and machines rather than a mere textbook. Following the tradition, N.I. Mertsalov used Willis's classification but added also the Reuleaux's ideas.

The introductory section, however, presenting the basics of kinematic geometry was a novelty. The course "Kinematics of Mechanisms" [19] considered the theory of mechanisms based on the geometric methods of analysis and synthesis. The presentation was excellent, elegant and occasionally virtuosic. Such virtuosity and elegance remind us of the best classic works of the French geometers of the 19th century. The course "Kinematics of Mechanisms" considered very lucidly the basics of kinematic geometry and their applications to the research of mechanisms using the methods of centroids, head swivels, etc. The same book included an extended formulation of the problem concerning the synthesis of plane mechanisms, which became the basis for further work in this area; the synthesis of non-circular wheels; the transfer of motion by centroids and mutually enveloping curves, etc. The book gave also the solution of some special problems of the synthesis of link mechanisms and, in particular, the problem of the synthesis of link guidance mechanisms using the head swivels.

To appreciate properly the importance of the second fundamental work of Mertsalov, namely, his course "Dynamics of Mechanisms" [20], it is necessary to point out that, in fact, there were no systematic courses of the dynamics of mechanisms in the world literature at that time. It was not until 1923 that the "Graphic Dynamics" of Prof. Vittenbauer appeared. So, we can consider the Mertsalov's book to be the first major composite work on the fundamentals of the dynamics of mechanisms. The main part of the book presented the methods of the dynamic study of mechanisms. To solve this task, Mertsalov widely used d'Alembert's principle. While solving the problems of controlling the machines and studying the motion of machines under the influence of given forces, Mertsalov used his original method for studying based on the "Zhukovskii's lever". Mertsalov expanded the applicability of Zhukovskii's theorem when he assigned the mechanism link masses to segments. He gave a new solution to the problem of calculating the fly wheel using his own methods. Prof. E.M.Gut'yar indicated later that this solution was more accurate and convenient than the solution of Vittenbauer.

In conclusion, we would like to say that the works of the above noted scientists had a great effect on the development of the national schools dealing with the theory of mechanisms and machines as well as other areas of mechanics in Russia in the 20th century.

Acknowledgement. The authors thank A.Vukolov and D.Saschenko, members of the Bauman Moscow State Technical University staff, for their valuable help in the design of the text, discussions.

References

[1] Niks, N.: The mission of professor is great and noble (Life and work of Moscow professors in second part of the XIX – beginning of the XX century). Working paper WP6/2004/01. State University – Higher School of Economics, Moscow (2004) (in Russian)

[2] Tyulina, A.: Development of Mechanics at Moscow University in the 18th and 19th century. In: The Scientific Works of History and Methodology of Science (IMI), vol. VIII, pp. 489–536. AN SSSR, Moscow (1955) (in Russian)
[3] Volchkevich: Esse of Moscow Higher Technical School History, Moscow (2000) (in Russian)
[4] Hvesiyk, V. (ed.): Nikolay Mertsalov. Bauman University Press, Moscow (2008) (in Russian)
[5] Golovin, A., Tarabarin, V.: Russian Models from the Mechanisms Collection of Bauman University. Springer Science + Business Media B.V (2008)
[6] Golovin, A., Mkrtchyan, D.: A. Yershov. In: Ceccarelli, M. (ed.) Distinguished Figures in Mechanisms & Machines Science, vol. 1, pp. 135–149. Springer Science+Business Media B.V (2007)
[7] Golovin, A., Tarabarin, V., Tarabarina, Z.: Feodor Orlov. In: Ceccarelli, M. (ed.) Distinguished Figures in Mechanisms & Machines Science, vol. 2, pp. 175–198. Springer Science + Business Media B.V (2008)
[8] Shevirev, S.: History of Imperial Moscow University, Moscow (1855) (in Russian)
[9] Tarle, E.V.: Soul, King and God of Sevastopol. In: Almanac "Arabesque history", 8th edn., Russia of seas, Moscow, pp. 463–472 (1997) (in Russian)
[10] Radovsky, M.I.: Verner Siemens (1816-1892). The founder of electrotechnical manufactory. In: Mitkevich, V.F. (ed.) Compilation "Pioneers of Machine Industry", pp. 239–294. Leningrad, Moscow (1937) (in Russian)
[11] Yanovskaya, S.A.: Two documents from the history of Moscow University, vol. 8. Bulletin of Moscow State University (1952) (in Russian)
[12] Chinenova, V.N.: A. Yershov - a representative of Applied Mechanics at Moscow State University. In: The History and Methodology of Natural Sciendes, Issue XX. Mathematics, Mechanics, pp. 214–222. MSU, Moscow (1978) (in Russian)
[13] Bogolyubov, A.N.: Theory of mechanisms and machines within the historical development of ideas. Science, 248 (1976) (in Russian)
[14] Zhukovskii, N.E.: The Complete Works. T. 1, 4, 8 (1937) (in Russian)
[15] Mathematical Collection, vol. 6 (1874) (in Russian)
[16] Tiulina, I.A., Smirnov, N.N.: Mechanics in Moscow University. Iris Press, Moscow (2005)
[17] Zernov, S.: Applied mechanics. Part 1. Theory of mechanisms, Moscow (1896) (in Russian)
[18] Naumov, S., Tyulina, I.A.: The research and teaching of N.I. Mertsalov. In: History and Methodology of Science, vol. 4, pp. 150–164. Perm' State University, Perm' (1997) (in Russian)
[19] Mertsalov, N.: Mechanism's Kinematics. Lithographic course of lectures, read IMTS. M. Phelinsky, Moscow (1916) (in Russian)
[20] Mertsalov, N.: Mechanism's Dynamics. Lithographic course of lectures, read IMTS. M. Phelinsky, Moscow (1914) (in Russian)

Royal Manufactures Promoted by the Spanish Crown during the 18th and 19th Centuries: An Approach to European Industrialisation

Javier Echávarri Otero[1], E. de la Guerra Ochoa, E. Chacón Tanarro, J.L. Muñoz Sanz, E. Bautista Paz, J.R. Álvarez Redondo, A. Díaz Lantada, and J.M. Munoz-Guijosa

[1] Universidad Politécnica de Madrid
José Gutiérrrez Abascal, 2. Madrid 28006, Spain
jechavarri@etsii.upm.es

Abstract. This work presents the Royal Factories promoted by the Spanish crown in the 18th and 19th centuries in the context of European industrialisation. They are classified together with the main types of factories and some specific examples are described. The article includes a discussion on the conditions that promoted their proliferation as the precursors of present-day state-owned industry, as well as the main causes that led to their decline.

1 Industrialisation in Europe

It is well-known that with the onset of the Industrial Revolution, in the second half of the 18th and in the 19th century, a rapid transformation took place in Europe from an economy based on manual work to an economy dominated by industry and manufacturing. As a consequence of this, radical changes came about at a socio-economic, technological and cultural level.

It may be said that this industrialisation had a prior and even parallel development in the so-called European Royal Manufactures in the 17th century, which formed the basis of a first approach to present-day state-owned industry. The systematic interference of the State in industry has its origin in the France of Luis XIV whose Secretary of the Treasury, Jean-Baptiste Colbert (Figure 1.a), transformed the "Manufactures Royales" into a key piece of his economic policy [1, 2].

During the years when Colbert was in the government (1661 -1683) more than a hundred Royal Factories were built, mainly in the luxury items and military sectors, among which it is necessary to highlight the Fine Cloth Mill at Abbeville, the Foundries at Nivernais and Delfinado, and the mirror works at Reuilly and Tourlaville that would later merge to become Saint-Gobain Glass [3] (Figure 1.b).

The prestige of the French Monarchy in Europe and the repercussions of the industrial policies embarked upon by Colbert caused other less advanced European countries to follow its example [2]. Such was the case of Russia, Prussia and the Austrian Empire, although with a different intensity [4].

T. Koetsier, M. Ceccarelli (Eds.): Explorations in the His. of Machines & Mech., HMMS 15, pp. 55–68.
springerlink.com

a)  b) 

Fig. 1. a) Jean-Baptiste Colbert (1619-1683); b) Glassware casting in an early Saint-Gobain factory.

The attempts at industrialisation in Russia had their origin in the initiative of Czar Peter the Great (Peter I) (1689 -1725) and his successors who used forced labour instead of free labour as in France. One of his outstanding successes was the metallurgical complex in the Urals, originally developed for military reasons. Such was the growth in this sector that Russia became the first European producer of iron, which it exported to countries like England. Nevertheless, the obsolescence of its technology meant that this supremacy was to be short-lived.

Something similar took place in Prussia, when King Frederick II (1740 -1786) took determined measures towards industrialisation. Its pattern followed that of the iron and steel works in Silesia and the industrial complex in Berlin that included cloth, silk, porcelain and weapons factories. In spite of the great expectations, the results were very poor due to the lack of buyers, with the king himself bearing witness to the disappearance of a large part of the industrial complexes created under his protection.

Finally, we should mention the efforts of Charles VI (1711 - 1740) to implement an industrial policy similar to Colbert's in the Austrian Empire. Porcelain manufactories were established in Vienna, cloth factories in Linz, and glass and mirror factories in Neuhaus. All these factories developed in the same way to those of other European countries; they were precariously run and ended up disappearing in the course of time.

2 The Royal Manufactures in Spain

The Royal Factories were brought to Spain by King Philip V, and became established from the beginning of the 18th century, and framed in the industrial policy of Enlightened Despotism. Philip V, the grandson of Louis XIV, promoted a type of company that was based on those in France and very specific to its age. Although continuity was not achieved from the time of the Royal Factories up to the present, they are considered to be the forerunners of present-day state-owned industries [5].

The reasons behind the founding and promotion of the Royal Factories were very diverse but the following are particularly significant:

- Strengthening of the State and reaffirmation of the monarchies.
- Self-sufficiency of the internal and colonial market. It was hoped to decrease the dependence on imported arms and imports in general, with the purpose of strengthening defence requirements.
- To promote industrial development. The traditional factories facing a crisis needed revitalising while new state factories needed to be set up. It was often a question of sowing the seed for the future development of private industries.
- To increase the fiscal revenues of the state by collecting indirect taxes that applied to manufacturing activities.
- To satisfy the luxury needs of the Court, regarding tapestries, glassware, mirrors, porcelain, etc.

To achieve such objectives, a series of industrial policy measures were adopted to produce a framework for running the Royal Factories:

- Customs barriers and protectionist policies to favour domestic products in the internal and colonial market.
- To increase the level of Spanish technology through a better knowledge of the industry of the surrounding countries by importing experts and technology, and even by carrying out espionage.
- Taking measures to eliminate social prejudice against the mechanical trades. The Royal Order of 18th March 1783 declared all manufacturing activities to be legal and honourable.
- To restrict the privileges of the craft guilds that blocked the growth of urban industries with their ordinances.
- Actions to stimulate and help private initiative regarding industry: tax exemptions, monopolistic privileges, interest-free credits and non-refundable subsidies.

The fate of state industries in Spain was no different from the rest of Europe and little by little there were changes in policy regarding them. The Crown gradually freed itself of the Royal Factories with the greatest losses, their enormous privileges were curtailed and tax exemptions were extended to private enterprise.

During Charles III's reign (1759 -1788) some Royal Factories were opened with the purpose of strengthening the state industrial sector, while at the same time other industries were privatised, mainly in the textile sector. Later on, with Charles IV (1788 -1808), the privatisations continued, the theories of Colbert tended to be abandoned and liberal economic thinking began to come to the fore. In spite of that, luxury industries were subsidised like the San Ildefonso Royal Glass Works and the Buen Retiro Porcelain Factory, and even military industries were created [5].

An important influencing factor on the destiny of the Royal Factories in Spain was the French invasion and the ensuing War of Independence (1808 -1814). Its consequences of death, destruction and famine were large-scale and led to an enormous drop in population, a loss of skilled workers and serious damage to manufacturing infrastructures. At the same time, the process of emancipation of the colonies made the crisis worse, leading to a narrowing of the of the goods market.

With the reign of Ferdinand VII (1814-1833) and Isabella II (1833-1868) the ideas of privatising state companies were gradually imposed [5], and most of them were sold, leased or simply abandoned.

3 Classification of Royal Factories in Spain

The Royal Factories can be classified according to the purpose and type of production they were intended for [4]. Four groups can be distinguished: those dedicated to the industrialisation of Castille, the ornamental industries for the internal consumption of the Crown, the arms industries and, lastly, the various State-controlled monopolies.

Figure 2, shows a map of Spain with the location of the different Royal Factories and their type. As can be seen, the ornamental industry is concentrated on the outskirts of Madrid, very near to the Court, which was the final consumer of the production. The arms industry, closely tied to the navy, is in the areas near the coast. The industries of Castille are concentrated in the middle part of the Peninsula. On the contrary, the monopolies are scattered throughout the whole country.

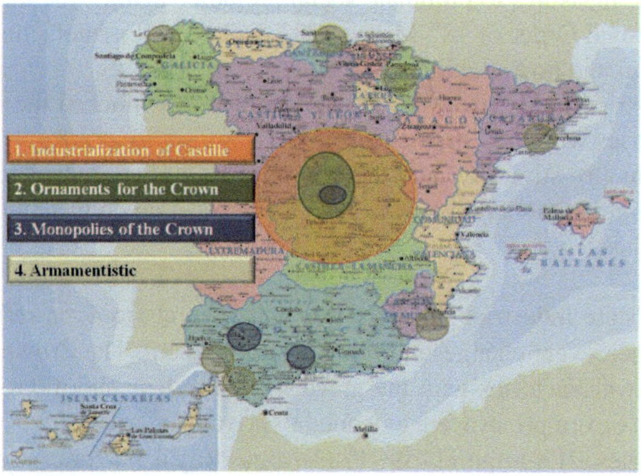

Fig. 2. Distribution map of the Royal Factories in Spain.

Industrialisation of Castille

It was hoped that the Royal Factories intended to industrialise Castille would endow the plateau of Castille with an industrial framework. So, it was attempted to bring together all the small craft industries scattered throughout the country, as well as provide them with new techniques to improve their production. Part of this group are the following factories: the Royal Cloth Factory in Guadalajara, the Royal Textile Factory in Avila, the Royal Textile Factory in Talavera, the Royal Cloth Factory in Segovia, the Royal Cloth Factory in Ezcaray or the Royal Cloth Factory in Alcoy.

Ornaments for the Crown

This type of factory was closely linked to the Crown since it supplied it with luxury items and objets d'art such as tapestries or mirrors. The main purpose of this type of Factory was to substitute imported goods for domestic products. Part of this group are the Royal Tapestry Works of Santa Bárbara, and the Silverwork and Clock Making Factory in Madrid, the Buen Retiro Porcelain Factory, the Moncloa Porcelain Factory, the Royal Mint and the Glass and Mirror Works in San Ildefonso.

Crown Monopolies

Forming part of this group are the industries that were state monopolies. These types of Royal Factories were of a very wide variety. The most important were the Tobacco Factories in Madrid and Seville, the Tin Works in San Miguel and the Playing Card Factory in Macharaviaya.

Arms

The purpose of the Royal Ordnance Factory was to strengthen the defence industry [6]. It was mainly centred on the arms programme (cannons and ammunition) and on the naval industry. A major part of this group are: The Royal Arms and Ammunition Factories in Eugui, Orbaitzeta and Muga, the Royal Bronze Cannons Factory in Barcelona and Seville, the State Artillery Foundries in Ximena and Trubia, the shipyards in Cartagena, Ferrol Guarnizo and Cadiz and the Cannons and Artillery Factory in Lierganes and La Cavada (Cantabria). Many of these Royal Factories formed the foundations of the steel industry in Spain.

4 Royal Cloth Factory in Guadalajara

This was founded in 1718 by Baron de Ripperdá. It was the first company to be modelled on the Colbert pattern of industrial state promotion and was founded to meet the urgent need to raise both the quantity and quality of woollen cloth production. Among the reasons for locating the factories in Guadalajara were their proximity to the market of the court and the natural resources surrounding the facility (wood, agricultural resources, water).

The factory comprised the building housing the looms, hammers, the wool laundry and the dying building. It also had dry lands in the proximity where the raw materials for the production of dyes were sown. Figure 3 shows an outline of the different hydraulic and hand-operated machines that were used for these tasks.

The cloth production process was the following one: the wool was washed to eliminate the grease, the carders worked to eliminate the hair, it was spun, and then it was knitted. Then the hammer was again used where the cloth was compacted and once more any grease was removed. Then, the nap of the cloth was cut and equalled. Finally the cloth was dyed.

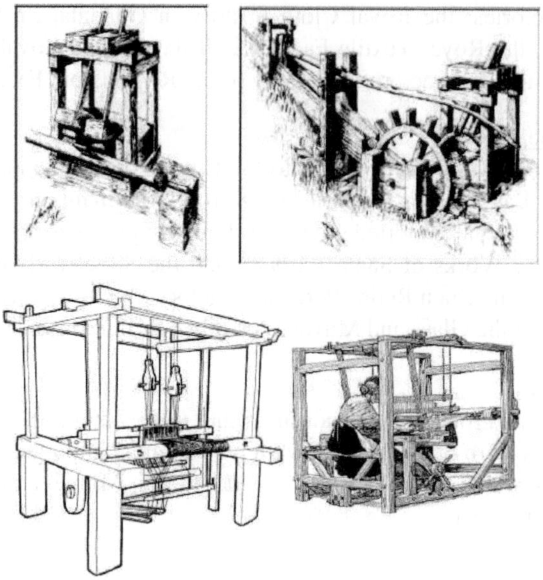

Fig. 3. Hydraulic hammers and loom.

From the very beginning (1718 -1731) problems arose through the management becoming politicised and through the workers' difficulties to adapt to the climate and language as many of them were foreigners. All this, on top of the lack of any tradition in the production of cloth in the area led to high costs and low quality products.

1731 saw a reorganisation of the administration in an attempt to put an end to the labour problems, improve management and quality and increase sales. Two branches were set up in nearby areas, in San Fernando de Henares (1746) and Brihuega (1750). The Royal Factory in Guadalajara became the most important factory in Spain and one of the biggest in Europe for its turnover; it ended up employing almost 15,000 spinners [7, 8]. Nevertheless, throughout its existence, it produced millionaire losses for the Royal Treasury.

5 The Royal Glass and Mirror Works of San Ildefonso

In 1727 the State decided to establish the Royal Works with the dual purpose of breaking the monopoly of French mirrors and to perfect the qualities of the glass. At that time, quality mirrors and glass were a symbol of the wealth and power of the Spanish Crown.

To meet these objectives, specialised manpower was brought to Spain and tasks of industrial espionage were carried out together with a heavy investment in technology. Major infrastructures were built in San Ildefonso to house all the technology and imported machinery [9].

The mirror production process was slow and complex. Several stages were required, which are detailed below.

Once the glass has been obtained, it must be rolled to be able to make it equal in thickness and to avoid imperfections in the reflection of the light. Once it has been levelled, both sides of the sheet are carefully polished. For the polishing process different stones and emery are used, which when mixed with water are able to give the mirror a very good finish. On completion of the polishing process, the glass sheet is given a mercury coating to ensure a full reflection of the light.

In order to save time and costs in the manufacturing process and to save on manpower, different machines were developed throughout the history of the Royal Works, some powered by hydraulics and others by animal traction (in dry seasons) which enabled the entire polishing process to be simplified.

The first hydraulic machine (1743) may be ascribed to Ventura Sit (Figure 4.a). It was capable of operating up to sixteen benches for the polishing process. In parallel to this, hydraulic mills were developed to obtain the emery required for polishing. In 1754, a machine was built powered by horses with six polishing benches designed by Alonso Garzón (Figure 4.b).

a) b)

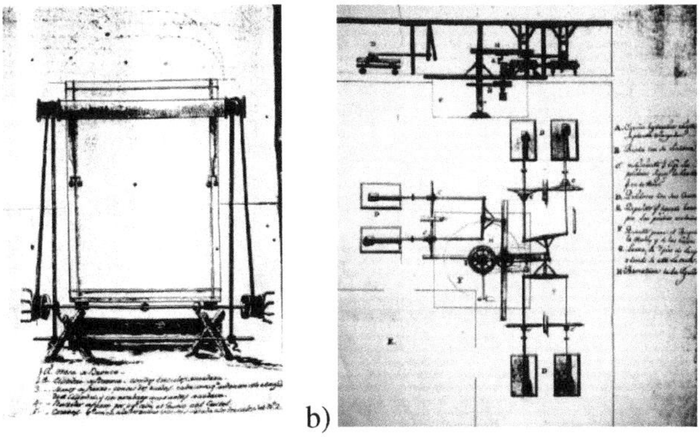

Fig. 4. Machines by: a) Ventura Sit, b) Alonso Garzón.

As the years passed these machines were improved, firstly by Juan Dowling (1761) and later by Demetrio Crow and Tomás Pérez (1786). Juan Dowling's hydraulic powered machine was capable of simultaneously powering a hundred polishing benches. The Encyclopaedia of Diderot and D'Alembert illustrates the plans of the machine in full detail (Figure 5) and presents it as a major step forward for the time.

In 1786 Demetrio Crow, together with Tomás Pérez, designed a more complex machine than the previous ones, with twelve polishing benches, as well as various millstones, which meant that the raw materials could be obtained by coupling the whole assembly to the same hydraulically powered drive shaft. The National Glass Centre Foundation has constructed a scale model of the machine (Figure 6).

Fig. 5. Juan Dowling's polishing machine.

Fig. 6. Demetrio Crow and Tomás Pérez's polishing and milling machine.

The Royal Factory continued working up to 1833, the year of Ferdinand VII's death, when production came to a halt. Queen Isabella II attempted to re-start production in 1853, but due to the scarce profits and the delicate political situation of the time, it was finally decided to lease the factory to private enterprise.

6 Royal Mint

We will now describe how the different techniques for minting evolved during the 16th and 19th centuries in the different locations where the coinage was minted in Spain. Figure 7 shows some of the presses used [10].

The techniques for minting coins at that time were as follows: by hammering, casting and mechanised minting (rolling on a Taschenwerk press, a flywheel press and already in the 19th century the mechanised press). The introduction of new techniques constituted a race against the ever increasing number of counterfeit coins.

Fig. 7. Rolling press with detail of the rollers (in Segovia) and a Taschenwerk press with detail of the die.

In 1661 production using hammers was brought to an end and the construction of mills and minting machines was begun so as to be able to carry out rolling in all the Mints, as Figure 8 illustrates. The purpose of introducing rolling was to reduce the number of fakes made with a hammer in amateur workshops.

Fig. 8. Flywheel press, automatic steam press and friction press.

The most common coinage at that time was made of copper lightly alloyed with silver. In spite of the measures taken to make it difficult to mint counterfeit coins (made of pure copper) the counterfeiting business was booming. Counterfeiting was carried out with a hammer using clandestine homemade equipment or in legal minting houses in Holland and Flanders. Finally in 1664 milled coinage was discontinued.

Throughout the following years new techniques were introduced, and made compulsory in 1730 for all metals, flywheel coinage and coinage with a milled edge. In the workshops equipped with mills, these were only used to roll the metal. Due to these changes, the number of legal mints was reduced to three: Segovia, Madrid and Seville.

It was not until 1808 when the next innovation reached Madrid; this was the flywheel press with a coin collar or the Droz press that can be seen in Figure 9.

Fig. 9. Coin collar and flywheel with a die.

Finally, in 1859 a steam machine, Watt's machine, was built for minting. This was one of the first applications of the steam machine in Spain.

7 Royal Tobacco Factories in Madrid and Seville

From 1636 tobacco was an exclusive product, that is to say, the state controlled its production and sale, exercising a monopoly over its trade. Construction of the Seville Factory began in 1728 with the purpose of grouping together the different factories that were scattered around the city. However, production would not begin until 1758, at which date the factory was still not completely finished [4].

Due to the exclusive nature of tobacco, the Factory was built along the lines of an old military fortress with a main building (workshops, stables, warehouses, etc.) with a chapel and prison as annexes. It was surrounded by a moat on three sides and a wall on the other. It had a drawbridge and guards that carried out compulsory searches on the way in and out.

During the more than thirty years that it took to build, the tastes in tobacco changed; snuff gave way to rolled tobacco and, so, the Seville Factory became practically obsolete right from its beginning. The snuff production process consisted of five stages described below.

The process begins cutting the leaves of the tobacco plant for drying in the sun and wind. In a next stage a first grinding of the tobacco is carried out and later it is sifted. Subsequently the tobacco is mixed with water containing different aromas that vary its appearance and flavour. It is then subjected to a drying process on floors by spreading the tobacco over a large surface to reduce the moisture and air it. Finally, in the finishing process, the tobacco is ground and refined. Figure 10, depicts the different machines used throughout the process.

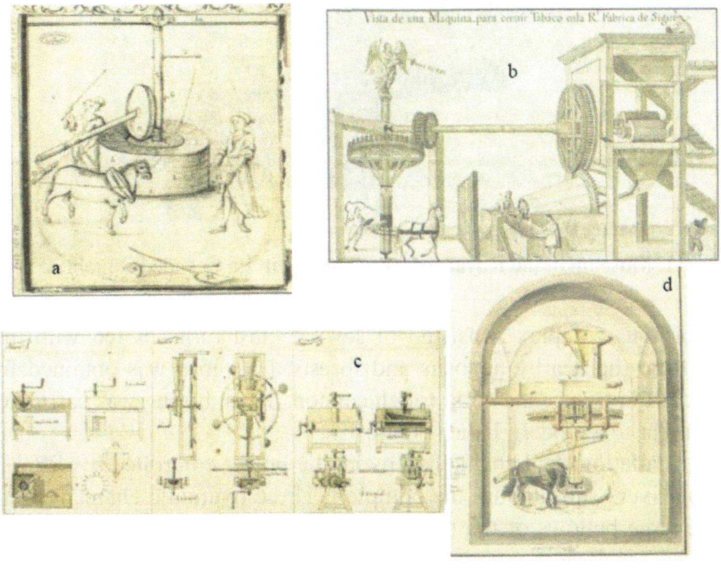

Fig. 10. a) Tobacco mill, b) Machine for sieving tobacco, c) Machine for cutting tobacco leaves d) Regrinding the tobacco

As the years passed, the consumption of cigars gradually increased. Towards 1800 the price of tobacco fell due to smuggling and changes were made to the exclusive nature of tobacco. New factories were built (Madrid, Alicante and La Coruña), the Seville factory was reorganised and the production of cigarettes and cigars was increased.

The changeover in the consumption of tobacco simplified the production process since it was no longer necessary to carry out the different stages previously mentioned. Production became a totally manual process. The process of rolling the cigarettes required no great physical effort and was habitually done by "cigarette women". The Royal Tobacco Factory in Seville continued operating up to 1950.

8 Cannons And Artillery Factory in La Cavada

The Factory began its work in 1637 manufacturing iron cannons and artillery (Figure 11), as well as other civil components like pipes or ships parts., Throughout their two hundred years of operation they manufactured twenty-six thousand cannons and nearly a hundred thousand tons of iron was cast [11].

Fig. 11. Museum of the Royal Artillery Factory in La Cavada (Cantabria, Spain)

This enormous complex consisted of several blast furnaces fed with the mineral and wood from the nearby deposits and forests. Cast iron was obtained from these blast furnaces to produce the solid rough cannon. Then, boring was carried out with a machine similar to a modern lathe (Figure 12).

All the installations had channelled hydraulic power controlled by different sluice-gates on the rivers surrounding the complex. These hydraulic channels were used to operate the large bellows of the blast furnaces and to operate the different machines like the boring machine.

Fig. 12. Scale model of the boring machine. Museum of the Royal Artillery Factory in La Cavada (Cantabria, Spain)

The Royal Factory in La Cavada continued working up to 1835, when the production of cannons practically ceased after Spain's defeat at the Battle of Trafalgar against the English Navy.

9 Reasons for the Decline of the Royal Factories

As already mentioned, the Royal Factories constituted a very specific type of industry at the time, which gradually fell into decline due to the different problems that will now be explained.

On the one hand, the multiplicity of purposes behind the creation of the factories led to a certain confusion and conflict among them (revitalising traditional factories, improving the quality and solid nature of the arms industry, reducing dependence on foreign imports, etc.). Consequently management was not very efficient, because frequently, political interests took priority over economic interests and the smooth running of the factory.

On the other hand, on the part of the government there was an ignorance of the technical complexity and instead of perfecting the existing technologies foreign technology was often imported as well as specialised technicians. If this is added to the scarce investment on the part of the bourgeoisie, it can be well understood that the Royal Factories were unable to transmit their technology to the surroundings in which they were located with the required effectiveness.

Also, the Royal Factories employed a large number of craftsmen, but they often worked in the same way as in the small workshops. They therefore had all the drawbacks of traditional production and few of the advantages of modern industry.

The internal market was scarce, a circumstance that was exacerbated by the colonies seeking freedom. Also, the attempts to industrialise Castille met with difficulties due to the bad communications and the increasing costs of transport.

As a result of all this, there was hardly any competition and the cost of the goods manufactured was high. There were defects in the organisation, low productivity, a shortage of raw materials, a high number of holidays and the factories were badly located.

10 Conclusions

The article describes the European and Spanish context in which the Royal Factories were created. A classification of the type of factories that were set up in Spain during the 18th and 19th centuries is shown and some of the most important ones are shown in more detail with several examples of the machinery used. The production activity carried out during the golden age and its subsequent decline is set out. However, it should be pointed out that in spite of the little success of the Royal Factories, these attempts at industrialisation served as a basis for the industrial boom of the 19th century in the different regions.

Acknowledgements. The authors would like to thank the following people and institutions for their valuable advice when preparing this article: Paloma Pastor (Fund. Centro Nacional del Vidrio), Rafael Feria (Real Fábrica de la Moneda), José Manuel Maza. (Fábrica de Cañones de La Cavada), Antonio Sama (Real Fábrica de Tapices).

References

[1] Coornaert, E.: Les manufactures de Colbert. L'Information Historique II, 1–50 (1949)

[2] Deyon, P., Guignet, P.: The Royal manufactures and Economic and Technological Progress in France before the Industrial Revolution. The Journal of European Economic History IX(3), 611–632 (1980)

[3] Pris, C.: La manufacture royale des glaces de Saint Gobain. Une grande Entreprise sous l'Ancien Régime (1665-1830), Lille (1975)

[4] Comín, F., Martín, P.: Historia de la empresa pública en España. Espasa Calpe, 51–87 (1991)

[5] González, A., Merino, J.P.: The Public Sector and Economic Growth in Eighteenth Century Spain. The J. of European Economic History (3), 553–592 (1979)

[6] Ruiz Cañete, O.: La empresa pública industrial en España: aproximación histórica y estrategia actual. PhD Thesis (1992)

[7] González Enciso, A.: Estado de la industria en el siglo XVIII: la Fábrica de Guadalajara, Madrid, pp. 558–567 (1980)

[8] Villaverde, M.D.: La Real Fábrica de Paños y la ciudad de Guadalajara. Wad-al-Hayara 8, 453–468 (1981)

[9] Pastor Rey de Viñas, P.: Historia de la Real Fábrica de Cristales de San Ildefonso durante la época de la Ilustración (1727-1810). CSIC (1996)

[10] Murray Fanton, G.S.: El Real Ingenio de la moneda de Segovia. Cámara de Segovia (2008)

[11] Rego, S.: En defensa de un Imperio. La Revista de Cantabria (July-September 2007)

The Influence of the Society of Jesus on the Spread of European Mechanical Knowledge in China in the XVIth and XVIIth Centuries

Michela Cigola

Laboratory of Documentation, Analyis, Survey on Architecture and Territory. DiMSAT. University Cassino, via G. Di Biasio 43, 03043 Cassino, Italy

Abstract. This article aims to investigate the role played by various missionaries of the Society of Jesus in the development and spread of European scientific and mechanical knowledge in China between the XVIth and XVIIth centuries.

1 The Society of Jesus

The Society of Jesus (in Latin *Societas Iesu*) is a male religious institution governed by papal law. The members of this order of clerks regular, called Jesuits, place the letters S.J. after their names.

The order was founded by the Spaniard Ignatius of Loyola who, along with several companions, professed vows in Paris in 1534 to preach in the Holy Land (this project was later abandoned in 1537) and swore obedience to the pope. Ignatius' plans were approved by Pope Paul III in 1540. After being expelled from several European countries in the second half of the XVIIIth century, the order was suppressed by Pope Clement XIV in 1773, but was restored by Pope Pius VII in 1814.

The Jesuits, whose motto is *Ad maiorem Dei gloriam*, indicating their search for mystical perfection, observe a vow of total obedience to the pope and are particularly involved in missions and education.

The spread of the order was rapid and its activities in several countries were extraordinarily vigorous. On the death of St. Ignatius (1556), it had more than 1000 monks with around 100 colleges and other houses spread out over 12 provinces, in Europe, Africa, the Indies and South America [1].

Some of the order's cultural centres were promoted by St Ignatius in Rome: the Collegium Romanum founded in 1551, and the Collegium Germanicum founded in 1552.

The history of Jesuit missions is integral to the history of relations between China and the West.

The first attempt made by a Jesuit to reach China was in 1552 by St Francis Xavier, a Spanish priest and missionary and founding member of the Society of Jesus. Francis Xavier, however, never actually reached China, dying a year after his arrival on the Chinese island of Shangchuan.

T. Koetsier, M. Ceccarelli (Eds.): Explorations in the His. of Machines & Mech., HMMS 15, pp. 69–79.
springerlink.com

Thirty years later, in 1582, the Jesuits began another mission to China, whose participants included Matteo Ricci. This mission brought western science and technology to the attention of the imperial court, and achieved significant results in intercultural dialogue with their Chinese pupils [2].

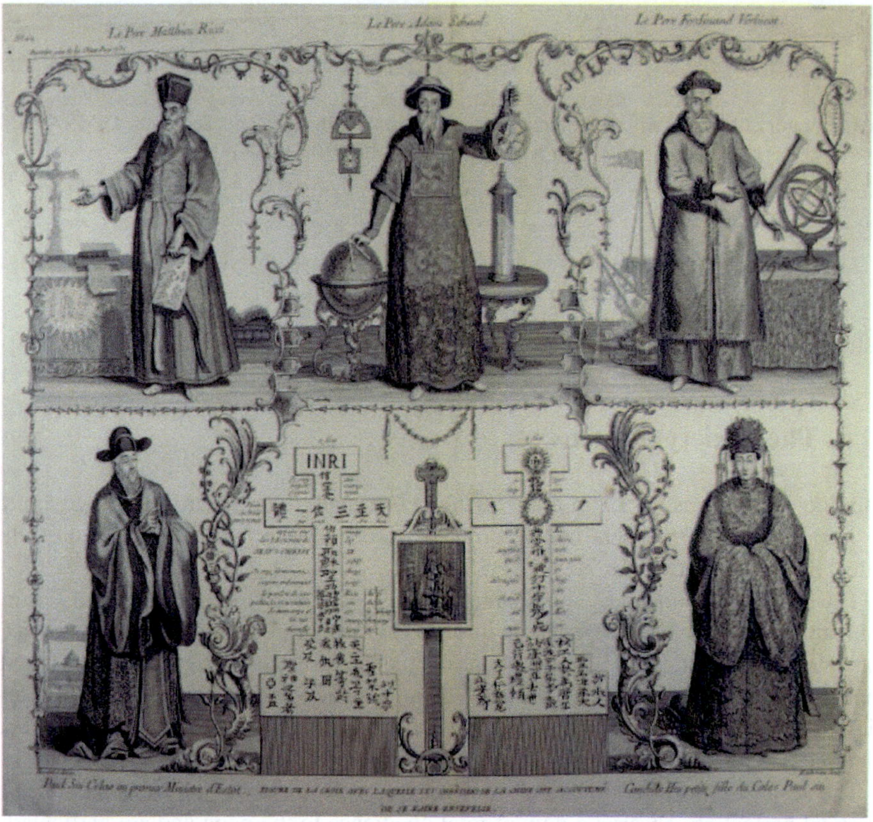

Fig. 1. Etching from the 17th century of the great Christians of China. Top left, M. Ricci and next to him F. Verbiest. Lower left, Xu Guangqi [3].

2 Missionaries in China

Matteo Ricci (1552 Macerata, Italy – 1610 Beijing, China; Chinese name Li Madou)

Arriving in Roma at the age of seventeen to study law, in 1571 he entered the novitiate of the Jesuits at Sant'Andrea al Quirinale, and the following year at the Collegium Romanum, remaining there until 1577, when he left for the East Indies. After reaching Goa, he was ordained into the priesthood and finished his theology studies. He then left to evangelise China. In Macau he began learning Chinese and in 1583 he reached Zhaoqing in Guangdong.

He presented himself as a scholar and man of science rather than as a priest, and began his work spreading western scientific knowledge. His evangelisation work was based on the utmost respect for Chinese cultural values and a relationship with the erudite class.

Fig. 2. Left, Emmanuel Yu Wen-Hui (known as Pereira) Portrait of Matteo Ricci, 1610. Rome, Church of the Gesù. Right, title page of De cristiana expeditione…" by Matteo Ricci.

He reached Beijing in 1601 and quickly won the esteem of the Wanli Emperor and many court personalities. He operated an extensive apostolate (2000 conversions, of which 400 in Beijing). He was buried in Beijing (he was the first foreign non-ambassador to be buried in China) [4].

His friendship with many of his disciples was a great help to him and the entire Society of Jesus. With some of them he compiled works in Chinese on theology and wrote books on science and technology based on European knowledge. Some of his pupils become Christians, the most important being Xu Guangqi, with whom Ricci translated *Euclid's Elements* and wrote many other works [5] (see below).

His collection of works is extremely vast: in addition to works on mathematics, astronomy, moral philosophy and music, with Li Zhao he also wrote: 1614 – *Tongwen suanzhi qianbian* (Treatise on practical arithmetics); 1614 - *Huanrong jiaoyi* (Treatise about geometry). Furthermore, as he was an expert cartographer, in 1584 he published a map of the world in Chinese, based on the cosmography of Christopher Clavius', his former master at the Collegium Romanum [6].

Xu Guangqi (1562 Shanghai, China – 1633 Beijing, China; Christian name Paulus Xu)

He was one of the most important Chinese to convert to Christianity. He was a minister of the Ming Dynasty, doing much for the Jesuits and promoting western scientific culture. He persuaded the emperor to assign the reform of the Chinese calendar to the Jesuits; they began in 1630 to work at this project under the supervision of Johann Schrek [7] (see below).

Fig. 3. – Left, etching featuring Matteo Ricci and Xu (A. Kircher, China illustrata, 1667). Right, portrait of Xu Guangqi (Guangqi Park, Shanghai).

From 1604 to 1607 he worked with Matteo Ricci studying, translating and writing several scientific texts. His most important work from this period was *Jihe yuanben* (Six books of Euclid). Other works with Matteo Ricci include: 1607 – *Celiang fayi* (On Trigonometry), *Gougu yi* (On rectangular triangles) [5].

Xu also worked with another Jesuit, Sabatino de Ursis, with whom he wrote *Taixi shuifa* (On western hydraulics) in 1612, a fundamental text which introduced European mechanical knowledge to China through the work of Agostino Ramelli (see below)

Sabatino de Ursis (1575 Lecce, Italy – 1620 Macau; Chinese name Xiong Sanba)

He was part of the first Jesuit mission to China arriving in 1603. In 1607 Matteo Ricci called him to Beijing on account of his knowledge of astronomy, hydraulics and mechanics [8].

He began collaborating with Ricci and Xu on the Chinese translation of Euclid's Elements. After Ricci's death, his collaboration with Xu grew stronger and in 1612 Sabatino orally translated the text of Agostino Ramelli's *Le diverse et artificiose machine* [9], which was then drafted in Chinese by Xu with the title *Taixi shuifa* from the notes made while studying with the Jesuit. The text was divided into four parts which describe the use of fluvial waters, groundwater and sources, rainwater and snow and includes an appendix covering miscellaneous subjects [10].

Together with two other texts on measuring instruments, this text was included in the collection *T'ien-hsüeh chu-han* (First Anthology of the Doctrine of Heaven) in 30 volumes which was published in Beijing in 1629 [11].

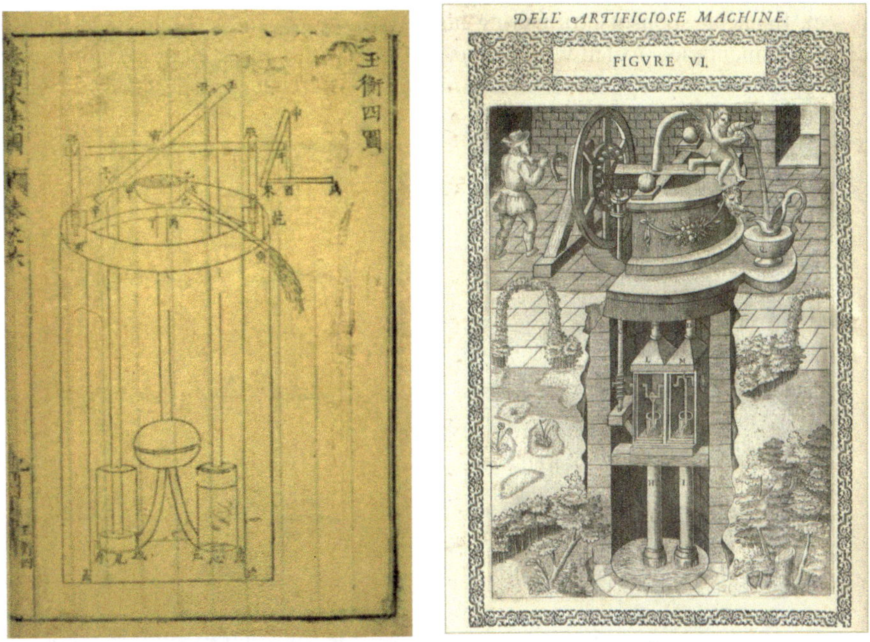

Fig. 4. Comparison of two machines for pumping water: left, image from Taixi shuifa by De Ursis-Xu, right, etching by Ramelli.

Figure 4 compares two tables taken from the works of Xu-De Ursis and Ramelli. Both illustrations present the machines in perspective, since a three dimensional view provides a better understanding of the overall aspect; however, in terms of graphic representation, the Chinese text, which was only 24 years more modern than its European counterpart, is extremely succinct, far removed from the refined luminist effects of Ramelli's version, which place the machines in their natural environments with several vanishing points.

In the European work, editorial flourished get the better of scientific and promotional intentions, whereas the Chinese version appears to concentrate on the desire to explain the machine and how it functions.

Nicolas Trigault (1577 Douai, Flanders – 1628 Hangzhou, China; Chinese name Jin Nige)

He joined the Society of Jesus and in 1610 landed in Canton, where he began his intense missionary activity, alternating periods in China with long trips to Europe.

The work by Trigault was the first draft and publication of Matteo Ricci's "Commentaries". On his death, Ricci left numerous written works in Italian – letters and diaries of his extraordinary experience in China – which he had been trying to systematise. The erudite Trigault was familiar with Mandarin and undertook to publicise these writings.

In August 1612 he received orders from Niccolò Longobardi, successor to Ricci, to return to Rome to deal with some important liturgical issues, recruit new missionaries and seek financial sponsorship. During his travels, he began the Latin translation of Ricci's Commentaries, which were printed in 1616 with the title *De Christiana expeditione apud Sinas suscepta ab Societate Iesu ex P. Matthaei Ricci* [12] (Fig. 2 right).

Fig. 5. Two portraits of N. Trigault by Peter Paul Rubens in 1617 during his travels in Europe. Left, drawing on paper (Metropolitan Museum of Art, New York, USA); right, Oil painting (Douai, Musée de la Chartreuse).

Trigault's trip to Europe began at the papal court in Rome and continued with visits to the major courts of the old continent. As testified in Rubens' portraits, he travelled wearing Chinese garb. He enrolled 22 young priests for the Chinese missions and obtained considerable funding.

Our study found Trigault's most important result to be the fact that he took a large number of scientific volumes back to China, along with numerous mathematical and astronomical instruments, including a Galilean telescope given to him by Cardinal Borromeo, which he in turn gave to the Emperor upon arrival in China.

Trigault was not really an expert in scientific and technological matters, but his works were of prime importance for the transmission and development of Western scientific knowledge in China [13].

Johann Schreck (1576 Bingen, Baden-Wurttenberg - 1630 Beijing; China; Chinese name Deng Yuhan)

A scholar with a wide range of interests and vast scientific and literary culture, he studied in Germany, France and Italy, obtaining a degree in medicine from Padua, where he became a disciple of Galileo. After meeting Galileo he Latinised his name to Johannes Terrentius [14].

He arrived in Rome in 1610 and became an associate of Federico Cesi, a scientist and naturalist from an aristocratic Italian family and founder of the Accademia dei Lincei. The objective of the Academy was to promote free and independent scientific research, and many of Italy's and Germany's best minds joined it. Galileo became a member of the academy in 1611, followed immediately by Johann Schreck in May of the same year [15].

Just a few months later, in November 1611, Schreck decided unexpectedly to join the Society of Jesus and left the academy, while maintaining contacts with his Academy friends.

During his theology studies, he met Nicolas Trigault, who persuaded him to become a missionary in China and travel with him to the courts of Europe.

On his arrival in China, he began collaborating on the project to reform the Chinese calendar begun by Xu Guangqi. While working on this task, Schreck promoted and publicised the scientific revolution undertaken by Galileo in Europe.

He and other confrères began working on the systematic translation of European scientific knowledge into Chinese [16].

One of the most important of these translated works is the *Qiqi tushuo* (Diagrams and explanations of the wonderful machines of the Far West) with Wang Zheng in 1627, a kind of encyclopaedia of European mechanical knowledge in Chinese. This was the first volume to present this kind of device to the Chinese public, since, despite being written in 1612, the work of Sabatino de Ursis and Xu Guangqi was not published until 1629.

The volume contained commented reproductions taken from the works of the Italians Ramelli [9] and Zonca [17] and the Frenchman Besson [18]. Figures 6 and 7 compare two tables taken from the works of Ramelli (fig. 6) and Zonca (fig. 7) with tables from the work of Schreck. A faithful reproduction of the original drawings in every detail, although the graphic style is completely Chinese.

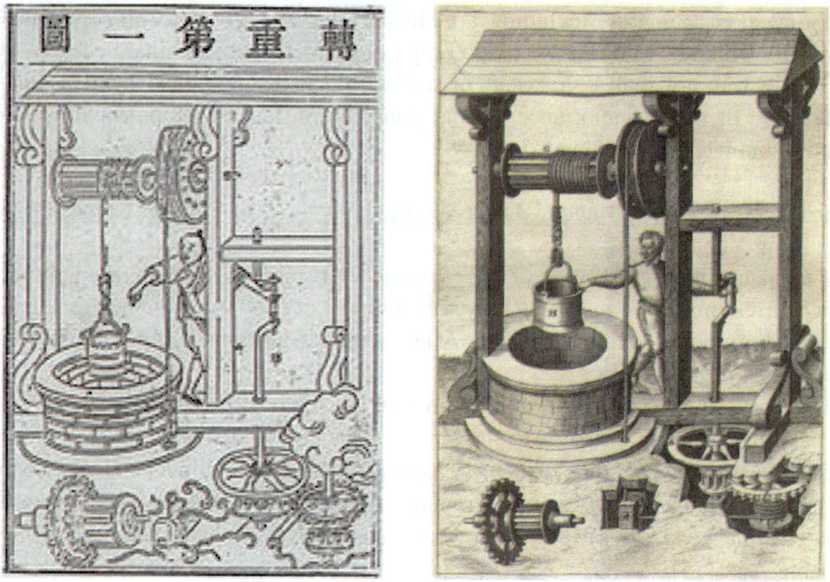

Fig. 6. Comparison of two wells: left, image of work by Schreck; right, etching by Ramelli.

The well taken from Ramelli's work is copied identically in terms of the overall layout of the table, the depth of perspective and the tiniest of details, although it completely lacks the graphic touches of the materials (such as the wood supports and the ground surrounding the well) [19].

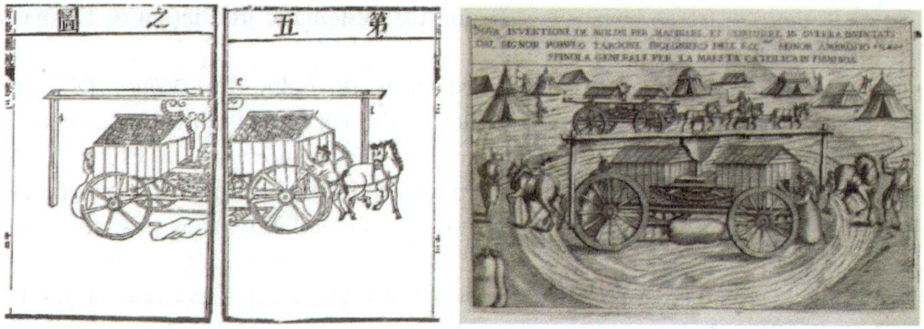

Fig. 7. Comparison of two carriages: left, image of work by Schreck; right, etching by Zonca.

As for Zonca's carriage, the Chinese text completely eliminates the natural setting and gives the carriage surroundings which are completely empty and ascetic. This concession to succinctness is not applied to the representation of the device, which slavishly replicates the Italian version in terms of overall three-dimensional space and the details presented of the machine and the animated objects surrounding it.

Ferdinand Verbiest (1623 Pitthem Flanders - 1688 Beijing; China; Chinese name Nan Huairen)

He operated in China during the Manchurian Qing Dynasty, which took over from the Ming Dynasty in 1644. He was a successful astronomer and mathematician and demonstrated the precision of the European astronomical model at the court of the Kangxi Emperor.

He began his mission in Shanxi, staying there until 1660, when he was called to be assistant astronomer to Adam Schall von Bell, Jesuit Director of the Imperial Observatory of Beijing and head of the Board of Mathematics, succeeding him on his death and rebuilding and re-equipping the observatory [20].

These offices led to a solid friendship between the Kangxi Emperor and Verbiest, who taught the emperor much about geometry, philosophy and music. He was often invited to the palace and accompanied the Emperor on his travels throughout the Empire.

He translated the first six of Euclid's books into Manchurian and, with the help of Chinese collaborators, redesigned the geographical map of the Empire. To recompense him, the Emperor bestowed on him the highest Mandarin title.

Fig. 8. Left, Portrait of F. Verbiest [21]; right, 19th century graphic reconstruction of steam carriage built by Verbiest for the Emperor.

In addition to his astronomy studies, Verbiest also experimented with steam. In around 1672 he designed – as a toy for the Emperor – a steam-propelled trolley which was, quite possibly, the first vehicle model to be powered by steam. We know the characteristics of this vehicle, thanks to the description given of it by Verbiest himself in his book *Defence of European Astronomy* [22]. However, the prototype was only 60 cm long, so was a scale model unsuitable for transporting passengers or a driver and, therefore, cannot be defined as an actual car. We do not know whether the model was ever actually built.

Notable works include in Manchurian translation: 1673/1674 - *Yi xiang tu* (Picture of 97 astronomical and mechanical instruments), 1674 – *Xin zhi ling tai/Yi xiang zhi* (Theory, construction and use of astronomical and mechanical instruments), 1681 - *Shenwei tushuo* (On cannon and military science).

3 Conclusion

The travels of the Jesuit missionaries in the XVIth and XVIIth centuries played an important role in the transmission and continuance of scientific knowledge in particular and culture in general between China and Europe, and had such an important impact on Christian culture and Chinese society that the effects are still felt today.

The importance of mechanical knowledge was even greater, as it was thanks to the Jesuit missionaries who arrived in China that many western texts and scientific instruments contributed – particularly through translations of important works by European authors - to transmitting, disseminating and developing European mechanical knowledge in China in the XVIth and XVIIth centuries.

Acknowledgments. The research work reported here was made possible partially by Grant n.250485 CIP-ICT-PSP2009.2.3 for Thinkmotion project by the European Community.

References

[1] Sommavilla, G.: La Compagnia di Gesu; prefazione del cardinale Carlo Maria Martini, Milano (1985)

[2] Masini, F. (ed.): Western humanistic culture presented to China by Jesuit missionaries (XVII-XVIII centuries). Proceedings of the Conference, Rome, October 25-27 (1996)

[3] Du Halde, J.B.: Description geographiqe, historique, chronologique, politique et physique de l'Empire de la Chine et de la Tartarie Chinoise, enrichie des car tes generales et particulières de ce pays, de la Carte générale et des Cartes particulières du Thibet et de la Corée, et ornée d'un grand nombre de Figures et de Vignettes gravées en Taille douce... P. G. Le Mercier, Paris (1735)

[4] Aleni, G., Criveller, G. (eds.): Vita del Maestro Ricci, Xitai del Grande Occident, Brescia (2010)

[5] Shangyang, S., Yijie, T.: Li Madou yu Xu Guangqi, Beijing (1993)

[6] Baldini, U. (ed.): Christoph Clavius e l'attivita scientifica dei gesuiti nell'eta di Galileo, proceedings of the International Congress (Chieti 28-30 aprile 1993) (1995)

[7] Jami, C., Engelfriet, P., Blue, G. (eds.): Statecraft and intellectual renewal in late Ming China: the cross-cultural synthesis of Xu Guangqi (1562-1633), Leiden (2001)

[8] von Collani, C.: Biography of Sabatino de Ursis SJ, China missionary, `http://encyclopedia.stochastikon.com`

[9] Ramelli, A.: Le Diverse et artificiose machine del capitano Agostino Ramelli dal Ponte della Tresia, Ingegniero del Christianissimo Re di Francia et di pollonia. Nelle quali si contengono varij et industriosi muvimenti, degni di grandissima speculatione, per cavarne beneficio infinito in ogni sorte d'operazione composte in lingua italiana e francese.., Paris (1588)

[10] Wang, C.: Origini familiari di Xu Guangqi. Shangai University press (2009)
[11] T'ien-hsüeh chu-han, photographic reprint in 6 volumes, Taipei (1966)
[12] Ricci, M.: De Christiana expeditione apud Sinas suscepta ab Societate Jesu... auctore P. Nicolao Trigautio Belga ex eautem Societate (About Christian expeditions to China undertaken by the Society of Jesus), Augsburg (1615)
[13] Dehaisnes, C.C.A.: Vie du Père Nicolas Trigault de la Compagnie de Jésus, Paris (1864)
[14] von Collani, C.: Biography of Johannes Schrecks SJ, China missionari and scholar, `http://encyclopedia.stochastikon.com`
[15] Battistini, A.: Galileo e i gesuiti: miti letterari e retorica della scienza, Milano (2000)
[16] D'Elia, P.: Galileo in Cina: relazioni attraverso il Collegio romano tra Galileo e i gesuiti scienziati missionari in Cina, 1610-1640, Roma (1947)
[17] Zonca, V.: Novo Teatro di machine et edificii per varie et sicure operazioni. Cò le loro figure tagliate in Rame é la dichiaratione, e dimostratione di ciascuna. Opera necesaria ad Architetti, et a quelli, che di tale studio si dilettano. Di Vittorio Zonca Architetto della Magnifica Comunità di Padova. In: Padova appresso Pietro Bertelli (1607)
[18] Besson, J.: Theatre des Instruments Mathematiques & Mechaniques de Jacques Besson Dauphinois, docte mathematichien. Avec l'interpretation des Figures d'iceluy, par Françoise Beroald, A Lyon, Par Barthelemy Vincent, Avec Prililege du Roy (1579)
[19] Zhang, B., Renn, J. (eds.): Transformation and transmission: chinese mechanical knowledge and the jesuit intervention, Preprint 313, Max Plank Insitute for the History of Science, Berlin (2006)
[20] Witek, J.W. (ed.): Ferdinand Verbiest (1623-1688): Jesuit missionary, scientist, engineer and diplomat, Nettetal (1994)
[21] Hamy, A.: Galerie Illustrée de la Compagnie de Jésus, Paris (1893)
[22] Bu de yi bian (1669) also as: Lifa bu de yi bian (Defense of the European Astronomy)

Section II

Biographical Contributions

Jang Yeong-sil: Inventor of the Striking Clepsydra during the Reign of King Sejong in Joseon

Nam Moon-hyon

Konkuk University and Jagyeongnu Research Institute, Seoul, Korea (South)

Abstract. Jang Yeong-sil was the Chief Royal Engineer in charge of the engineering and construction project for re-equipping the Royal Observatory during the reign of King Sejong (世宗 r. 1432-1438). Jang devoted himself to making five kinds of astronomical instruments and ten timekeepers. In this paper the life and contributions of Jang are introduced along with his major invention: the striking clepsydra. This paper also discusses how the clepsydras were handed down and utilized by Jang's successors in the making of astronomical clocks in Joseon. Not only did Jang assimilate the techniques of his Korean, Chinese and Islamic predecessors, but he also was a creative and innovative engineer. Jang is in some ways the most articulate among those in a long tradition of "automation" in Asian science and technology, where automation is a matter of submission to the laws of nature rather than the means of controlling nature, as seen in our time. This paper positions Jang's work as a catalyst for new directions in automation and robotic arts. He was one of those mechanical geniuses who contributed to the advancement of pre-modern science and technology

1 Introduction

In the Confucian cultures of East Asia, the doctrines of the legendary Chinese sage-emperors Yao (堯) and Shun (舜), which placed a priority on agriculture, became both a royal duty and a royal prerogative. To be a legitimate ruler, one must receive the Mandate of Heaven (天命), and only the recipient of this Mandate was deemed capable of observing the movements of the heavenly bodies. The successive dynasties and kings inherited the doctrines and were informed by these classical teachings, and kings made armillary spheres to observe the movements of celestial bodies, particularly the Seven Regulators (七政 Chiljeong), which include the sun, moon and five planets (Mercury, Venus, Mars, Saturn and Jupiter). Moreover, they employed astronomers to monitor movements of the heavenly bodies, and astrologers to divine portents in order to determine the appropriate imperial response to celestial anomalies. Under the Neo-Confucian tenet of "Observing Heaven and Granting the Seasons to the People," the production of calendars for farming and the reckoning of the time of day remained among the ruler's most important tasks since China's Song Dynasty (960-1279).

Neo-Confucianism was adopted on the Korean Peninsula by the founders of the Joseon dynasty (1392-1910) as state political ideology. The fourth ruler, Sejong the

T. Koetsier, M. Ceccarelli (Eds.): Explorations in the His. of Machines & Mech., HMMS 15, pp. 83–105.
springerlink.com

Great, was strongly influenced by the Neo-Confucian education he received during his formative years, and his upbringing encouraged him to study natural phenomena and human society in an effort to bring Heaven and Earth into harmony. Many of Sejong's accomplishments, from the invention of the Korean alphabet *hangeul* to re-equipping the Royal Observatory, were reflections of his Neo-Confucian values and their implementation in accordance with the doctrines of Yao and Shun. He oversaw the production of a water-clock with his scientifically-inclined father, King Taejong (太宗 r. 1400-18), before being invested as crown prince. That project was probably carried out with the aid of Jang Yeong-sil (蔣英實 1380? – 1450?).[1] Sejong was a political leader as well as a cultural hero whom many Korean revere for his fostering of cultural creativity, revitalizing intellectual life and encouraging technological and scientific advances. The king was particularly interested in astronomical and calendric matters: seasonal changes, times of sunrise and sunset, and rainfall and draught in his realm. These were vitally important for a state almost totally dependent upon agriculture.[2]

In 1420, Sejong reorganized the Hall of Worthies (集賢殿 Jiphyeonjeon) and assigned talented scholars and officials to spend their time researching and producing important writings on the classics, history, geography, linguistics, law, music, agriculture, astronomy, and medicine. The king wanted to make his own government calendars for farming independent from the Chinese calendar and to build ingenious machines that could tell time and show heavenly movements automatically. These devices would serve as a symbol of prestige for the Joseon government. Chief Royal Engineer Jang Yeong-sil ("Jang") devoted himself to making state-of-the-art timekeepers and astronomical clocks in response to the wishes of the King Sejong.

In this paper the life and works of Jang are introduced with his major inventions: striking clepsydras as well as how the traditional Chinese clock-making technologies were applied and innovated by Jang. It discusses the genealogical nexus employing Chinese and Islamic horology, which were believed to have inspired Sejong and his court scholar-officials. Also covered is how the features of the clepsydras were inherited by Jang's successors in the making of astronomical clocks. Jang is in some ways the most articulate of what is a long tradition of "automation" in Asian science and technology, in which automation is *a manner of submission* rather than the *means of control* seen in modern times. Not only did Jang assimilate the techniques of his Korean, Chinese and Islamic predecessors, but he also was a creative and innovative engineer. This paper positions Jang's work as a catalyst for new directions in automation and robotic arts and one of mechanical genius who contributed to the advancement of pre-modern science and technology.

2 Life of Jang Yeong-sil

Astronomical Instruments and Clocks made under King Sejong's Reign

Before embarking on a presentation of Jang's biographical sketch and achievements, it might be useful to contextualize the Korean science and technology of astronomy and timekeeping that informed and substantiated his work. Importantly,

King Sejong emphasized and encouraged the systematic development of science and technology as mentioned above. Immediately after his ascending the throne, Sejong worried that his astronomers were incapable of predicting eclipses with precision and had them study computational methods related to mathematical astronomy. The king decided to correct the Great Concordance System of Calendric Astronomy (大統曆 Datongli), the new calendar of Ming China, which was in fact identical to the Season-Granting Astronomical System (授時曆 Shuoshili) developed in China in 1279 and adopted on the Korean Peninsula during Goryeo. Sejong's action was to give his people an independent calendar. Consequently the first Korean calendar system, the Inner Volume of Computation of Seven Regulators (七政算內篇 Chiljeongsan-Naepyeon), referenced to 38 *tu* 1/4 (present 37deg 41min.) the latitude of Hanyang (漢陽 old name for the Joseon capital and present-day Seoul), was compiled in 1432 and disseminated in 1442.[3]

After the compilation of a new calendar, Sejong appointed Jeong In-ji (鄭麟趾 1396-1478) to serve as Supervisory Council Controller for Calendar Computation, with the task of re-equipping the Royal Observatory in the fall of 1432. The project involved researching the classics, creating the instruments and clocks, and providing the observation and examinations. Jang was in charge of engineering and construction under the guidance of Yi Cheon (李蕆 1376-1451), the Supervisory Council Controller for Constructing the Observatory. Sejong wanted to equip his observatory with precise astronomical instruments and clocks inside the palace complex and to use them to calculate celestial time and thereby determine human time. Following the royal decree, the government undertook the construction of the Royal Observatory (簡儀臺 Gaunuidae) between 1432 and 1438. The king recognized the need for determining the precise length of the tropical year. Work began on building "simplified instruments" (簡儀 *jianyi*) and a platform first referenced to the latitude of Seoul, a 45-foot-high measuring-scale and gnomon (圭表 *guibiao*), and square table (正方案 *zhengfangan*) used to find the precise north-south direction table. These devices were modeled after that developed by Guo Shou-jing (郭守敬 1231-1316) in Yuan China. Also, a water-operated armillary sphere (渾天儀 *huntianyi*) and celestial globe (渾象 *hunsang*), referred to the Chinese History Books and writings. These five instruments were arranged around the Banquet Pavilion (慶會樓 Gyronghoeru) inside the main palace complex, called Gyeongbokgung (景福宮) in Hanyang.

Ten kinds of timekeepers were created: the "Clock of Annunciating Clepsydra Pavilion" (報漏閣漏 Borugaknu), "Clock of Respectable Veneration Pavilion" (欽敬閣漏 Heumgyeonggaknu) and "portable water-clock" (行漏 *haengnu*); scaphe-sundial (仰釜日晷 *angbuilgui*), plummet-sundial (懸珠日晷 *hyeonjuilgui*), horizontal-sundial (天平日晷 *cheonpyeongilgui*), and south-fixing-sundial (定南日晷 *jeongnamilgui*); and "smaller time-determining instruments" (小定時儀 *sojeongsiui*), "time-determining instrument" (日星定時儀 *ilseongjeongsiui*) and "smaller simplified instruments" (小簡儀 *soganui*). Among these timekeeping devices, two striking clepsydras along with the scaphe-sundial, time-determining instruments, smaller time-determining instruments and smaller simplified instruments have been evaluated as Korean inventions, while the others were modified and re-engineered from older Islamic and Chinese models. The major part of inscription with an introduction on the creating of "Time-determining Instruments" was written by Sejong himself. Most of the above timekeepers were equatorial-types derived from the simplified instruments and armillary spheres. All the instruments and clocks were employing the Korean version of the Zhou-*foot* (周尺 *jucheok*) equivalent to 207.0 [mm].[4] In addition to these instruments, a rain gauge (測雨器 *cheukugi*) was invented by the crown prince, who later became King Munjong (r. 1451-2), in 1442, and rainfall was measured in *jucheok*. The principal driving force behind the undertakings was King Sejong, who played a pivotal role in supporting the research efforts of scholar-officials and the creation of instruments and clocks by engineers and craftsmen.[5]

The following aspects of the science and technology in Sejong's reign can be inferred from the project to re-equip the Royal Observatory:

1. To promote the makings of astronomical instruments and clocks, the government undertook state-sponsored R&D projects, led by senior scholar-officials, and the king was instrumental in providing support. He even joined the research team in person.
2. Talented scholar-officials, scientists, engineers and craftsmen took part, and the state invested in the research environment.
3. The most advanced Chinese and Islamic models of astronomical and timekeeping devices were studied.
4. The process began by researching the classics. Then new instruments were devised through modeling and experimentation.
5. To enhance their practicality, the instruments and timekeepers were designed to be smaller, automated, standardized and documented, leading to their adoption.
6. The astronomical and calendric undertakings of Sejong laid the groundwork for a new astronomical tradition in Joseon.[6]

The king was really a pioneer and originator of Korean Practical Learning (實學 Silhak).

Biography of Jang

Jang's family background is not exactly known, except that his mother was an official *kisaeng* serving Dongnae sub-county in Left Gyeongsang Province, where a naval headquarters was located. This is in the southeastern part of the Korean Peninsula, the area closest to Japan. As Jang's mother had low-born status, he also worked as a government slave during his youth.[7] He served as Chief Royal Engineer under two kings, Taejong and Sejong, from the early 1400's to 1442. He was put in charge of handcrafts and engineering affairs at the royal court, and promoted from a supervisory official (Rank 5A) at the Royal Guard Headquarters to a royal guard (護軍 *hogun*, Rank 4A) in 1433. In 1438, Jang became a grand royal guard (大護軍 *daehogun*, Rank 3B). The dates of Jang's birth and death are not known, and most of his later years are shrouded in mystery.

During Joseon, the biographies of the king as well as the "ministerial-class officials" (Rank 2B or higher) were recorded in the dynastic annals, known literally as the *Veritable Records* (實錄 *Sillok*). The annals covered the important events that occurred during the reign of each king and brief entries were included on important achievements by the ruler's subjects such as Jang.[7] More than ten entries on Jang appear times in the *Annals of Sejong* (世宗實錄 *Sejong Sillok*) during his twenty-year service to King Sejong. In addition to the official records, extant writings on Jang include the *Genealogy of the Asan Jang Clan* (牙山蔣氏世譜) and documentation on his works. The content of the available information varies greatly.

The *Genealogy of the Asan Jang Clan* states that the first settler (1st generation) at *Asan*, on the central west coast of the Korean Peninsula (across the Yellow Sea from China), was a Chinese General named Jiang Xu (蔣壻), the Commander-in-Chief of Army who was exiled from Northern Song at Keifeng around 1120 after the war between Song and the Jin tatars. From that time his descendants lived at Asan and Gaeseong, the Goryeo capital, serving the Goryeo government as higher officials in mostly military posts. Jang Sung (蔣崇), the 5th generation, served concurrently as director of the Royal Observatory Bureau (書雲觀 Seoungwan) and the Weaponry Office. Five brothers of Jang Yeong-sil's father (8th generation) all served as ministers in the government just before the fall of Goryeo, so they resisted the new Joseon regime. Four brothers survived by escaping from the capital to the countryside in Gyeongsang Province. Only Yeong-sil's family remained in Gaeseong; his father was executed and the social status of the remaining members was lowered. His mother was forced to serve as a government *gisaeng* at Dongnae sub-county.

According to the *Annals of Sejong*, the progenitor of Jang's family (General Jiang Xu) came from the Suzhou-Hangzhou area of the Great Yuan (actually Northern Song). This confirms the above-mentioned *Genealogy of the Asan Jang Clan*. Jang's biography and writings on his works have been revised as follows by bringing together all extant writings and records:

1. Jang Yeong-sil was born as a son of Jang Seong-hwi (蔣成暉), a Goryeo government minister, at Gaeseong around 1380.
2. In 1392 the Goryeo dynasty fell and the new Joseon dynasty was founded. In the chaos of dynastic change, Jang's father remained loyal to Goryeo, resisted the new government and was finally executed. The status of his family was lowered, and his wife and children had to serve Dongnae sub-county, the former as a government *gisaeng* and the latter as government slaves. (At that time people officially registered as lowborn, as Jang's family was, lived in their own house, worked independently in their own workshops, and participated in government projects during a certain period of each year.) We can guess that under the circumstances at that time, the young Jang received a Confucian education and was attached to Dongnae sub-county as an officially registered craftsman, along with his mother, after the fall of Goryeo. Dongnae was a frontier with Japan, and the Joseon government set up a navy headquarters there. Most of registered craftsman in that area had to make and repair weapons, build warships, print books and so on. The Joseon government treated as government slaves most craftsmen registered as

lowborn. Although these craftsmen were government slaves, they were quite different from private slaves and similar to agricultural slaves. They could be freed in some cases; their social status could be raised, and they were allowed to enter government service. Jang was a very talented and intelligent craftsman and it is said that the sub-county magistrate recommended him to King Taejong when the government was looking for talented man in the early fifteenth century. By law, Jang's registration was transferred to the capital to serve at the Royal court.

3. After the enthronement of King Sejong, Jang was freed from lowborn status and assigned as a supervisory official (別坐 *byeoljwa*, Rank 5A) at the Bureau of Royal Attire (尙衣院 Sang-uiwon) in 1423. The next year he was sent to Yanjing (now Beijing) for on-the-job training in astronomy with two imperial astronomers. Upon his return to Joseon, he made an improved night-watch clepsydra (更點器 *gyeongjeomgi*) to replace the night- clepsydra (更漏 *gyeongnu*) made in 1398, and was transferred with the same rank to the Royal Military Headquarters (行司直 Haengsajik). Around this time, Sejong commissioned Pak Yeon (朴堧 1378-1458), the supervisory council controller for traditional music and dance, to improve the classical court music (雅樂 *a-ak*), and Jang developed native musical instruments with him. Pak Yeon also developed armillary spheres with him later.
4. In 1432, he was assigned to participate in the Royal Observatory construction project, where he was put in charge of engineering and construction. He developed astronomical instruments and an automatic annunciating water-clock, as well as made improvements to metal moveable type.
5. In the ninth moon of 1433, he completed the Striking Palace Clepsydra (自擊宮漏 Jagyeokgungnu). The king determined that new water-clock was more refined than the Palace Clepsydra of Shun Di (順帝) of Yuan (China), promoting Jang to royal guard and assigning him to take charge of engineering affairs at the royal court.
6. In the seventh moon of the following year, the new clepsydra was put into service as the standard time-annunciator for the kingdom. And he developed high strength copper-zinc and lead-tin alloys for casting the metal type of the Gap-in Year (甲寅字 *Gapinja*), 1434.
7. In 1437, he learned metal refining techniques from a Chinese metallurgist and developed a new astronomical clock, the Striking Heavenly Clepsydra, for King Sejong.
8. In the first month of 1438, the new astronomical clock was installed in the Respectful Veneration Pavilion. Jang was promoted to a grand royal guard.
9. In 1438, he was assigned as special supervisor for collecting copper, iron and soft iron produced in Gyeongsang Province.
10. In 1442 Jang oversaw the construction of an open palanquin (sedan chair) for the king. The vehicle broke down during a royal procession, and Jang was punished. His appointment letter was revoked, and he was forced to resign his post. Jang's judgment was subsequently suspended by special royal decree. At that time they said that "Jang Yeong-sil and Pak Yeon were born to meet the needs of the times, achieving King Sejong's astronomical and musical undertakings." The

wholehearted devotion of these talented two figures enhanced the implementation of Neo-Confucian rites and music in early Joseon.

3 Major Works of Jang Yeong-sil

The Striking Palace Clepsydra: Borugaknu

In the Joseon period, the Royal Observatory Bureau was in charge of astronomical, calendrical, meteorological and clepsydral matters. Founder King Taejo (太祖 r. 1392-8) ordered the construction of a bell tower (鐘樓 *jongnu*) and had a big bell hung in it in on Jongno (鍾路 Bell Street) in downtown Hanyang. The bell was rung to announce the time provided by the newly cast night-watch clepsydra from 1398. It was the first standard time-annunciator in the new capital. Sejong commissioned Jang to cast a new night-watch clepsydra (*gyeongjeomjigi*) in 1424 to announce the curfew alert (人定 *injeong*, "people at rest") in the evening and morning lifting of curfew (罷漏 *paru*, "quit the clepsydra"), employing the Chinese calendar based on the Great Concordance system. The bell tolled 28 times for *injeong* at the third point of the first night-watch, symbolizing the 28 constellations in the heavens and also was a request to Heaven for a peaceful night. As soon as the tolling ended, the main gates of the capital (Great South, East, West and North Gates) and major palace gates were all closed. The curfew went into effect and club-wielding night-watchmen patrolled the streets.

At dawn, the bell was rung 33 times to signal the lifting of the curfew at the third point of the fifth night-watch, and the gates to the city were opened. On this signal, the king awoke with the rest of the people in the palace complexes and among the capital populous. The number 33 was a comparison of daybreak to the Buddhist 33rd Heaven, overseen by by Sakra-devanam Indra, and was also a supplication for a peaceful day. Time signals at the times for *injeong* and *paru* generated by the annunciating clepsydra pavilion were relayed to Gwanghwamun, the main (south) gate of Gyeongbokgung, and to other gates of the main palace compound. The signals were then conveyed by drum to the bell tower on Jongno. An official dedicated as the bell-striker at the bell tower would strike the big bell upon receiving the drum signal from the nearest drum relay post. At noon the midday drum sounded at the main gate as well as at the other major gates of the palace, announcing the time to thousands of people residing inside the palace complexes, including the king and queen, and to officials inside and outside the palace. The midday drum was a signal to quit morning activities and begin preparations for the midday meal.

The king worried that the old clepsydra was imprecise and unsatisfactory for operating the night-clepsydra, which employed the Chinese calendar. Therefore, he commissioned Jang to build a new one capable of automatic time-announcement without human intervention. This constituted a major aspect of the project to re-equip the Royal Observatory. Sejong wanted to run a new standard time-annunciator that would employ local times for sunrise and sunset, calibrated according to each fortnightly period in the Joseon capital. Upon receiving the order, Jang developed a new clepsydra capable of automatically announcing the time in 1433. This was followed by an astronomical clepsydra that demonstrated celestial movements with time-announcing jacks in 1437. The making of these clocks was documented in the *Annals of King Sejong* and supplemented by accounts in the *Revised and Expanded Edition of the Comprehensive Study of Civilization* (增補文獻備考 *Jeungjeong munheon bigo*).[4, 8]

In 1433, Jang invented an automatic time-annunciating water-clock called the "Striking Palace Clepsydra" under an order from King Sejong. On the first day of the seventh moon in 1434, the new clepsydra was installed at Annunciating Clepsydra Pavilion (報漏閣 Borugak), and employed as the standard time-annunciating time-keeper. The clepsydra was officially named the Clock of Annunciating Clepsydra Pavilion (報漏閣漏 Borugaknu) and also known as the palace water-clock (禁漏 *geumnu*) or self-striking water-clock (自擊漏 *jagyeongnu*). According to the *Memoir and Inscriptions with Preface* (報漏閣記銘幷序)written by Kim Don (金墩 1385-1440) and Kim Bin (金鑌 ? – 1455), scholars at the Hall of Worthies, the new water-clock consisted of three parts: a ball-falling inflow-clepsydra and the ball-operated dual time-announcing striking machines, and the interface with ball-traveling routes connecting these two mechanical systems. The uniqueness of this clock was its capability to announce twelve double-hours (one double-hour, or *si* 時, corresponds to two of today's hours) and five night-watches and their points and automatically with visual and audible signals.[9] (Each night-watch, *gyeong* 更, was divided into five *joem* 點, resulting in twenty-five time subdivisions per night. Each night-watch corresponds to about two hours today.)

The Striking Palace Clepsydra was made about eighty years after the Palace Clepsydra *Gong Lou* of Shun Di (順帝, Togan Timur, r. 1333-67). Sejong admired its novelty and specialized functions. The Joseon monarch told his high-ranking court officials, "Although he (Jang Yeong-sil) made the Striking Palace Clepsydra (自擊宮漏 Jagyeokgungnu) according to my instructions, nobody could make it except him. I have heard about the self-striking Palace Clepsydra (自制宮漏) made at the time of Shun Di, which was less ingenious than the craftsmanship of Jang's new clepsydra. Now that he has made a new clepsydra capable of transmitting through all the ages permanently, I propose you recognize his service on inventing new clepsydra. I would like *haengsajik* (Rank 5A) Jang to be promoted to royal guard (*hogun*, 4A) ….".[7] From this dialog, we can find that Sejong already knew about Shun Di's Palace Clepsydra from the dynastic annals of Yuan and wanted to make a version in Joseon by commissioning Jang to develop the automatic time-annunciating clepsydra. Also at that time the court scholars, including Jang, knew about Chinese and Islamic time-keepers as well as calendars, with the history books and writings collected by the kings of Goryeo, who had maintained close relationships with the Yuan (Mongol) emperors.

It is evident that the making of the Striking Palace Clepsydra in Joseon originated from and was inspired by Shun Di's Palace Clepsydra, which in turn, was influenced by Guo Shuo-Jing's knowledge of Islamic horology to some degree. Islamic technology might have seemed exotic to those who were accustomed to traditional Chinese water-clocks from Han and Tang onwards. Shun Di's clepsydra might have encouraged Sejong and his court scholars to innovate the traditional time-annunciation by employing inflow-float clepsydras such as the night-watch clepsydra (*gyeongjeomjigi*) made in 1424. To support the project to re-equip the Royal Observatory, Sejong commissioned Lee Sun-Ji (李純之 ? – 1464) to edit source texts on the astronomy, calendars, armillary spheres and celestial globes, sundials and water-clocks by compiling selected historical records and writings collected by the past and present governments into books. The result was the publishing of four volumes entitled *Collections of the Chinese Articles on the Astronomy, Calendar, Armillary Spheres and Celestial Globes, Sundials and Water-clocks* (諸家曆象集 *Jegayeoksangjip*) in 1445. Thus extensive information on clock-making was prepared and provided to the scientists and engineers. Jang and his supporters might well have received relevant scientific and technical data from the scholars and surveyed the historical documents on traditional Chinese and Islam timekeepers before developing the ball-falling clepsydra with ball-operated striking machinery for accurate time-measurement and time-annunciation. The ball-operated jack-work mechanisms sound a bell, a drum and a gong to announce twelve double-hours and the night-watches audibly, while the twelve double-hours are displayed with visible time-indicators. Traditional Chinese clepsydra technology was improved upon by adopting the Islamic ball-operated jack-work mechanism, creating an automatic striking water-clock in Seoul in the early fifteenth century.

To reconstruct Jang's Striking Palace Clepsydra there followed several studies in the history of science led by Needham from 1960's in England,[10] and control systems engineering led by Nam in 1984 onwards.[11] A major technical feature was presented to HMM2000.[12-13] The National Palace Museum of Korea initiated a reconstruction project during 2004-5 under contract with Konkuk University, with which the author has been affiliated, and the reconstructed model was installed for daily operation from November 2007 as shown in Figure 1.[14] The details of this ambitious project were reported to relevant academic societies.[15-16]

The technical details can conveniently be considered sectionally by modularity (M) following the orders of the time-ball travel and time-announcement signals as follows:

M-1. A ball-falling inflow-clepsydra: (1) a timekeeping inflow-clepsydra which consisted of three water-delivering vessels – large (reservoir), medium (compensator) and small (regulator) - and one receptacle, plus two inflow-float vessels for daily changeover; (2) eleven floating-indicator rods for sequential use from winter solstice to summer solstice and vice versa during the year, and the twelve double-hours with one-hundred intervals and five night-watches divided into twenty-five subunits were graduated on the surface of each float-rod depending on seasonal changes of the day-night apportionment; and (3) the time-ball dropping mechanisms above the inflow vessels where the float-rods operated to tipping buckets the evenly spaced twelve (for twelve double-hours) and/or twenty-five (for five night-watches and twenty-five night-points) ball-racks and drop the small (musket-ball-sized) copper balls, which are held in buckets, to outside the time-ball dropping mechanism. The clepsydra

Fig. 1. Completed replica of the Striking Palace Clepsydra (south elevation); the ball-falling clepsydra in the left, striking machinery in the right, and these two are connected by an interface (falling ball catch funnels and ball travel route are on the slantwise board) in the middle. The whole system measures approximately 6 [m] in width, 2 [m] in depth, and 5.7 [m] in height. (Courtesy: National Palace Museum of Korea, Seoul)

generates discrete time-signals – twelve times for double-hours and twenty-five for five night-watches and their points. The adoption of a vertical linear ball-rack in a (wooden-casing) ball-dropping mechanism above the inflow-vessels as the time-announcing signal-generator removed the complicated dynamic-control mechanisms and simplified the overall appearance. The flow rate into the inflow vessel became the sole factor that governed the time measurement accuracy of the clock. This enabled the clepsydra warden to measure and annunciate time accurately.

M-2. An interface connecting the clepsydra (M-1) and the striking machinery (M-3) provide the funnels to catch the falling balls from the ball-dropping mechanisms and the routes for the balls to travel to the ball-shift mechanisms (M-4) installed in the striking machinery M-3. This enables transmission of the dual discrete time-signals to two time-announcing machines.

M-3. The striking machinery is a two-story structure consisting of (1) the three pairs of time-announcers - a bell-striker and bell, a drum-beater and drum and a gong-sounder and gong – occupied on the second story, and twelve time-tablet bearing jacks occupied in the upper part of the first story, respectively; (2) the ball-shift mechanisms (M-4) placed under the second story platform; (3) the double-hour annunciating machines (M-5); and (4) the mechanisms for announcing the night-watch-points (M-6). The falling large (hen-egg-sized) iron balls from the ball-shift mechanisms (M-4) transmit energy to two time-announcing mechanisms M-5 and 6.

M-4. The ball-shift machine is a power converter consisting of (1) two ball-relay mechanisms – one for double-hours and the other for night-watches' small copper balls; (2) two ball-rest-release mechanisms – one for double-hours and the other for night-watches' large iron balls. The machine consists of two discrete energy converters that drop large iron balls to the striking mechanisms M-5 and 6 in conjunction with the time-balls' falling from the clepsydra to the ball-relay mechanisms.

M-5. The machine for announcing double-hours consists of (1) a bell-striking mechanism to sound the bell and (2) the time-tablet display control systems that work in conjunction with the bell-striking mechanism by rotating the horizontal carousel wheel. At each double-hour, the bell-striker sounds the bell once, and the time-tablet bearing jack, which has displayed the former double-hour, is changed to next one simultaneously by employing an on-off escapement mechanism. Thus visual and audible time-announcing signals are given.

M-6. The machine for announcing night-watches and their points in the night-watches consists of (1) a mechanism that beats a drum and gong to announce by the numbers, respectively, the night-watches and their first-points; (2) the mechanism announcing night-watch points by sounding the gong by the numbers. Announcing of the first point of each night-watch is followed by the announcement of each night-watch. E.g., to announce the first point of the 5th night-watch, the corresponding iron ball for the 5th night-watch operates the night-watch mechanism, which sounds the drum five times. The ball then operates the first night-watch mechanism, which sounds the gong once, and so on. Thus the machine operates according to logic similar to that employed by the modern digital counter.

M-7. The fabrication of above six modules makes up the time-annunciating machinery for the Striking Palace Clepsydra. To install the machinery inside the Annunciating Clepsydra Pavilion (**M-0**) requires accompanying facilities to array the clepsydra vessels and working space for operation and maintenance.

Figure 2 shows schematically how the seven modules of the Striking Palace Clepsydra at the annunciating clepsydra pavilion (M-0) were arranged, with all of the works except for the building (M-0 is omitted). L (left) and R (right) shows the directions of the components installed. The circled numbers are major sub-assemblies of each module.

Technical features and their origins appearing in the Striking Palace Clepsydra can be traced and considered sectionally as follows:

The ball-falling inflow-clepsydra consisted of a traditional Chinese three-stage inflow-type clepsydra with linear ball-rack mechanisms operated by the rising float-rod. The rising indicator-rod shows time and tips over the ball-bucket at evenly spaced intervals dropping the respective time-balls in the bucket. Thus the clepsydra measures time and falling time-balls to operate the time-announcing mechanisms concurrently. The three-stage inflow-type clepsydra, which was initially developed by Wang Fu (王普) of Song, was developed widely through Yuan and Ming onwards and introduced to Joseon, probably by Jang, for use as the basic timekeeper of the striking clepsydra. The clepsydra made use of an overflow hole at the regulating-vessel maintaining a static head.

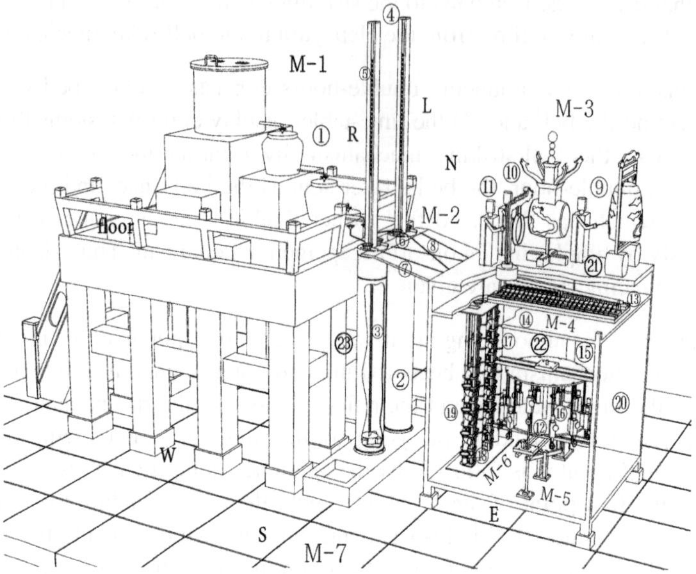

Fig. 2. Illustration of the Striking Palace Clepsydra with explanatory diagrams showing how to fabricate six modules into the time-annunciating machinery (south elevation) **M-1**: ① Water-delivering vessels are arranged on the three-stage platforms, and ㉓ a receptacle (hidden behind the column) and ② water-receiving (inflow) vessels are arrayed on the lower platform. The ③ float-rods are in the inflow-vessels ④ ball-dropping mechanisms. A ⑤ pair of copper plates secures the tipping ball-rack mechanisms. **M-2:** ⑥ four funnels, ⑦ inclined wide board (interface), ⑧ four ball routes extend to M-3. **M-3**: ⑨ the bell striker and bell for announcing double-hours, ⑩ drum-beater and drum for announcing night-watches, ⑪ gong-sounder and gong for announcing night-points, ⑫ 12 figures bearing time-tablets, ⑳ striking machinery, ㉑ upper-story, ㉒ lower-story. **M-4**: ⑬ a pair of ball-relay copper tubes to hold the ball-shift hole mechanisms, ⑭ four ball-rest channels and ball-release with spoon mechanisms. **M-5**: ⑮ the bell striking mechanism (hidden behind horizontal wheel), ⑯ horizontal wheel mechanism carrying the figures bearing the twelve time-tablets. **M-6**: ⑰ the night-watch mechanism, ⑱ the night-watch initial-point mechanism, ⑲ the night-watch point mechanism. **M-7**: interior of the building (M-0) for the Striking Palace Clepsydra and accompanying facilities such as staircase (west) and floor are built for clepsydra operations.

This is a simpler and more robust system than using the valve-operated feedback control technique, e.g. Ctesibian float-valve controller. Installing the linear ball-rack mechanisms above the inflow-vessel tipping the ball-bucket to drop the balls inside was a novel technique for dissecting (counting) time into discrete intervals corresponding to each time-announcing instant. It is comparable to the steelyard clepsydra-escapement mechanisms to count "One-hundred-intervals' in 'The Water-Powered Armillary (Sphere) and (Celestial) Globe Tower' of Su Song (蘇頌) & Han Gong-Lian (韓公廉)[17] and the ball-dropping mechanism driven by a submersible bowl called *tarjahar*, counting every 'half-an-hour' in the 'Elephant Water-clock' of

Al-Jazari.[18] They were all operated like modern analog to digital converters (A/DC). This signal conversion technique developed by Jang made it possible to measure analog time and announce digital time simultaneously as well as to separate the water mechanisms from the ball-operated striking mechanisms. To generate more accurate time-signals, small balls were chosen for insertion in the ball-bucket. Doubling the number of inflow-vessels was necessary to ensure uninterrupted time-keeping, and daily shifting from two inflow-float vessels was an unprecedented way of operating a water-clock. To operate the striking devices with sufficient power, larger balls were employed to take over the energy transmission mechanisms. Various pathways, spoon-shaped levers, detents and chain devices played key roles in ball-relaying and releasing. Ball-shift mechanisms work as energy converters, or mechanical amplifiers, employing the ball-relay hole mechanisms and the ball-rest-release mechanisms. The small copper balls dropped from the clepsydra are used to release large iron balls resting in their respective channels, operating the time-announcing machines. Thus the inventor separated energy sources for water machinery generating time and time intervals from the ball-operated jack-work mechanisms and generating time-announcing signals via ball-shift mechanisms.

A unique feature in the double-hour apparatus is the mechanism for changing the figures that indicate the double-hours. The double-hour display system was very simple and utilized an escapement mechanism making use of the figure's own weight without requiring any gearing. The night time-announcing mechanism is far more complex. The five night-watches and their points-announcing mechanism are similar to a modern digital 5-digit counter consisting of various automatic switches to close and open valves, change flow direction or reset. Falling balls enabled the time-announcing mechanisms to work in a digital fashion and secure effective energy transmission and accurate time-announcement. A copper tube with hole openings in the ball-relay mechanism is a mechanical device that works like a modern electronic shift register. Delayed-action mechanisms postponed the opening or closing until a set period had elapsed. Importantly, the night-watch announcements by pre-programmed float-rods enabled the Striking Palace Clepsydra to operate as a programmable computer. This clepsydra operated on a closed-loop system whereby it continued to work as long as balls were loaded into the bucket at the ball-dropping mechanisms above the inflow-vessels and into the channels of the ball-shift mechanisms at the striking machinery similar as the Elephant Clock of al-Jazari. As Needham argued, the elaborate time-annunciating mechanism using ball-operated jack-works was inspired by al-Jazari's 1206 clocks, which were not copied but instead were adapted to fit the Korean time-keeping tradition by the eminent mechanical engineer Jang Yeong-sil and his supporters.[10]

The Striking Palace Clepsydra was a key machine in the history of clock-making during Joseon. We can find vestiges of it in the Striking Heavenly Clepsydra, an astronomical clock made by Jang in 1437,[19] and various astronomical clocks made in Joseon during the ensuing sixteenth and seventeenth centuries.[20] The Striking Palace Clepsydra deserves to be called the first mechanically engineered time-keeper in Korean history, and the construction of a full-scale replica only served to reveal its structure, novelty and creativity, and innovative genius of its creator, Jang Yeong-sil.

In 1536, a replica of Jang's was made with some improvements by King Jungjong (中宗 r.1506-44) at another palace complex, Changdeokgung (昌德宮). Pak Se-ryong (朴世龍), the chief royal horologist, was in charge of construction and repair. [21] Jang's original Striking Palace Clepsydra was burned and the new one was greatly damaged during the Japanese invasions between1592 and 1598. After the war the government repaired the surviving clepsydra and installed it at the newly built Annunciating Clepsydra Pavilion in 1616. The *Record on the Repair of the Annunciating Clepsydra Pavilion* (報漏閣改修儀軌 *Borugak gaesu uigwe*) was compiled for the king's review.[22] (The text was lost during the French invasion of Ganghwa Island in 1866.) From the mid-eighteenth century, the striking parts of the annunciating clepsydra were dismantled and the clock operated manually until the abolition of the *injeong* and *paru* regulations in 1895.[5] Figure 3 shows a replica made in 1536 by after Jang Yeong-sil's 1433 production of the Striking Palace Clepsydra.

Fig. 3. The relics of 1536 replica are currently on the grounds of Deoksugung. From the upper left, the water-delivering dragon-mouthed reservoir (large), compensator (medium) and regulator (small) vessels, on the floor bed two dragon-ornamented inflow-float measuring vessels 2 [*m*] high, 32 [*cm*] in diameter. (Korean National Treasure No. 229 Striking Clepsydra (*Jagyeongnu*) of *Borugak* preserved in *Deoksugung* since 1938.)

<u>The Striking Heavenly Clepsydra : Heumgyeonggaknu</u>

In 1437 Jang invented another automatic striking astronomical clock demonstrating heavenly motion as well as twelve double-hours and five night-watches for King Sejong. On the eighth day of the first moon in 1438, the new heavenly clepsydra was put into service at the Respectful Veneration Pavilion (欽敬閣 Heumgyeonggak), so the clock was officially named the Clock of Respectful Veneration Pavilion (欽敬閣漏 Heumgyeonggaknu). As a reward for the invention, Sejong promoted Jang to a grand royal guard. According to the *Memoir of Respectful Veneration Pavilion* (欽敬閣記 *Heumgyeonggakgi*) by Kim Don in 1438, "Everything was more ingenious and excellent than the traditional designs were. All of these operations are performed automatically, without any human help. It looks like the immortals operate the mechanisms; onlookers were struck with admiration and can't guess how it works."[23]

Although several studies have been made on the clock, most centered on translations of source materials with basic reconstructions related with its origins. Detailed, operable structures have not been fully realized because of the dearth of technical information in the *Memoir* and later records. This clock partly came from the water-operated armillary sphere and celestial globe (水運渾天 *sunun honcheon*) and the Striking Palace Clepsydra, both made in 1433, plus the Lantern Clepsydra (燈漏 Deng Lou) made by Guo Shou-jing in 1270. It had been reconstructed in 1554 and again in 1614 after destruction by fire. The last version was destroyed in 1656.[24] It was maintained for 218 years, and the technical details might have been handed down and innovated upon to make subsequent armillary clocks from the seventeenth century onwards in Joseon. Information related to make astronomical clocks made it possible to conjecture plausible technical features.

According to the Kim Don's *Memoir*, the overall view of the clepsydra was as follows: Beside the West Council Hall (千秋殿 *Cheonchujeon*) at Gyeongbokgung was built a small structure called the Respectful Veneration Pavilion. A papier-mâché mountain more than 1.5 meters high (seven *jucheok*) was constructed inside the pavilion. Inside the mountain is a waterwheel linkwork rotating by means of trickling water from a jade clepsydra (玉漏機輪 *oknugiryun*). There were five-colored clouds surrounding a golden image of the sun, which rose and set once a day passing over the mountain. Below the sun-model stood four jade female immortals (玉女 *oknyeo*), each holding a golden bell in hand. At the four cardinal corners there stood the four guardian gods (四神 *sasin*): the Blue Dragon (East), the White Tiger (West), the Red Bird (South) and the Black Snake-Turtle (North) - facing the center of the mountain. There stood four hour-figures of immortals in uniform with their backs turned to the mountain, while three warrior figures – the bell-striker, drum-beater and gong-sounder – sound their respective instruments under the signals from the main hour-figure, on a high platform at the south foot of the mountain. On the level ground are the twelve gods of the Zodiac (十二神), standing at their respective positions, and holes are placed behind them. At the Rat double-hour (子時 *Jasi*) the hole behind the Rat opens; the Rat-hour time-tablet bearing jade female immortal comes out and the Rat stands still. Similar performances take place throughout the day. South of the mountain is another platform where the figure of a uniformed official pours water into an inclining vessel (欹器 *uigi*). On the four sides of the mountain there are country scenes displaying the works and days of four seasons, and wooden carvings of men, birds and plants according to the *Odes of Bin* (豳風) to remind viewers of the labors undertaken by peasants. Figure 4 shows an artist's concept of the Striking Heavenly Clepsydra.

Fig. 4. The artist's concept of the Striking Heavenly Clepsydra shows a replica of the yellow sun revolving the mountain, four jade female immortals standing on the multicolored cloud, each facing in one of the four cardinals directions, striking a golden bell with a wooden mallet to announce the double-hours. At same time, the four guardian gods (Blue Dragon, Red Bird, White Tiger and Black Snake) face the cardinal directions. At the southern foot of the mountain are an hour-jack and three warriors that strike the double-hours, night-watches and points with respective instruments similar to those in Figure 1. On the ground are twelve gods of the Zodiac, occupying their respective positions, and behind them are holes. At midnight, the Hour of the Rat, the hole behind the Rat opens, a Rat double-hour immortal with a time-tablet comes out, and the Rat figure stands still. The process continues with the other hours in succession. South of the mountain a man figure with a basin pours water into an inclining vessel that lies on its side when empty, stands upright when half full of water, and falls over again when full. Rural scenery depicts peasants working around the mountain in the four seasons.

The annotated translation of the *Memoir* found in Nam,[3] Needham, et al.[10] enabled modern scholars to infer the technical details of the clock in depth as well as its origins. Most of the mechanisms have been reconstructed recently by technology historians including the author.[24] The technical details can conveniently be considered sectionally as follows:

First, the heavenly motion mechanisms consist of a sun-model and clockwork to rotate the model. The sun replica, the size of a crossbow-bullet, crosses alone over the middle of the mountain, which looks like a round island surrounded by a plain. The sun is surrounded by five-colored clouds while keeping the same polar distance and directions of the rising and setting to the real sun for each season. The sun-model might be moved by an ingenious mechanism driven by water supplied from an inflow-clepsydra similar to the armillary sphere and celestial globe mechanism operated by a hidden waterwheel. Jang made a revolving sun-model employing the ecliptic ring from the armillary sphere that has five-layered rings at the top of the mountain. He left the elliptic ring and removed the other four. The ecliptic ring was moved by the celestial wheel, which was employed by the same members who made the water-operated armillary sphere and celestial globe in 1433.[25] To arrange the sun-model corresponding to the degrees of the respective heavenly rotations at the top of the mountain, the time-announcing mechanisms could be arranged separately under the sun-drive mechanism.

Second, the time-announcing mechanisms consisted of three main parts:

1. Four jade females and four guardian gods stood under the track of the sun-model. The four jade females are positioned at the corner of the four cardinal directions, and hold bells struck with wooden mallets at the beginning of each double-hour. The four guardian gods change the direction in which they face sequentially by 90 degrees for each double-hour, depending on the bell sound, resuming their original positions after four double-hours.
2. Under the four guardian gods, four female hour-figures stood on a platform in front of the mountain. Three other figures hold, respectively, a bell-mallet, drumstick, and gong-stick in front of their respective instruments. At the beginning and in the middle of each double-hour, the bell-striker sounds the bell, and the main hour-figure would turn to the east. At the beginning of each night-watch, the main-hour figure turned to the west, which sounded the drum and gong simultaneously. At the next four points (from second to fifth point of night-watches) only the gong would sound, in the same manner as the Striking Palace Clepsydra. These time-announcing features are similar to those of the Striking Palace Clepsydra except for the addition of a fourth hour-figure. The mechanisms are moved by a waterwheel linkwork rotating by means of water flowing from a jade clepsydra set up inside the mountain. The waterwheel would rotate by water flowing from the jade clepsydra in the same way that steelyard clepsydras do with the escapement. Thus the water-supplier jade clepsydra would operate as an inflow-type – a reservoir, a compensator and a regulator - without receiving vessels. The reservoir and the compensating vessels are arranged outside the mountain in consideration of their size and for ease of operation. Only the regulating vessel is inside the mountain, which is turned as water is received by the scoops attached to the rim of the waterwheel. Twenty-four balls are needed to move the four immortal figures to sound the bell at the beginning and in the middle of each of the twelve double-hours, and twenty-five more balls are needed for the night-watches and their points. A clockwork mechanism linked with the waterwheel is required to release the resting balls into their respective channels.
3. The twelve gods of the Zodiac of the double-hours occupy positions in front of slots on the ground level, while the jade female bearing the time-tablets would rise into position behind the respective Zodiacal god (animals like the Rat, Ox, etc.) at each double-hour. This mechanism would run linked with clockwork as described in 2 above.

These mechanisms might be combined properly with waterwheel rotation and ball-dropping through channels introduced from the Islamic tradition, something never employed in traditional Chinese astronomical clocks. This was an ingenious way to operate the time-announcing mechanisms.[24]

In addition to the time-announcing mechanisms, an inclining vessel was installed on the small platform south of the stationary twelve Zodiacal gods of the double-hours. An official dressed in a uniform and holding a jar pours the water into the inclining vessel, which lies on its side when empty, stands upright when half full, and falls over when full to the brim. The water that overflows from the inflow clepsydra is poured into the vessel, which empires to create a waterfall. The uniqueness of the clepsydra was its capability to demonstrate celestial movements and to announce the twelve double-hours and five night-watches and their points automatically with visual and audible signals. The system also reminded viewers of their responsibility to venerate Heaven and work hard for the people, and presented the principle of waxing and waning as standard of conduct."

In 1656, King Hyojong (孝宗 r. 1650-9) abolished the Heumgyonggak which was re-created by Prince Gwanghae (r. 1608-1623) in 1614. After two decades, King Hyeonjong (顯宗 r. 1660-74) commissioned Yi Min-cheol (李敏哲) and Song I-yeong (宋以穎) to revive the Striking Heavenly Clepsydra that Jang made for King Sejong. In 1669 Yi Min-cheol made a water-driven armillary clock (渾天儀 *honcheonui*) according to the traditional method, while Son I-yeong made a weight-driven armillary clock called the "self-sounding bell" (自鳴鐘 *jamyeongjong*) by employing Western clockwork. Details of these clocks could be found elsewhere.[10, 20, 26] Thus, the East Asian tradition of astronomical clock-making was carried on successfully in Joseon. Figure 5 shows the armillary clock of Song I-yeong, which was designated as Korean National Treasure No. 230, the Armillary Clock, in 1985. Regarding this clock, Needham asserted:

"It would be an instructive thing to have a replica of this whole instrument, as well that of Su Sung, together with suitable historical explanations, in every great museum of the history of science and technology in the world."[27]

4 Innovation and Succession in Seoul

In ancient East Asia, elaborate timekeeping mechanisms, especially demonstrating heavenly motions and time-announcing jack-work operations, symbolized the prestige of rulers and governments, just as great clocks did in medieval Europe. Shun Di's Palace Clepsydra *Gong Lou* built around 1350s in Dadu (now Beijing, China) might well have inspired King Sejong to commission Jang to make an equivalent or even more complex device. To fulfill the king's wishes, Jang developed an elaborate automatic striking water-clock while maintaining a quite ordinary, traditional float-rod clepsydra capable of dual time-measurement and announcement by employing automatic ball-operated jack-work mechanisms. Further to this was an astronomical clock consisting of a revolving sun-model, three kinds of time-announcing mechanisms employing a waterwheel-linked clockwork and falling balls, rural scenery and an inclining vessel. All functions were fully completely automated.

From the technical details and features on the striking clepsydras reviewed above enables us to draw a timeline of clock-making in fifteenth century Korea with inspiration from Chinese and Islamic sources. Figure 6 Shows how the Chinese and Islamic science and technology influenced Joseon court scholar-officials, including those in the reign of King Sejong, to make astronomical instruments and clocks. The immediate inspiration for the Striking Palace Clepsydra was probably the elaborate one produced by Shun Di. However, the brief descriptions of that clock do not

Fig. 5. Armillary Clock: clockwork-linked armillary sphere (left); weight-driven clockwork for the striking- train and the bell striking (middle); time-annunciator displaying twelve double-hour time-tablets, going-train and striking-release mechanisms (right). Length: 120 [cm], Breadth: 52 [cm], Height: 98 [cm] (Courtesy of Korea University Museum, Seoul, Korea)

provide enough information to specify its power-transmission mechanisms.[28] In addition to this, writings on the water-driven armillary (sphere) and celestial (globe) tower of Su Song & Han Gong-lian carry traditional astronomical clock-making information utilizing waterwheel linkwork and various time-announcing jacks as well as a revolving armillary sphere and celestial globe. Those records were reviewed by Joseon officials the process of making two clepsydras.[17] The Lantern Clepsydra of Guo Shou-jing, which employed time-announcing jack-work propelled by water and the balls (pearls) used as a part of time-announcing system, also directly influenced the making of the 1437 astronomical clock.[24] Besides the Lantern Clepsydra, Jang and his associates re-created the astronomical instruments of Guo Shou-jing—the "simplified instruments," 40-foot-high measuring-scale and gnomon, and direction-determining table—and these became key instruments in Sejong's Observatory.

Needham argued that Shun Di's Palace Clepsydra may be related to two mechanical clepsydras with striking jack-work operated by a waterwheel escapement made for the first Yuan Emperor, Kubilai Khan (r. 1260-94): the Precious Mountain Clepsydra (*Baoshan Lou*), c. 1262 and the Lantern Clepsydra of Guo Shou-jing in 1270.[29] Guo constructed not only clocks with fanciful jacks, but also celestial globes rotated by mechanical means exactly in the tradition of Zhang Heng and Su Song.[30] The general similarity of the Lantern Clepsydra to the water-clocks can be found in the Castle Clock described by al-Jazari.[31] Our detailed reconstruction of the Striking Palace Clepsydra's time-announcing jack-work allows us now to suggest more definitely from the ball-operated power-transmission mechanisms described in al-Jazari's *Book of Knowledge of Ingenious Mechanical Devices* (1206).[32]

As noted above Shun Di's Palace Clepsydra inspired Joeson court scholars, including Sejong and Jang, in which its ancestry can be traced, through the clepsydras of Guo Shou-jing, back to the al-Jazari Elephant Clock. Jang inherited traditional

Chinese horology while innovating the float-rod clepsydra, creating ball-operated power-transmission mechanisms in 1433, and re-creating a waterwheel linked, ball-operated, jack-work time-announcing astronomical clock in 1437 in the Joseon capital. Thus he inherited a long tradition of astronomical clock-making from Zhang Heng onwards. Jang's astronomical clock, in turn, was inherited by Lee Min-cheol and Song I-yeong to make water-driven and weight-driven armillary clocks, respectively, in 1669, two decades of Jang's replica was destroyed in 1656.[24, 33-5] Jang's successors inherited a long tradition of Chinese astronomical clock-making from Zhang Heng onwards, and carried it on in the Joseon capital. It is noteworthy to recall Needham's statement:

"Whatever may have been the precise route of transmission of these techniques from the Islamic world through China to Korea, the fact that in the fifteenth century automatically sounding clepsydras employing falling balls were in use across the entire span of the Old World, from Fez to Seoul, testifies eloquently to the ability of ideas to travel far and quickly among the greatest civilisations of premodern times."[34]

Fig. 6. Timeline of the striking clepsydras. This shows Jang's mentors and inspirations: Guo Shou-jing's Lantern Clepsydra borrowed astronomical clock-making technology from Zhang Heng and Su Song and influenced Shun Di's Palace Clepsydra , which inspired in turn King Sejong and Jang. The Lantern Clepsydra was very similar to al-Jazari's waterwheel escapement jack-work mechanisms used in the Castle Clock. Two striking clepsydras: the former was inspired from the Shun Di Palace Clepsydra, which was influenced from the Islamic tradition, and the latter was made up of the waterwheel linkwork of the water-driven armillary (sphere) and celestial (globe) tower made by Su Song & Han Gong-lian, and ball-operated time-announcing mechanisms of the first striking clepsydra and Lantern Clepsydra of Guo Shou-jing. Jang inherited and innovated "Chinese automation: in astronomical clock-makings, and his works were applied by Yi Min-cheol and Song I-yeong when making the Joseon armillary clocks in 1669.

5 Concluding Remarks

In summary, Jang, an outstanding mechanical engineer, inherited the long tradition of Chinese astronomical clock-making and created two superb striking clepsydras by making innovations while keeping intact the tradition of employing Islamic ball-falling power transmission mechanisms. The clepsydras operated on a closed-loop system that continued to work as long as balls were loaded into the magazines and the reservoir was filled with water throughout the day. These two monumental clocks were symbols of the pre-modern Korean horology, and marked the beginning of the history of Korean mechanisms, mechanical engineering, automation and robotics. Jang's innovations were then inherited by two other mechanical geniuses in Joseon, resulting in water-driven and weight-driven armillary clocks in 1669, carrying on the long tradition of astronomical clock-making in East Asia. This shows a genealogical nexus among the machines, automation and robotics of China, Islamic states and Korea. Jang is in some way the most articulate of what is a long tradition of "automation" in Asian science and technology in which automation is a manner of submission to nature rather than means of controlling nature, as in our time. Jang not only assimilated the techniques of his Korean, Chinese and Islamic predecessors, but he was also creative and innovative. Hill argued, "No really satisfactory definition of the term 'engineer' has yet been produced, …… Nowadays we are apt to think of engineering in terms of the application of mathematical processes to the solution of problems, together with rigorous attitude towards cost control. Here I intend to define the term 'engineer,' somewhat arbitrarily I admit, as follows: A mechanical engineer is a man who designs and constructs machines of some complexity, which when completed fulfill the function for which they were intended. He should pay due regard to previous practices, and make the best use of the tools and materials at his disposal."[36] As defined by Hill, Jang Yeong-sil of fifteenth century Joseon takes his place among the great mechanical engineers and horologists, who include Han Gong-lian of the eleventh century Song, al-Jazari of thirteenth century Islam and Guo Shou-jing of thirteenth century Yuan. The contributions of all these great men helped advance modern science and technology in the world.

Acknowledgements. The author is grateful to Professors Timothy Atkinson at Seoul University of Foreign Studies and Dr. Yoonkey Nam at KAIST for reviewing the manuscript.

References

[1] Rufus, W.C.: Trans. Korea Branch Royal Asiatic Soc. 26, 1–48 (1936)

[2] Yi, T.J., Jeon, S.W.: King Sejong the Great - The Light of 15th Century, Korea, pp. 233–237 (1998)

[3] Nam, M.H.: Korean Water-Clocks: Jagyongnu, The Striking Clepsydra & The History of Control & Instrumentation Engineering. Stud. Korean History, vol. 101, pp. 75–114. Konkuk University Press, Seoul (1998)

[4] Annals of Sejong (Sejong Sillok), ch. 77, pp. 7a–11b, 1437. National Institute of Korean History, Seoul (1986)
[5] Nam, M.H.: JangYeong-shilgwa Jagyeongnoo (Jang Yeong-Sil and the Striking Clepsydra – Revival of the Time Measuring History of the Chosun Period). Seoul National University Press, Seoul (2002)
[6] Nam, M.H., Jeon, S.W.: From Guo Shoujing to King Sejong. In: Nha, I.-S., Stephenson, F.R. (eds.) Proceedings of the First International Conference on Oriental Astronomy, pp. 305–324. Yonsei University Press, Seoul (1997)
[7] Annals of Sejong (Sejong Sillok), ch. 61, p. 55b, 1433. National Institute of Korean History, Seoul (1986)
[8] Comprehensive Study of Civilization, revised and expanded (Jeungjeong munheon bigo), ch. 2, pp. 27b–30a, 1907. The King Sejong Memorial Soc., Seoul (1980)
[9] Annals of Sejong (Sejong Sillok), ch. 65, pp. 1a–3b, 1434. National Institute of Korean History, Seoul (1986)
[10] Needham, J., Lu, G.-D., Combridge, J.H., Major, J.S.: The Hall of Heavenly Records (HHR):Korean Astronomical Instruments and Clocks, 1380-1780. Cambridge University Press, Cambridge (1986)
[11] Nam, M.H.: Ind. Tech. Res. Report Konkuk Univ. 10, 75–89 (1985)
[12] Nam, M.H.: Korea J. 30(7), 9–21 (1990)
[13] Seo, M.H., Hahn, Y.H., Lee, J.H., Nam, M.H.: In: Ceccarelli, M. (ed.) Int. Symp. Hist. Machines and Mechanisms Proceeding HMM 2000, pp. 341–350. Kluwer Academic Pub., Dordrecht (2000)
[14] National Palace Museum of Korea, Jagyeongnoo (Self-striking Water Clock). In: Catalog of National Palace Museum of Korea. National Palace Museum of Korea, Seoul (2007)
[15] Nam, M.H., Hahn, Y.H., Seo, M.H., Lee, J.H., Seo, J.: Chin. J. Hist. Sci. Tech. 31(4), 461–470 (2010)
[16] Nam, M.H.: Mapping the Oriental Sky. In: Nakamura, T., Orchiston, W., Sôma, M., Strom, R. (eds.) Proceedings of the Seventh International Conference on Oriental Astronomy, pp. 116–125. National Astronomical Observatory of Japan, Tokyo (2011)
[17] Su, S.: Xin-yixiang Fayao. Shang Wu Inshuguan, Shanghai (1937)
[18] Al-Jazari, I: Kitab fi mar'ifat alhyal al-handasiya (The Book of Knowledge of Ingenious Mechanical Devices). Reidel, Dordrecht and Boston (1974) (1206) translated and annotated by Hill, D.R
[19] Annals of Sejong (Sejong Sillok), ch. 80, pp. 5a–6a, 1438. National Institute of Korean History, Seoul (1986)
[20] Nam, M.H., Hahn, Y.H., Lee, S.W., Yang, P.S.: Academic Report Konkuk Univ. Res. Cen. Kor. Hist. Tech. 39, 519–543 (1995)
[21] Annals of Jungjong (Jungjong Sillok). ch. 82, p. 33, 1536. National Institute of Korean History, Seoul (1986)
[22] Annals of Gwanghaegun (Ganghaegun Ilgi), ch. 103, 1616. National Institute of Korean History, Seoul (May 1986)
[23] Annals of Sejong (Sejong Sillok), ch. 80, p. 6a, 1438. National Institute of Korean History, Seoul (1986)
[24] Hahn, Y.H., Nam, M.H., Lee, S.W., Yi, M.K.: Tech. Hist. (J. Korean Soc. Hist. Tech. Ind.) 1(1), 99–140 (2000)
[25] Annals of Sejong (Sejong Sillok), ch. 77, p. 9b, 1437. National Institute of Korean History, Seoul (1986)

[26] Comprehensive Study of Civilization, revised and expanded (Cheungbo MunheonBigo), ch. 3, p. 2a, 1907. The King Sejong Memorial Soc., Seoul (1980)
[27] Needham, J., Wang, L., Solla Price, D.J.: Heavenly Clockwork: The Great Astronomical Clocks of Mediaeval China, 2nd edn., p. 163. Cambridge University Press, Cambridge (1960)
[28] Yuang shi.: ch. 48, pp. 7ab. Chung-hua Shu-chu, Beijing (1976)
[29] Needham, J., Lu, G.-D., Combridge, J.H., Major, J.S.: The Hall of Heavenly Records (HHR): Korean Astronomical Instruments and Clocks, 1380-1780, pp. 40–41. Cambridge University Press, Cambridge (1986)
[30] Yuang shi: ch. 5, p. 2a. Chung-hua Shu-chu, Beijing (1976)
[31] Needham, J., Wang, L., Solla Price, D.J.: Heavenly Clockwork: The Great Astronomical Clocks of Mediaeval China, 2nd edn., pp. 135–138. Cambridge University Press, Cambridge (1960); with a supplement by Combridge, J.H. (1986)
[32] Nam, M.H., Hahn, Y.H., Seo, M.H., Lee, J.H., Kim, S.K.: Konkuk J. Ind. Sci. Tech. 31, 67–101 (2006)
[33] Annals of Hyeonjong (Hyoenjong Sillok), ch. 21, pp. 55a, 1669. National Institute of Korean History, Seoul
[34] Needham, J., Lu, G.-D., Combridge, J.H., Major, J.S.: The Hall of Heavenly Records (HHR): Korean Astronomical Instruments and Clocks, 1380-1780, p. 44. Cambridge University Press, Cambridge (1986)
[35] Hahn, Y.H., Nam, M.H.: J. Korean Hist. Sci. Soc. 19(1), 3–19 (1997)
[36] Al-Jazari, I: Kitab fi mar'ifat alhyal al-handasiya (The Book of Knowledge of Ingenious Mechanical Devices). p. 279 Reidel, Dordrecht and Boston (1974) (1206); translated and annotated by Hill, D.R

Kurt Rauh: Saxon Mechanism Theory Brought to Aachen

Burkhard Corves

RWTH Aachen University, Germany

Abstract. In 2009, the biannual German colloquium on mechanism science took place in Aachen, Germany organized through the Department of Mechanism Theory and Dynamics of Machines of RWTH Aachen University in order to celebrate the 50th anniversary of its foundation. For this purpose the author of the current paper investigated into the history only in order to find out that the 50th anniversary was only partly true and "institutional" mechanism theory could even be traced back further. But this was not yet the end of the line reaching back in history, because Prof. BRAUNE could even trace back certain roots of mechanism theory to FRANZ REULAUX, sometimes seen as the "father of kinematics" in Germany.

This paper intends to shed some light on early mechanism theory at RWTH Aachen University especially considering the activities of Saxony-born KURT RAUH, Fig.1, starting in the twenties of the last century and lasting until shortly after WWII. In the following it will be shown that mechanism theory in Aachen has already played a major role before the official founding of the Chair of mechanism theory in Aachen in 1959.

Fig. 1. Kurt RAUH (1897-1951

1 Early Mechaniim Theory at RWTH Aachen University

In the middle of the Franco-Prussian War 1870/1871 on October 10th, 1870, the then "polytechnic university" in Aachen was officially opened. As shown in [1], FRANZ REULEAUX as the then undisputed doyen of mechanism and machine theory in Germany was involved during the planning and development phase.

This involvement can be deduced from the following evidence: The first elected rector of the "polytechnic university" in Aachen was ADOLF VON GIZYCKI who acted as professor of *"Descriptive Machine Theory"*, *"Theoretical Machine Theory"* and *"Theory of Kinematics (Reuleaux System)"* from the very beginning in Aachen

T. Koetsier, M. Ceccarelli (Eds.): Explorations in the His. of Machines & Mech., HMMS 15, pp. 107–121.
springerlink.com

among just three other mechanical engineering professors. He was educated at the Royal Industrial Institute in Berlin and acted there as assistant and lecturer in contact with FRANZ REULEAUX. Another evidence of FRANZ REULEAUX'S influence is the fact, that an exhibition catalogue [2] produced by RWTH Aachen University lists the Inventory of the "Kinematic Collection" containing *"195 demonstration models, produced in the workshops of the Royal Industrial Institute of Berlin, costing 15200 Mark in total."*. Unfortunately, these today very valuable, so-called REULEAUX models are lost. Fortunately many other REULEAUX models have survived in other collections [3].

Besides the two above mentioned evidences, the development of mechanism science at the end of the 19th century at RWTH Aachen University is somewhat vague and in the chronicles of the university "Kinematics" appears as documented subject for the last time in 1906. But in 1909 a dissertation thesis within the field of mechanism science was finished by EUGEN ESSICH [4].

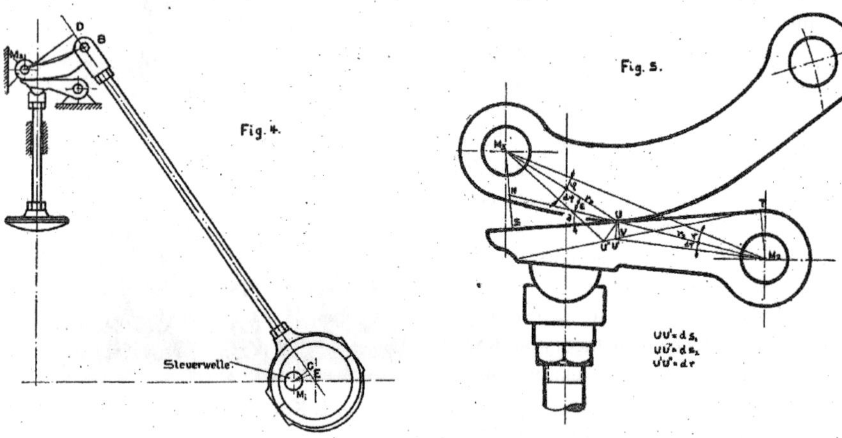

Fig. 2. Control mechanisms with rolling cams, according to [11]

Fig. 2 shows on the left side a typical control mechanisms with rolling cams which are shown in detail on the right side as investigated in [4]. In his introduction ESSICH stated that for rapid oscillating motions the dynamics of mechanisms must be considered. He also remarks that in this respect the crank mechanism has already been exhaustively treated, an estimation that appears somewhat pre-emptive from today's perspective.

Other particular evidence of research and teaching of mechanism science at RWTH Aachen University shortly before and after WWI was not available to the author of this paper and it seems that someone else had to come to re-establish mechanism science at RWTH Aachen University.

2 Education in Saxony

In fact it was someone from Saxony who would bring new life to mechanism and machine science at RWTH Aachen University. Born in Plauen, which in the late 19th century became a textile manufacturing center, specializing in lace as a kind of industrial 'boomtown' in the western part of Saxony, KURT RAUH attended the Gymnasium in Plauen before he had to join the military service during WWI. Afterwards he started his studies at the then "Technische Hochschule Dresden" now Technical University of Dresden in 1919 after having attended preparatory courses in machine drawing at the vocational school in his hometown Plauen. As explained in detail in [5], TH Dresden was one of the leading institutions in the field of mechanism theory since the late 19th Century. Therefore, it is probably no coincidence that KURT RAUH in 1919 enroled as a student of mechanical engineering at TH Dresden. Already after one year he had earned his "Vordiplom" (intermediate examination for a diploma) and just after another year he finished as "Diplomingenieur" (graduate engineer) in November 1921. Immediately after finishing his diploma he started working as a design engineer for mechanical handling devices in the company "Mühlenbauanstalt und Maschinenfabrik vorm. Gebrüder Seck" (mill manufacturer and machine company, formerly brothers Seck) in Dresden-Zschachwitz. Fig. 3 shows the machine hall in 1965, but from today's point of view it appears as if nothing much has changed since the time when KURT RAUH started working as a design engineer. But after less then a year it seems he was already lured back into academia starting a new job as scientific assistent at the "Institut für Maschinentechnologie" (institute for machine technology) of Prof. RUDOLF HUNDHAUSEN again at TH Dresden.

Fig. 3. Machine hall of the mill manufacturer and machine company, formerly brothers Seck, February 1965

3 From East to West

But also this engagement did not last long and after another job in Saxony in 1923 KURT RAUH finally moved to the west of the Weimar Republic to Bonn, where he initially worked as the "*Leiter des Neuheiten- und Konstruktionsbüros und der Versuchsabteilung der Firma F. Soennecken Bonn, dazu seit 1924 Leiter des Patentbüros und seit Sommer 1925 Assistent des Betriebsdirektors derselben Firma*" (head of innovation and design offices and experimental department of the company F. Soennecken, Bonn, and then, since 1924 director of the Patent office, and since the summer of 1925 assistant of the operating director of the same company), [6]. In fact we will see later that his interest in patent issues kept on even after he again quit this job in 1926 in order to pursue a job as scientific assistant and later as scientific chief assistant at the "Landwirtschaftliche Hochschule Bonn-Poppelsdorf" (Agricultural College).

At that time the Agricultural College in Bonn, which was founded in 1847 and only in 1934 became part of the Rheinische Friedrich-Wilhelms University in Bonn, did not have the permission to promote PhD candidates which is why KURT RAUH submitted his dissertation at the former TH Hannover where he received his Dr.-Ing.-title on July 12th, 1927, fig. 4. Comparing this date with the above mentioned dates of his professional career it seems obvious that at least after having quit his first academic position at TH Dresden he never quit academia at least in his mind. Otherwise it seems a very short time of just about one year to prepare his doctoral thesis.

DIE TECHNISCHE HOCHSCHULE HANNOVER
UNTER DEM REKTORATE
DES ORD.PROFESSORS Dr.FRIEDRICH QUINCKE
VERLEIHT MIT DIESER URKUNDE DEM HERRN
DIPL.-ING. KURT RAUH
AUS PLAUEN I.V.
DIE WÜRDE EINES
DOKTOR-INGENIEURS
NACHDEM ER BEI DER FAKULTÄT
FÜR MASCHINENWIRTSCHAFT
FACHRICHTUNG „MASCHINENINGENIEURWESEN"
IM ORDNUNGSMÄSSIGEN PROMOTIONSVERFAHREN
UNTER VORSITZ DES
ORD. PROFESSORS Dr.-ING. E.H. GEORG DETTMAR
UND UNTER MITWIRKUNG DER REFERENTEN
ORD.PROFESSOR Dr.-ING. ARTHUR PRÖLL
UND ORD. PROFESSOR Dr.-ING. WALTHER KAUFMANN
DURCH SEINE DISSERTATION
„UNTERSUCHUNG UND WEITERENTWICKLUNG
DER GETRIEBE MIT PERIODISCHEM HIN- UND
RÜCKLAUF UND BESCHLEUNIGUNGSFREIEM
ARBEITSGANG"
SOWIE DURCH DIE VORGENOMMENE MÜND-
LICHE PRÜFUNG SEINE WISSENSCHAFTLICHE
BEFÄHIGUNG ERWIESEN UND HIERBEI
DAS PRÄDIKAT „BESTANDEN"
ERWORBEN HAT.
HANNOVER, DEN 12. JULI 1927
REKTOR UND SENAT
DER TECHNISCHEN HOCHSCHULE

Fig. 4. Doctor certificate of Kurt Rauh

In the introduction to his dissertation entitled "Investigation and development of a periodic mechanism with forward and reverse and acceleration-free operation," he notes the key points of his dissertation [6]: "*In mechanical engineering there is often the need for a mechanism, that realized a periodic forward and return stroke with temporarily uniform velocity. For example such a mechanism can be used advantageously in automatic production machines of modern construction, in which work pieces are moving with uniform velocity through the machine. For certain operations such as punching, pressing, cutting etc. it is necessary that the required tools temporarily are moving with the same velocity when the operation takes place, as the work piece itself. Then the operation takes place at the time, during which there is no relative motion between tool and work piece in the direction of movement of the work piece. In that case, highest precision must be placed on the synchronicity of the corresponding mechanism, and even more so with increasing duration of the synchronicity between tool and work piece.*" In fact, with these words he describes a task still valid today in many applications even if increasingly servo-controlled drive systems are used for this purpose. Interestingly, throughout the history of mechanism theory in Aachen in the second half of the last century, mechanism with temporarily synchronous velocity has been a special topic. At least five research reports [7-11] which originated under the auspices of WALTER MEYER ZUR CAPELLEN, the successor of KURT RAUH as head of mechanism theory in Aachen, can be mentioned here among other publications. Still the successor of WALTER MEYER ZUR CAPELLEN, GÜNTER DITTRICH kept the topic alive as can be seen from the dissertations of SHI and LEUSCH [12,13]. And even the author of this paper himself as chairman of a VDI guidelines committee supported by GOTTLIEB LEUSCH dealt with the same topic leading to the VDI guideline 2740. Part 3 "Mechanical devices for automation equipment - Mechanisms for partially synchronised motion". Based on the dissertation of Leusch. this guideline should support the mechanical engineer in selecting appropriate solutions for a given task by defining specific properties for the comparison and evaluation of suitable solutions, which can be selected from a collection of different solutions [14], which has been adapted from the well-known IGM mechanism descriptions [15,16].

Since the clarity of principles and solutions in mechanism theory shown through pictures and models is of utmost importance, it is certainly worthwhile to look at the detailed and very informative illustrations of the doctoral thesis of KURT RAUH (Fig. 5).

Interestingly, these illustrations display a great similarity in the representation of examples known from LUDWIG BURMESTER's illustration in his textbook "Lehrbuch der Kinematik" shown in fig. 7, [17]. But this is no coincidence, since KURT RAUH in his literature review explicitly refers to BURMESTER particularly with respect to his figures 1 through 4, above right in Fig. 5, which find their counterparts in the figures 422-425 in Fig. 6.

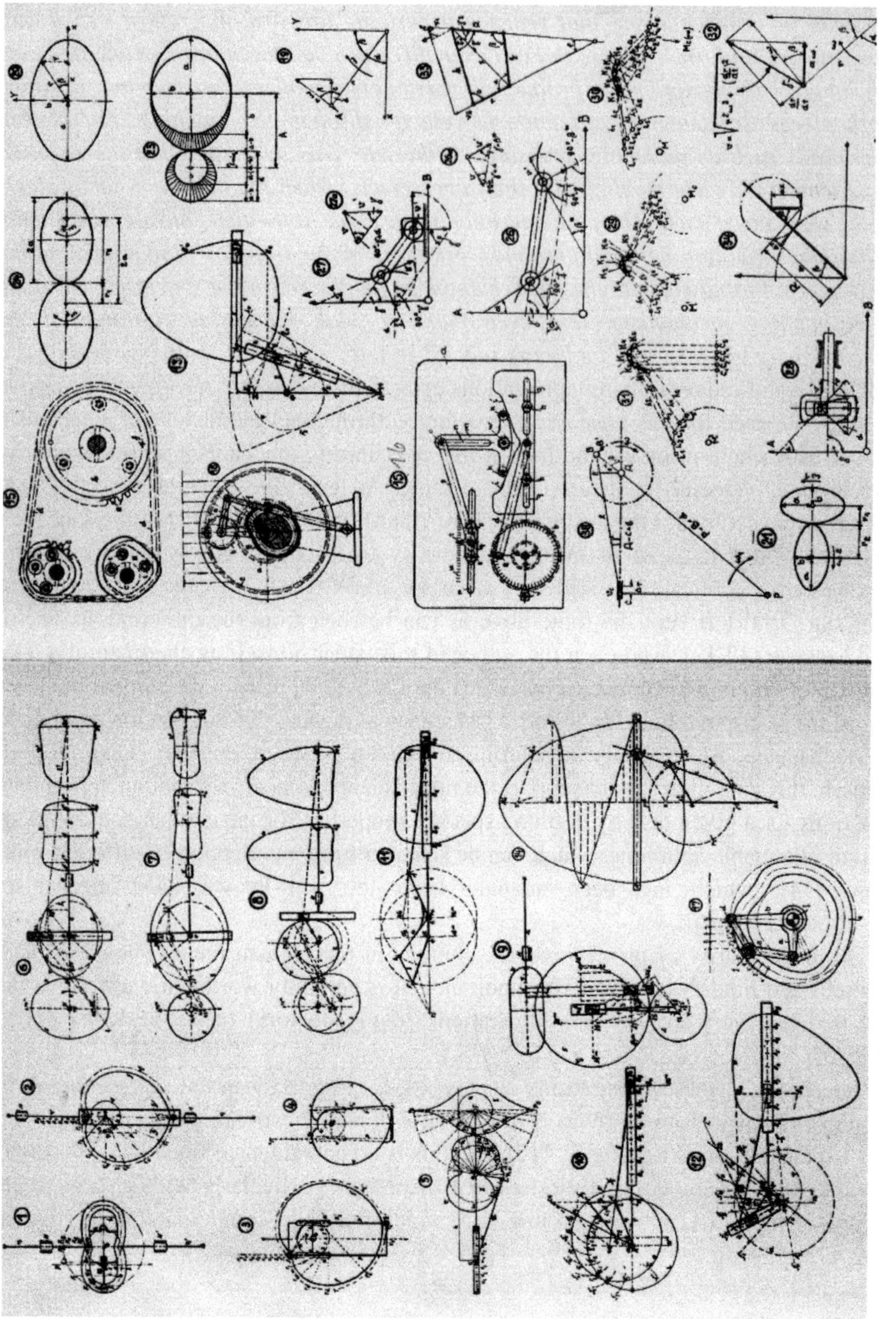

Fig. 5. Mechanism illustrations from [6] (rotated counter clockwise for improved enlargement)

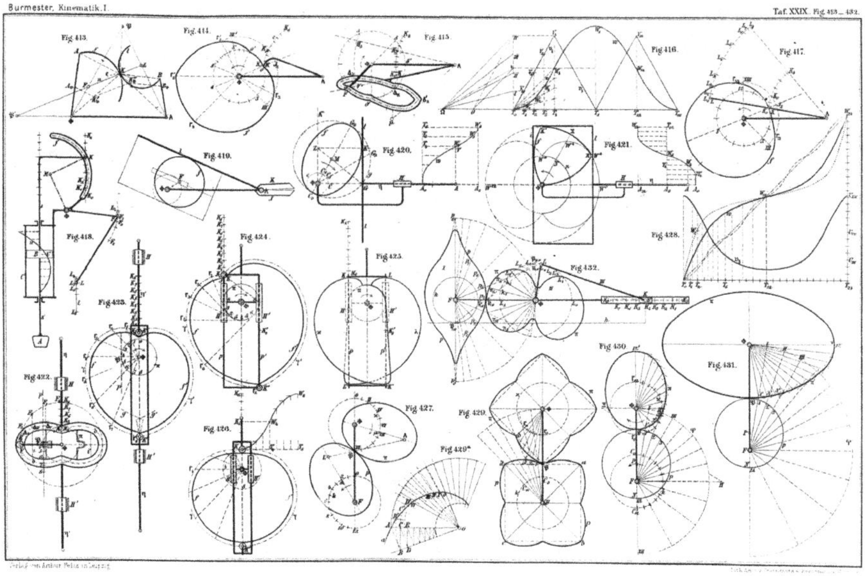

Fig. 6. Table 29 from the Atlas accompanying BURMESTER's illustration in his textbook "Lehrbuch der Kinematik"

A letter of the Institute Director of the Agricultural College in Bonn Prof. Dr.-Ing. KARL VORMFELDE dated April 17th, 1928 to the Minister for Agriculture, Domains and Forestry in Berlin addressing the following words:

"In order that agricultural engineering gets a firm footing also at RWTH Aachen University which is in the interest of German agriculture, I have caused my assistant, Dr.-Ing. Rauh to habilitate in Aachen with the topic "Kinematics and Agricultural Engineering." RWTH Aachen University will agree and I would therefore respectfully want to ask to approve to the habilitation of Dr.-Ing. Rauh in such a way that he will hold for 1 day each week during the winter term a 2 or 3 hour lecture in Aachen. The work of the Agricultural Machinery Institute in Bonn-Poppelsdorf will not be negatively influenced by this habilitation. The scientific work and progress of Dr. Rauh will ensure an honourable success."

marks the re-beginning of mechanism theory through a dedicated lecturer in persona KURT RAUH at RWTH Aachen University.

"*At the request of the Faculty of Mechanical Engineering*" the Prussian Minister for Science, Art and Education gave "*a teaching assignment for mechanism theory to KURT RAUH with effect from the start of the winter term 1929/30*". So only a few years after finishing his dissertation and receiving the doctoral degree as Dr.-Ing., KURT RAUH has received the habilitation at RWTH Aachen University and the position of a private lecturer.

Already in 1931 his first textbook on mechanism theory [18] appeared with a large number of illustrations, including about twenty pages with color illustrations of various mechanisms (Fig. 7). In the beginning thirties of last century this was quite a special feature where all the other publications were simply in black and white print. The assumption of the author of this paper is, that this textbook to a certain extend also reflects the experience that KURT RAUH gained through his current research still partly at the Agricultural College in Bonn and through the lectures and courses at RWTH Aachen University.

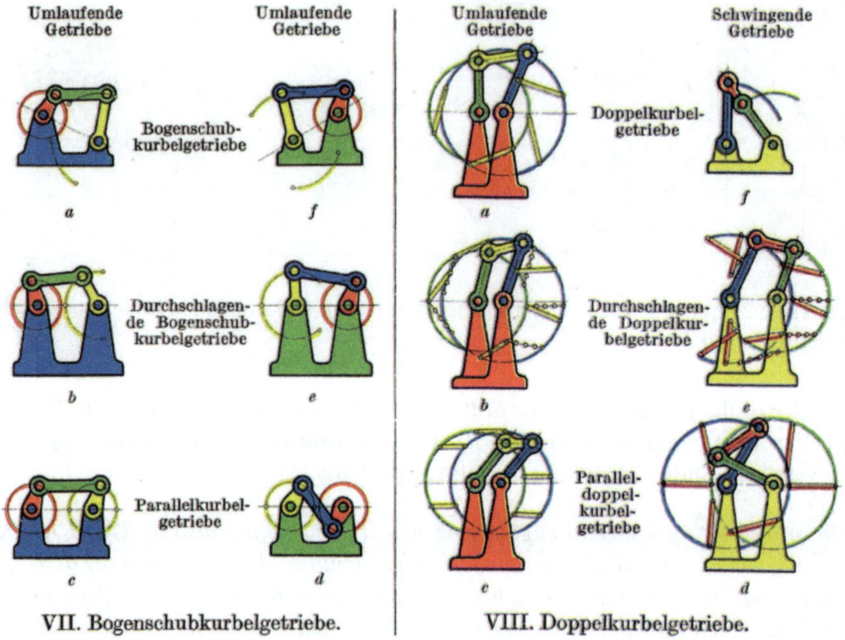

Fig. 7. Table 7 und 8 from "Praktische Getriebelehre" by Kurt Rauh [18]

The title "Praktische Getriebelehre" (Practical mechanism theory) already indicates that he wants to address the mechanical engineering designer in the design office and intends to support him in his daily work. In the preface he makes it clear that he wants to put the designer in a position to not only analyze an existing mechanism, but also to find a suitable mechanism for a given motion task.

As already mentioned, KURT RAUH also acted as director of the patent office during his assignment with the company F. Soennecken, Bonn. Obviously his interest in patent issues kept alive even after he left the company in order to pursue his academic career. This can be proven by the fact, that besides his lectures and courses in mechanism theory, he also held lectures regarding patent administration, which also were broadcast through the West German Radio (WDR) in 1932. The title of the broadcast sounded somewhat striking "I register my own invention myself" and was announced as being based on lectures on patent administration at the RWTH Aachen University, "grouped into five radio lectures". A printed version was issued in 1935 [19]. Even after WWII KURT RAUH continued with lectures about patent administration as can be seen from [20] which is built on the former paper [19] but broadens the scope such, that it deals with two main chapters: Chapter A repeats [19] headed "I register my own invention myself", whereas chapter B carries the heading "What does the engineer have to know about the patent process?" giving explanations about patent issues regarding patent infringement, preliminary injunction, infringement claims, patent expertise, patent trials and not to be forgotten: How to fix the fees for a patent expertise?

From the historical perspective it is interesting to take a closer look at the second page of [19] with references to other publications. These references are provided with the heading „Veröffentlichungen aus dem Institut für Getriebelehre der Technischen Hochschule Aachen" (Publications of the Institute for Mechanism Theory at RWTH Aachen University). This means that already in 1935 there existed an Institute for Mechanism Theory. This assumption is somehow confirmed by the fact that according to the personnel records of RWTH Aachen University KURT RAUH was appointed as a full-time lecturer at the Technical University of Aachen as of December 1st, 1934. These personnel records also contain a publication list of KURT RAUH, which starts with his dissertation in 1927, [6] and ends exactly in 1934, which supports the assumption that he submitted this publication list prior to be appointed as a full-time lecturer. This publication list contains publications mostly dealing with agricultural engineering, most of them published in German agricultural journals, among them also reports about national and international agricultural exhibitions, for example in Paris and Newport as well as test reports on different agricultural machinery. But also publications in mechanism theory are listed especially about coupler curves and dwell mechanisms that can be derived from them.

After war-related interruption from 1939 to 1945 and the "denazification" by the corresponding "Einreihungsbescheid" (notice of classification) of the Military Government of Germany within the British control area, KURT RAUH was fully active again at RWTH Aachen University since 1947. In the 1949 yearbook of RWTH Aachen University one can find a short report [21] of KURT RAUH about the "Dozentur für Getriebelehre, Maschinenaufbaulehre und Patentlehre" (Lectureship in mechanism theory, machine design theory and patent theory). Obviously the scope of KURT RAUH's lectureship now also included patent theory and machine design theory. The latter subject is explained in the same report be KURT RAUH as the theory of machine design and construction "ohne notwendiges Vorbild" meaning without necessary model, i.e. machine design from scratch. Furthermore KURT RAUH reports that even though the rooms and teaching material were completely destroyed during the war, already new research activities had started as could be seen from two new dissertations and 6 published research reports. Additionally KURT RAUH reports about activities to incorporate agricultural engineering within his scope probably with the idea to build upon his expertise in this field. In fact as a result of his efforts to establish agricultural engineering in 1949 he publishes a book "Entwicklungslinien im Landmaschinenbau" (Design lines in agricultural engineering), [22], which as he explains in the acknowledgement of this book resulted from an advice of Prof. Dr.-Ing. KARL VORMFELDE under which he started his academic career in Bonn telling him in 1934, when he left for Aachen, to write a book about machine design theory for agricultural machinery. Looking at this book it is interesting to see that even though is shows a lot of technical details in picture, see fig. 8, and text, it starts with statements about cost effectiveness and economic efficiency in agriculture.

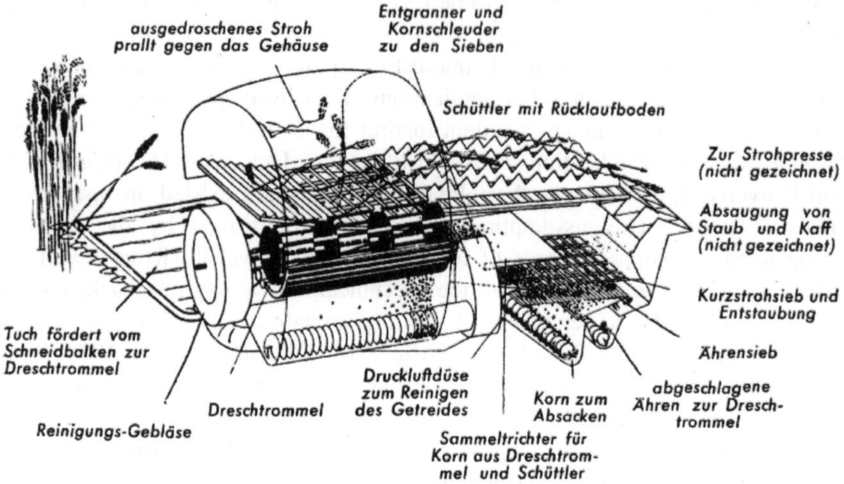

Abb. 7. Grundsätzlicher Aufbau und Arbeitsweise des Claas-Mähdreschers (Nachkriegsmodell). Es ist zu beachten, daß das geschnittene Getreide vom Schneidwerk ohne Bandwechsel bis in die Dreschtrommel gefördert wird.

Fig. 8. Post-war model of a Claas-combine according to [22]

Thus the first chapter of [22] is titled: "*The allowable construction cost. (How much can you charge for the agricultural machine?)*" and the second chapter is devoted to the operating costs of some agricultural machinery.

If [22] can be seen as the result of integrating agricultural engineering into his research portfolio, then his activities within the area of machine design are reflected by his third textbook [23] devoted to the machine design of processing machinery in general, published in 1950, which, as he notes in his preface, is adressed to "*every design engineer*" and is "*accordingly separated in one volume of text and a figure book for clear and easy handling*". The figure book contains 453 figures. These figures not only depict processing machines or machine parts but also process related information as well mechanism solutions, as shown in fig. 9.

In 1951 KURT RAUH embarks on a research journey within the ERP (European Recovery Program) for a roundtrip through the United States. The European Recovery Program better known under the name of the "Marshall-Plan" after the then

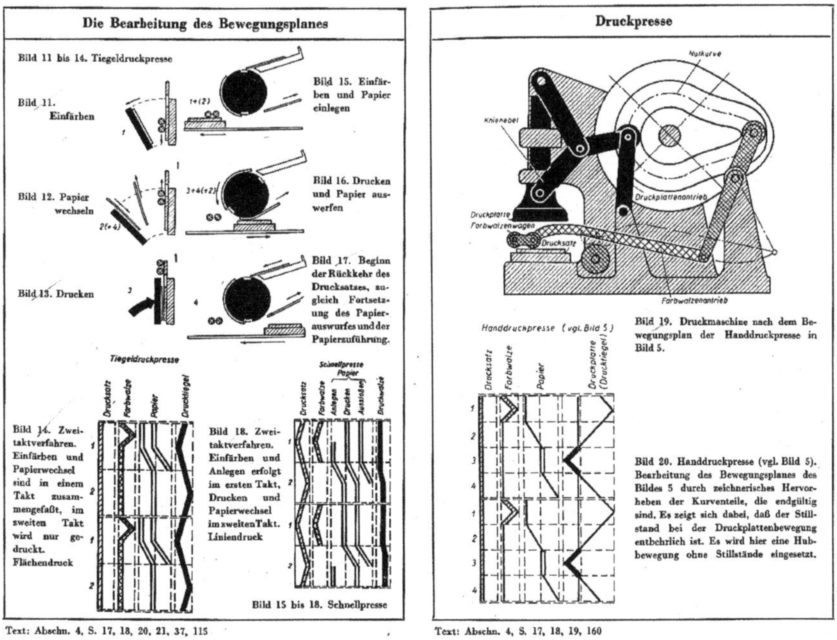

Fig. 9. Manually operated printing press and process according to [23]

U.S. Secretary of State GEORGE C. MARSHALL, who proposed this plan, started on April 3rd 1948. It was aimed at helping the countries of Europe, which had been destroyed by the war [24]. The money funneled into Germany through this program was meant as a loan, but the money was not payed back to the U.S. but into a fund which exists still today and is administered by German Authorities. Today in a way similar to KURT RAUH's journey to the U.S., German students who plan to take up a one or two-year Master's course at a distinguished university in the U.S. or plan a research visit after completion of a degree in Germany can apply to the well-endowed "ERP Fellowship Programme" which provides financing: fellowships for subsistence from 12 to 21 months and subsidies towards academic fees up to a maximum of 25,000 dollars per year. Since 1994, the program has been financed by funding from ERP special assets which are administered by the German Federal Ministry of Economics and Technology (BMWi) [25].

KURT RAUH kept a four volume diary about his journey, [26]. In the beginning of this diary he also explains the motivation behind this journey: "*The trip takes 2 months, and is planned under the general title agricultural machine operation and testing and should – in general terms – encompass the study of the organization and methods at universities and technical schools in teaching courses for agricultural machinery, testing and repair. Special requests will be considered.*" Fig. 10 shows

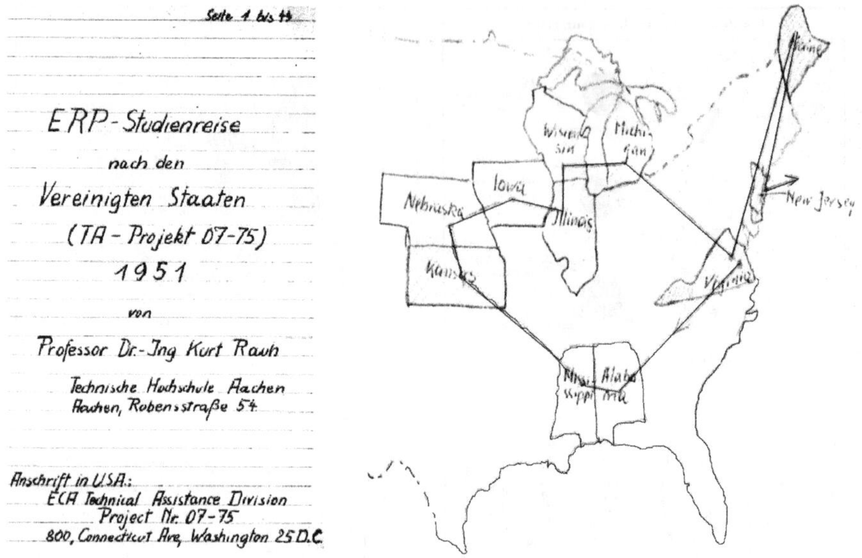

Fig. 10. ERP roundtrip to the U.S. [26]

on the left the cover of his travel diary and on the right a sketch of his trip map drawn by himself.

Besides information about travel details considering places he visited and people he met, also quite detailed sketches of agricultural machinery and equipment, as shown in Fig. 11, can be found in the diaries together with detailed descriptions of the same machinery and equipment but also about the underlying processes, procedures and test methods in particular.

Less than a year after returning from his study trip to the U.S., KURT RAUH died from a serious illness. His last appearance at a scientific conference was probably the VDI conference on mechanism science in Berlin in October 1951 [27]. As formulated in the preface to the proceedings of this conference issued in 1953 as the first VDI Proceedings Volume this was the first German conference on mechanism science after WWII. HERMANN ALT as then chairman of the committee of mechanism science within the VDI also notes that choosing Berlin and especially the facilities of the Technical University in Berlin-Charlottenburg for the conference should also be seen as an appreciation of FRANZ REULEAUX as "*the founder of mechanism science*" who taught at the then "Königliche Gewerbeinstitut" (Royal Industrial Institute) between 1864 and 1896. The papers collected in the proceedings reflect that almost all well-known German scientists in mechanism theory active at that time were gathered just to name amongst others HERMANN ALT, WILLIBALD LICHTENHELDT, GERD KIPER, KARL-HEINZ SIEKER, PAUL LOHSE, KURT HAIN, RUDOLF BEYER and ROBERT A. KRAUS.

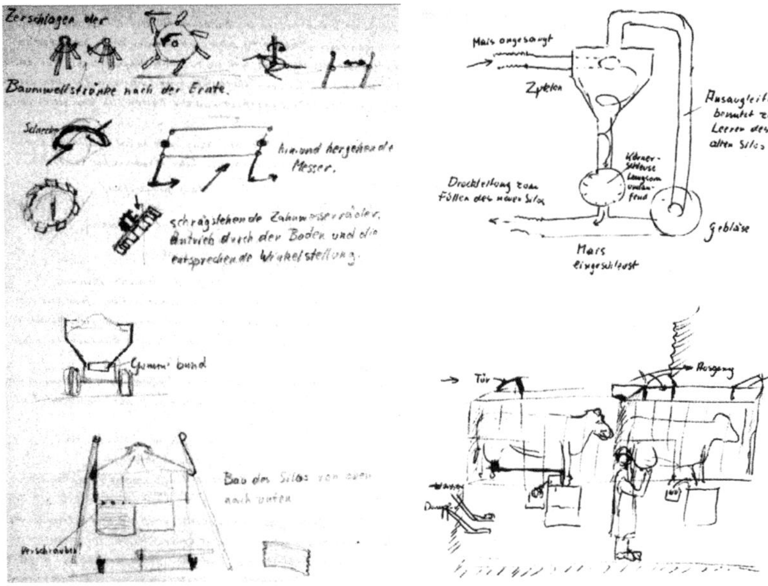

Fig. 11. Sketches of agricultural machinery, equipment and processes. [26]

In his obituary in honor of KURT RAUH, the Rector of RWTH Aachen University says, "*His great achievement was the development of teaching in mechanism science, which was performed by him in his own way. In several important works, especially in the two-volume textbook on mechanism theory, he has documented his research. The deceased was a man of high ideals and aspirations. To all his staff, he was not only an inspiring teacher, but also a companionable consultant.* "

4 Summary and Acknowledgements

With this paper the author intended to show the strong roots that mechanism science has in the Institute of Mechanism and Machine Dynamics at RWTH Aachen University. These roots go far beyond the 50th anniversary celebrated in 2009. Especially from the beginning of KURT RAUH's activities in Aachen, it was mainly publications regarding systematics, synthesis and selection of mechanisms of any kind, which should allow the engineer working in practice to find, to select and synthesize in a very simple and clear way an optimal solution for a given motion task.

The paper has been compiled by the author, who is not a "learned" historian, but judges his sources as a mechanical engineer specialized in the field of mechanism and machine science. A lot of the publications on which this paper is based could be retrieved with the help of the DMG-Lib internet portal. Therefore acknowledgement is given to the German Research Foundation (DFG) which supported the DMG-Lib project and also to the European Union which supported the thinkMOTION project. Both projects are aimed at digitally preserving past and present knowledge in Mechanism and Machine Science.

References

[1] Braune, R., Corves, B.: Franz Reuleaux and Rhenisch roots of mechanism theory. In: Proc. of HMM 2012, International Symposium on History of Machines and Mechanisms, Amsterdam (2012)

[2] RWTH Aachen University Library, Gewerbe- und Kunstausstellung in Düsseldorf 1880 – Ausstellungskatalog, `http://darwin.bth.rwth-aachen.de/opus/volltexte/2009/2792/`

[3] Moon, F.C.: Mechanical Engineering Heritage Collection, Reuleaux Collection of Kinematic Mechanisms at Cornell University 1882, American Society of Mechanical Engineers (2004)

[4] Essich, E.: Über Steuerungsgetriebe mit Wälzhebeln. Dissertation bei Prof. Langer, RWTH Aachen. RWTH Aachen University Library, Signature S2305 (1909)

[5] Mauersberger, K.: Theory of Mechanisms and Machines at Dresden University of Technology - A Historical Review. In: Third International Workshop on History of Machines and Mechanisms, Moscow (2005)

[6] Rauh, K.: Untersuchung und Weiterentwicklung der Getriebe mit periodischem Hin- und Rücklauf und beschleunigungsfreiem Arbeitsgang. Diss. TH Hannover, Rhenania-Verlag G.m.b.H., Bonn (1927)

[7] Meyer zur Capellen, W.: Eine Getriebegruppe mit stationärem Geschwindigkeitsverlauf. Forschungsbericht Nr. 606 des Landes Nordrhein-Westfalen. Westdeutscher Verlag, Köln (1958)

[8] Meyer zur Capellen, W., Lehn, F.: Kinematische Kenngrößen der ebenen elliptischen und der räumlichen Schleifen. Forschungsbericht Nr. 1593 des Landes Nordrhein-Westfalen. Westdeutscher Verlag, Köln (1966)

[9] Meyer zur Capellen W., Schreiber, E.: Getriebe mit stationärem Geschwindigkeitsverlauf. Forschungsbericht Nr. 1851 des Landes Nordrhein-Westfalen. Westdeutscher Verlag, Köln (1967)

[10] Meyer zur Capellen.W., Schreiber, E.: Raumgetriebe mit stationärem Geschwindigkeitsverlauf. Forschungsbericht Nr. 2096 des Landes Nordrhein-Westfalen. Westdeutscher Verlag, Köln (1970)

[11] Meyer zur Capellen, W., Willkommen, W.W.: Sphärische Viergelenkgetriebe als Proportionalgetriebe. Forschungsbericht Nr. 2386 des Landes Nordrhein-Westfalen. Westdeutscher Verlag, Köln (1970)

[12] Shi, Z.: Maßsynthese von Getriebekombinationen aus ebenen oder sphärischen Doppelkurbeln und anderen Getrieben zur Verwirklichung bereichsweise vorgegebener Übersetzungsverhältnisse. Dissertation RWTH Aachen (1982)

[13] Leusch, G.: Ein Beitrag zur Systematisierung und Maßsynthese ebener ungleichmäßig übersetzender Getriebe mit bereichsweise annähernd konstanter Übersetzung. Dissertation RWTH Aachen, VDI Fortschrittbericht Reihe 1, Nr. 248. VDI-Verlag, Düsseldorf (1995)

[14] VDI 2740, Part 3: Mechanical devices for automation equipment - Mechanisms for partially synchronised motion. VDI, Düsseldorf (1999/2007)

[15] Corves, B., Kloppenburg, J.: History and Future of the IGM-Mechanism Collection. In: History of Machines and Mechanisms 2006, IFToMM Workshop Lectures, Ithaca, NY, USA (September 9-10, 2006)
[16] Corves, B., Niemeyer, J., Kloppenburg, J.: IGM-Mechanism Encyclopaedia and the Digital Mechanism Library as a Knowledge Base in Mechanism Theory. In: Proceedings of DETC 2006: ASME 2006 Int. Tech. Conf. & Comp. and Inform. Eng. Conf., Philadelphia, Pa., USA (September 10-13, 2006)
[17] Burmester, L.: Lehrbuch der Kinematik, Erster Band, Die ebene Bewegung. Verlag von Arthur Felix, Leipzig (1888)
[18] Rauh, K.: Praktische Getriebelehre. Erster Band. Verlag von Julius Springer, Berlin (1931)
[19] Rauh, K.: Ich melde meine Erfindung selbst an: Vorlesungen über Patentlehre an der Technischen Hochschule Aachen zusammengefasst zu fünf Rundfunkvorträgen (1932 übertragen vom Westdeutschen Rundfunk). In: 2. Erweiterte Ausgabe als Handschrift Gedr. Borowsky, Köln (1935)
[20] Rauh, K.: Vorlesungen über Patentlehre an der Technischen Hochschule Aachen. In: 3. Erw. Aufl., Als Hs. gedr., Soennecken, Bonn (1947)
[21] Rauh, K.: Dozentur für Getriebelehre, Maschinenaufbaulehre und Patentlehre. In: Jahrbuch der Rheinisch-Westfälischen Technischen Hochschule Aachen, pp. 176–177, Buchverlag W. Girardet Essen (February 1949)
[22] Rauh, K.: Entwicklungslinien im Landmaschinenbau. Girardet, Essen (1949)
[23] Rauh, K.: Aufbaulehre der Verarbeitungsmaschinen, Textband und Bildband. Verlag W. Girardet, Essen (1950)
[24] `http://www.bundesfinanzministerium.de/nn_39820/DE/BMF__Startseite/Service/Glossar/E/011__European__Recovery__Program___28ERP_29.html`
[25] `http://www.studienstiftung.de/erp.html`
[26] Rauh, K.: ERP-Studienreise nach den Vereinigten Staaten. Tagebuch, `http://www.dmg-lib.org/dmglib/main/portal.jsp?mainNaviState=browsen.docum.meta&id=4869009&id=487000,&id=487100,&id=487200`
[27] VDI(Hrsg.): Getriebetechnik: VDI-Tagung in Berlin, VDI-Tagungsheft 1. Dt. Ingenieur-Verl., Düsseldorf (Oktober 1951) (1953)

About Karl Hoecken and Some of His Works on Mechanisms

Hanfried Kerle

Formerly Technical University of Braunschweig, Germany
Institute of Machine Tools and Production Technology,
Langer Kamp 19b, D-38106 Braunschweig
h.kerle@t-online.de

Abstract. After the death of Wilhelm Hartmann in 1922, the direct successor of Franz Reuleaux on the Chair of Kinematics (Lehrstuhl für Getriebelehre) at the Technische Hochschule (TH) Berlin-Charlottenburg, there was a row of intermediate professors until Hermann Alt from Dresden finally got the chair in 1939. Karl Hoecken belonged to this row of intermediate professors between 1930 and 1934. Looking through Hoecken´s estate we can find some new aspects of the famous kinematic model collection established by Reuleaux. This paper also takes up three publications of Hoecken on mechanisms, models and their kinematic equations, and thus gives a short historical overview over Hoecken´s contribution to the field of kinematics at that time. Incidentally, Hoecken himself added 26 own models to the then Reuleaux collection.

1 Introduction

Karl Hoecken was born on May 31, 1874 in Berlin. To the end of World War II he moved to Braunschweig with his family where he died latest in 1963 at the age of 89. Following some documents from Berlin and Braunschweig, we learn that he studied Applied Mathematics and Geodetics in Berlin, Bonn and Darmstadt [1]. Since 1909 and at least till 1914, but most probably till 1930 Hoecken had a leading technical position as senior engineer with the company "Optische Anstalt C. P. Goerz AG" in Berlin-Friedenau. Between 1920 and 1924 he also worked in Amsterdam.

Between 1922 and 1928 there was no real holder of the Chair of Kinematics at the TH Berlin. When Franz Reuleaux (1829-1905) had left the TH in 1896, Wilhelm Hartmann (1853-1922) became his successor in the chair and at the same time head of the famous and well-known Kinematic Model Collection of Reuleaux in the western part of the ground floor in the main building of the TH (Fig. 1).

Six years after Hartmann´s death, the TH management asked Rudolf Franke (1870-1962) from Hanover to take over the chair temporarily. At that time Franke was already full professor since 1922 at the TH Berlin in the Chair of Telecommunications and Precision Engineering, a field between electrical and mechanical engineering, and had successfully established his own collection of device examples in this field. In 1930 – at the age of 56 – Karl Hoecken got his chance and became Lecturer of

T. Koetsier, M. Ceccarelli (Eds.): Explorations in the His. of Machines & Mech., HMMS 15, pp. 123–134.
springerlink.com

Fig. 1. Main building of the TH Berlin around 1900

Kinematics and head of the Reuleaux collection. He did this job only four years till 1934, then he left the TH Berlin. Rudolf Franke came back in the Chair of Kinematics and could retire only when Hermann Alt (1889-1954) from Dresden followed in 1939 [2]. The author of this paper does not know exactly what Hoecken did and where he lived in the meantime till the end of World War II. There are documents at the then TH Braunschweig which prove that in 1946 Hoecken – now living in Braunschweig – applied for a position to give lectures on kinematics, his request was declined.

2 Karl Hoecken and Franz Reuleaux

Most of Hoecken´s scientific papers were written and published in the time between 1930 and 1934. There are a lot of patent papers existing before 1930, a fact that demonstrates Hoecken´s practical orientation and is typical for a working life in industry.

Immediately after his appointment as lecturer and head of the Reuleaux collection, Hoecken took inventory of the collection. He made many photos of the kinematic models, by means of a typewriter he wrote down the names of all the models in the collection, together with the numbers assigned to them. In 1930/31 the Reuleaux collection consisted of 818 different models. Probably already Reuleaux, but certainly Hartmann gave numbers to the models. The system of numbers followed a special classification scheme, cf. Table 1.

Table 1. Classification of Reuleaux models in Berlin

Group number	Model numbers	Model types
0	001-043	Element pairs and four-link mechanisms
1	101-156a	Drawing auxiliary devices and mathematical instruments
2	201-234	Simple machine tools
3	301-348	Straight-line linkages
...	...	...
8	801-833	Gear trains
...	...	...
19	1901-1915	Belt and rope trains
20	2001-2051	Degrees of freedom

There is an additional list of 26 Hoecken models, partly belonging to the Reuleaux collection and partly distributed among chairs at the polytechnics in Karlsruhe, Danzig and Munich.

3 Three Works on Mechanisms of Karl Hoecken

In the following three papers of Hoecken are presented and discussed to give an idea of his mathematical skills, knowledge of kinematics and sense for practical applications.

Conchoidal straight-line mechanism

From the kinematic point of view the conchoidal straight-line mechanism is a four-link centric angle-slider mechanism (Fig. 2a). The coupler link as the straight line gg constantly slides through the fixed sliding point O, whereas the movable sliding point A is led along the fixed straight line through Q having the distance a from O. The sliding direction of point A is perpendicular to the symmetry axis passing through O and Q, the latter being also the horizontal axis x of a rectangular coordinate system. Every point P on the coupler straight line gg follows a conchoidal curve. In case P is located on the opposite side of OA, the conchoidal curve contains a closed crunode. If this crunode is partly similar to a circle, it may be approximately replaced by a circle arc with centre point M on the x-axis and radius r (Fig. 2b). Thus a centric inverted slider-crank mechanism is found as a substitute mechanism, where the slider in A can be dropped, because its path partly consists of an approximately straight line, perpendicular to the x-axis [3].

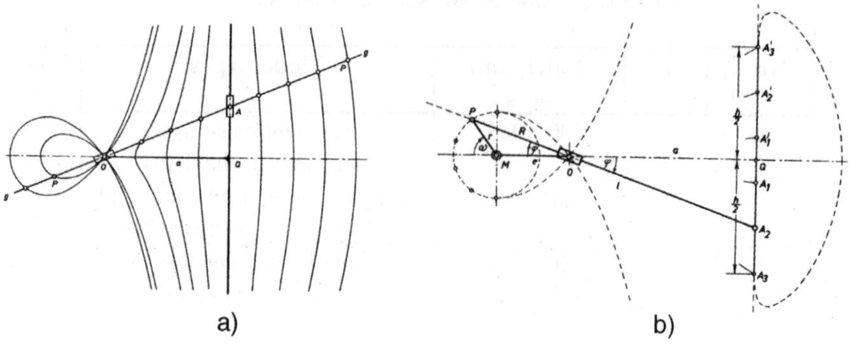

Fig. 2. Centric angle-slider with conchoidal coupler curves (a) and a centric inverted slider-crank (b) which approximately replaces (a)

Hoecken equates the polar equation of the crunode with that of a circle around M and takes into account that all coupler curves of the inverted slider-crank are symmetric related to the x-axis. He determines the dimensions of this substitute mechanism which approximately leads the coupler point A along a straight line through six points in finite neighbouring order with a distance of h/5 between two points each. Taking a = OQ = 100 length units for given, Hoecken gets the following data: e = OM = 38.6378, h = $A_3A_3{}'$ = 120.7658, l = AP = 161.9516 and r = MP = 23.3341. Hoecken writes that the maximal deviation of the coupler curve within h = $A_3A_3{}'$ from an exact straight line results in 0.0203. Moreover, Hoecken built two compact models in order to show the usefulness of his ideas and the validity of his calculations (Fig. 3).

One should keep in mind that the mathematical aids at the time of Hoecken were limited and mainly consisted of tables of logarithmic and trigonometric functions. Slide rules were also in use, but not very efficient and/or precise. Solving Eqn. 1 for example, i.e.

$$2el \cdot \cos^3(\varphi) - (l^2 + 2ae + e^2 - r^2) \cdot \cos^2(\varphi) + 2al \cdot \cos(\varphi) - a^2 = 0 \quad (1)$$

to calculate e, l, r, where three angles φ are given above or below the x-axis (cf. Fig. 2b), requires multiple interpolation and iteration procedures to gain results with four digits after the decimal point.

The author used the commercial program "Mathcad 11" and followed a sequence of equations published in [4]. Thus he could show that Hoecken´ calculations were correct (Fig. 4).

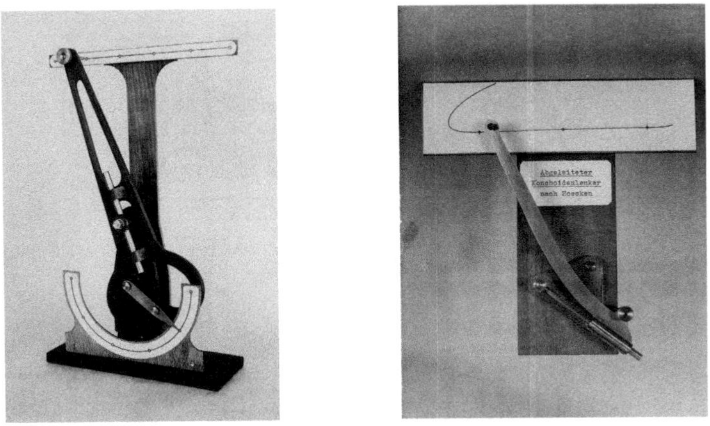

a) b)

Fig. 3. Two Hoecken models representing conchoidal straight-line mechanisms: Original form (a) and derived form (b)

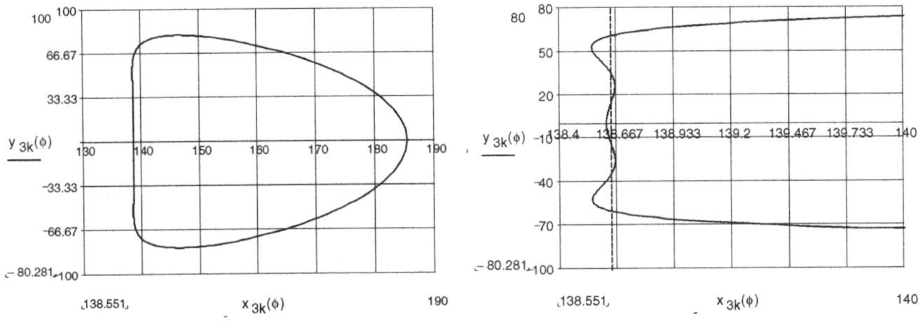

a) b)

Fig. 4. Coupler curve of Hoecken´s conchoidal straight-line mechanism: a) General view, b) zoomed view around the vertical straight line $x_{3k} = a + e$

Six-link dwell mechanism

With dwell mechanisms having one d.o.f. the driven link temporarily comes to a standstill, while the driving link moves continuously. The standstill or dwell position is characterized by zero values of the velocity and acceleration of the driven link, related to the fixed link. In return positions, i.e. in so-called "dead positions", the velocity of the driven link is also zero, but not its acceleration. The dwells generated by linkages are normally only approximate ones, i.e. the zero values of velocity and acceleration of the driven link mentioned above are reached only approximately or on an average [5, 6].

In 1980 Kurt Hain (1908-1995) edited a booklet on dwell mechanisms [7]. In the beginning of this booklet Hain dealt with the systematics of dwell mechanisms based on six links of the type "Watt" and "Stephenson" and found nine structures, Fig. 5.

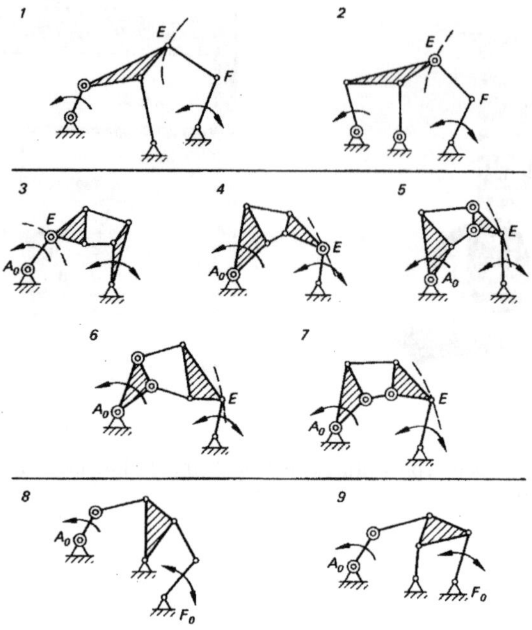

Fig. 5. Systematics of six-link dwell mechanisms with simple turning pairs by Kurt Hain

The driving link is marked by a circle segment with one arrow, the driven link with the possibility to have a dwell position by a circle segment with two arrows. Turning pairs with a full rotation of 360 deg are marked by a double circle, those only swinging are marked by a simple circle. So, when we bear in mind Hain´s systematic overview and look at Hoecken´s six-link dwell mechanism (Fig. 6), we find out at once that Hoecken´s choice corresponds with the number 1 of Hain´s catalogue: there is a four-link crank-rocker with a two-bar E-F coupled to it. It is very interesting to discover that Hoecken put his linkage onto a typical Reuleaux frame and assigned to it the model number 2201. There is also a paper of Hoecken on dwell mechanisms [8], but it does not describe his model shown in Fig. 6.

Before manufacturing the model "six-link dwell mechanism", Hoecken made a technical drawing of it on February 11, 1932 and named it "Rastgetriebe", Fig. 7. The drawing shows details of the links, its dimensions and even a counterweight on the opposite end of the rocker for static balancing.

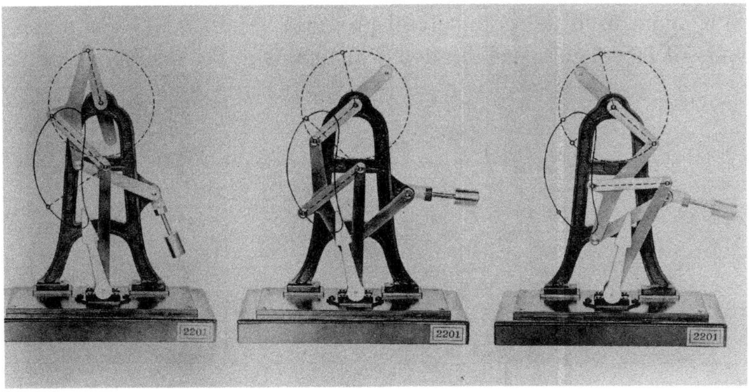

Fig. 6. Hoecken´s six-link dwell mechanism as a model mounted on a Reuleaux frame

The dimensions of Hoecken´s six-link dwell mechanism are as follows:

$n_0 = B_0C_0 \approx 124.0$	$A_0C_0 \approx 215.0$
$n_1 = A_0B_0 \approx 127.0$	$n_2 = A_0A = 61.3$
$n_3 = AB = 131.9$	$n_4 = B_0B = 94.2$
$n_5 = CP_{3k} = 83.9$	$n_6 = C_0C = 139.5$
$AP_{3k} = 84.8$	$\angle P_{3k}AB = 24.45$ deg

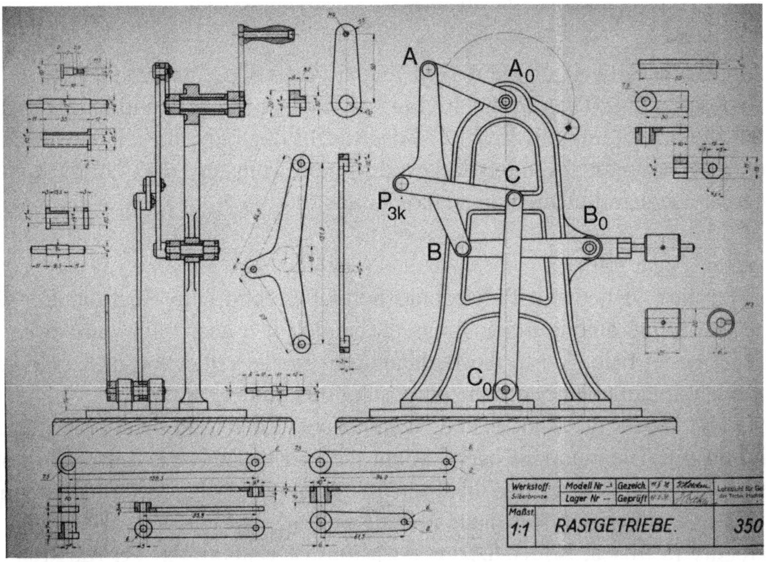

Fig. 7. Technical drawing of Hoecken´s six-link dwell mechanism showing details for its manufacturing

The author again used the commercial program "Mathcad 11" to make a kinematic analysis of Hoecken´s six-link dwell mechanism, the results are presented in Fig. 8.

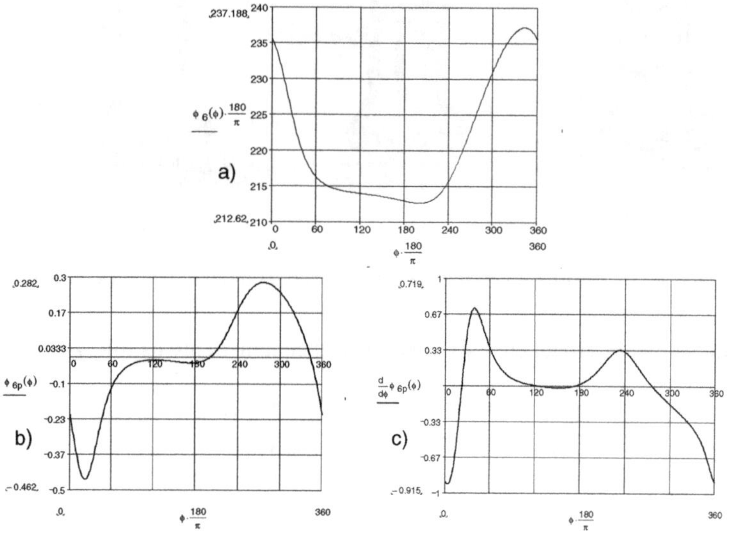

Fig. 8. Kinematic analysis of Hoecken´s six-link dwell mechanism: Angle (a), angular velocity (b) and angular acceleration (c) of the driven link 6 versus crank angle φ (driving link)

The approximate dwell of the driven link 6 (C_0C, Fig. 7) is realized within the range 120 deg $\leq \varphi \leq$ 180 deg of the driving link 2 (A_0A, Fig. 7). Within this range the angle φ_6 of the driven link falls from 214 deg to 213 deg only, the angular velocity of the driven link shows an absolutely largest deviation from zero of 0.022 rad/s, the one of the angular acceleration amounts to 0.018 rad/s^2.

Redtenbacher´s gear train

In 1926 Hoecken visited the Polytechnic School (Technische Hochschule, TH) in Karlsruhe. He found among other things the model of a gear train with seven gearwheels. The model belonged to the Redtenbacher model collection in Karlsruhe, but nobody could explain to Hoecken the purpose of this model [9].

Fig. 9 (left) shows the back of the Redtenbacher gear train on a photo made by Hoecken himself. The author is happy about the fact that there is a second photo on the internet [10], now revealing the front of the model, Fig. 9 (right) with model number UK083. This photo was made in 2001 or 2002 by Francis C. Moon from Cornell University in Ithaca, NY (USA) on occasions of visits to the TH Karlsruhe. It is not necessary to write about the significance of Ferdinand Jakob Redtenbacher (1809-1863) in the evolution of mechanical engineering to a science in Germany, because there is detailed and excellent description of his lifework, cf. [11, 12].

Fig. 9. Redtenbacher´s gear train, back view (left) and front view (right)

Hoecken explains that three of the seven gearwheels are not necessary for the correct function of the mechanism, but only serve to have the input rotary axis with the driving crank coincident with the output rotary axis ("returning" mechanism). Both axes have metal pointers in form of arrows for demonstration, similar to those of a clock.

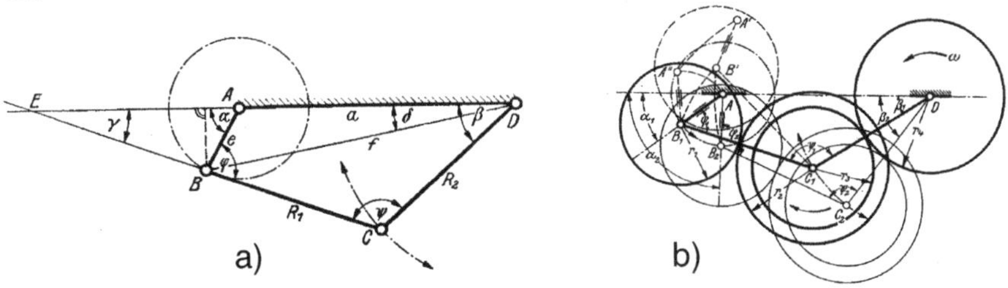

Fig. 10. Hoecken´s sketch of Redtenbacher´s gear train

The base mechanism is that of a crank-rocker ABCD (Fig. 10a) with four two-by-two mating gearwheels with their centres in B, C and D (Fig. 10b). The gearwheel centred in B forms the driving crank and rotates around A (eccentric wheel). Dependent on the dimensions of the crank-rocker and the radius r_1 to r_4 of the gearwheels it is possible to vary the output angle ω of the gearwheel 4 related to the input angle α of the crank or gearwheel 1, i.e. the transmission ratio can be partly positive, zero (standstill of gearwheel 4) or negative ("pilgrim´s step" motion of gearwheel 4), while the crank 1 rotates continuously with constant angular velocity. The rotary motions of the gearwheels originate from the variable difference angles φ, ψ and β between the adjacent bars of lengths e-R_1, R_1-R_2 and a-R_2 [13, 14].

Hoecken used his own extensive geometric-mathematical equations to calculate the output angle $\omega = \omega(\alpha, t)$ (time t) and output angular velocity $d\omega/dt = (d\omega/d\alpha)\cdot(d\alpha/dt)$

for given dimensions and given input quantities α and $d\alpha/dt \equiv 1$ rad/s of the gear train [9]. Fig. 11 shows Hoecken´s results in form of his original diagrams for three different gearwheel ratios (cf. Table 2).

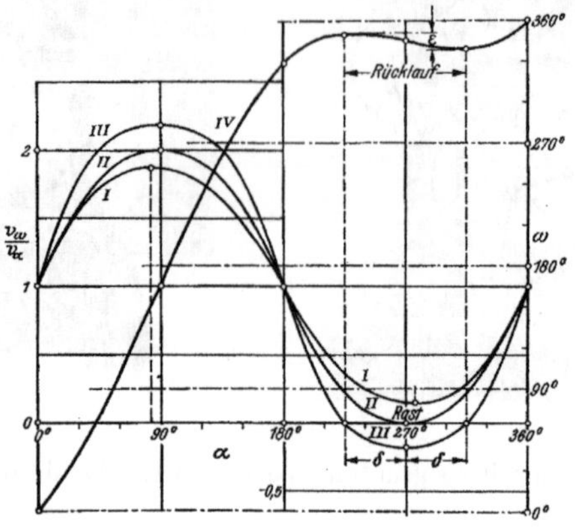

Fig. 11. Hoecken´s diagrams for Redtenbacher´s gear train: Diagrams I – III: output angular velocities $d\omega/d\alpha$, diagram IV: output angle $\omega(\alpha)$

To save time and pains, the author put again into action the commercial program "Mathcad 11". Given the dimensions a = AD, e = AB, R_1 = BC, R_2 = CD of the crank-rocker and the radius r_1 to r_4 of the gearwheels, where $r_1 + r_2 = R_1$ and $r_3 + r_4 = R_2$ and setting $r_1/r_2 = \mu$ and $r_3/r_4 = \upsilon$, we first make use of a set of kinematic equations for four-bar linkages and calculate the angles $\beta = \beta(\alpha)$ and $\varphi_{BC} = \varphi_{BC}(\alpha)$, both related to the horizontal axis AD (x-axis). Then we care for the wheel rotations and follow the equations

$$\psi(\alpha) = \phi_{BC}(\alpha) - \beta(\alpha) - \pi \tag{2}$$

$$\phi(\alpha) = 2 \cdot \pi - \phi_{BC}(\alpha) + \alpha \tag{3}$$

$$\omega(\alpha) = \omega_0 + \beta(\alpha) - \beta_0 - \upsilon \cdot [\psi(\alpha) - \psi_0] + \mu \cdot \upsilon \cdot [\phi(\alpha) - \phi_0] \tag{4}$$

$$\omega_0 = \omega(\alpha = \alpha_0 = 0) = (1 + \upsilon) \cdot \beta_0 + \upsilon \cdot (1 - \mu) \cdot \phi_{BC0} = 0 \tag{5}$$

In order to demonstrate the effectiveness of these equations, we take Hoecken´s example [9] and set the lengths of the four-bar links and the gearwheel radii as follows: a = 50, e = 15, $R_1 = R_2 = R = 35$. We distinguish between three cases concerning different gearwheel ratios, i.e. case I: $r_1 = 25$, $r_2 = 10$, $r_3 = 20$, $r_4 = 15$; case II: $r_1 = 20$, $r_2 = 15$, $r_3 = r_1 = 20$, $r_4 = r_2 = 15$; case III: $r_1 = 10$, $r_2 = r_3 = r = 25$, $r_4 = r_1 = 10$ (Hoecken made a slight mistake when he wrote $r_1 = r_4 = r = 25$.). Case III is equivalent with the fact that there are only three instead of four gearwheels. The results are shown in Table 2.

The columns represent the output angle $\omega = \omega(\alpha)$ and its derivatives $d\omega/d\alpha$ and $d^2\omega/d\alpha^2$. Because of the fact that the input angular velocity is constant and of unit value, the derivative functions $d\omega/d\alpha$ and $d^2\omega/d\alpha^2$ are equivalent with the angular velocity and angular acceleration with respect to the output gearwheel 4.

Table 2. Output motion of gearwheel 4 versus input crank angle in Redtenbacher´s gear train showing three different cases

Case I Positive ratio	**Case II** (Hain) “Point dwell”	**Case III** (Hoecken) “Pilgrim´s step”

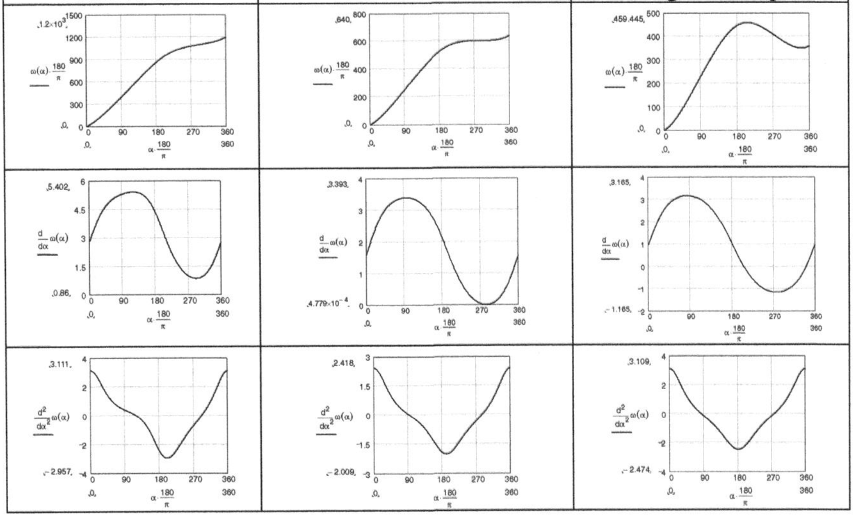

Case I reveals a continuous positive ratio between the output angle ω and input angle α, the output angular velocity varies within the positive range without reaching zero; in case II there is a standstill of the output gearwheel momentarily at point α = 285 deg, a condition that Kurt Hain already named a “point dwell” position in coherence with a special six-bar linkage [15]; case III stands for a “pilgrim´s step” motion of the output gearwheel in the range 205 deg $\leq \alpha \leq$ 340 deg, i.e. gearwheel 4 rotates partly in opposite direction related to input gearwheel 1.

4 Conclusions

Karl Hoecken was one of the intermediate professors teaching kinematics in the time between Wilhelm Hartmann and Hermann Alt at the Technische Hochschule Berlin-Charlottenburg. Moreover, from 1930 till 1934 he had the unique chance to work as a curator of the famous Reuleaux collection. He seized this chance and took an inventory of the collection – for the sake of coming generations of kinematicians in Germany and elsewhere. By describing and explaining some of his works on kinematics, the author of this paper also wants to thank Karl Hoecken for his initiative at that time.

References

[1] Kerle, H.: Über drei Getriebe von Karl Hoecken aus Berlin – im Nachgang betrachtet und neu berechnet. In: Berger, M. (ed.) Proceedings of the 9th Kolloquium Getriebetechnik, September 07-09, pp. 313–332. Universitätsverlag Chemnitz, Germany (2011)

[2] Hain, K.: Prof. Dr. phil. Rudolf Franke zum Gedächtnis. Konstruktion 15(2), 66 (1963)

[3] Hoecken, K.: Rechnerische Ermittlung eines Konchoidenlenkers. Reuleaux-Mitteilungen – Archiv für Getriebetechnik 6(8), 421–423 (1933)

[4] Luck, K.: Kinematische Analyse ebener Grundgetriebe in Matrizenschreibweise. Wiss. Zeitschrift d. TU Dresden 19(6), 1467–1474 (1970)

[5] Alt, H.: Koppelgetriebe als Rastgetriebe. Zeitschrift VDI 76(19), 456–462 (1932); (22) 533-537

[6] Meyer zur Capellen, W.: Zur Theorie der Bahnkurven-Rastgetriebe. Konstruktion 15(10), 389–392 (1963)

[7] Hain, K.: Entwicklungsstand der Gelenk-Rastgetriebe. VDI-Forschungsheft (597), 5–7 (1980)

[8] Hoecken, K.: Die Erzeugung von Rasten durch Koppelgetriebe. Zeitschrift VDI 74(19), 594 (1930)

[9] Hoecken, K.: Das exzentrisch angetriebene Räderknie zur Vermittlung ungleichförmiger Drehung mit Rast und Rückwärtsgang. Zeitschrift VDI 74(16), 509–512 (1930)

[10] (2011), `http://kmoddl.library.cornell.edu`

[11] Wauer, J., Mauersberger, K., Moon, F.C.: Ferdinand Redtenbacher (1809-1863): Pioneer in scientific machine engineering. Mechanism and Machine Theory 44(9), 1607–1626 (2009)

[12] Wauer, J., Mauersberger, K., Moon, F.C.: Ferdinand Jakob Redtenbacher (1809-1863). In: Ceccarelli, M. (ed.) Distinguished Figures in Mechanism and Machine Science – Their Contributions and Legacies, Part 2. History of Mechanism and Machine Science, vol. 7, pp. 217–245. Springer Co. (2010)

[13] Hain, K.: Winkelzuordnungen in Räderkurbelgetrieben der Viergelenkkette. Konstruktion 5(8), 257–263 (1953)

[14] Hain, K.: Das Übersetzungsverhältnis in Räderkurbelgetrieben. Werkstatt und Betrieb 87(10), 625–628 (1954)

[15] Hain, K.: Das sechsgliedrige Punktrastgetriebe für Werkstück- und Werkzeugführungen besonderer Art. Werkstatt und Betrieb 114(5), 333–337 (1981)

Gioacchino Russo – A Versatile Scientist

Francesco Petrone[1] and Viviana Russo[2]

[1] Full Professor of Mechanics of Machines, Dipartimento di Ingegneria Industriale e Meccanica, Università degli Studi di Catania, Italy
fpetrone@diim.unict.it
[2] PhD in Maritime Engineering
russo.viviana@libero.it

Abstract. The Italian scientist Gioacchino Russo made several important contributions to the development of machines and mechanisms. We briefly discuss his life and then give a detailed explanation of the principles behind the navipendulum, one of his most famous mechanical inventions.

1 Biographical Notes about Gioacchino Russo

Gioacchino Russo (Figure 1) was born in Catania, Italy, on September 8th, 1865.

He was first educated by his uncle, the priest Gioacchino Russo, rector of the Archiepiscopal Seminary. In 1887, he was awarded a diploma in Civil Engineering from the 'Scuola di Applicazione' of Turin. In the same year he began his military career with the 'Royal Navy Engineers'. In 1889, he was awarded another diploma in Naval and Mechanical Engineering from the 'Scuola Superiore di Ingegneria Navale' of Genoa.

From 1890 to 1892, he taught 'Naval Architecture' at the Naval Academy of Leghorn. During this period, he wrote a valued treatise, 'Manuale di Architettura Navale', the first of its kind in Italy, which was published in 1905 by the international publishing house Roux & Viarengo of Turin.

He also presented two important scientific inventions: the cleptoscope and the navipendulum.

The cleptoscope, a useful instrument for submarines which can observe above the water surface before surfacing (as suggested by Eng. Laurenti), is a precursor of the modern periscope. There was a model of the cleptoscope at the 'Museo della Scienza e della Tecnica' of Turin, but it was destroyed by bombing during the Second World War.

The Navipendulum is G. Russo's most known invention (Figure 2). Built in Florence by 'Officine Galileo', the instrument was exhibited at the Institution of Naval Architects (I.N.A.), London, in 1902, at the behest of the British Marine Minister. After a practical demonstration in front of a packed house, the 'Gold Medal' – rarely awarded to foreign scientists – was given him by the Executive Council of the I.N.A. After a similar demonstration in Paris during the same year, he received the 'Palmes Académiques'.

T. Koetsier, M. Ceccarelli (Eds.): Explorations in the His. of Machines & Mech., HMMS 15, pp. 135–144.
springerlink.com

Fig. 1. Gioacchino Russo (1865-1953).

During his military service for the Marine Ministry in Rome, he worked on designs for the Dreadnought battleships: 'Cavour' and 'Dante Alighieri' with General Masdea in the Executive Department for New Constructions. He completed these designs after the death of General Masdea (1909). One of his most interesting ideas for the shooting towers was placing two cannons one upon the other, so as to shoot in two different or opposite directions at the same time. From 1910 to 1912, he was Director of the Royal Building Site in Castellamare di Stabia, and then in Venice.

In 1915 and in cooperation with the industrial company 'Ansaldo', he created an experimental model basin with elastic walls to reproduce wave-motion.

Fig. 2. The Navipendulum.

In 1927, he began service with the 'Navy Reserve' and was elected a member of Parliament from 1919 to 1921. He was then re-elected from 1924 to 1929. Subsequently commissioned Senator of the Kingdom for his contribution to science, he was Under-Secretary of State for the Navy up to 1933.

During these years, he designed an experimental camera, capable of coloured output. Photographic polychromy was obtained with a triple-object glass and three primary colour filters (Red, Green and Blue). The image in black and white - fixed by a triple-object glass, filtered through the three primary colours and then exposed on the same plate (Figure 3a) - was projected onto the same three coloured filters used before (Figure 3b). A fully polychromatic output of the image was obtained by overlapping the three coloured images (Figure 3c).

a) b) c)

Fig. 3. Preliminary steps for obtaining a coloured output from Chromac : image in black and white filtered through the three primary colours (a), image projected onto the three primary colours (b) and the polychromatic image (c) [25].

Initially, the Korista company of Milan was entrusted with the construction of the experimental camera, but was subsequently transferred to the Optische Anstalt C.P. Goertz company of Berlin (Figure 4), which had already been taken over by Carl Zeiss, because Korista was over-burdened with work and the project manager (Mr Carlo Koristka himself) suddenly died. The first Chromac camera was made in 1927 (Figure 5).

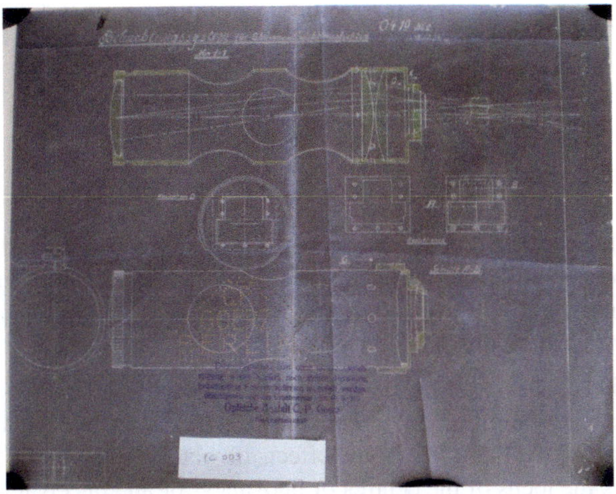

Fig. 4. Chromac experimental camera, design made by C.P. Goertz – Berlin [25].

Fig. 5. Chromac experimental camera [25].

At the end of the Second World War when the scientist was already eighty years old, he appeared at the Chancery Court to answer about his governmental positions during the previous twenty years. He was one of two isolated cases, together with Senator Sirianni, to be re-elected Senator among all the ministers of that period.

During his latter years, he actively contributed to designing the Town Planning of Catania, giving significant inputs and suggestions, which unfortunately were only applied after his death.

He was also:

- Member of C.N.R.
- Member of the National Optical Institute
- Secretary of the Corps of Engineers and Architects
- President of the Corps of Engineers and Architects of the Province of Catania
- President of the 'Unione Industriali' of the Province of Catania
- Board member of Salmoiraghi
- Board member of SGES ('Società Generale Elettrica per la Sicilia')
- President of the Board of Directors of SAIVO ('Società per il vetro e l'ottica') of Florence

He died in Catania on May 7th, 1953.

2 The Navipendulum

The Navipendulum designed by G. Russo was capable of experimentally reproducing the behaviour of ships hit by transverse waves, with a wide range of amplitudes and frequencies. Some of the scientific considerations contained in the papers of G. Russo have been reported within this section.

Theoretical basis

It is well known that a ship floating in an upright or inclined position can be statically compared to a supported body free of the effects of water. Furthermore, the geometric surface, which is the locus of the centres of buoyancy for all conceivable inclinations, becomes a material surface, on which the ship rests in a horizontal plane [1-4].

Considering only transverse inclinations, the surface of the centres of buoyancy may be replaced by a cylindrical surface, with cross-section the line of the centres of buoyancy in a transverse section of the ship (plane of inclination). In addition to this line and the corresponding cylindrical surface there are an infinite number of other lines, which equally posses the property of exciting a reaction of the same value and line of action as the buoyancy at any angle of inclination (Figure 6).

Among the infinite number of lines which may be drawn parallel to the line of the centres of buoyancy, preference can be given to a line with all its points as near to the centre of gravity as possible.

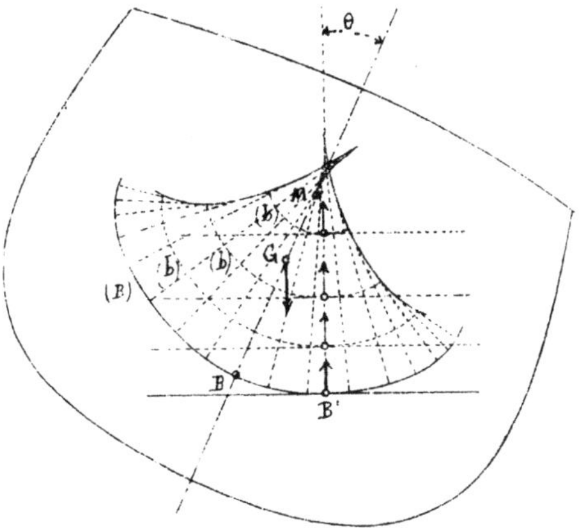

Fig. 6. Lines of the centres of buoyancy in the plane of inclination.

It is easy to understand that the central areas of these curves (for angles of inclination between the upright and 10° or 12°) can be considered as circumferential arcs, with their centres at points *M* which represent the metacentrics of the ship in various inclinations. Moreover, tracing the metacentric evolutes, and consequently the desired curves, can be performed very easily, starting from the ordinary stability curves.

Reverting to the dynamics of rolling in still water, it can be seen that the rolling motion of the supported ship is practically equal to the free rolling of the floating ship. This is true if we neglect the influence of the instantaneous axis of rotation of the supported ship on the longitudinal axis passing through the centre of gravity, and, secondly, the influence of the dipping oscillations accompanying the transverse free oscillations.

The oscillatory motion of rolling surfaces cannot, of course, be produced for real ships although there is no difficulty in applying it at a reduced scale using the general principles of mechanical similitude [5].

Take a body (Figure 7) oscillating by rolling upon a cylindrical surface, the cross section of which is similar to curve *b* of Figure 6, passing near the centre of gravity; assume the similitude ratio be arbitrarily fixed as λ; assume the body's centre of gravity is in the plane of symmetry of the cylindrical surface at distance $d = \lambda H$ from the central straight line, *H* being the corresponding distance for the ship. Similarly, arbitrary similitude ratios may be attributed to the masses, as μ. So, if I_G is the moment of inertia of the ship, $\mu\lambda^2 I_G$ will be the moment of inertia of the model. Because it is impossible to modify the acceleration due to gravity, the similitude ratio for time must be: $\tau = \lambda^{1/2}$.

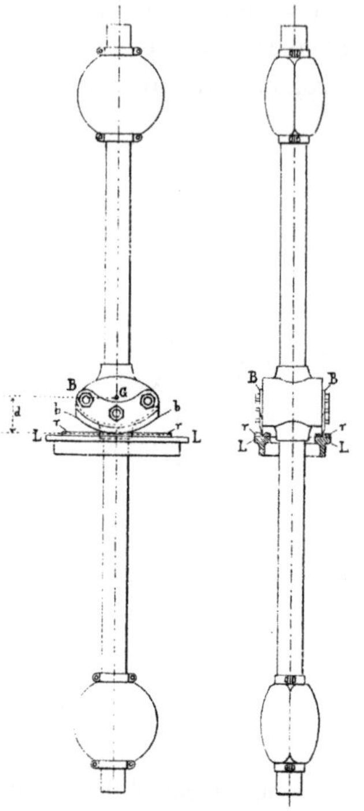

Fig. 7. The oscillating body of the navipendulum simulating the behaviour of a ship

With these approximations a body (as in Figure 7) can reproduce the oscillatory behavior of a ship.

The apparatus components

The navipendulum in these experiments was constructed by Officina Galileo of Florence.

The navipendulum was aligned as in Figure 8. Its most important part is plate *L* on which it rolls. This plate moves because it is connected to rod *k* which is coupled to a double crank linkage with axis of rotation in *O* and O_1; physically, it is perpendicular to the plate, and acts as a governing lever to regulate its inclination. The radii values *r* and r_1 of the cranks can be set.

In addition to plate *L,* the apparatus includes a registering device mounted on pin *C* around which the central axis of plate *L* oscillates. A small board, *z,* is also mounted on pin C which, during the apparatus' motion remains horizontal. Plate *z* is connected to rod *j*, which is coupled to another double crank parallel linkage, with *O* and O_2 the axes of rotation.

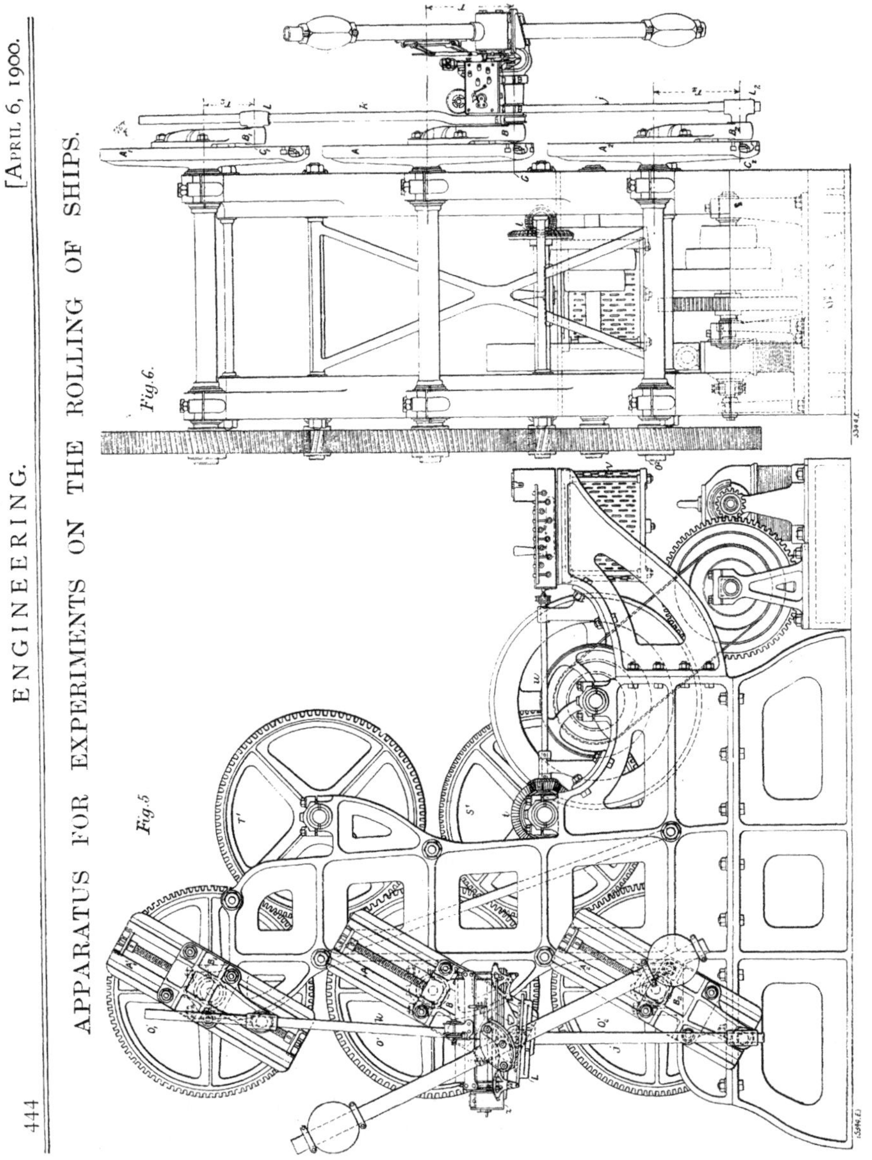

Fig. 8. The apparatus of navipendulum [8].

The registering device traces five lines on a piece of paper: 1) a toothed line which is a scale of times in seconds; 2) a sinuous line crossed by a straight line (3), giving the amplitude of the navipendulum's rolling relative to the vertical; 4) a sinuous line, crossed by a straight line giving the inclinations of the perpendicular of the wave to the vertical.

Motion is given by an electric motor at different speeds regulated by a potentiometer. By means of a couple of cogwheels and a pair of stepped pulley cones, motion is reduced and transmitted to a train of five identical cogwheels so the three wheels on the axes O, O_1, and O_2 turn at the same velocity. By regulating the motor rotation with the pulley pair, the speed of the gear train may be regulated between 10 and 80 rpm. To restrict motion, a Valessie indicator has been set.

<u>Results</u>

Figure 9 shows the experimental curves for a given ship in three different conditions of still water oscillation and five different conditions of wave motion [6-7].

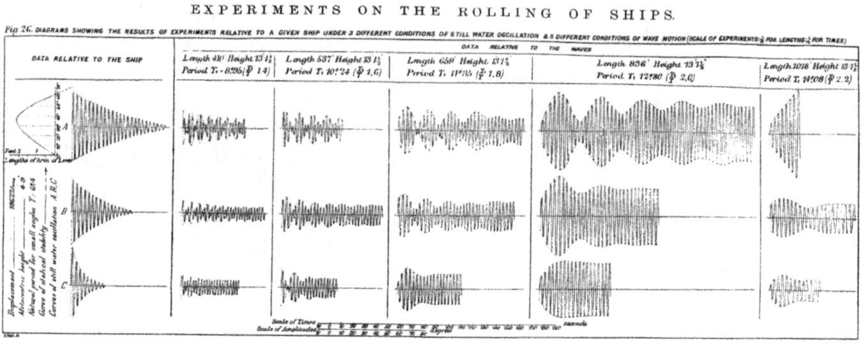

Fig. 9. Navipendulum diagrams for a given ship [14].

3 The Reception of the Navipendulum

Several papers [8-13] contain the description and theoretical basis to the complex machine – the navipendulum – designed by G. Russo. The paper published by I.N.A. [3] stands out from the rest. G. Russo presented this paper at a meeting at I.N.A. in April 1900, cited in [8], with drawings of the machine.

Comments by I.N.A. conference attendees have been reported below.

E. FROUDE (Associate Member of Council): *"I do not know whether most admire the boldness of his conception, or the acuteness and good judgment with which he has carried it out His instrument may be described as a machine for calculating rolling, almost instantaneously on the assumptions usually made for rolling calculations, with this important exception, that the curves of stability and resistance used are the correct ones for the ship, instead of more or less rough approximations, such as have to be used in any but very laborious calculations."*

MR. B. MARTELL (Vice-President): *"I think we are greatly indebted to Captain Russo for bringing forward this paper. ..... This is a paper that requires reading through very carefully, and, the more we read it, I am quite sure the more we shall be gratified with the ingenuity displayed, and the great scientific skill that has been shown by Captain Russo in the production of this paper".*

MR. ARCHIBALD DENNY (Member of Council): *" I think that Captain Russo is a benefactor to naval architecture in the production of this machine.....All Captain Russo's conclusions appear to me to be perfectly correct, and this study of rolling will be a most interesting study for naval architects, and especially for warship naval architects".*

Subsequently, on March 20th 1902, Captain G. Russo was invited to exhibit the navipendulum results, published in [6], at the I.N.A. meeting.

The navipendulum was sent to London, at the behest of the Navy Minister, to show how it works. A report of the meeting was published in [14] (a photo of the machine was also shown) where some experimental diagrams, obtained on board some Italian warships, were presented.

The comments, all of them enthusiastic, have been reported below [6].

MR. ARCHIBALD DENNY (Member of Council): *"I should like to have the opportunity which Captain Russo has had of studying rolling by this machine.....I will just add that I covet that machine, and I think that the Institution owes Captain Russo a very great debt of gratitude for the drouble he has taken in getting that machine brought here and showing us his experiments."*

MR. MACFARLANE GRAY (Member of Council): *"I cannot express to the full the admiration I have for this marvellous mathematical automation. While looking at it, I said to Lord Glasgow, "What ability in initiation, what patience in perfecting, and how excellent the workmanship!" And is the work of a foreigner. These are the men we have to contend with; let every Englishman, Scot, and Irishman realize this".*

The Right Hon. the EARL of GLASGOW, G.C.M.G. (the President): *"I should communicate to the Minister of Marine in Italy our thanks for his kindness in allowing Captain Russo to bring his wonderful machine here for us to see. Now gentlemen, I think you will all agree with me when I ask you to give Captain Russo a most hearty vote of thanks for his interesting paper, and for allowing us to see his most beautiful apparatus".*

This meeting caused quite a stir within the scientific community [16-22].

The navipendulum, the theory behind it and the experimental results were inserted into the book published by John Harvard Biles in 1908 [23]. This work, essential for naval design, was recently reprinted by Salzwasser Verlag, Bremen [24].

Acknowledgements. The authors are grateful to the other members of the Russo family for kindly providing all the necessary information and documents, and to Ms Chiara Puccio for providing support - within her Master's thesis - by retrieving and cataloguing the material needed to publish this work.

References

[1] Russo, G.: Appareils et moyens mécaniques pour executer des experiences de cabinet sur le roulis des navires, manuscript with patent request (1898)

[2] Russo, G.: Apparatus and Means for ascertaining by Laboratory Experiments the Behaviour of Ships on Waves, Fairfax & Wetter, patent no. 2387, London (1899)

[3] Russo, G.: An Experimental Method of ascertaining the Rolling of Ships on Waves. Institution of Naval Architects, London (1900)

[4] Russo, G.: An Experimental Method of ascertaining the Rolling of Ships on Waves, Transaction of the Institution of Naval Architects. Greenhill, London (1900); discussion by Froude, R.E., Martell, B., Denny, A., Wingfieldm, C.H.

[5] Russo, G.: Esperienze navipendulari sul rollìo delle navi in mare ondoso. Rivista Marittima, Roma (1900)

[6] Russo, G.: The Navipendular Method of Experiments, as Applied to some Warships of Different Classes, Transaction of the Institution of Naval Architects, the R. H. the Earl of Glasgow, London (1902); discussion by Froude, R.E., Denny, A., Biles, J.H., Gray, M., Barnaby, N., Yarrow, A.F., Thornycroft, J.T.

[7] Russo, G.: Il metodo degli esperimenti navipendulari applicato ad alcune navi da guerra. Rivista Marittima, Roma (1902)

[8] An Experimental Method of ascertaining the Rolling of Ships on Waves. Engineering, 442–445, 493–496 (1900)

[9] Esperienze navipendulari sul rollìo delle navi in mare ondoso. La Corrispondenza – Giornale di scienze militari, Roma (1900)

[10] Abstract of [3]. The Engineer, London (1900)

[11] The Rolling of Ships. Engineer's Gazette XIV(148), 102 (1900)

[12] Captain Russo's "Navipendulum". The Mariner XX(8), 262–263 (1900)

[13] Rolling of Ships on Waves. The Shipping World XXII(358), 380–382 (1900)

[14] The Navipendulum - The Navipendular Method of Experiments, as Applied to some Warships of Different Classes. Engineering 402–403, 520–524, 590–593 (1902)

[15] Abstract of [6], The Engineer, London, pp. 310–312, 330 (1902)

[16] Schiffspendel – Versuche Von Capitan Russo. Schiffbau, (20 21), 859–861, 900–906 (1902)

[17] Le Navipendulum, Revue Géneral de la Marine Marchande, Paris, p. 346 (1902)

[18] The Marine Engineering, London, vol. XXIV(277), p. 2 (1902)

[19] Le Yacht – Journal de la Marine, Paris, p. 235 (1902)

[20] Le "Navi Pendulum", Le Moniteur de la Flotte, Paris, p. 10 (1902)

[21] Der Apparat Russo zur Bestimmung des Rollens der Schiffe, Ueberall – Illustrierte Wochenschrift fur Armee und Marine, Berlin, pp. 747–748 (1902)

[22] The Navipendulum Method of Experiment. Scientific American, LXXXVII(23), 374–375 (1902)

[23] Biles, J.H.: The Design and Construction of Ships, Stability, Resistance, Propulsion and Oscillations of Ships, vol. II. Charles Griffin and Company, London (1911)

[24] Reprint of [23]. Salzwasser Verlag, Bremen (2009) ISBN: 978-3-86195-094-3

[25] Puccio, C.: Ing. Gioacchino Russo - Analisi e Restauro dei Progetti "Chromac", Master's thesis, Accademia delle Belle Arti, Catania (2011)

Applied Mechanics: The Information Sheets of Professor Leonid Smirnov

Alexander Golovin[1] and Anastacia Golovina[2]

[1] Professor, Theory of Machines and Mechanisms Department,
Bauman Moscow State Technical University
aalgol@mail.ru
[2] Undergraduate Student, Laser Treatment of Materials Department,
Bauman Moscow State Technical University
a.a.golovina@gmail.ru

Abstract. In the first half of XXth century one of the most well-known scientific and pedagogical schools of Applied mechanics was the Imperial Moscow Technical School (IMTS), led by Leonid Smirnov. A number of its graduates became prominent scientists and founded their own scientific and pedagogical schools. A significant segment of prof. Simonov's activity took place in an extremely difficult period of life in Russia. In the conditions of transition from empire to the Soviet Union the concept of learning had to be repeatedly changed.

This paper describes the evolution of the course of Applied Mechanics, which was devoted to TMM, by Professor L. Smirnov in the 1st half of the XXth century. The books «Kinematics of Mechanisms and Machines» and «Kinetics of Mechanisms and Machines», published in 1926, «Applied Mechanics Info sheets» and «Methodical Text-books for Carrying-out Graphical Works in Applied Mechanics» were developed by Professor L. Smirnov in collaboration with his students. Basing itself on these works, the TMM Department of Bauman university prepared and issued a series of lecture notes for all sections of the TMM course, transformed laboratory practice, produced a series of compilations of tasks for course projects and subsequently published a textbook on the TMM in the 1950-60s.

1 Professor Leonid Smirnov (1877-1954)

In 1903 Leonid Smirnov (fig. 1) graduated from the Imperial Moscow Technical School (IMTS, now Bauman University), where he was student of professor Nikolay Zhukovsky - a famous Russian mechanic. Then from 1903 until 1906 he lived and studied abroad (Dresden, Polytechnic School, the study of steam engines and compressors; Stuttgart, Polytechnic School, Laboratory of Engineering; Goettingen lectures by Klein, Hilbert and Prandtl). In 1906 he was invited to teach on Theoretical Engineering (in the Department of "Applied Mechanics") and do the organization of the laboratory of steam engines upon the recommendation of Professor Nicolay

T. Koetsier, M. Ceccarelli (Eds.): Explorations in the His. of Machines & Mech., HMMS 15, pp. 145–158.
springerlink.com

Mertsalov. All of his subsequent life was closely connected with the Departments of Applied Mechanics and Steam Engines in the Bauman University.

From 1929 to 1949 he headed the Department of "Applied Mechanics" and since 1943 - the "Theory of Mechanisms and Machines" Department [1]. The time of his activity coincided with big changes in Russian society. This has left a definite imprint on all his activities. Life was not easy. Yet under those conditions with his graphical methods of solving the "Applied Mechanics - TMM" problems he succeeded in meeting the demands of modern engineering activities.

Fig. 1. Professor Leonid Smirnov.

2 A Characterization of Smirnov's Activity Period

In Russia in the first half of 20th century one can distinguish three periods: before the Great October revolution (until 1917), the 1920s and the1930s.

Until 1917 universities had received students that were well prepared. The courses were built considering this fact. This is particularly noticeable in the course of Professor N. Zhukovsky and N.A. Mertsalov [2, 3, 4, 5].

The revolution made it possible for workers and peasants to get an education of which they had actually been deprived before 1917. However, their training level was inadequate. So called working faculties (abbreviated "rabfak") were founded in order to get them ready. In these faculties adults studied who had life and work experience but not much more. Accordingly, the level of education obtained by the "Workers' Education", was lower than the level of education received in public and real schools before the revolution. In the 1920s, most of the students at Bauman University were graduates of the "workers' faculties". This required a lower academic level of courses and only tests in spite of examinations to be introduced [1, 6].

Fig. 2. Title of "Info sheets…"(a); title of one of "Methodical text-books…"(b).

Characteristic of that period was the use of visual teaching methods and techniques - graphic or graphic-analytical methods with strong algorithmic transformations using as little calculations as possible, and also strictly limited amount of educational material. The development by Smirnov of the "Applied Mechanics Info Sheets" [7] was a natural response to this challenge. Six editions of the sheet appeared from 1922 to 1935. Their total volume was 40 pages.

In 1926 two books were published by Smirnov: "Kinematics of Mechanisms and Machines" and "Kinetics of Mechanisms and Machines" [8, 9], that reflected the content of the course "Applied Mechanics" in the 1920s.

In the 1930s young people, who had graduated from normal schools and had the level of knowledge, comparable with one given in pre-revolutionary schools, entered the higher educational institutions of Russia. This required regular of courses, including courses of "Applied Mechanics".

The restructuring of the course was reflected in the production of special Info Sheets [7], the development of practical laboratory work, the creation of scientific films, the development of new home work tasks and the compilation of the "Methodological guidelines for the implementation of graphic works on Applied Mechanics" that was published 1940 [10]. Smirnov's disciples were involved in work with the material. The title of Info Sheets is shown in the fig.2a; the title of one of the "Methodological guidelines ...." is shown in fig.2b.

3 Prof. Smirnov's "Kinematics of Mechanisms and Machines" and "Kinetics of Mechanisms and Machines" [8,9]

In 1926 the State Publishing house published two books - "Kinematics of mechanisms and machines" and "Kinetics of mechanisms and machines" which were recommended as guidance for high technical colleges by the Scientific Technical Section of State Science Council.

In the preface to "Kinetics of ...," the author notes that proposed work represents the main methods of investigation for mechanisms and machines (and its separate parts) movement as well as their more complex combinations summary.

From the first pages, the author introduces the so-called scale in all the formulas and graphical representations. Currently, complex geometric transformations and the associated scale are unnecessary because of the presence of modern computing. However, during this period, these techniques allowed to solve problems only on the basis of elementary geometric transformations and calculated proportions. "All the exposition held only for the plane case, as the study of motion that occurs in three-dimensional space, is not available for elementary exposition," [7].

“Kinematics of mechanisms and machines”

In book 10 chapters consisted of L. Smirnov rendered the idea for presentation of Ampere’s kinematics as a kind of geometry. In chapters 1-3 graphical methods are presented for the simplest cases of a point and a body, covered in a course in Theoretical Mechanics. In chapters 4-8 these methods are applied to the complex motion of a point and a body, including the rolling of one curve on another and point trajectory curvature radius definition. The author turns to the direct questions of the theory of mechanisms only in the 9th and 10th chapters. Accordingly, in the 9th chapter the kinematics of a highest pair is examined as an extension of Chapter 5. In 10th chapter the kinematics of Assure’s structural “double lead” groups-based linkages. However Smirnov did not use "structural group" insignia within his studies.

“Kinetics and mechanisms of machines”

In the preface the author stressed that the meaning of the book considered the dynamics of machines, but giving the name "Kinetics of ..." A. Smirnov emphasized the dominant role of the fact that the amount of kinetic energy is the result of external forces. It should be noted that the book provides only positional systems.

The book consists of 10 chapters. Some chapters of the book are minor and prepare the material for the study of motion on the basis of experimental data or set parameters.

The simplest types of motion studied in theoretical mechanics (rectilinear, curvilinear, which is the result of two rectilinear motions, and rotating around a fixed axis) are considered in chapters 1-3. As in "Kinematics of ...", the explanations lie at the heart of graphic techniques and methods. In addition, it describes the instruments for studying the motion parameters: the device for determining the numerical value of acceleration - deceleration in a straight horizontal motion, the instrument for studying forces in curvilinear motion (diagram of the device is shown, for example, in [1, 6]), the device for studying the forces in the case of rotation of a mass axis round a fixed axis.

Chapter 4 discusses methods for determining the mass and moment of inertia of the body of arbitrary shape relative to the selected axis. As a result of formal and geometrical transformations the mass and the moment of inertia appear to be proportional segments in elementary calculated scales. The principle of one of the methods is given in [1, 6]. Additionally, an experimental method for determining the moment of inertia - as the result of physical pendulum oscillations exploration - is presented there. The coupler of a steam locomotive which was used by L. Smirnov for carrying out laboratory work on the definition of the moment of inertia during the period of oscillation [6] is still kept in the laboratory of TMM department.

Chapter 5 is devoted to the study of unbalanced rotors and methods of their balancing.

Chapter 6 discusses the problem of rotation of the flywheel with moment of inertia given under the action of given forces, motion under constant force (acceleration till achievement of the given speed), the movement in the event of a given constant engine power, acceleration at a given relationship between angular velocity and acceleration time, the movement in case of engine power changes during controllable acceleration.

Chapter 7 describes the method of forces and moments reduction. The practical review for the problem of riveting machine power pneumatic cylinder was given as well.

Chapter 8 is dedicated to issues of friction features in the Coulomb's law-based joints and guides. The conditions of friction coefficient changing with increasing pressure are marked and a graphical method for the determination of friction in this case is given. The problem of the losses from friction in a linkage is examined on the base of the method of forces reduction.

In chapters 9, 10 the method for determining the reduced moment of inertia is given. The problem of determining the law of motion for given reduced moment of inertia, work of forces and driving resistance, as well as certain initial velocity, is solved.

4 Applied Mechanics Info Sheets [7]

Currently, these sheets are a rarity. Authors used their own archives largely while working on this article. Unfortunately, the given collected set of sheets proved to be incomplete. There were references to the publication of 1922 (Bauman university) and the publication of 1935 (RSL) in the library catalog of Bauman university and the RSL - The Russian State Library. The copy found in the RSL of «Applied Mechanics Info sheets» contained only 5 pages on the "wedge" theme. The materials were not mentioned in the museum of Bauman university, the Department of TMM, in the library catalog of the Museum and the Polytechnic library. Fortunately, significant part of the material was preserved in the author's archives. For possibility of extension the missing part of material was maintained to be restored using books of former Smirnov's students [11, 13] and links from "Methodical Text-books ...." [10] sources list.

In fact, info sheets were lecture notes of Professor L. Smirnov. He called a number of his students - members of the department, "Applied Mechanics" Baranov, L. Reshetov, B. Shitikov - to work on these sheets with him. Subsequently, that staff grew to professors, well-known experts in the field of TMM. Baranov had become the author of a well-known textbook on TMM which withstood a number of publications. L. Reshetov [6] left his mark in all sections of TMM, founded a scientific school of mechanisms without redundant links, became the author of more than 200 models of mechanisms for the collection of the Department of TMM; his analysis of the methods of evolvent gears calculation [11] formed the basis of the gear geometric theory that was designed by Gavrilenko who was the student of prof. Smirnov [12].

The main topics of the info sheets are following:

- The scale and symbols;
- Kinematics of linkage, including determination of curvature radius of any point of the curve;
- Wedge and its application, including a double wedge (a fragment of these sheets is shown in Fig. 3)
- Grashof's theorem;·
- Hook's cardan joint;· Rectilinear motion.
- Graphic methods of differentiation and integration;· Rotary movement, the triangle of linear speeds;
- Higher pair, gearing, gears, cam mechanisms;
- Rectilinear oscillatory motion (detail of these sheets is shown in Fig. 4);
- Inertial forces of a conrod placed in four-crank kinematic chain;
- Identifying external forces needed to accelerate a four-crank kinematic chain;
- Identifying the "kinetic energy" and "reduced mass" for a four kinematic chain (Fig. 5a);
- Determination of forces that are necessary for value changing of the kinetic energy of machine (fig.5b);
- Calculating procedure of inertial torque for flywheel (fig.5v);
- Flywheel motion under the action of applied forces (fig.5g);
- Determination of reduced friction.

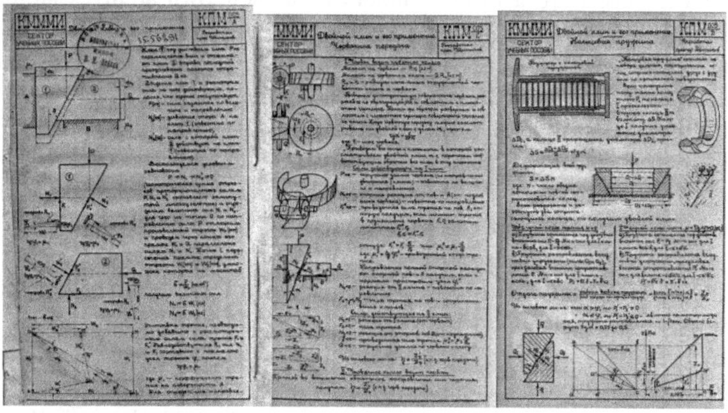

Fig. 3. The fragment reference sheets (subject - a double wedge).

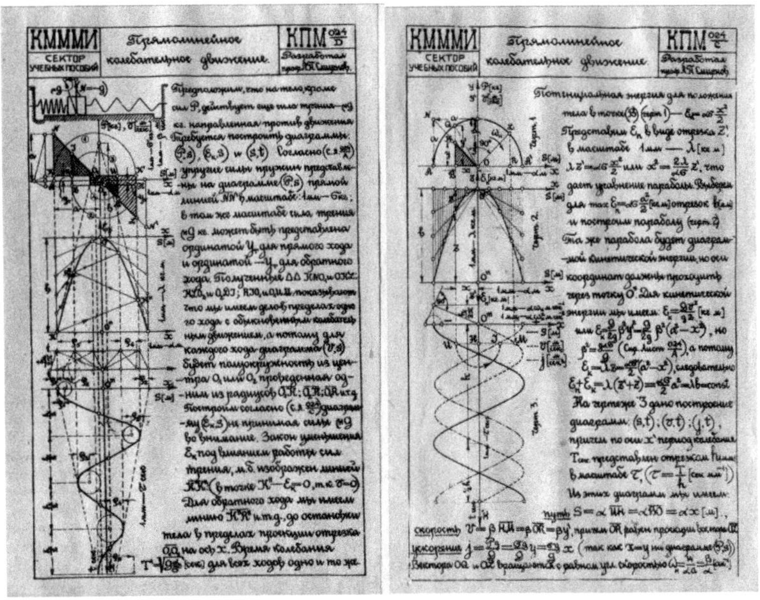

Fig. 4. The fragment reference sheets (theme - straight vibratory motion).

5 "Methodological Guidelines for the Implementation of Graphic Works on Applied Mechanics," Edited by Professor L. Smirnov

This work was completed 1940. Methodological guidelines were prepared by the Associate of the TMM department Assistance Professor N. Vzorov, who, according to the memories of his contemporaries, was a brilliant lecturer. Model sheets (shown in Figure 6 - 13) were prepared by N. A. Savelova, who had been the Head of TMM department from 1949 to 1951, under the leadership of professor N. Vzorov. The reference to "Kinematics of ...", "The kinetics of ...", "Info sheets of ..." were used in this article, as well as to the Soviet and foreign translation (USA, Germany) editions, that had been published in the period of 1931- 1935. Scrupulousness of the authors should be noted. For example, there is a reference to the use of unpublished reception, offered by L. Reshetov, in the instructions to the 5th work.

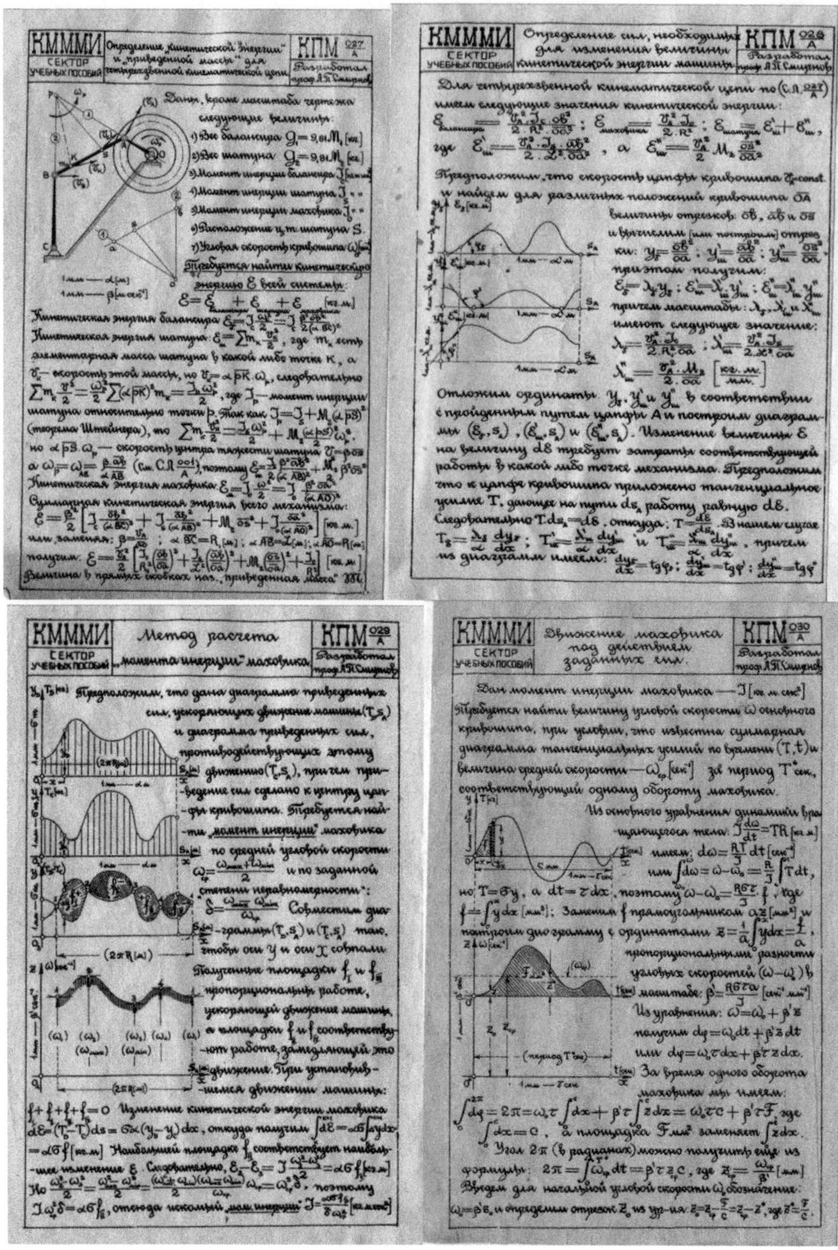

Fig. 5. Info Sheets fragment ("The Dynamics of Mechanisms")

Eight themes of graphic works were prepared in the 1st edition. They are explored while going down.

Work №1.Research of the absorbing device for automatic coupling of the "Peerless" (Fig. 6).

Materials on the theme “Wedge and its application” were used in the guidance to this work (author: Shitikov). Fig. 5. Fragments of the Info Sheets(“The dynamics of mechanisms“ theme).

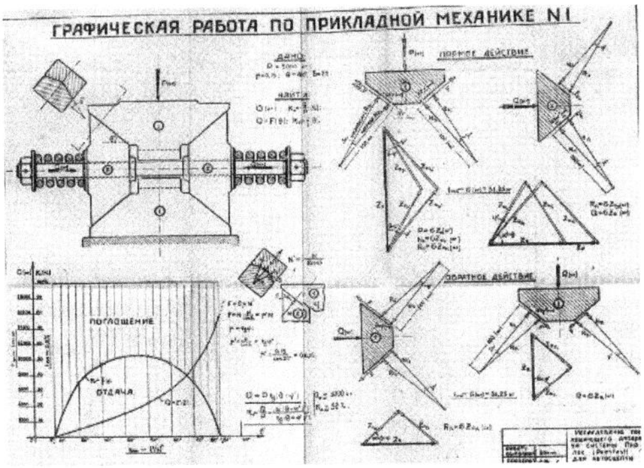

Fig. 6. Graphic work on theme "Absorbing device for the automatic railway wagon coupler"

Work №2. Evolvent gearing of spur gears with straight teeth (Fig. 7).

This work bases on studies of L. Reshetova [11]. This book is the completion of the work on engagement, launched for the All-Union Exhibition for gear-cutting business in 1932. The Exhibition committee suggested to the author as the leader of links to review and organize existing correcting systems. On the advice of his teacher (prof. Smirnov) the author chose the path of self-dependent research. The required mesh had been designed by the opening of the exhibition, the theory was presented in the spring of 1932 in the course of Applied Mechanics, that was read by him in MMMI in the "Metal Treatment” department. Implementation of the link was managed only in the beginning of 1934 due to the lack of equipment.

Work №3. Research of planetary gears with cylindrical wheels (Fig. 8).

The graphic techniques developed by prof. L. Smirnov in [7], [8] were laid in the basis of this work. Some techniques were written with the help of L. Smirnov lectures and published by G. Baranov, who was his student, [13] in 1932 and became a part of the latest edition of info sheets. In particular, the issues of friction, wear, efficiency were reviewed. The kinematics and forces in the planetary gear were analyzed in that work.

Work №4. The kinematic research of rectilinear motion (graphical integration and differentiation; Fig. 9).

The graphic techniques developed by prof. L. Smirnov in [7, 13] formed the basis of that and were outlined in the relevant info sheets.

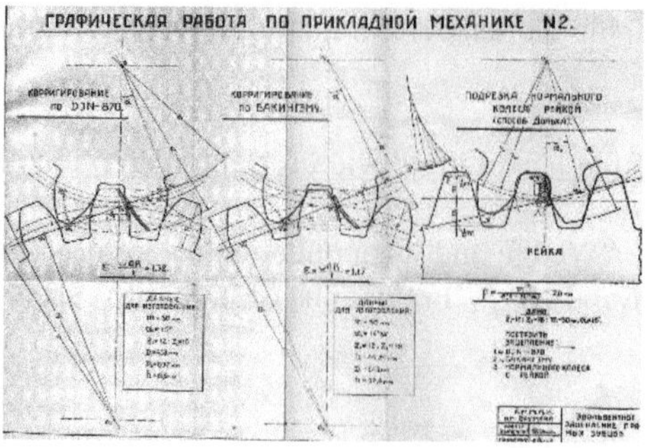

Fig. 7. Graphic work on "Evolvent gearing spur gears".

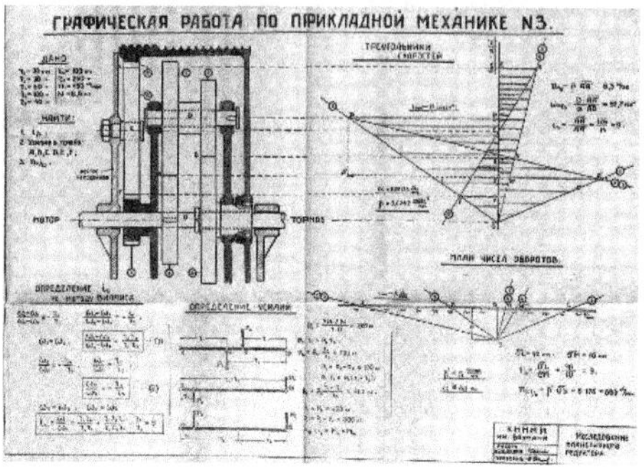

Fig. 8. Graphic work on "Research on the planetary gear".

Work №5. The kinematic research of the "Junkers" mechanism (Fig. 10).
The method of velocity and acceleration triangles underlines the research [7], including the research on the kinematics crank point, and set out in the relevant info sheets.

Work №6. Kinematic research of a cam using Hartmann's method. (Fig. 11).
The study is based on sections approximation of the profile with circular arcs or straight line segments and transition to replacement of relevant sections of cam mechanism with equivalent one with four links. Construction of the kinematic characteristics is procured with the methods of graphic integration and differentiation (work №4).

Work №7. Defining of flywheel moment of inertia (Fig. 12).
The problem of dynamics of mechanisms, described in [8] and reference sheets: "a four-rod inertia forces of kinematic chains", "Identification of external forces that are necessary to accelerate the four-crank kinematic chain", "The definition of "kinetic energy' and 'reduced mass" for a four-link kinematic chain", "Determination of forces required to change the value of machinery kinetic energy", "The method of calculating the "moment of inertia of a "flywheel", "Movement of a flywheel under the influence of applied forces".

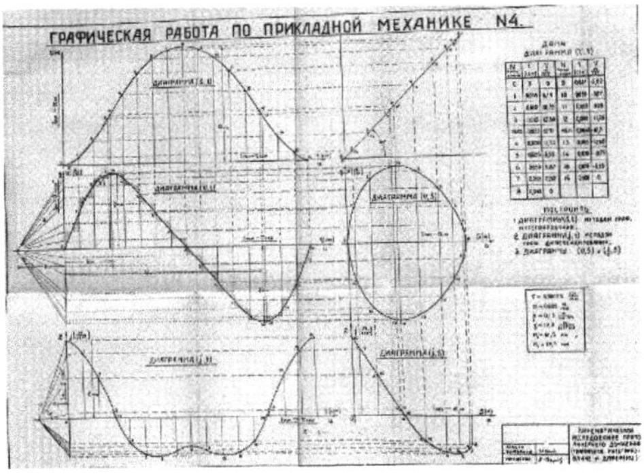

Fig. 9. Graphic work on "Graphical integration and differentiation"

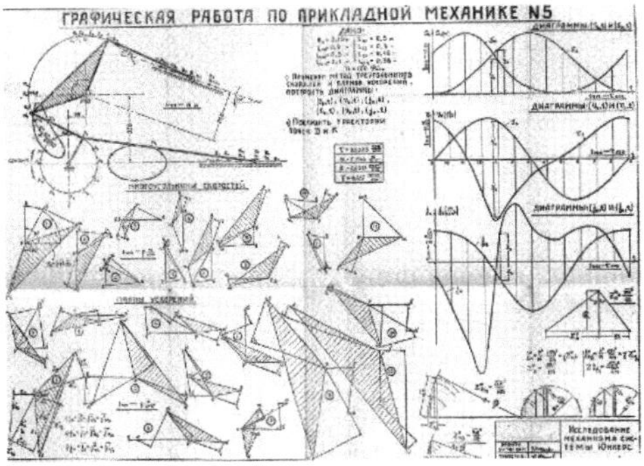

Fig. 10. Graphic sheet of work on the theme "The kinematical analysis of "Junkers" mechanism"

<u>Work №8. Balancing of rotating masses (Fig.13).</u>

The problems of static and dynamic balancing of rotors are analyzed in that part on the basis of the material presented in [8] and reference sheets [7, author - B. Shitikov].

Conclusion. In the period of 1920-30s the Applied Mechanics Course, created by Professor L. Smirnov and his disciples, corresponded to the level of students' training at all stages of the Soviet Union and reflected the needs of the industry at that time. High level of course development and perfection of graphic techniques for solving problems of mechanisms design created high authority on the discipline in that phase, which lasted until the 1960s. Later, during the transition to modern numerical methods, these advantages have become a hindrance in presenting the subject of TMM and, consequently, led to the downfall of its authority.

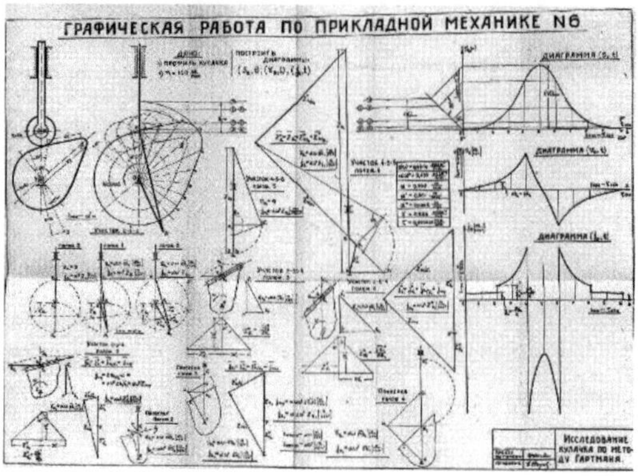

Fig. 11. Graphic work on the theme "The kinematical analysis of a cam using Hartmann's method"

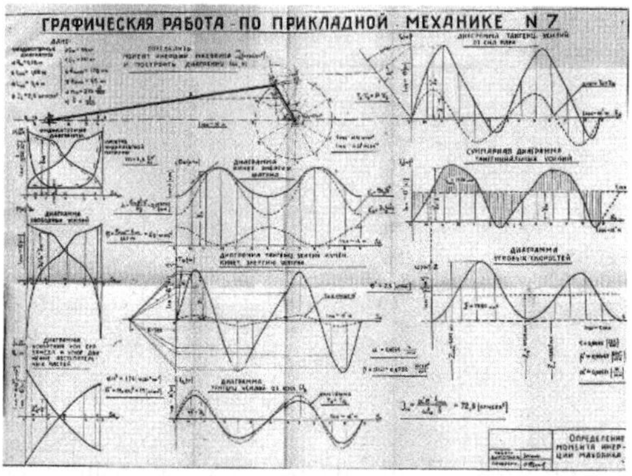

Fig. 12. Graphic work on theme "Moment of inertia determination for a flywheel"

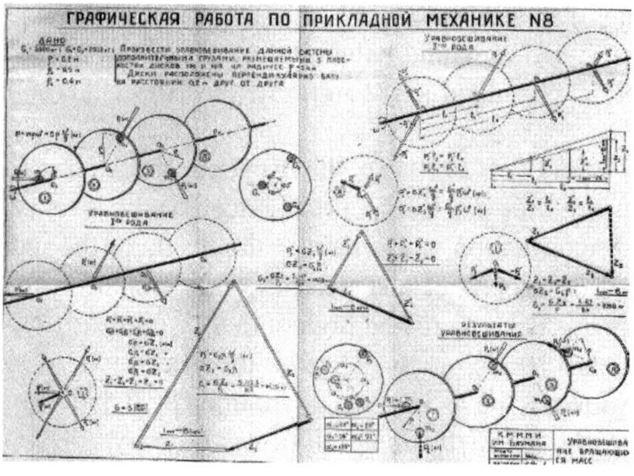

Fig. 13. Graphic work on "Balancing of rotating masses"

In the 1950-60s of the XXth century the TMM Department staff prepared and published five issues of lecture notes for all sections of the course [14 - 18], transformed laboratory practice, produced Transactions of tasks for Theory of Mechanisms term-project [19] and finally published a TMM aid-text book [20]. "Info sheets of ..." [7] and "Methodical Text-books ..." [10], developed under the guidance of Smirnov, became the basis of them. Five students of prof. Smirnov, who were involved in his work, became well-known scholars, the founders of their science schools - Baranov, V. Gavrilenko, Petrov, L. Reshetov, B. Shitikov.

References

[1] Golovin, A.: Bauman University TMM Department in the Former Half of the XX-th Century (Professors N. Merzalov & L. Smirnov). In: Proceedings of International Symposium on History of Machines and Mechanisms, pp. 137–148. Kluwer Academic Publishers (2004)

[2] Joukovsky, N.: Analytical mechanics/ Lithographic edition of lectures. IMTS edition, Moscow (1898/1899) (in Russian)

[3] Joukovsky, N.: Theoretical mechanics/ Edition of lectures from 1874, Moscow (1921) (in Russian)

[4] Mertsalov, N.: Mechanism's Dynamics/Lithographic course of lecture, read in IMTS. Processed and published by M. Phelinsky, Moscow (1914) (in Russian)

[5] Mertsalov, N.: Mechanism's Kinematics/Lithographic course of lecture, read in IMTS. Processed and published by M. Phelinsky, Print-Lithographic Company I.N. Kushnerev and Co., Moscow (1916) (in Russian)

[6] Golovin, A., Tarabarin, V.: Russian Models from the Mechanisms Collection of Bauman University. Springer Science and Business Media B.V (2008)

[7] Smirnov, L.: Applied Mechanics Info sheets (in Russian)

[8] Smirnov, L.: Kinematics of Mechanisms and Machines. State Publishing House, Moscow (1926) (in Russian)

[9] Smirnov, L.: Kinetics of Mechanisms and Machines. State Publishing House, Moscow (1926) (in Russian)

[10] Smirnov, L.: Methodical Text-books for Carrying-out Graphical Works in Applied Mechanics, Moscow (1940) (in Russian)

[11] Reshetov, L.: Correlation of Evolvent Gearing. Methods and Foundation. ONTY Moscow-Leningrad (1935)

[12] Gavrilenko, V.: Geometrical Theory of Evolvent Gearings. Mashgiz, Moscow (1949) (in Russian)

[13] Baranov, G.: Kinematics and Dynamics of Mechanisms, Part 1. The State Power Publishing House, Moscow (1932) (in Russian)

[14] Skvortsova, N.A., Malysheva, Z.S., Musatov, A.K.: Theory of mechanisms and machines/Lectures (Linkages and Cams), vol. 1. Publishing House of Bauman Moscow Higher Technical School, Moscow (1968)

[15] Skvortsova, N.A., Malysheva, Z.S., Musatov, A.K.: Theory of mechanisms and machines/Lectures (Evolvent gears), vol. 2. Publishing House of Bauman Moscow Higher Technical School, Moscow (1956)

[16] Savyelova, A.A., Lukychev, D.M., Musatov, A.K., Nikonorov, V.A.: Theory of mechanisms and machines/Lectures (Dynamics of Mechanisms and Machines), vol. 3. Publishing House of Bauman Moscow Higher Technical School, Moscow (1959)
[17] Malysheva, Z.S., Glukhov, N.A., Minut, S.B., Petrov, G.N.: The Theory of mechanisms and machines edited by L. Reshetov/Lectures (Dynamics of Mechanisms and Machines), vol. 4. Publishing House of Bauman Moscow Higher Technical School, Moscow (1959)
[18] Reshetov, L.: Theory of mechanisms/Lectures (Gears), vol. 5. Publishing House of Bauman Moscow Higher Technical School, Moscow (1962) (in Russian)
[19] Transactions of tasks for Theory of Mechanisms term-project. Bauman Publishing House of Moscow Higher Technical School, Moscow (1962) (in Russian)
[20] Gavrilenko, V.: Theory of mechanisms. Aid-Text Book, Moscow (1973) (in Russian)

Konstantin Frolov Continuing the Soviet School Traditions on MMS (Mechanism and Machine Science)

Olga V. Egorova* and Gennady A. Timofeev**

Moscow State Technical University Named after N.E. Bauman (MGTU), Moscow, Russia

Abstract. Konstantin Vasilevich Frolov was an internationally appreciated scientist. He was one of the pioneers of vibration studies dealing with the influence of vibrations on machines and man. He devoted his life not only to scientific research but also to the promotion of the Russian (Soviet) Engineering School. He received a lot of high honors and awards in his Motherland and a great recognition abroad. He contributed to modern Mechanism and Machine Science through his work as an engineer, scientist, and as Director of the Institute for Engineering Science (IMASH RAS) named after A.A. Blagonravov in Moscow. The aim of the present text is to attract attention to this extraordinary personality in the year of the 80th anniversary of his birth birth.

1 Introduction

«For everything that I have managed to reach ... I am obliged to all those wonderful people who were working together with me, I am obliged to my friends, and I am obliged to my remarkable family» [1].
Konstantin V. Frolov

Huge scientifically-organizing and public activity, wide scientific interests and ability for cooperation demand the very special qualities from a person. The basis of all of them can be put only in the childhood and youth, in the family of a person. That is why it is so important to study the biography and lifework of the well-known scientists if we want to understand the reasons which have allowed them to become successful. On an example of outstanding people the new generation of scientists and engineers grows.

2 Childhood and Youth

The future academician of the Russian Academy of Sciences (RAS) Konstantin Vasilevich Frolov (Fig.1) was born on July, 22, 1932 in Kirov city of the Kaluga region (Russian Federation) in a family of office employees [2].

---

* Prof., Doctor in History Science.
** Prof., Doctor in Engineering Science.

T. Koetsier, M. Ceccarelli (Eds.): Explorations in the His. of Machines & Mech., HMMS 15, pp. 159–169.
springerlink.com

Fig. 1. Konstantin Vasilevich Frolov (1922-2007)

His mother, Frolova Alexandra Sergeevna (Fig.2, a), was a doctor and worked in X-rays-offices of military and of municipal hospitals. His father, Frolov Vasily Ivanovich (Fig.2, b), unfortunately, in 1937 was subjected to repression.

From Konstantin Frolov's personal memoirs [1]:

"I should remember my mum who has endured together with us the consequences of our father reprisal, of the war, and of the beginning of the very difficult post-war period. Her surprising optimism and soft humour, her kindness not disappearing under no circumstances, almost religious goodwill to everyone, her quiet and self-possessed character were a constant example in our family. Surprisingly, despite my often made faults, I was never punished physically, never! My greatest sin was if I did not yield in something to my twin sister Elena or, even worse, if I have offended her.

... the large family library was creating an atmosphere of some mystery and desire (curiosity is more) to learn what all these names and figures under interesting pictures and illustrations mean. Perhaps, for this reason we, I with my twin sister (Fig.3), learned to read very early».

a) b)

Fig. 2. a, b. K.V. Frolov's parents: a – mother, b – father

Fig. 3. Twins: Konstantin Frolov and Elena Frolova

In 1956 Konstantin Frolov graduated with honour from Bryansk University of Transport Machinery (Russia). It is important that being yet a student enlisted on the most prestigious specialty "Turbine construction" he has already actively participated in research works of cathedrals. After the graduation the young Frolov was directed «on student distribution base» to Leningrad Metal Plant (Saint-Petersburg today), one of the largest power engineering enterprises in the Soviet Union at that time. There he began his activity in the design bureau of steam and gas turbines. Later he passed to work to the laboratory "Vibration Study of Steam and Gas Turbines" [3].

From K. Frolov's personal memoirs [1]:

«Once the "Turbine shop floor" superintendent Mr. E. Dolinsky came and told us that he had to reduce time for our vibration tests in order to ship the finished goods from the plant in due time and to get paid a bonus payment. Luckily, together with

us there was our laboratory chief Mr. A. Zilberman who asked Mr. Dolinski to write with his own hand: «Today I have forbidden to the future academician K.V. Frolov to carry out vibration tests» that in his prospective memoirs Mr. Frolov would remember this and publish. Mr. Dolinsky burst out laughing and, to my astonishment, allowed me to go on with the tests.
... In the late fifties of the last century Leningrad Metal Plant was visited by a group of scientists from the Institute for Engineering Science (IMASH RAS) and among them there were Prof. F.M. Dimentberg, Prof. S.V. Serensen, Prof. V.O. Kononenko, and others. They familiarized with the researches of our design bureau and advanced development laboratory and with the works on durability and strength in particular. Consequently of our acquaintance and the detailed analysis of my own research results I was offered a postgraduate study in IMASH ... »

3 Institute for Engineering Science Named after A.A. Blagonravov (Imash Ras)

And already in autumn, 1958, Konstantin V.Frolov has arrived to Moscow and become a postgraduate student of IMASH. From the first day in the Institute and up to his last days he has been working in IMASH, connected forever his life with it. It is interesting that one of outstanding Soviet scientists, Doctor in Engineering Science, Professor Victor O.Kononenko (1918-1975), who met him first in Leningrad (Saint-Petersburg today) became his thesis advisor.

In IMASH Frolov started a research of influence of energy sources properties on stationary and non-stationary vibration of mechanical systems. In 1962 he has successfully defended his first thesis with a theme "Influence of energy source properties on vibration of autonomous systems". In his thesis the problem of stability of various modes for a usual nonlinear, a parametrical, and a self-oscillatory systems were studied, and also the link between the nonlinear system parameters and the energy source characteristics was shown. In parallel with postgraduate study in 1960 Konstantin V.Frolov carried out the works on research of resonance states of solid bodies with nonlinear elastic links. As a result the dependence of resonance state on an energy source was shown and experimentally proved, and also some new objective laws reflecting the energy source influence on vibration of solid bodies were found out and studied.

Successes of the young scientist have been noticed, and in 1964 Frolov was given a title of a senior research assistant and he was appointed to be the chief of vibration engineering laboratory. He managed to combine his theoretical work and scientific publications with organizational activity [4].

In October, 1975, after the death of academician A.A. Blagonravov, the Director of the Institute for Engineering Science, Konstantin V.Frolov has taken his place and up to his tragic death he was the irreplaceable head of IMASH, now named after A.A. Blagonravov.

It is greatly to his credit that several new scientific fields were developed in IMASH: machinery and equipment vibroacoustics, power engineering objects strength, robotic technology, biomechanics, etc.

From K. Frolov's personal memoirs [1]:

"Here I have done a way from the post-graduate student to the academician and the director of institute. It became my life, my daily care, my pleasure, and sometimes disappointment ... I am happy that within more than 30 years up to this day I am heading this wonderful collective of remarkable people, outstanding scientists, and employees of Institute for Engineering Science of Russian Academy of Sciences».

In the late sixties Frolov has formulated the problems of a new scientific field – complex system vibration protection "Man - Machine - Environment». The case is that as a result of technical and industrial progress the speed and power of modern machines and technological equipment increased and created new external conditions for a Man. That is why previously insignificant for a human life certain individual environmental factors took on enormous significance [5]. The problem of the harmful vibration loads influence has become acute. The more serious problem was to protect a man-operator from the influence of a special type of machines and equipment, the working process of which was based on the use of vibrating principle of action. Frolov proved that vibrations, acting on a Man, can reduce the productivity of labour and its quality. It was therefore quite important to study the influence of various factors of the efficiency of an operator as a live link of "Man-Machine-Environment" system, as we now call - the integrated biotechnical system. For the first time ever Frolov proposed to consider dynamic features of a body of a Human being as an oscillatory system. Under Frolov's direct participation various dynamic models of the body of a man-operator for various working positions under the influence of vibrations were developed and studied.

Also under Frolov's direct control there was done a number of researches devoted to studying of destructive vibration influence connected with the load increase on mechanism links and machine components, which leads to stability loss, fatigue breakages, desired motion law changes. New problems of machine dynamics were solved with the help of classical methods of analytical mechanics, vibration nonlinear theory, theory of variable weight mechanics, elasticity theory, and etc.

In 1970th in IMASH by Frolov's school the problems of self-synchronization of oscillatory systems with two energy sources have been successfully solved, as well as the problems of theoretical results application for a design of new machines of vibrating principle of action.

Results of numerous researches and experiments spent in the late sixties by Konstantin Frolov have received a full reflection in many scientific publications and were summarized in his thesis for a doctor's degree "Vibration in Machines with Variable Parameters in Application to a Hydraulic Power Drive Dynamics". In 1970 K.V. Frolov has been awarded a scientific degree of a Doctor en Engineering

Sciences (Dr.Sci.Tech). The main results were published in magazines first and then in 2007 in his "Selected works" [5].

Later in the vibration technics and technology laboratory, as well as in the biomechanics department of IMASH with personal Frolov's participation there were worked up very interesting researches on a man-operator behavior as a live link of integrated biotechnical system. These researches have caused the transition to the idea of vibration field rearrangement, and as a result allowed not only to exclude a harmful vibration influence on a human being, but in contrary, to raise labor activity of a person with the help of vibrostimulation. A number of essentially new and courageous experiments on vibrostimulation and vibration therapy were put, in the field of bone fabrics surgery in particular [5].

Under the great ability of Frolov for cooperation the very interesting and unique collaboration of IMASH with well-known USA firms "Boeing" and "Mac Donnel Douglas" on a contract basis was established. As a result the whole complex of problems in the field of aviation and space technics, including the use of composite materials for manufacturing of rocket tanks for fuel, the suppression of micro gravitational influences in space vehicles, and creation of comfortable conditions for work of cosmonauts, was solved.

Frolov's activity was always aimed to practical applications and to a teaching of both the solutions and approach for the formation of clever professionals. Thus, he contributed considerably to make IMASH RAS to become one the most famous and successful scientific institutes not only in Russia but all over the world.

4 Moscow State Technical University Named after N.E. Bauman: Pedagogical Talent

Konstantin V.Frolov was an excellent teacher. His lectures were always full of the latest information about the achievements in the international science and technology. He continued the Russian traditions and skillfully organized the educational process, promoting the introduction of modern methods of teaching. More than 45 years K. Frolov combined his research and pedagogical activities. Within 1961-1976 he has been working at Moscow Technology Institute of Light Industry and in 1973 he was elected there a Department Chairman of "Theoretical Mechanics and Theory of Mechanisms and Machines" chair. He has written a number of lecture courses, methodical study guides, and a new set of laboratory operations manuals. As a Chairman he insisted that each professor-lecturer should have very active scientific researches together with students. In 1971 K. Frolov has been awarded an academic status of professor.

In 1978 Konstantin Vasilevich has headed a chair «Theory of Mechanisms» (TM) at Moscow State Technical University (MGTU) named after N.E. Bauman and up to his death in 2007 he was in charge of it, as well as he was a "Board of Studies" member and a "Dissertation Council" member of the University.

His textbook (Fig. 4) "Theory Mechanisms and Mechanics of Machines" [6] based on a new concept of teaching on "Theory of Mechanisms" course has sustained four editions. Frolov was also the editor of the original manual for higher technical schools «Mechanics of industrial robots» [7], which has been published in 1988-1989. Even today, in 21 century it can be used as a management for designers and experts in the field of industrial robotics.

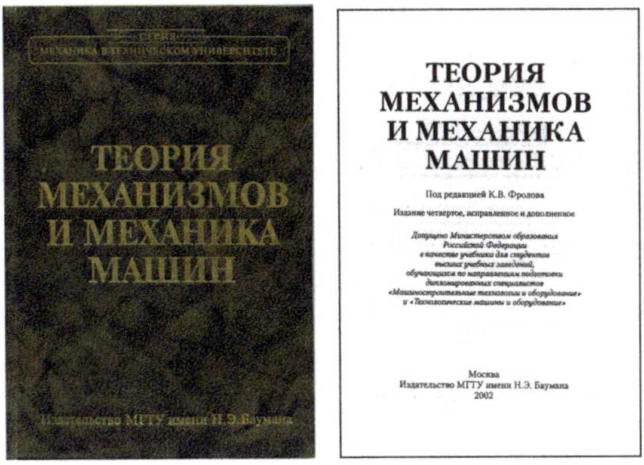

Fig. 4. Textbook "Theory of Mechanisms and Mechanics of Machines"

Everybody who could ever be in touch with Frolov recognizes his unusual open mind. He appreciated not only analytical or experimental, but also organizational skills of his colleagues and students. He thought that such qualities as business knowledge and ability to understand people were necessary for a modern science leader. He tried to push people with these talents and to move them up to the career ladder.

Not coincidentally, that today almost all his former post-graduate students became doctors of sciences and professors.

5 Main Frolov's Scientific Works

Frolov's bibliography includes more than 1000 published works. The most interesting results of research works were published in fundamental monographies: "Applied Theory of Vibration Protection Systems" (1980) [8], "Theory of vibration technics and technology" (1981) [9], "Membrane Vibration in a liquid" (1983) [10]. In 1978-1981 he participated in six-volume scientific edition "Vibrations in Technics" [11] issued by publishing house "Machinostroenie", Moscow, and personally was the scientific editor of two volumes of this edition (v.1, 1978; v. 6, 1981).

Great interest among scientists and experts was caused by the monography "Methods of Machine Development and Problems of Mechanical Engineering" (1984) [12] published in frames of a scientific series "Fundamentals of Machine Design" initiated by Frolov. No doubts, that this edition promotes considerable expansion of modern methods of machine design and helps designers, researchers, experts of mechanical, instrument and tool engineering, and etc.

It is necessary to note the academician activity connected with the scientific literature edition as Frolov was the editorial board chairman of the multivolume encyclopedia “Machine Building” [13]. Now 44 volumes are let out already. The main aim of this edition is to promote essential improvement of quality of machines and efficiency of a machine-building complex in total. The encyclopedia should highlight all the progressive methods and technologies of mechanical engineering and the best decisions in new technics design and construction.

Since 1977 Konstantin Vasilevich was the editor-in-chief of academic magazine "Engineering Science". Under his initiative in 1990 the edition was transformed to a new magazine "Problems of Mechanical Engineering and Machine Reliability" for the purpose of increase of the topics range and the information support level of fundamental and applied scientific researches. Also Frolov was an editor-in-chief of international magazine “Mechanical Engineering and Automation Problems” which received the recognition among scientists and experts all over the world. Engineers and specialists from USA, Finland, Hungary, Poland, Bulgaria, Czechia, Slovakia and many other countries are presented in this magazine editorial board. The most important task of the international magazine is to inform on the advanced research-and-production experience, on new constructional materials, progressive forms and mechanical engineering high technologies, as well as the assistance in multinational contacts of scientists.

K. Frolov was supervisor of a multivolume edition “Security of Russia. Legal, social-economic, and scientific and technical aspects”, published by the International Humanitarian Fund “Znanie” [14]. There issued already 24 volumes.

Konstantin Vasilevich headed the editorial council of "Scientific and Technical Progress in Mechanical Engineering" books.

In 1991 the fundamental monography “Science in the strategy development” [15] based on the results of many-sided creative Frolov’s activity was published.

6 Recognition and Awards, International Activity

In 1976 Konstantin Vasilevich was selected a member-correspondent, and in 1984 – the full member of Academy of Sciences of the USSR (now RAS). In 1985 as the academician-secretary he has headed a new scientific field – Machine Building, Mechanics and Process Control Systems. Within seven years (1985-1992) he promoted the organization of the perspective fundamental researches in that field and active participation of the Russian academic institutes in the decision of the most important applied problems. He also stimulated the results of scientific researches to be economically effective.

Frolov achieved a well earned international fame due to his interdisciplinary studies in the field of mechanics, biomechanics, vibration, physics and ergonomics. These studies made and developed under his lead enabled to solve a number of urgent problems dealing with a human being as an operator of machinery and equipment vibration environment.

The very special attention Frolov gave a problem of risk decrease and safety maintenance of complicated technical systems, human beings and environment at occurrence of technogenetics and natural calamities. Under his initiative the nationwide scientific and technical "Safety" program was generated. Also with the purpose of cooperation all over the world the International Safety Institute was created.

The realization of the "Safety" program was based on the assistance with the United Nations (UN) Organization and on close international cooperation with many countries: the USA, Japan, Great Britain, the Netherlands, France, the Peoples Republic of China, South Korea and etc. As a result the creation of uniform international norms and requirements in the field of safety were developed.

According to his own, proved to be fatidic words:

"Today, there is an urgent need to develop an understanding for the systems behaviors and vulnerabilities of interacting infrastructure networks in order to protect that infrastructure from naturally occurring hazards and man-made acts of terrorism. Infrastructure is the underlying foundation or basic facilities, services, and installations needed for the functioning of a community or society, such as transportation and communications systems, water and power lines, and public institutions. The NATO-Russia Workshop focuses on the creation of a risk-informed analysis capability for modeling and predicting the behavior of complex infrastructure networks; applying emerging technology to the problems of designing, constructing, monitoring, and operating critical infrastructure systems; and building an understanding of the social, economic, and environmental factors that affect, and are affected by, critical infrastructure. The problem posed by modern, complex infrastructure is that society has become completely dependent on the reliable functioning of these systems, yet, we do not fully understand their systems behavior, which is subject to perturbations by natural hazards and malicious threats. Further, the characteristics of these hazards and threats are changing, driven both by global change and land use, and by changing political conditions. The disruption of the smooth functioning of civil infrastructure has important engineering and public policy implications, as well as social and environmental impacts [16].

Konstantin Frolov carried on a great international activity. He held the position of the President of the International Fund "Znanie" (Knowledge), named after S.I. Vavilov and he was also a member of many foreign academies, active and emeritus professor of leading national and foreign universities.

For the great services to his Motherland academician Frolov was given the title of Hero of Socialist Labour, awarded two Orders of Lenin, "Gold Star" medal of Hero of USSR, and many other very prestige awards. (Fig. 5). He also became the winner of USSR Council of Ministers and Russian Federation Government prizes. Frolov was given the "Stodola gold medal". Among foreign prizes and awards we can see the gold medal of Czech Academy of Science and the "Mikhail Pupin gold medal" (Yugoslavia).

Frolov regarded knowledge as means of communicating with the others. He contributed much to develop the International Federation for the Promotion of Mechanism and Machine Science and he was IFToMM honorary member that is a unique recognition from the IFToMM Federation.

From K. Frolov's personal memoirs [1]:

«The constant strong belief in good deeds and its fulfillment is the main source of optimism that always inspires me on finding the difficult problems solutions».

Fig. 5. Academician Konstantin V. Frolov

7 Conclusions

Academician Konstantin Vasilevich Frolov is considered to be one of the greatest scientists on modern TMM (now MMS). He published more than 1000 scientific works [5-16], etc. Modern MMS widely applies achievements of all innovative sciences, from elementary particle physics and nanotechnology, genetics, conquest of outer space to engineering and etc.

No doubts, his enormous followers, young researchers, members of IFToMM and of the international scientific community, will ensure that his memory lives on, the memory of a considerate, warm human being and prominent scientist.

To the main achievements of Frolov's scientific school in the field of vibro-pulsatile technics (percussive machines) there belonged:

a) theory of periodic modes of movement of multi-mass systems of the above mentioned type;

b) researches of transitive modes, complicated periodic modes of movement of the simplest systems;

c) optimization results of systems with one and two degrees of freedom;

d) studies of synchronization phenomenon, including synchronization of unbalanced vibrators, and etc.

The heritage left by Konstantin V. Frolov, the patriarch of national (Russian) engineering science, can be considered as a source of ideas for many generations and everyone who works for Machines and Mechanisms Science. His role in the development of mechanical engineering, vibration studies, biomechanics, Man-Machine-Environment Systems, and safety problems is inestimable. We, his post-graduate students and followers, are assured that in the heart of our success there is his belief and carry. We all appreciate what he has done.

References

[1] Frolov, K.V.: Selected works, Vibration Technics, Biographical sketch, vol. 1, pp. 9–39. (2007) (Russian)

[2] Timofeev, G.A., Frolov, K.V.: The continuer of TMM School Traditions. In: Polet, Special issue, pp. 121–130. Machinostroenie, Moscow (2005) (Russian)

[3] Fortov, V., Mahutov, N., Bessonov, A., Petrov, V. (eds.): Achievements and Problems of Machine Engineering. Devoted to the 70th anniversary of academician Konstantin Vasilevich Frolov, Znanie, Moscow, Russia, pp. 17–45 (2006) (Russian)

[4] Konstantin Vasilevich Frolov. In: Documents for Scientists Bibliographies, RAS, Nauka, Moscow, Russia, vol. (4), pp. 10–39 (2001) (Russian)

[5] Frolov, K.V.: Modern problems of vibration in the systems "man - machine - environment"// Man under vibration: Suffering and Protection. In: Proceedings/CISM - IFToMM - Symp., Udine, Italy, pp. 1–41 (1979) (English)

[6] Frolov, K.V. (ed.): Theory of Mechanisms and Mechanics of Machines. Textbook for technical universities, Vicshaya Shkola, Moscow, Russia (1987)

[7] Frolov, K.V., Vorobiev, E.I. (eds.): Mechanics of industrial robots: manual for institutes in 3 volumes. Vicshaya Shkola, Moscow (1988-1989) (Russian)

[8] Frolov, K.V., Furman, F.A.: Applied Theory of Vibration Protection Systems, Machinostroenie, Moscow, Russia (1980) (Russian)

[9] Frolov, K.V., Goncharevich, I.F.: Theory of vibration technics and technology, Nauka, Moscow, Russia (1981) (Russian)

[10] Frolov, K.V., Antonov, V.N.: Membrane Vibration in a liquid, Nauka, Moscow, Russia (1983) (Russian)

[11] Frolov, K.V. (ed.): Vibrations in Technics in 6 volumes. Machinostroenie, Moscow (1978-1981) (Russian)

[12] Frolov, K.V.: Methods of Machine Development and Problems of Mechanical Engineering, Machinostroenie, Moscow, Russia (1984) (Russian)

[13] Frolov, K.V. (ed.): Machine Building: Encyclopedia in 40 volumes. Machinostroenie, Moscow (1995) (Russian)

[14] Frolov, K.V. (ed.): Security of Russia. Legal, social-economic, and technical aspects, MGF, Znanie, Moscow, Russia, vol. 24 (1992-2002) (Russian)

[15] Frolov, K.V.: Science in the strategy development, Nauka, Moscow, Russia (1991) (Russian)

[16] Protection of civilian infrastructure from acts of terrorism. In: Frolov, K.V., Baecher, G.B. (eds.) Proceedings of the NATO Advanced Research Workshop on Protection of Civilian Infrastructure from Acts of Terrorism, Moscow, Russia, May 27-29. Springer, Dordrecht (2006) (English)

Francisc Viliam Kovács – Leader in the Development of Mechanisms and Robotics in Romania

Erwin-Christian Lovasz, Dan Perju, Inocenţiu Maniu, Corneliu Rădulescu and Valentin Ciupe

"Politehnica" University of Timisoara, Department of Mechatronics

Abstract. This paper presents a short biography and the academic career of Professor Francisc Viliam KOVÁCS. During the academic carrier he occupied administrative leadership positions at the "Politehnica" University of Timisoara. Professor Francisc KOVÁCS is a road opener of the teaching and research field of Robotics in Romania and made several important contributions to the theory of the machines and mechanisms.

1 Introduction

Francisc Viliam Kovács was born November 26th 1929 in Timisoara. Starting at a young age he manifested remarkable intellectual qualities and capacities. He followed the courses of the Piarist Private Highschool in Timisoara from 1940 to 1948, being the prize-winner of his class every year and the first classified at the final exam in the summer session of 1948.

During the fall of 1948 he was admitted at the Mechanical Engineering Faculty of the Politehnica Institute of Timisoara, as "chief of year". He took courses of mechano-energetics and rolling stock (railroad engines) and passed the final exam in February 1953, obtaining a diploma of mechanical engineer.

The remarkable professional qualities proven by Kovács during his student years led to a republican scholarship, then in 17.11.1951 –as a student in fourth year– to the nomination as tutor at the department of machine elements and mechanisms. He is promoted as assistant in 1.10.1952 and lecturer from 1.10.1953 with the course of Theory of mechanisms. In February 1963 he wins a competition for the post of university professor at the same discipline. In December 1970 he is nominated as head of Machine elements and Mechanisms Department in Timisoara, function held until his retirement in 1.10.1995, with a break from 1976 to 1984, when he was dean of the Mechanical Engineering Faculty.

Starting with 1980 Prof. Kovács (s. Figure 1) holds several courses in robotics, being one of the trail blazers in teaching this modern scientific domain in Romania.

Prof. Kovács had also a fruitful scientific research activity. He published at renowned publishers 16 books, 20 books for student use, 269 scientific articles in journals and volumes of conferences in the Romania and abroad, having 22 recorded inventions and innovations and participated in 63 research contracts and grants, to the majority of them as head of research staff [8], [14], [16-17], [23-24] .

T. Koetsier, M. Ceccarelli (Eds.): Explorations in the His. of Machines & Mech., HMMS 15, pp. 171–182.
springerlink.com

Fig. 1. Francisc Viliam Kovács Prof. Dr. Eng. mult. Dr.h.c. (1929-2009)

At the beginning of his university activities he was involved in the study of the geometry and durability of gears and the correlation between them. His work in this field was awarded with a Prize of Education Ministry in 1966. At the beginning of the sixties he started working in mechanisms synthesis, a domain in which he prepared his Ph.D. thesis named "Contributions to the elaboration of a unitary method for mechanisms synthesis", thesis alleged in September 1969 at "Traian Vuia" Politehnica Institute of Timisoara. In this work an original notion of "connection" was introduced, a true "foundation brick" for designing any mechanism, idea developed later in other papers, among others in the dissertation "The use of the notion «connection» in theory of mechanisms and robotics", prepared with the occasion of the awarding of the scientific title of Dr.h.c. at the University of Craiova.

At the advice of acad. Anton Ioan, rector of the "Traian Vuia" Politehnica Institute, Francisc Viliam Kovács started in 1979 to work in robotics. In this domain, Prof. Kovacs had obtained important results in theoretical research, organized and led, starting with 1979 and alongside Prof. Tiberiu Muresan, the dean of the Electrotechnic Faculty, a strong multidisciplinary research and development team that helped obtaining many results in applied research. He also founded, organized and held the post of honorary president the Romanian Society of Robotics (SRR) in 1990 along with the past Prof. Cristian Pelecudi from the Politehnica University of Bucharest.

As a result of the robotics' community activity, led by Prof. Kovács, was realized the first Romanian industrial robot REMT-1 and the first Romanian flexible manufacturing cell, for machining the shafts of electric motors (in 1982, results awarded with the "Traian Vuia" prize of the Romanian Academy – Figure 4.b); 12 other types of robots and synchronic positioners – two of them awarded at TIB (Bucharest International Technical Fair) with gold medal (1987 and 1988) and 7 robotized flexible manufacturing systems.

In the field of theoretical contributions, Kovács produced an original method of robot optimization, based on the approximate iterative computation of a robot's dynamic performance, and also a method of optimization for the synthesis of robotized flexible manufacturing systems. Starting with 2000 he introduced the theory of the reference pairs systems (PeSiR), a mathematical modelling way of mechanical systems, of technological systems based on relative movements and even of some economical systems [18-20], [25], [26].

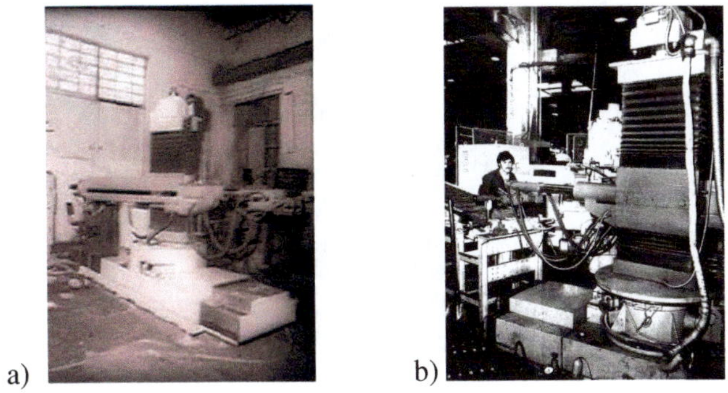

Fig. 2. REMT-1 in the first Romanian flexible manufacturing cell (8 feb. 1982) a) Design of REMT-1 b) REMT-1 in the manufacturing cell

Starting with 1990 Robotica & Management Journal appears, initiated and coordinated by Kovács together with Vela Ioan from "Eftimie Murgu" University of Resita [27], [29].

Prof. Kovacs was the founder of the higher technical education in robotics at the "Politehnica" University of Timisoara and other universities in the country [11], [12], [15], [22]. Since the '80 he started teaching at the Mechanical Engineering Faculty of Timisoara an optional course in robotics. Later, he initiated the specialization of graduates in robotics by holding post-university courses in robotics, for beginning at Timisoara, then Arad and Oradea. In 1990 at the Mechanical Engineering Faculty of Timisoara was founded the "Industrial Robots" specialization in Romanian language, and in 1992 a similar specialization in German language. Many former and actual courses held at these specializations had Prof. Kovács as titular. He also held courses in robotics, after 1995, at the University of Oradea, "Aurel Vlaicu" University of Arad and "Eftimie Murgu" University of Resita. He was the author of several books in robotics (including the first book in the country from this field, "Manipulators, robots and their industrial applications", 1982 – s. Figure 3).

Prof. Kovács' scientific and academic activity was appreciated by awarding numerous prizes, distinctions, decorations, titles and diplomas, by institutions in his country and abroad (s. Figure 4 a) Diploma of Doctor Honoris Causa of the "Eftimie Murgu" University of Resita - 12 oct. 2006, b) Traian Vuia Prize of the Romanian Academy - 1982, c) Excellence Diploma in Robotic Development)

Prof. Kovács was honorary member of many scientific societies in the country and abroad, like: International Federation of Robotics; International Federation for the Theory of Machines and Mechanisms; The Public Corporation of the Hungarian Science Academy etc., and was also member of the scientific committee of several scientific journals in the country and abroad.

He past away in 29.04.2009.

Fig. 3. The first book on robotics in Romania (1982)

Fig. 4. Awards of Kovacs scientific and academic activity

2 The Concept of Connection

In order to generalize the kinematic groups by Assur [1], which used only the joints with one degree of freedom, Kovács has introduced the concept of connections. The definition of the connection, given by Kovács, means all the restrictions in the relative movement between the elements in order to provide constraint motion of the mechanism.

The connections can be of geometric/kinematic type if the imposed restrictions are made geometrically or of dynamic type if the restrictions occur due a certain system of forces acting on mechanism in its movement.

The kinematic/geometric connection is defined as an open kinematic chain which is interposed between 2 elements with known or imposed relative movement.

The connections are classified upon their number of degrees of freedom (DoF) and the number of elements and kinematic pairs. They are divided in 3 types: connection type A, connection type B and connection type C (s. Figure 5).

Tipul conexiunii	Numarul de elemente	Cuple cinematice		Schema cinematica a conexiunii	Gradul de libertate L_k	Notare
		Nr.	Clasa			
A	$n_k = 0$	1	$c_{5k}=1$		-2	$K_A(-2)$
			$c_{4k}=1$	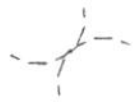	-1	$K_A(-1)$
B	$n_k = 1$	2	$c_{5k}=2$	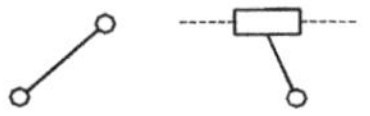	-1	$K_B(-1)$
			$c_{5k}=1$ $c_{4k}=1$		0	$K_B(0)$
C	$n_k = 2$	3	$c_{5k}=3$	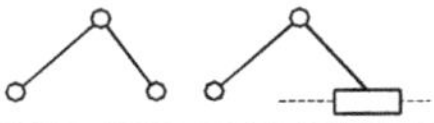	0	$K_C(0)$
	oarecare				L_k	$K_C(L_k)$

Fig. 5. Classification of the connections by Kovacs

At planar mechanism a connection of type A is just a kinematic pair which can be of V or IV class, according with Artobolevsky's classification, and are denoted $K_A(-2)$ and $K_A(-1)$ respectively, where in bracket is given the L_k degree of freedom of the kinematic chain.

A connection of type B has one element and two kinematic pairs. If both of them are of V class the connection has DoF=-1 and is denoted $K_B(-1)$. For the connection with one kinematic pair of V class and the other of IV class DoF=0 and is denoted $K_B(0)$.

A connection of type C may have any number of elements and kinematic pairs. The Assur groups (kinematic/structural groups according to Artobolevsky's classification) represents a particular case of the connection of type C denoted $K_C(0)$. The degree of freedom of the connection´s kinematic chain is calculated with the formulae:

$$L_k = 3 \cdot n_k - 2 \cdot c_{5k} - c_{4k} \tag{1}$$

where: n_k - number of the connection´s elements,

c_{5k} - number of the connection´s kinematic pairs of class V,

c_{4k} - number of the connection´s kinematic pairs of class IV.

Prof. Kovács developed in his Ph.D. thesis [4] a unitary method for mechanisms synthesis based on the concept of connections (Figure 6) and published them in the books [5], [6], [7] and [13].

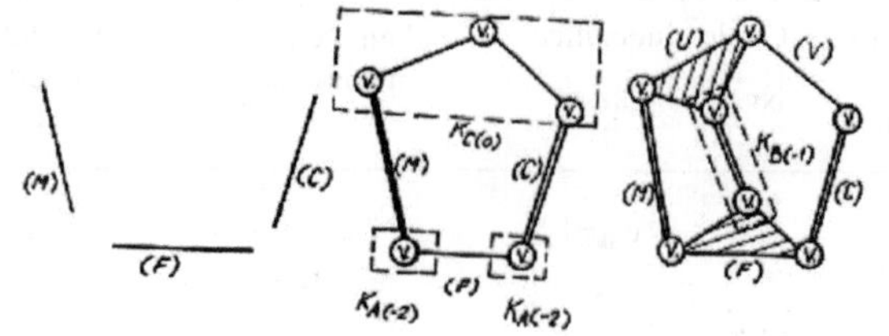

Fig. 6. Unitary method for mechanisms synthesis based on the concept of connection

The concept of connection will be developed in 1973 by Crudu in his Ph.D. thesis [2] considering the dynamical connections. Rehwald used in his mechanisms theory [30] and in the self developed program for analysis of the planar mechanisms KOSIM a similar concept "elementary groups", which consider also the kinematic chains with kinematic pairs of class IV or with DoF=2. The concept of connection is presented in many books on the mechanisms theory edited at the "Politehnica" University of Timisoara [28], but also was proposed to be used in robotics [21].

3 The First Romanian Flexible Manufacturing Cell

During the eighties the Robotics Multidisciplinary Team led by Prof. Kovacs developed an intense activity for conceiving and designing a large number of manipulators and industrial robots as requested by the industrial applications of some companies across the country.

The first real problem solved by the Robotics Multidisciplinary Team referred to the study, research, design and manufacturing, in collaboration with Electromotor Factory in Timisoara, of the electrically actuated industrial robot REMT-1 (Figure 2).

The design of this first robot and actually the entire conceiving activity of this collective were coordinated by Prof. Nicolae Gheorghiu. Among the first members of the collective there was Prof. Ioan Văcărescu, who participated at the conception of the REMT-1 robot. REMT-1 robot characteristics:

- mechanical system structure T ⊥ R ‖ T ⊥ T ‖ R (s. Figure 7);
- actuation: electric;
- control system : ECAROM 800 adapted for robot control;
- payload: 63 daN;
- work space: parallelepiped + cylinder;
- v_{max}: 0.2 m/s;
- ϖ_{max}: 0.5 rad/s;
- rotational encoders: TIRO 100

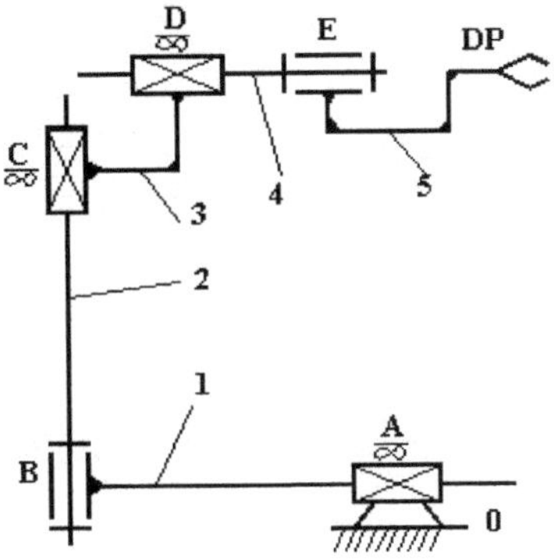

Fig. 7. Kinematic Schema of the REMT-1 robot

In the archive of the Department of Mechatronics were found a few of the original blue-prints containing mechanical assembly drawings of some REMT-1 modules

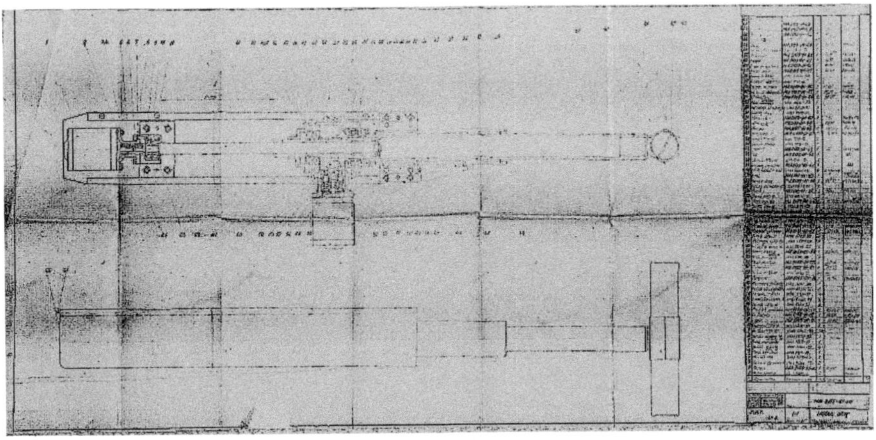

Fig. 8. a) REMT-1 arm module

(Figure 8 a, b, c). These drawings stand as true proof of the amount of work provided during the REMT-1 industrial robot design by the Robotics Multi-disciplinary Collective.

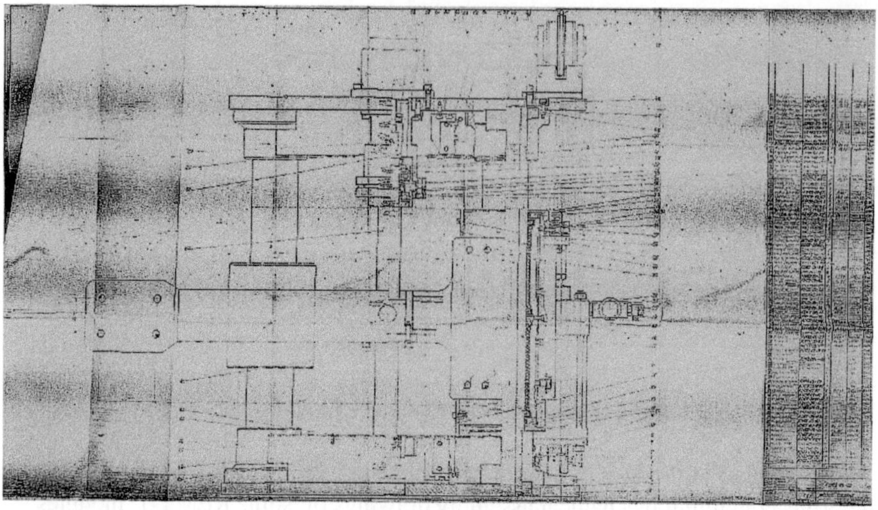

Fig. 8. b) REMT-1 columns module

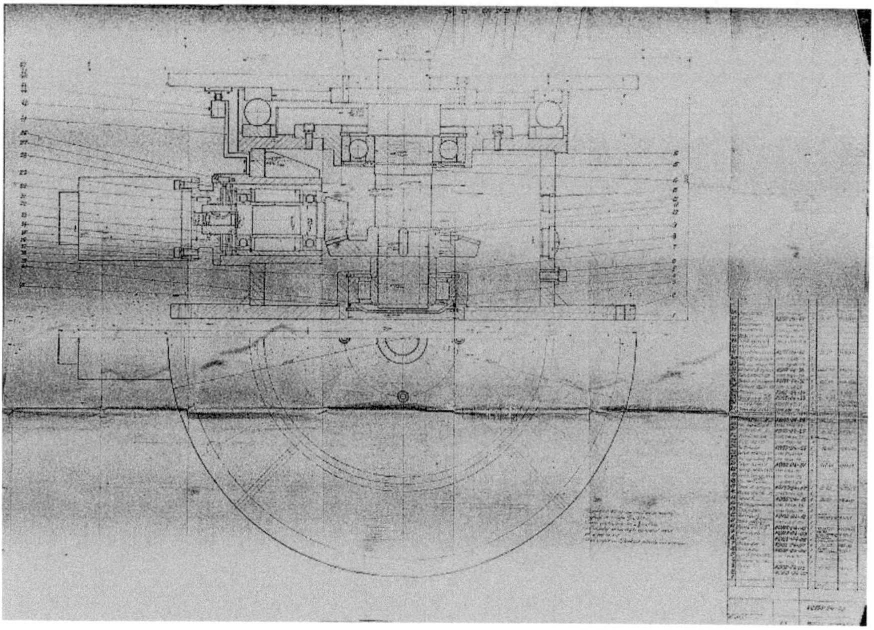

Fig. 8. c) REMT-1 rotational module [3]

By 8th February 1982, at the Electromotor factory in Timisoara was conceived and put into operation a flexible manufacturing cell used for the machining of 12 typo-dimensions of shafts for the rotors of the middle series electrical motors, cell that was served by the REMT-1 robot (Figure 2 a, b).

In Figure 9 the layout of the manufacturing cell is presented, and in Figure 10 a, b, c there are some pictures of the machining manufacturing cell for the shafts family, served by the REMT-1 robot.

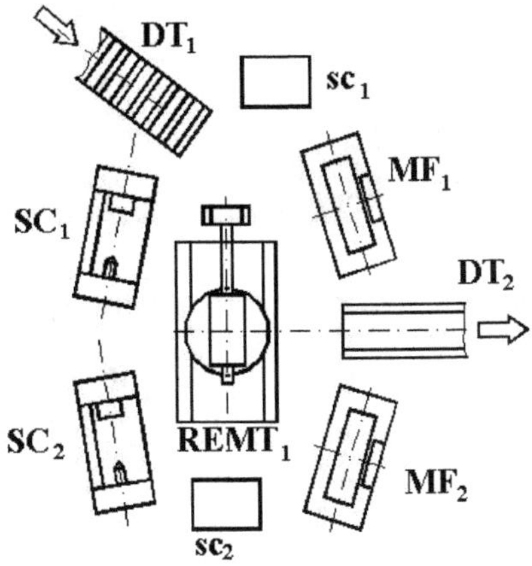

Fig. 9. Layout of the manufacturing cell for machining the shafts

The components of the manufacturing cell for stepped shafts:

- copying lathes SC_1, SC_2 2 pcs;
- automated milling machines MF_1, MF_2 2 pcs;
- transfer device for parts DT_1 1 pcs;
- gutter type transfer device for finished parts DT_2 1 pcs.

Inside the cell, the REMT-1 industrial robot was gripping the part presented as a laminate bar from the stepper gravitational transfer device, shown in Figure 10a, and transferred it to one of the two numerically controlled copying lathes.

When the lathe finished machining the stepped shaft, the REMT-1 industrial robot was gripping the part from the lathe and transferred it to one of the two available milling machines, which were machining the keyway in an automated cycle (Figure 10 b).

After the milling operation ended, the finished shaft was transferred to the evacuation device (Figure 10 c), device that was placed between the two milling machines.

By approaching the construction of manipulators and robots in a modular way, during 1982-1988, in collaboration with Electromotor factory in Timisoara, there were conceived and executed five variants of robots in the REMT family [9], [10].

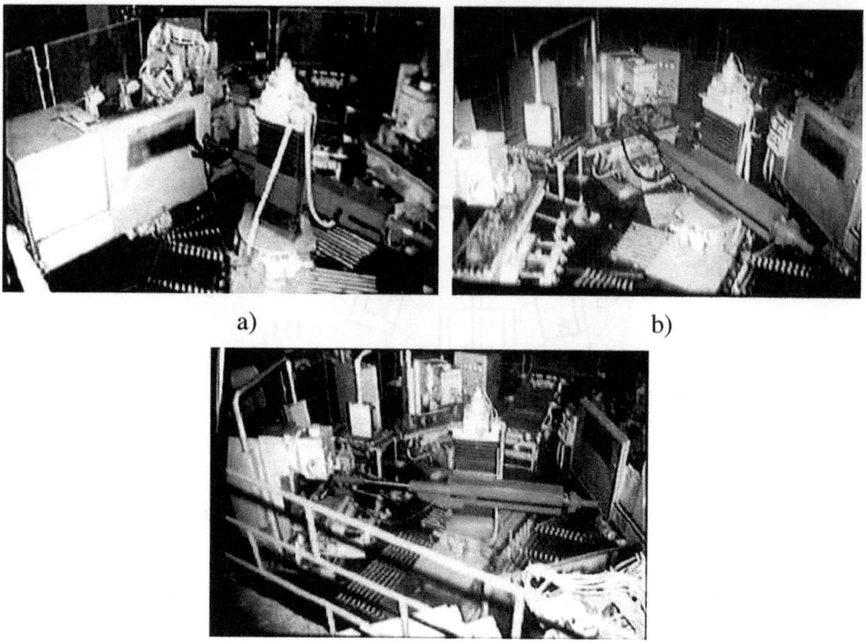

a) b) c)

Fig. 10. Manufacturing cell for machining the shafts: a) REMT – 1 robot, feeding a part to the copying lathe; b) REMT – 1 robot, feeding a machined shaft to the first milling machine; c) REMT – 1 robot, transferring a finished shaft to the cell evacuation device

References

[1] Assur, L.V.: Issledovanie ploskih cterjnevih mehanismov s nizsimi parami s tociki zrenia ih strukturi i klassifikatii. In: Izdatelstvo Akademii Nauk, CCCR, Moskva (1952)

[2] Crudu, M.: Mecanisme cu conexiuni dinamice (Mechanisms with dynamic connections), Ph D. Thesis, Politehnica University of Bucharest (1973)

[3] Gheorghiu, N., Rădulescu, C., Văcărescu, I., Dolga, V., Tanvuia, V.: Modulul de rotaţie al robotului industrial REMT-1 (Rotation modul of industrial robot REMT-1). In: The 3rd National Symposium MTM, Timişoara, October 24-26, vol. 4, pp. 123–128 (1980)

[4] Kovács, F.V.: Contributii la elaborarea unei metode unitare de sinteza a mecanismelor (Contribution to the elaboration of a unitary method for mechanisms synthesis), Ph. D. Thesis, Politehnica Institute "Traian Vuia" Timişoara (1969)

[5] Kovács, F.V., Perju, D., Savii, G.: Metode noi in sinteza mecanismelor (New methods in mechanisms' synthesis). Editura Facla, Timişoara (1976)

[6] Kovács, F.V., Perju, D.: Sinteza mecanismelor (Synthesis of the mechanisms) – Lito Politehnica University of Timişoara (1977)

[7] Kovács, F.V., Perju, D., Crudu, M.: Analiza mecanismelor (Analysis of mechanisms) – Lito Politehnica University of Timişoara (1978)

[8] Kovács, F.V., Cojocaru, G.: Manipulatoare, roboti si aplicatiile lor industriale (Manipulators, robots and their industrial applications), Editura Facla, Timisoara (1982)

[9] Kovács, F.V., Mureşan, T., Gheorghiu, N., Micşa, I., Şelariu, M., Rădulescu, C.ş.a.: Robotul industrial REMT-2 (Industrial Robot REMT-2). In: Proceedings of the 2nd National Symposium of Industrial Robots, Bucureşti, October 29-31, pp. 416–422 (1982)

[10] Kovács, F.V., Groşanu, I., Mureşan, T., Rădulescu, C.ş.a.: Robot industrial REMT-2S (Industrial Robot REMT 2S). In: Proceedings of the 6th National Symposium of Industrial Robots "ROBOT-1986" Braşov, December 11-13, vol. 3, pp. 79–86 (1986)

[11] Kovács, F.V.: Trecut, prezent şi perspective în cercetările de robotică în Timişoara (Past, present and future in the robotic research in Timisoara), Presentation at the Romanian Academy, Technical Section (November 20, 1987)

[12] Kovács, F.V., Mureşan, T., Gheorghiu, N., Rădulescu, C., Dragomir, T.: Contributions to Research, Construction and Industrial Applications of Robots, Mechanics Seminary, Politehnica Institute "Traian Vuia" Timisoara, pp. 114–183 (1987)

[13] Kovács, F.V., Perju, D., Crudu, M.: Mecanisme (Mechanisms) Lito Politehnica University of Timişoara (1992)

[14] Kovács, F.V., Radulescu, C.: Roboţi industriali (Industrial Robots), vol. 1, 2. Litografia Universităţii "Politehnica" din Timişoara (1992)

[15] Kovács, F.V., Mureşan, T., Gheorghiu, N., Rădulescu, C.: Fifteen Years of Robotics Reasearch at the Timisoara Technical University, Heavy Machinery TM 1993, Kraljevo, vol. 2, pp. 254–261 (October 1993)

[16] Kovács, F.V., Radulescu, C.: Flexible Manufacturing Line as part of "Robotized Flexible Manufacturing Systems" Laboratory of "Politehnica" University of Timisoara. In: Proceedings of the 10th National Symposium of Industrial Robots, Timisoara, November 16-19, vol. 1, pp. 75–82 (1994)

[17] Kovács, F.V., Grigorescu, S., Radulescu, C.: Sisteme de fabricatie flexibilă robotizată (Robotized Flexible Manufacturing Systems), vol. 1, 2. "Politehnica" University of Timisoara (1994)

[18] Kovács, F.V.: On Social Impact of Robotics in Romania. In: Proceeding of the First International Conference on Advanced Robotics and Intelligent Automation, Athens (1995)

[19] Kovács, F.V.: Robot's Guiding Devices. In: New Geometrical Model, Proceedings of the 11th Robotics National Symposium, October 10-12, vol. 2, pp. 13–20. Reşiţa-Văliug (1996)

[20] Kovács, F.V.: Un nou model geometric al dispozitivului de ghidare al roboţilor (A new geometrical model of the guiding mechanism of the robots). In: Proceedings of the 11th National Symposium of Robotics, October 10-12, vol. 2, pp. 20–26. Reşiţa-Văliug (1996)

[21] Kovács, F.V.: Using the Concept "Connection" for Structural Analysis/Synthesis of Parallel Topology Robots". In: Proceedings of the 7th IFToMM - International Symposium on Linkages and Computer Aided Design Methods - Theory and Practice of Mechanisms – SYROM 1997, Bucharest, August 26-30, vol. 2, pp. 189–194 (1997)

[22] Kovács, F.V.: The "Robotic Family". In: Proceedings of the 7th IFToMM International Symposium on Linkages and Computer Aided Design Methods – Theory and Practice of Mechanisms – SYROM 1997, Bucharest, August 26-30, vol. 2, pp. 185–188 (1997)

[23] Kovács, F.V., Tusz, F., Varga, Ş.: Fabrica viitorului (The plant of the future). Editura Multimedia Internaţional Arad (1999)

[24] Kovács, F.V., Varga Şt., Pau, V.: Introducere în robotică (Introduction in Robotic). Editura Printech, Bucureşti (2000)

[25] Kovács, F.V.: Noţiunea "Perechi de Sisteme de Referinţă" (PeSiR) şi unele utilizări în domenii de ştiinţe tehnice (The concept of "reference system pairs" (PeSiR) and some applications in the technical science fields). Robotica & Management 6(1), 22–29 (2001)

[26] Kovács, F.V.: Structural and Kinematical Modelling of Mechanical Systems of Mobile Robots using the concept "Pairs of Reference Systems"(PRS). In: Proceedings of the 2nd National Workshop on Mobile Robots, Craiova, pp. 38–44 (2001)

[27] Kovács, F.V., Rădulescu, C., Dolga, V., Maniu, I.: Activity of the multidisciplinary robotic team in Timisoara. Robotica & Management 7(2), 4–21 (2002)

[28] Perju, D.: Mecanisme de mecanică fină (Mechanisms for precision mechanics), vol. 1, 2. Lito Politehnica University of Timişoara (1990)

[29] Radulescu, C., Maniu, I.: 1979–2009 - 30 Years of Robotics at "Politehnica" University of Timisoara. Robotica & Management, 14(2), 14–21 (2009)

[30] Rehwald, W.: Analytische Kinematik von Koppelebenen. Otto Krausskopf Verlag, Mainz (1972)

Section III

History of Technical Systems

An Examination of the Mechanisms of Movement in Heron of Alexandria's *On Automaton-Making*

Finlay McCourt

Classics, School of Humanities,
University of Glasgow, Glasgow, G12 8QQ
f.mccourt.1@research.gla.ac.uk

Abstract. In the first book of *On Automaton-Making* a number of mechanisms are described which move a 'mobile-automaton' around a performance space. These mechanisms are presented in the form of new translations from the original Greek. The feasibility of each mechanism is discussed, and other technical issues in each plan are considered along with those applicable to all. From this some conclusions as to which mechanisms had actually been constructed or witnessed by the author are made, and the theoretical nature of the remaining devices is considered. Additionally, a short introduction to the author Heron of Alexandria is given, along with an introduction to the text *On Automaton-Making* as a whole.

1 Introduction

Although Heron (Hero) of Alexandria is a well-known author in the field of ancient mechanics and mathematics, the text *On Automaton-Making* remains relatively obscure due to the lack of an appropriate and accessible English translation of the work. There are previous translations in German [1] and French [2] (though the latter deals only with the second book of the work). There is also a more recent English translation [3], which is less than ideal, as it does not engage with the technical issues present in the text.

Therefore, before presenting and analysing the mechanisms of movement from the first book of *On Automaton-Making,* I will provide a brief introduction both to the text itself and automata in the classical world. I will also provide a short but comprehensive introduction to Heron of Alexandria and in so doing clear up some common misconceptions about his life and work.

Heron of Alexandria

Heron (Hero) of Alexandria is a mathematician and engineer of the first century AD, and the author of a number of technical texts, of which we know of around eleven. The majority of these are extant in Greek, with the *Mechanica* surviving fully only in an Arabic translation. He is certainly most famous for his record of the aeolipile (a simple steam-engine) in the *Pneumatica*, which is the first full description of a steam-engine (Heron, *Pneumatica* 2.11 [1], though there is an earlier mention in Vitruvius

T. Koetsier, M. Ceccarelli (Eds.): Explorations in the His. of Machines & Mech., HMMS 15, pp. 185–198.
springerlink.com

1.6.2 [4]). His work has been treated as something of a footnote in the history of Greek science until recently [5], but its survival is undoubtedly due to its popularity and utility [6: 307-8]. Much of the material went on to have an influence in both the Byzantine period and the Renaissance (see Brett [7] for discussion of a Byzantine automata influenced by Heron with reference to Heronic Renaissance automata).

Heron is part of the tradition of technical writing on mechanics and related fields, which appears to have been somewhat disconnected from, and neglected by, other areas of contemporary thought [8] (I refer specifically to Heron's mechanical texts here as mathematics seems to have been considered a higher pursuit on a par with philosophy, see [9: `Mathematics'.II.A]) This mechanical tradition has many similarities, and may be considered congruent, with the tradition of medical writing as seen in the works of Galen and the Hippocratic corpus. There are a number of authors who are known to be have been working in the field of mechanics, beginning with Aristotle and encompassing Ktesibios, Philon and Heron up to Pappus of Alexandria. Heron's engagement with mechanics may extend to mechanistic philosophy, seen by some as a distinct branch from contemporary philosophical thought [10, 11], though Heron's engagement with contemporary Peripatetic theory is evident (See Heron, *Pneumatica* 1.proem [1] and Boas [12]). Heron's immediate predecessors in the field are Ktesibios (the one Heronic reference to Ktesibios comes from the title of the *Belopoeica*; see Heath [6] for discussion) and Philon, both of whom wrote treatises on pneumatics which served as the basis for Heron's own work on the subject.

The most common method for dating Heron is by the likely date of the eclipse mentioned in the *Dioptra* (Heron, *Dioptra* 35 [13]). Since his method of calculating the distance between Rome and Alexandria involved simultaneous observation of the eclipse from the two cities, it is likely that he observed this event in person (see Keyser [14] and Heath [6], but with important updates and corrections in Neugebauer [15]).

On Automaton-Making

Heron's *On Automaton-Making* (abbreviated as *Automata*) is based on an earlier work by Philon, (Heron, *Automata* 20.1&3 [1]). The attribution of him as his source serves as a point of departure, from which Heron develops his own work. He places himself firmly in this tradition of automaton-building, that is, he does not claim that the form is his own innovation. Associated with this stance is the acknowledgement and criticism of his predecessors (Heron, *Automata* 5.1 [1]), which informs much of the content of the work, in that he defines many aspects of his own treatise as improvements upon the work of Philon.

Automata make their first appearance in Homer, in a form that is strikingly similar to those described in *On Automaton-Making*. Hephaestus is depicted creating self-moving tripods to serve the gods on Olympus (Homer *Iliad* 18.373-381 [16]). This passage from the Iliad is the inspiration for the later automaton, when what was once impossible and divine becomes possible through the human application of mechanics [17]. Certainly, the form of automaton seen in Heron and Philon would seem to be a conscious emulation of this passage (Schneider [17: 201-207] gives a potted history of the development of ancient automata).

On Automaton-Making consists of two books dealing with the construction of two different kinds of automata, one mobile, the other stationary. Both automata utilise a falling weight as their power source. The descent of this weight is regulated by seeds or sand, and then this energy is transferred to the mechanisms by a cord and axle system. The mobile automaton consists of a miniature shrine of Dionysus containing a statue of the god, surrounded by dancing Bacchantes, and with an altar on either side. This is set on top of four pillars that are borne by a wheeled case. This automaton will travel to a pre-defined point, stop, and then the figures will carry out a number of actions before the automaton travels back to its point of origin. The stationary automaton consists of a miniature theatre (*pinax*) set on top of a single column. The theatre will open and close a pair of doors automatically, and the viewer will see a series of scenes from the legend of Nauplios play out before them (the sequence of which seems to be dictated by what was technically possible rather than narrative alone). The text does not present these automata only in isolation, but details techniques utilised in their construction as well as offering alternative methods for a number of the components.

The most likely context for the display of the automata described in *On Automaton-Making*, given their size, is that of a symposium, particularly in the case of the moving automaton described in the first book, especially considering its Dionysian theme. There is also a clear parallel created between the three wheeled automata of Heron and the tripods of Hephaestus[16] when they are presented in a sympotic context. There are a number of other points in the text that may provide insight into the display context of the automata described, but analysis of these is outside the scope of this paper.

The readership for whom the treatise *On Automaton-Making* was intended is still unclear. It purports to contain all the information needed for the implementation of the described automata and all variations thereupon (Heron, *Automata* 1.8 [1]) but many important details are missing, as will be discussed below. At certain times Heron gives precise measurements of the automata's elements that have less impact on its workings, and at others he misses out key information. The result of this is a text that could not be used without considerable technical knowledge.

2 Mechanisms of Movement

I define mechanisms of movement as those which Heron presents to effect the movement of the mobile automaton around the performance space. These mechanisms are specifically presented by him as innovations upon the work of his predecessors (Heron *Automata* 5.1-2 [1]). Heron here characterises the mechanisms of his predecessors as being both troublesome and difficult to effect, but does not detail their workings. He characterises his own designs for basic forward and reverse movement as easy and risk-free. In addition to this he claims further innovations that deal with movement in other patterns.

Heron states that his aim is to furnish the reader with all the tools necessary to implement whatever scenario they should wish (Heron, *Automata* 1.8 [1]). The addition of these other patterns of movement must be viewed in this light, as an attempt to fulfil this goal, but he does not attach any guarantee to these new mechanisms in the way that he does for forward and reverse motion.

I will demonstrate the theoretical, rather than practical, underpinnings of most of Heron's innovations in patterns of movement through the investigation of each of these mechanisms. While these would ultimately disappoint a prospective ancient automaton-maker, they demonstrate the application of known and observable mechanical concepts.

<u>Basic Forward and Reverse Movement.</u>

Heron begins by detailing his design for basic forward and reverse movement in a straight line. A cord is attached and wrapped around an axle, and as a falling weight pulls upon this cord, it is unrolled and the axle and the wheels turn. This moves the automata forwards. Reverse movement is achieved by looping the cord around a peg in the axle and wrapping the cord in the opposite direction.

(All translations of Heron's *On Automaton-Making* presented in this paper are my own, from the edition of Schmidt [1]. Necessary supplements or changes to the text are presented in square brackets. Explanatory notes and glosses are presented in parentheses.)

> 5.3. First of all I will describe [movement] upon a straight line. Let there be a base ΑΒΓΔ, in which let an axle EZ be fixed, turning on pins and fixed into settings in the walls of the base. Let there be two equally-sized wheels which have been worked into a convex (lit. 'lentiform') shape, ΗΘ & ΚΛ, attached to the axle; and let there be a cylinder MN attached in the middle of the axle, around which the cord will be wound up.
>
> 5.4. Let there be a peg, Ξ, attached to this, around which a loop of the cord will lie. Let there be another wheel, ΟΠ, in the middle of the side ΓΔ, turning in the frame ΡΣΤΥ, around a very small axle ΦΧ. Let the wheel's axles be adjusted so that the base stands level at all points. When the loop of cord has been put round the peg Ξ, let the cord be wound up around the cylinder.
>
> 5.5. After a rectangular tube has been placed upright in the middle of the base, let the other end of the cord be passed through a pulley at the top of the tube and fastened to a lead counterweight inside. Accordingly, if anything should let the counterweight travel downwards in the tube, the cord would be drawn tight. As this is unrolling from the cylinder it turns the wheels ΗΘ and ΚΛ; and rolling along the ground they will move the case until the loop falls from the peg, or the counterweight comes to rest upon something.

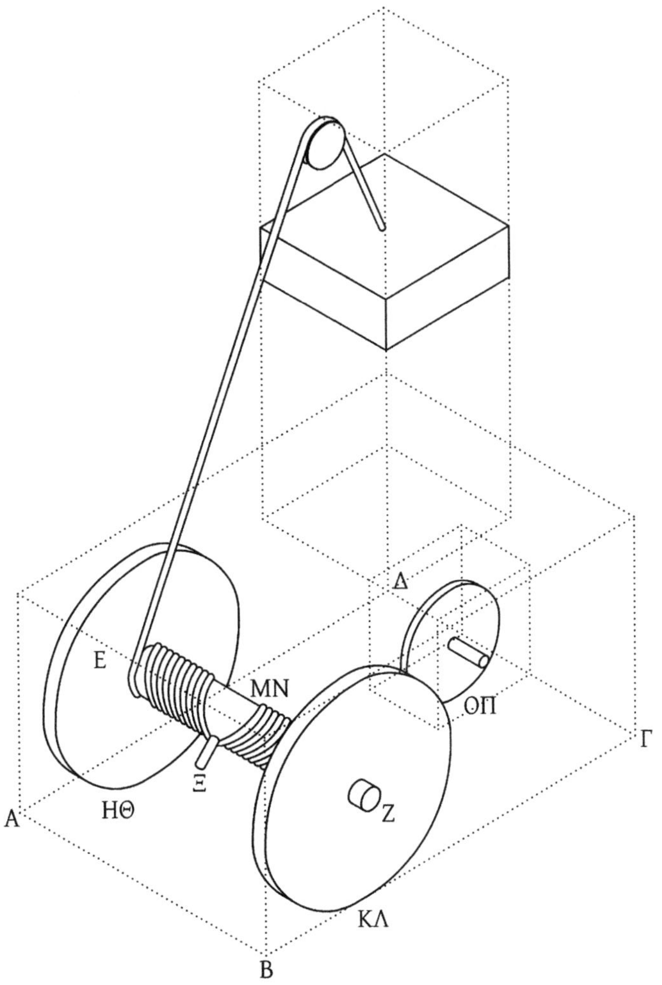

Fig. 1. Drive mechanism for forward and reverse movement

> 6.1. The movement is effected in the manner that has been described, but the return movement as follows: When the rope has been wound around the cylinder to a specific point, let it be looped around the peg Ξ, and wound in the opposite direction around the cylinder. Then let it be similarly attached to the counterweight, secured to it with a ring. So, when the counterweight again moves downwards it will unroll the first winding, and the base will move.
> 6.2. Then, when [the cord] drops from the peg, the wheels will turn in the opposite direction, and thus the return movement is effected.

Unfortunately, we cannot assess whether Heron's claimed superiority in design is justified as the texts by earlier authors in the automaton-building tradition are not extant. It is however a feasible method of producing forward and reverse motion in the automaton, and forms the basis for all other mechanisms of movement presented in *On Automaton-Making*.

The use of cords and axles, instead of gearing, for the entire drive system is undoubtedly influenced by two things, adaptability and scale. The use of cords and axles lends an adaptability to the device that could not be achieved by a gearing system, which would provide only a fixed pattern of movement. As described in chapter 11.6 of *On Automaton-Making*, the cord and axle system allows for different patterns of movement to be programmed easily and instantly for different contexts. Contrast this with the complex but relatively fixed nature of devices such as the Antikythera mechanism [18, 19, 20]. The scale of the device is such that it needs to be made as light as possible (Heron *Automata* 1.8 [1]), and the movements, and thus any gearing system, must have a degree of precision. Gears made of wood would be imprecise and heavy, and gears of bronze would be heavy and expensive, and so neither are appropriate for use in this context.

Circular Movement.

Heron proceeds to then describe mechanisms utilising the same drive system to generate different patterns of movement. He begins by describing the means by which the automata can be made to move in a circle of a fixed size. He proposes that two wheels of differing sizes be mounted upon an axle, in a similar way to the previous example. The two wheels would be cross-sections of a cone, the length of which is equal to the radius of the circle around which the automaton should travel. The effect of these two wheels rolling together will be the same as a cone rolling upon its side and thus the automata will travel in a circle.

> 7.1. Movement upon a circular course is effected in the following way. Let there be a circle, ABΓ, around which the base must be borne, the centre of which is Δ. Let the line AΔ be drawn across, and EAZ at right angles from it at A; and let EZ be the diameter of one of the three wheels, of whatever size we might choose. Let the centre of this be A and let ΔE and ΔZ join up at the top.
>
> 7.2. Let AH be equal to the size of the axle of the wheels and let [ΘHK] be parallel to EZ. Let the base be MΛNΞ, with NΞ being parallel to AΔ. Let another line, ΔO, be drawn, and ΠP which bisects it at right angles underneath O. The positions of the wheels will be in relation to the diameters EZ, ΘK and ΠP, but their axles will be TY and OX.
>
> 7.3. Thus let the wheels be arranged in such a position that the base, having be set upon them, is evenly balanced. Then the axle pins will be at the points T, Y, O, and X. Let the cylinder around which the cord is wound be placed again between A and H, and let these things happen as stated previously. Thus the base will be borne around the circle which has been described.

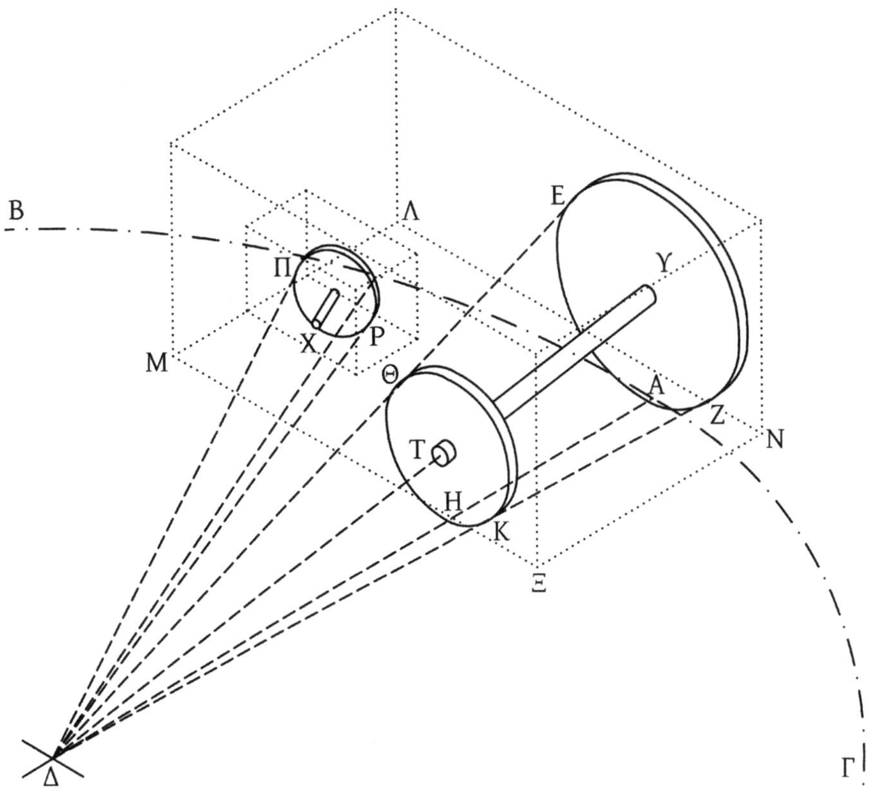

Fig. 2. Wheel and axle arrangement for circular movement based on cones

8.1. If a cone is rolled upon a flat surface, its base will describe a circle, the distance from the centre of which is equal to the side of the cone, but the tip remains motionless, being the centre of the aforementioned circle. The wheels EZ, ΘK, and ΠP are in two cones, the bases of which are the circles EZ and ΠP, and the tip of which is the point Δ.

8.2. It is clear that these isosceles cones rolling around describe circles, and the tip is motionless. For [a cone] lying upon a flat surface and moving upon its own side is evenly balanced with itself, for it is cut through by a plane perpendicular to the line produced by the side, dividing [the cone] in two. Whenever it is rolled, having been overcome by some other force, each of the semicircles on its surface bears upon the remaining semicircle of this circle with equal force, and thus it moves.

There are however some obvious problems with this design. Heron conceives of the wheels EZ and ΘK as circular cross-sections of the cone EZΔ mounted upon an axle passing through the centres of both circles. He states that the wheels EZ and ΘK must be parallel with one another, but not the angle at which they should be mounted upon the axle. If the wheels EZ and ΘK are mounted vertically, as with all other wheels in the text, then it follows that they are not mounted at right angles to the axle TY. An arrangement such as this is not workable as the rotation of the axle would create uneven movement and lateral 'wobble'. The wheels EZ and ΘK must therefore be mounted at right-angles to the axle TY (Fig. 3). If this arrangement is adopted, then the wheels EZ and ΘK must be mounted closer to the centre of axle TY to prevent any contact with the walls of the base. As the angles A and B become more acute (Fig. 3), the more unstable the automaton becomes. This is because the wheels have to be mounted closer to the centre of TY. Similarly, an increase in the acuteness of the angles A and B also weakens the axle and wheel assembly, as the weight of the automaton, instead of being vertically distributed upon the wheels bears down upon the wheels at an angle.

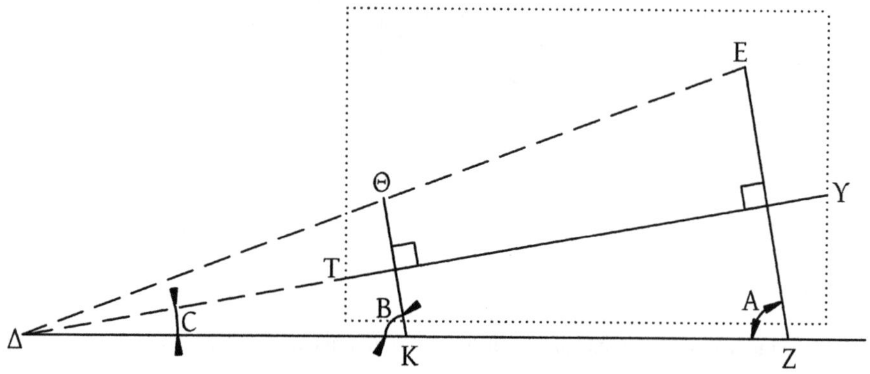

Fig. 3. Wheel and axle angles for circular movement based on cones

Thus, the closer the angle C is to zero, the more feasible this mechanism for circular movement becomes. It is preferable, as I will discuss below, for the wheels that are moving the mechanism to be as large as possible. Therefore, the outer wheel, EZ, must be made as large as is allowed by the size of the base. The only variable that can be changed to reduce the size of angle C is the radius of the circle around which the automaton will travel.

Besides describing the mechanism to effect this circular movement, Heron describes the theoretical underpinnings of the circular movement in 8.1-2. This seems to be supplied as evidence of the feasibility of this arrangement of wheels. The genesis of this method of generating circular movement seems to be the observation of the behaviour of cones, hence the inclusion of this information, rather than the construction of the device in question. Had Heron himself constructed this arrangement of wheels, or observed it in use, then he surely would have included additional information regarding the radius of the circle and the mounting of the wheels.

Rectilinear Movement.

Heron then describes how rectilinear movement will be effected through the use of two sets of wheels, one of which can be alternately raised and lowered. The first fixed set of wheels will bear the automaton along one side of a rectilinear form. The second set of wheels, mounted at right angles to these, will descend on a screw mechanism and lift the first set of wheels and the automaton off the ground. It will then proceed along the next side of the rectilinear form. Next, it will again raise the second set of wheels and the first set will again rest upon the ground. It will then move along the third side of the rectilinear form upon these wheels. It will proceed in this manner as many times as the automaton-builder may wish.

> 9.1. Rectilinear movement of the base will be effected in the following manner. Let there be the base, ΑΒΓΔ, inside which let there be the axle, ΕΖ, attached to which are the wheels ΗΘ & ΚΛ; and let there be a third wheel ΜΝ, by means of which both the forward and return movements will be effected, just as has been mentioned before. But let there also be another axle, ΞΟ, attached to which are the wheels ΠΡ & ΣΤ and likewise ΥΦ.
> 9.2. Let the axle ΞΟ be above the axle ΕΖ, and be a sufficient distance away from it. Let the wheels ΠΡ & ΣΤ along with the axle ΞΟ be able to be raised and lowered in height, as we will hereafter describe, and the wheel ΥΦ similarly. So, if we should make the wheels ΠΡ, ΣΤ & ΥΦ descend, so that they sit upon the floor, the wheels ΗΘ, ΚΛ & ΜΝ will be raised from the ground, and the base will make its movement upon ΠΡ, ΣΤ & ΥΦ.
> 9.3. After the axle ΞΟ is drawn up, so that ΗΘ, ΚΛ & ΜΝ sit upon the floor again, the base will be driven along another side of the rectangle upon these wheels. Then having stopped itself again, ΠΡ, ΣΤ & ΥΦ will be lowered, and the base will be borne again along another side of the rectangle. These things occurring alternately, the base will travel along the rectangle as often as we should choose.

In chapter 10, Heron goes on to describe his proposed mechanism for raising and lowering the second set of wheels. He intends that the axle ΞΟ and the wheel ΥΦ each be mounted in their own frames. Consequently, they can be raised and lowered by means of screw gears which are turned when cord wrapped around them is unrolled when pulled upon by the counter-weight.

It is, of course, impossible for these wheels to be alternately raised and lowered by the force generated by the falling counterweight. For it to do so would be a contradiction of Newton's first law of motion, as there is not an unbalanced force acting to raise the automaton upon the lowering set of wheels. The force generated by the falling counterweight would not be sufficient to overcome the overall weight (F_g) of the mechanism.

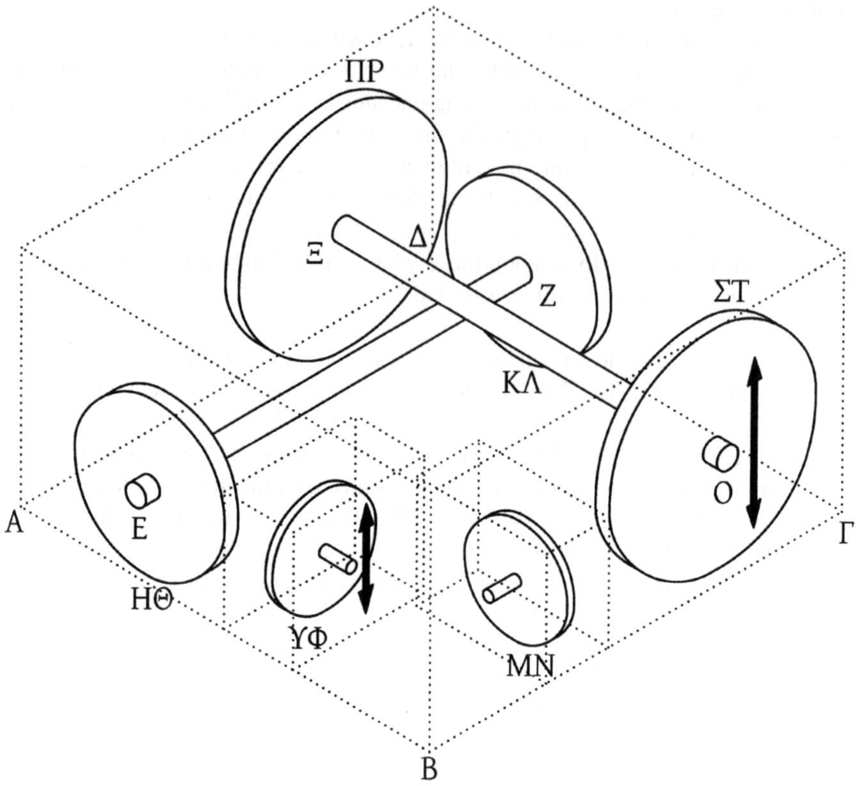

Fig. 4. Arrangement of wheels and axles for rectilinear movement

Screws were not a new technology [21], and were very familiar to Heron. He had described a machine for cutting internal screw threads in his *Mechanica* (3.21 [22]) and made use of screws in his *Dioptra* (3 [13]). Once again the description of this mechanism seems rather to be based on a conception of how it would ideally work than on a device that has been built and observed in action. It is telling that he says in 10.3 [1] when describing the raising and lowering mechanism, "If someone should turn about the screw ΦΧ the little bar ΗΘ will be raised and lowered by the peg Π." This would seem to relate more to observation of the use of screws in presses and other devices than the use of screws in automata. He mentions the screw being turned by hand, and goes on to assume that it will work in the same way when turned by a cord pulled by the falling counterweight.

Serpentine Movement and Other Forms.

Finally, Heron discusses a mechanism that he believes should allow for all patterns of movement. It utilises two independently mounted wheels that can be independently stopped and started, though not driven at different velocities. To effect a turn, the wheel on the inside of the turning circle will be stopped by slack in the cord pulling upon it from the counter-weight. The outer wheel will continue moving and the automaton will be turned until the inner wheel once again begins turning.

11.1. It is also possible to turn the case in other ways, not only in a rectangle, but also in all rectilinear figures; it is also possible to effect snake-like movement, and it is much more easily dealt with than the method previously written about.

11.2. Let there be a base, in which there are wheels, ΑΒΓΔ, and inside which let there be fixed two axles, ΕΖ & ΗΘ. Let ΗΘ turn easily on pins, with the wheel ΚΛ attached to it, but let ΕΖ be joined to the case, having been worked to an even thickness on a lathe. Let two axle-boxes, ΜΝ & ΞΟ lie around this, turning freely and closely around [the axle], with both the inside and outside surfaces having been worked upon a lathe. Let there be wheels of an equal size, ΠΡ & ΣΤ attached to the axle-boxes.

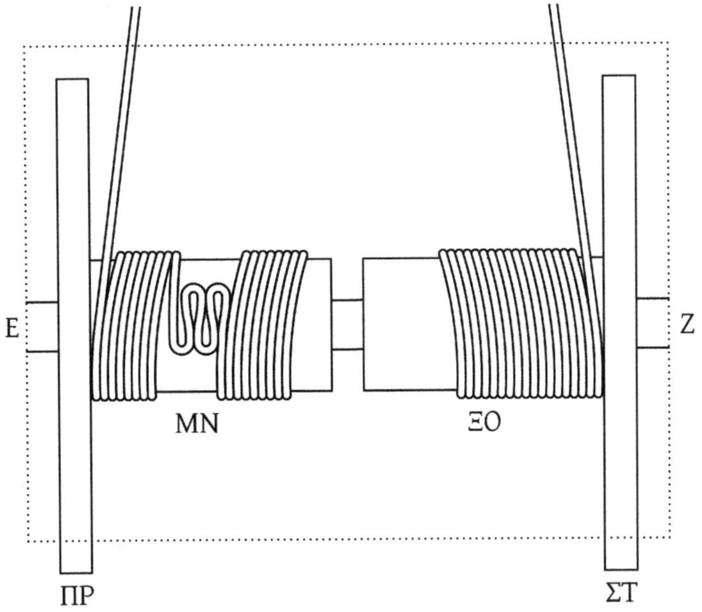

Fig. 5. Independently mounted wheels with programmed loops in the cord to effect a turn

11.3. So, if cord, having been wrapped around each of the axle-boxes should be attached to the counter-weight in the tube, it will happen that when the counterweight descends, and when the cord is unrolled the wheels will turn about with the axle-boxes, thus the base will move along a straight line, and the wheel ΚΛ will rotate along with it.

> 11.4. So, if ΠΡ, from the wheels ΠΡ & ΣΤ should remain motionless, its cord having some slack, the wheel ΣΤ will turn around, and ΚΛ will turn around with it, until the counterweight draws in the slack [created by] the little kinks around the axle-box MN. Then, when the cord has tension again, the wheels ΠΡ & ΣΤ will turn at the same time. The base will be borne along a straight line determined by the turn that has occurred.

This, unlike the previous two mechanisms, is a feasible and familiar method of turning a moving vehicle.

Practical Limitations

Beyond the basic way in which these mechanisms function there is another fundamental issue that affects the effectiveness of these mechanisms individually and the mobile automata as a whole, namely, the distance of travel that the automata could achieve. Insufficient information is given in the text to provide a complete analysis of this topic, but I will explore the issue in relation to the mechanism for linear forward motion alone.

The only dimensions that we are given of the automata are presented in chapter 3:

> 3.1. Let there be a base one cubit in length, four palms in width and three palms in height, with a groove running around both the upper and lower parts. Set upon its corners four small pillars, each with a height of ten (MSS: eight) palms and a width of two palms, with rounded mouldings at the base and corresponding capitals at the top. Upon these little capitals lies a kind of circular lintel with a height that is an eighth of that of the whole pillar, i.e five fingers.

These are the sole dimensions mentioned in the text. Therefore, to calculate the distance that the automaton is able to travel from these dimensions it is necessary to make a set of reasonable assumptions. We can assume that the wheels could have been, at most, around three palms in diameter (w), allowing for clearance between the bottom of the base and the floor. A smaller axle would allow for greater distance, but a larger one would provide more surface area for slack in the rope to be affixed, and thus more movements to be programmed. A reasonable size for a smaller axle would be two fingers in diameter (a), and a palm (a) for a larger axle. The maximum distance that the weight could drop is around 10 palms (h), allowing space above for the mechanisms of the little figurines, and below for the wheels. Given these figures the maximum distance (d) that the automata can travel is (roughly) determined by the following equation:

$$\frac{(h \times w)}{a} = d$$

Using this calculation we find that, with the large axle, the automaton could travel roughly 30 palms (around 2210mm, see [9: 'Measures'.II.A]), and, with the smaller axle, roughly 60 palms (around 4425mm). These distances are not particularly large given the actions that Heron expects them to complete, and they may have brought the device into a room, but no further.

Heron goes some way towards solving this issue in his description of the use of axles and wheels of varying size, as well as additional axles, to gain mechanical advantage in chapters 17 and 18. However, the presentation of this concept is extremely general, and does not provide any detailed information on the necessary arrangement and ratio of the axles to one another. This lack of information would make the system hard to implement for anyone who is not already familiar with the workings of similar devices. Additionally, Heron fails to make any demonstration of the advantage that a system such as this would provide to the other mechanisms of movement that he has presented, and so his inclusion of this information seems to be an afterthought rather than a response to issues presented by the other mechanisms.

3 Conclusions

The main conclusion that can be drawn from the examination of these mechanisms of movement is that Heron has not built all of the devices that he has described, but develops from observable principles mechanisms that he believes will produce the desired effects. He presents these as working improvements upon the tradition of automaton-building.

The mechanism that he describes for forward and reverse movement is practical and feasible, and probably has substantial similarities to mechanisms described by Philon for the same purpose. I would consider it likely that of all the described devices, this is the one that has been constructed. The improbability, given the physical limitations of the automata, that the mechanism for circular movement will work as intended means that it has not been constructed, certainly not on a working scale. The root of the design seems to be the observation of the behaviour of a cone upon a surface, and it is this behaviour that is emphasised rather than descriptions of the behaviour of this kind of wheel arrangement. The impossibility that the mechanism for rectilinear movement will work at all proves that it has not been constructed, and it seems to have developed from Heron's knowledge of other screw-mechanisms, and the belief that it would work as described. The mechanism that he describes as imparting "snake-like" movement is feasible, but the complexity of the movements that are possible is limited by the physical constraints of the automata.

Heron is attempting to expand upon the established body of knowledge on the construction of automata, at least partly to prove that his treatise is a worthwhile expansion or improvement upon that of previous authors. The designs which are new and unworkable are backed up with explanation that exposes their theoretical origins and the genesis of the concepts. Rather than expose the fact that they are not tried and tested, Heron attempts to prove the designs by other means. The mix of practical and impractical designs demonstrates in microcosm some of the constraints under which ancient technical writers were operating.

References

[1] Schmidt, W. (ed.): Heronis Alexandrini Opera Quae Supersunt Omnia, vol. 1. Teubner, Leipzig (1899)

[2] Prou, V.: Les théâtres d'automates en Grèce au IIe siècle avant l'ère chrétienne d'après les *Αυτοματοποιηκα* d'Héron d'Alexandrie, Paris (1884)

[3] Murphy, S.: Heron of Alexandria's On Automaton-Making. History of Technology 17, 1–44 (1995)

[4] Granger, F.: Vitruvius - On architecture, vol. 1. Heinemann, London (1931)

[5] White, K.D.: Greek and Roman Technology. Thames and Hudson, London (1984)

[6] Heath, T.: A History of Greek Mathematics, vol. I & II. Clarendon Press, Oxford (1921)

[7] Brett, G.: The Automata in the Byzantine "Throne of Solomon". Speculum 29(3), 477–487 (1954)

[8] Lloyd, G.E.R.: Methods and Problems in Greek Science. Cambridge University Press, Cambridge (1991)

[9] Brill's New Pauly: Encyclopaedia of the Ancient World: Antiquity, vol. 1-16. Brill, Leiden

[10] Berryman, S.: Ancient Automata and Mechanical Explanation. Phronesis 48(4), 344–369 (2003)

[11] de Solla Price, D.J.: Automata and the Origins of Mechanism and Mechanistic Philosophy. Technology and Culture 5(1), 9–23 (1964)

[12] Boas, M.: Hero's Pneumatica: A Study of Its Transmission and Influence. Isis 40(1), 38–48 (1949)

[13] Schöne, H. (ed.): Heronis Alexandrini Opera Quae Supersunt Omnia, vol. 3. Teubner, Leipzig (1903)

[14] Keyser, P.: Suetonius "Nero" 41. 2 and the Date of Heron Mechanicus of Alexandria. Classical Philology 83(3), 218–220 (1988)

[15] Neugebauer, O.: Uber eine Methode zur Distanzbestimmung Alexandria-Rom bei Heron. Kongelige Danske Viden-skabernes Selskab 26.2, 21–24 (1938)

[16] Murray, A.T.: Homer - Iliad, vol. 2. Harvard University Press, London (1999)

[17] Schneider, H.: Einführung in die antike Technikgeschichte. Wissenschaftliche Buchgesellschaft, Darmstadt (1992)

[18] Freeth, T., et al.: Decoding the ancient Greek astronomical calculator known as the Antikythera Mechanism. Nature 444, 587–591 (2006)

[19] Freeth, T., et al.: Calendars with Olympiad display and eclipse prediction on the Antikythera Mechanism. Nature 454, 614–617 (2008)

[20] Koetsier, T.: Phases in the Unraveling of the Secrets of the Gear System of the Antikythera Mechanism. In: Yan, H.S., Ceccarelli, M. (eds.) International Symposium on History of Machines and Mechanisms, Proceedings of HMM 2008, pp. 269–294. Springer, Heidelberg (2008)

[21] Oleson, J.P. (ed.): The Oxford Handbook of Engineering and Technology in the Classical World. Oxford University Press, New York (2008)

[22] Nix, L., Schmidt, W. (eds.): Heronis Alexandrini opera quae supersunt omnia, vol. 2.1. Teubner, Leipzig (1900)

The Mystery of the Defense Chain Mechanism of Constantinople

Junichi Takeno[1] and Yoshihiko Takeno[2]

[1] Meiji University
juntakeno@gmail.com
[2] Heuristics Science Research
kielfluegel@gmail.com

Abstract. Istanbul, Turkey, is rich in relics of the Roman Empire. We are particularly interested in the iron chain that is said to have closed off a huge gulf to entry by strange ships. The width of the strait across which the gigantic chain was allegedly laid is well over 500 m. The chain gate was closed to keep out unwanted ships and opened to let allies in. How did it work? Where was it actually located? We try to solve this mystery based on available materials and rational considerations.

1 Introduction

Istanbul was once the capital of the Eastern Roman Empire, and often called *Nova Roma* instead of its official name of Constantinople. The city was renamed to Istanbul when it fell to the Ottoman Empire on 1453.

Constantinople prospered for about 1,000 years until its fall. As many history books suggest, its long history of survival and prosperity was due to the secure life of its people protected by Roman statutes, peace of mind from belief in Christianity, and powerful, well-organized armaments[1-4]. The city was guarded by two rigid defense systems: the Defense Walls surrounding the entire city, and the Defense Chain blocking unfamiliar groups attempting to attack the city from the sea.

The 3-tiered Theodosian Defense Walls repelled enemies' land attacks well. The Defense wall along the Golden Horn bay, however, was difficult to fortify because many military and commercial ports existed there. The Defense Chain was contrived to reinforce protection of this area.

Constantinople was conquered by Mehmet II of the Ottoman Empire. During the war, the Defense Chain prevented the surge of the Ottoman fleet into the Golden Horn. Mehmet II implemented an astonishing strategy of having his fleet climb a hill and then go down to enter the bay on the far side to attack Constantinople from behind. It is a well-known fact that without the Defense Chain he would not have hit upon this strategy.

We focus in this paper on the mechanism of this defense chain, this unique military installation for guarding the city. The chain itself is preserved at the Istanbul Archaeological Museums and at the Istanbul Military Museum but details are

T. Koetsier, M. Ceccarelli (Eds.): Explorations in the His. of Machines & Mech., HMMS 15, pp. 199–211.
springerlink.com

unknown because of its nature of being a defense system, and many questions remain unanswered.

We attempt in this paper to solve some of the mysteries while remembering the characteristics of the Romans — reasonable with a propensity of using whatever engineering technologies that are feasible.

2 Mystery of the Floating Chain-Only a Chain or a Chain with Floats?

This section is dedicated to a study of the existing chain to explain how blocking the sea was possible.

Properties of the Chain

Several types of chain links may be observed. Fig. 1 and Fig.2 (a). Though not all links are visible, the sizes are nearly the same. There is also a device looking like large pincers (Fig. 2 (b)), which is seemingly used at the ends of the chain. These pincers may be part of the system to tie the chain to a shore structure.

The links and pincers must have been made by casting. The internal surface of each link is severely chafed on both sides, suggesting that they had been used for a long time period.

The calculated volume v_i of a link is about 3,000 cm^3 according to the measured dimensions of the link. The calculated total weight of each link m_i is about 20 kg, assuming that the specific gravity of iron C_i is 7.8.

Total Length and Weight of the Chain

Let us estimate the length of the defense chain from one end to the other. The starting point of the chain is known to be at Galata Castle (*Kastellion Ton Galaton*) in the Galata district on the northern shore of the Golden Horn bay.

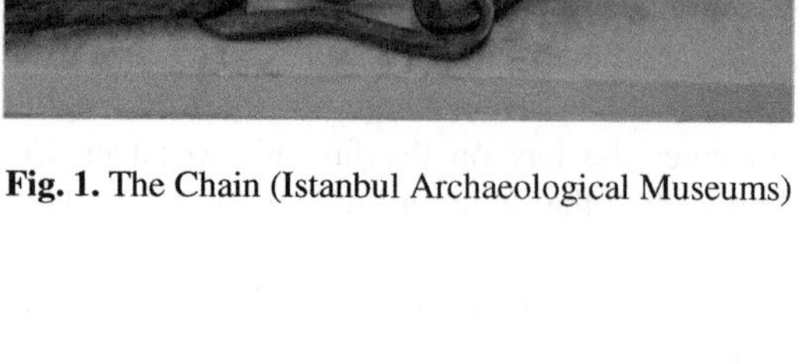

Fig. 1. The Chain (Istanbul Archaeological Museums)

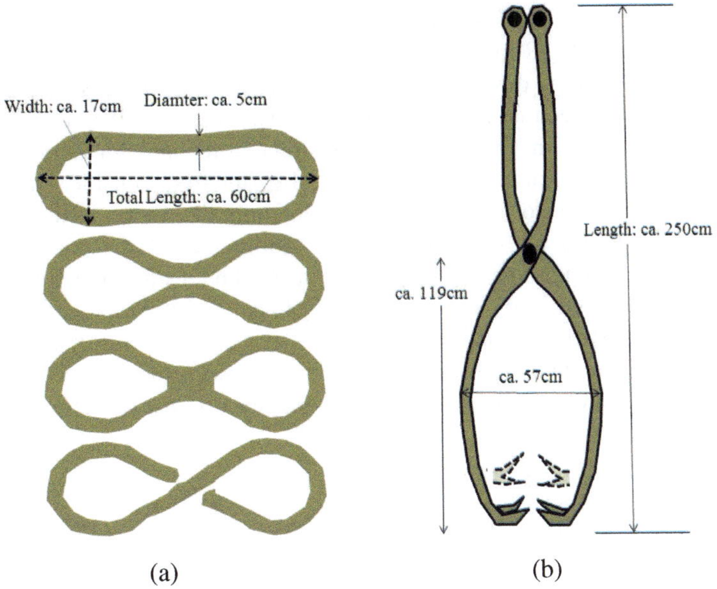

Fig. 2. Chain Links(a) and Pincers(b)

The castle is actually an underground mosque (*Yeralti Camii*), a Muslim place of worship now. This point, P0, is about 150 m in the direction of the Bosphorus from the foot of the Galata Bridge extending from Istanbul. The final end point of the defense chain is unknown, but many materials suggest it ended on the southern shore of the Golden Horn just below Topkapi Palace. We are doubtful of this assumption, and our findings will be introduced in the later section. For now, we assume that the conventional theory is correct, and call this end point P1. The total length of the defense chain is then calculated to be about 1,000 m, considering that the shoreline at that time was farther back inland than the present-day shoreline.

This length is enormous beyond our imagination. How many links were connected for the chain to cover this length? Taking overlap into consideration, the effective length of a link l_i is calculated to be about 50 cm (60 cm less 5cm *2). To cover a distance of 1,000 m, about 2,000 links (100,000 cm/l_i) must be connected. Extra length should be considered to account for the relaxing of the chain for opening the gate, but for now, we calculate the straight length of the chain as it is connected directly between two points.

What is the total weight M_c of the chain? It is calculated to be about 40 t (2,000* $m_1 \fallingdotseq$ 40,000 kg).

3 Float Design

The defense chain is intended to prevent the entry of enemy ships. While weighing about 40 t, the chain must be kept taut at a position near the sea surface. High-level technologies and a tremendous amount of energy would have been required to keep such a heavy chain stretched at the sea surface at all times without any auxiliary devices. The Romans were able to construct huge monuments as we know today, so their descendants in Constantinople must have been well acquainted with the technologies of their scientist ancestors. It is therefore easily conceivable that some mechanical and water-related technologies must have been used to make the defense chain practical. Let us first consider water-related technology. Assume use of raft-like floats, something like those used on lane ropes found in today's athletic swimming pools. Both ends of the rope are tied to fixtures on the ground to keep the rope tensioned. The tensile force applied to both ends of the rope is reduced by the use of the floats, and the rope is stably kept at the water surface by these floats.

A set of wooden floats is exhibited side by side with the defense chain in the Istanbul Military Museum. It is unknown if they are original objects. An outline is shown below (Fig. 3).

Realizing Zero Buoyancy

The volume v_f of this raft-like wooden float is calculated to be about 0.187 m^3. The total volume V_i of the 1,000m long chain is about 5.13 m^3 (M_i / C_f; where C_f is the specific gravity of iron). The buoyancy F_i is about 5.33 t (V_i *1.04), considering the properties of Mediterranean seawater. The total weight W_i of the iron chain submerged under the sea is about 34.7 t (40 t – F_i).

The weight of wood per 1 m^3 submerged under the sea is –0.59 t considering its buoyancy. The total volume V_w of the amount of wood required to balance out the weight W_i of the iron chain is therefore about 58.8 m^3 (W_i /0.59).

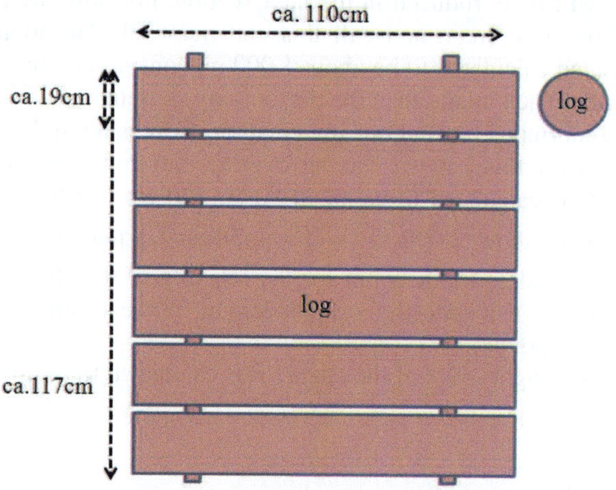

Fig. 3. Appearance of Wooden Floats

More than balancing out the total weight W_i of the iron using the floats, let us make the chain rise to the surface. To achieve this, we multiply the volume of the floats by 1.2, i.e., the total volume $V_w^{'}$ of the wood required is $V_w * 1.2 \fallingdotseq 70.6\,m^3$.

Since the volume of a single float is v_f, the number of floats K required is $V_w^{'} / v_f \fallingdotseq 378$ (sets). Assume these floats are installed on the 1,000 m long chain, then the distance between adjacent floats l_f is about 2.64 m (1000/(K+1)). This length proves that the above number of floats could have actually been installed on a 1,000 m long chain, considering the size of the float shown in the figure.

Experiments with a Small Model

The model chain used in the experiment is about 136 cm long and weighs about 25 g. The wood sticks used as the floats each weigh about 2.25 g. Fresh water was used in the experiment. Twenty-three wood sticks are tied to the model chain over a length of 100 cm. Each wood stick represents several floats. The results of the experiment generally support the reasoning of the author as described above. The floats were forcedly submerged under the water, and it was confirmed that there was enough force to buoy the chain to the water surface level.

4 Mystery of the Position of Traction–How Many Chains? From Where to Where Is the Sea Blockaded?

Mystery of the Number of Defense Chains

It remains a mystery how many defense chain systems existed in Constantinople. Three major old maps and pictures that depict the defense chains are introduced below. The first one is a woodcut named *Constantinople* printed in 1493 in Nurnberg [5]. Forty years had passed since the city fell to the Ottoman Empire when this map was printed. This old map therefore shows the city immediately after the attack. At the center-right of the map, we see the Hagia Sophia with its ground section securely guarded by ruggedly constructed defense walls. On the right side water looking like a river is depicted. The Galata district and its defense walls are visible on the right bank of the river.

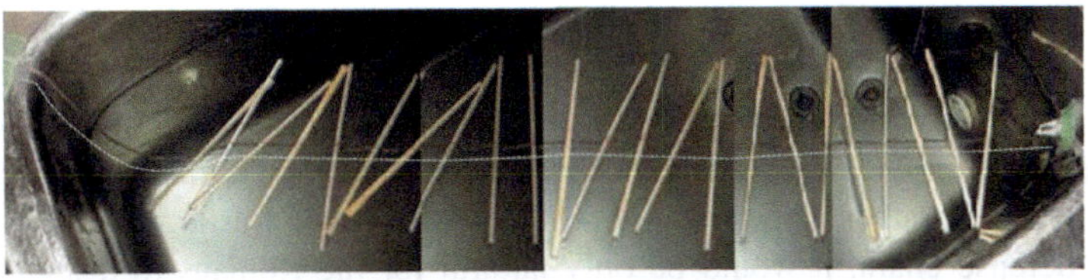

Fig. 4. Experiment to Keep Chain on Water Surface Using Floats

The river seems to be the Golden Horn bay. There are two objects which seem to be defense chains in the bay. The positions of all objects depicted in this map, excepting the Hagia Sophia, are not accurate.

In the map, one of the defense chains hangs down near the Hagia Sophia. The other is located on the deeper part of the bay. In this map, one defense chain closes the mouth of the Golden Horn while the other shuts down the deeper part of the bay. The connecting points of both chains are not depicted in this picture. Anything suggesting floats that should have kept the chain afloat on the sea surface are not visible. Nevertheless, if you look at this picture intently, you will see one defense chain at the entrance of the Golden Horn and another at the deeper part of the bay.

The second old map also shows a defense chain [6]. In this copperplate drawing, the defense chain extends from the Galata district to the north side of Topkapi Palace. This picture was printed in 1688 in Venice.

The third and last map is a lithograph printed in 1918 in Vienna [7]. A new defense chain, not appearing in the above two maps, is depicted, thus two chains are shown. One is the same as the chain shown in the old map printed in Venice. The second chain extends from the defense wall on the east side of Topkapi Palace to the Asian coast across the Bosphorus. A small islet, probably where *Kiz Kulesi* (lighthouse) stands, a short distance off the Asian coast, seems to relay the chain. The total length of this second chain would be about 1,800 m.

Having examined these three old maps, it appears that two defense chains once existed, but their locations are not definite.

The defense chain closing off the Golden Horn at its inlet existed without question. Although some materials and old maps suggest existence of a second defense chain, it is not certain, however, if a second defense chain actually existed at the deeper part of the Golden Horn bay or extended to the opposite shore across the Bosphorus.

It is a famous historical event that the Ottoman's fleet climbed the hill and then went down into the deeper part of the bay, detouring around the northern part of the Galata district to avoid the defense chain on the sea, and attacked the Byzantines from behind. This incident attests to the existence of the defense chain.

Mystery of Chain Locations

We now examine where the chains were located. One end of the first chain was tied to a small castle in the Galata district (P0, Fig. 5). Many materials suggest that the other end of this chain was tied at position P1 on the southern shore of the Golden Horn, just below Topkapi Palace. This place is supposedly located near the defense wall separating ancient Byzantium and the city, on the shore a little distance from the ancient Prosphorion Harbor in the direction of the Bosphorus.

Contrary to this generally accepted idea, We assume that the chain extended from points P0 to P2 based on several points of evidence. The first is the fact that the façade of the foundation of Galata Castle (P0) directly faces the Bosphorus from where point P2 can be reached, rather than facing point P1. Since Galata Castle was constructed to guard the defense chain, it would be only natural for the Romans to design the best traction route for the chain from the viewpoint of mechanical rationality. The second piece of evidence to support our assumption is the presence of traces of where the chain was dragged on the defense wall located at P2. A large square opening, never seen on other walls, exists at about 8 m above ground. This masonry was not retrofitted but seems to have existed from when the wall was first constructed. This particular wall was built relatively rigidly compared with neighboring walls. Traces of the dragged chain extend from the opening as shown in Fig. 6.

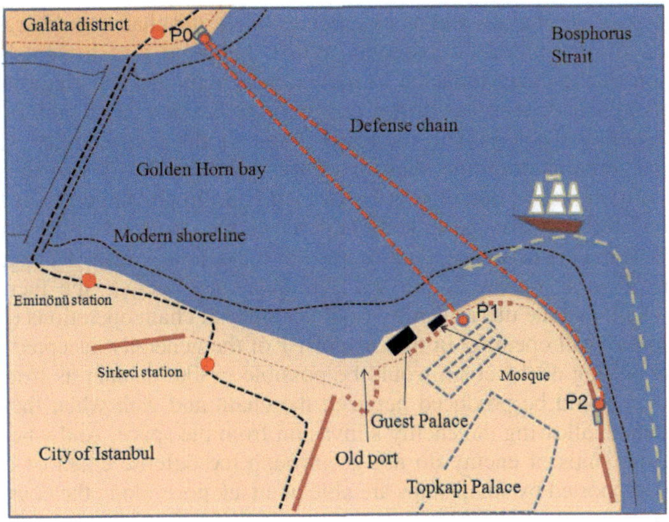

Fig. 5. Location of the Chain

Fig. 6. Wall Opening, Traces of Chain Traction, and a sub-Ditch (?)

These traces suggest that the chain ran along the curved defense walls below Topkapi Palace and then, at some point, left the walls and headed straight to Galata Castle on the opposite shore of the Golden Horn. There are several good points that support this supposition. The chain started from the opening on the wall and entered the sea gradually. It proceeded along the wall for about 300 m until it reached the Golden Horn (the coastline at that time was set back and was closer to the wall). From here, floats were used to set the chain afloat up to the point on the opposite shore where Galata Castle stood. This design offered a number of advantages when operating the chain to allow passage of allies and block enemies. First, the chain would scarcely leave the wall when it was pulled or loosened. When the chain was loosened, the float-coupled section of the chain retracted to reduce its length, and the chain-only section (without floats attached) that was drawn out onto the Golden Horn sank in the sea to allow ships to enter the bay. The retracting force of the float-coupled section was identical to the force to draw the chain from the wall opening. When the chain was pulled, the float-coupled section closed the Golden Horn to prevent entry of the enemy. One more good thing is that the linked chain along the curved defence walls is very effective to prevent destruction of a mechanism in the defense tower (P2) against direct attacks on the floating chains of the sea. Disadvantages include the increased total length of the chain and the unavailability of direct sight of chain operations (closing and opening) from the wall opening. In the case of P1 of the generally accepted theory, the loosening and pulling of the chain would be possible in like manner as from P2, but a large free space would be produced between the chain and wall when the chain was pulled or loosened, allowing the enemy's invasion from the space. And when the chain is laid at P1, the ships of enemy do not go through the defense chain of course. But ground troops supported by small ships are able to attack ports along the coast easily.

And the location which suggested as the terminal location of the defense chain at old paintings of Constantinople [8] has many point of similarity at the P2. Both have similar defense tower (A, A') which have comparatively big openings each, and there are suggested as a special watching tower (B, B') which have three narrow openings each on the left. And these towers are connected by a normal defense wall (C, C'). And the location of A' in the old paintings seems just the location P1. But we guess the painting was used as a navigation map of ships which approach a port of Constantinople because the position of directed sight is moving.

(a)

Fig. 7. Very similar constructions.
(a) a photograph of the location P2 and
(b) a part of "Panorama of Constantinople" in Benaki museum.
(c) Whole of the painting at the museum, coloured copper engraving, 0.32x3.83m.

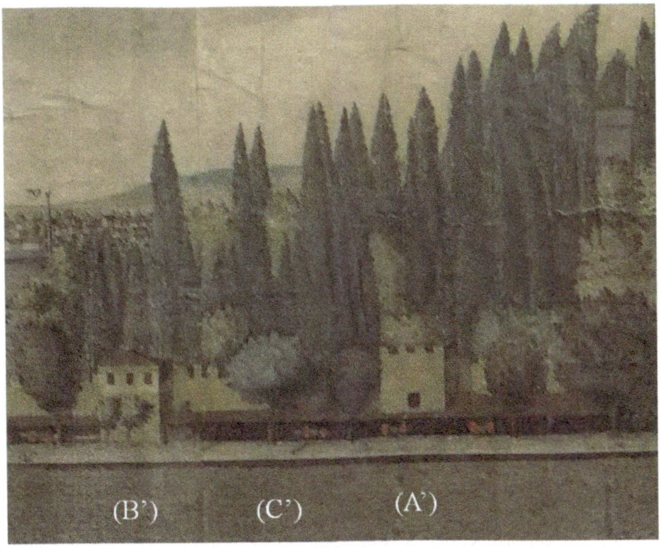

(b)

(c)

Fig. 7. *(continued)*

Therefore the location of A' is possible to consider just the location P2. Fig. 7

For these reasons, the author estimates that the defense chain extended from P0 to P2.

Experiments with a Small Model

The feasibility of operating the defense chain in this manner was experimented with a small model. The results of the experiment confirmed that when the chain is pulled out the float-coupled section retracted like a bellows and at the same time the chain-only section sunk deep into the water (Fig. 8) as foreseen.

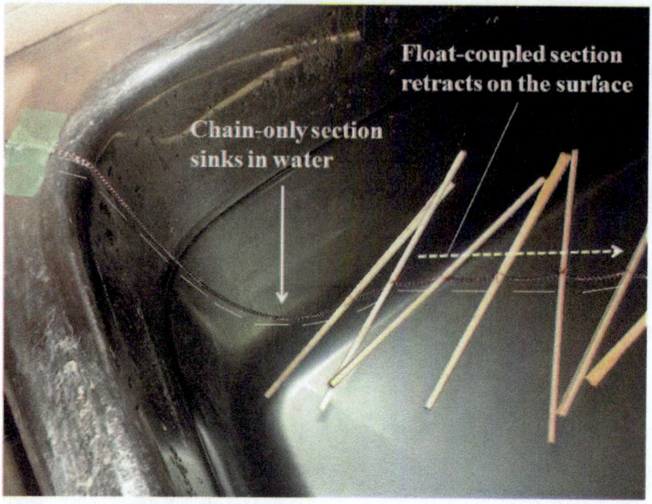

Fig. 8. Chain-only Section Sinks in Water and Float-coupled Section Remains Afloat

5 The Mystery of the Control Mechanism–Human-Powered or Mechanically Driven?

What was the mechanism for operating the chain depending on allowing allies in or keeping enemies out? It takes a longer time to tighten the chain to shut out enemies than to loosen it to admit allies. Assume that the chain was stretched between points P0 and P2, and the total length of the chain is about 1,300 m. Floats were assumed to be installed for a length of 1,000 m across the bay while the remaining part of the chain (300 m) ran along the walls without floats. Then the weight of the float-coupled section is nearly zero and the weight of the chain-only section is considered to occupy most of the weight of the entire chain.

The total weight W_i of a 1,000 m long chain submerged in the sea has been calculated to be about 34.7 t. Based on this, the weight W_i' of the 300 m long chain-only portion would be about 10.0 t ((300 m * 34.7 t)/1000 m). The estimated weight of the mechanism calculated from the experiment with the small model was, however, 7 t. In short, the floats are pulling to some extent.

We assume that the chain drive mechanism was housed inside the defense wall at point P2. -

Observing an aerial photograph of this point, we see something like a square-shaped defense tower, different from the neighboring ordinary defense walls. Complex geography can be seen inside the tower. There is also a small structure seeming to be a sub-ditch on the outer wall. Water could have been drained from above the upper layer of this structure through this ditch.

Simulation of the Expected Drive Mechanism

A waterwheel measuring about 2 m in diameter and 30 cm in width is assumed considering the internal size of the defense wall at point P2. Water required to run the wheel comes from the underground reservoir of Basilica Cistern (*Yerebatan Sarnici*), near the Hagia Sophia, via a duct. The elevation difference h is about 40 m. Now assume that water of 0.3 m * 0.3 m * 0.1 m, or 0.009 t by weight, per second, is constantly available. The potential energy E_h of this volume of water is calculated by $M * g *$ h to be 3528 (MKS), where the gravitational acceleration g is assumed to be 9.8 (M / s^2). The water runs down the duct, producing energy E for the wheel. Assuming the efficiency of the mechanism to be 5%, the effective energy E' is calculated by $E * 0.05$ to be approximately 176 (MKS).

Well, how does the object which has so fixed energy supplied continuously do any movement? It can make a calculation by using Newton's dynamics as follows.

Resistance by the seawater was ignored here. The reason is that the movement of the chain is very short (30 m).

A general answer. A speed $v(t)$ can be calculated as follows when energy E was given to the object of the mass m and all of that were changed into the movement energy.

Than
$$E = \frac{1}{2} m v^2, \quad v(t) = \sqrt{\frac{2E}{m}}. \tag{1}$$

The power supplied from the outside to the object of the mass m is put with A [W] = const. (fixed number).

The total of the energy supplied in the time t [s] is $t * A$ [J].

If all this energy becomes movement energy momentarily, a speed $v(t)$ is as the following.

$$v(t) = \sqrt{\frac{2A}{m}} \sqrt{t} = \sqrt{\frac{2A}{m}} * t^{1/2} \tag{2}$$

Then a distance $x(t)$ is found by the integral calculus.

$$x(t) = \frac{2}{3} \sqrt{\frac{2A}{m}} * t^{3/2} + C \tag{3}$$

At the initial condition, C is 0 when a movement distance $x(t)$ is 0.

Therefore,

$$x(t) = d = \sqrt{\frac{8A}{9m}} t^{3/2} \tag{4}$$

Therefore, the duration t(sec.) is as the following.

$$t = \sqrt[3]{\frac{9m}{8A}} * d^{2/3} \fallingdotseq 34.5 \text{ (sec.)} \tag{5}$$

$$(m = 7\text{t},\ A = 3528*0.05,\ d = 30 \text{ m})$$

If the above consideration is applied to an iron chain 7 tons, it is moved 30 m.

It can be calculated when it takes about 34 seconds.

It takes about 34 seconds to pull the chain for a distance of 30 m, meaning that the chain can be operated quickly enough by this mechanism without using human power during a war. The estimated waterwheel mechanism is described below (Fig. 9).

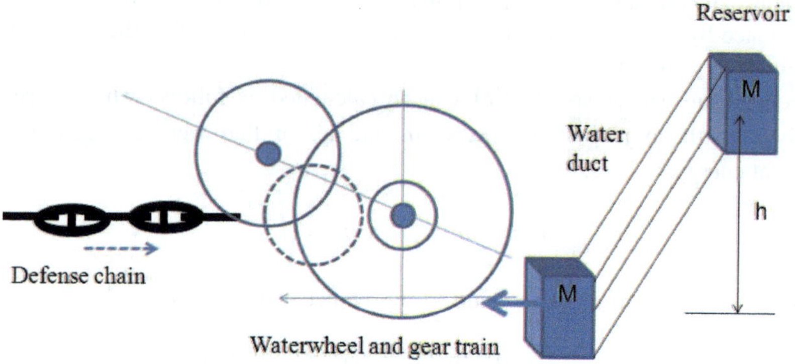

Fig. 9. Chain Traction Mechanism Using a Waterwheel

6 Conclusion

This paper discusses the mysteries of the defense chain of Constantinople; the number of chains, their locations, floats and traction, and the drive mechanism for operating the chains. The number of chains is seemingly two according to records, but this paper focuses on the one which existed without question and actually closed off the Golden Horn. The author estimates that this chain was connected between Galata Castle, at one end, and point P2 on the defense tower east of Topkapi Palace, facing the Bosphorus Strait, at the other. The chain was about 1,300 m long in total, of which about a 1,000 m long section was set afloat using floats and actually closed off the Golden Horn to entry by enemy ships. A waterwheel mechanism existed inside the defense wall at point P2 to haul the chain. Traction energy was supplied by water brought from an underground reservoir (Basilica Cistern) through a duct. This water was guided down to the waterwheel to drive the mechanism. It is shown in this paper that using this mechanism the chain could easily be opened and closed as required.

References

[1] Norwich, J.J.: A Short History of Byzantium. Penguin Books (1988, 1991, 1995)
[2] Herrin, J.: Byzantium: The Surprising Life of a Medieval Empire. Penguin Books (2007)
[3] Goldsworthy, A.: Roman Warfare. Phoenix Press (2007)
[4] Tate, G.: The Crusaders -Warriors of God. Harry N. Abrams (1996)
[5] Constantinople, Istanbul –Map of Istanbul 1422-1922, p. 35. Benizler Kitaberi (2010)
[6] Constantinople, Istanbul –Map of Istanbul 1422-1922, pp. 66–67. Benizler Kitaberi (2010)
[7] Constantinople, Istanbul –Map of Istanbul 1422-1922, p. 216. Benizler Kitaberi (2010)
[8] Panorama of Constantinople, Benaki museum in Athens, No. ΓΕ 25724 (18th c.), Greece
[9] History of Military Art and Science Literature During the Ottoman Period, Research Centre for Islamic History, Art and Culture, IRCICA (2004)
[10] Heiberg, L.J.: Naturwissenschaften, Mathematik und Medizin im klassishen Altertum. B.G. Teubner in Leipzig (1920)
[11] Lemerle, P.: Histoire de Byzance. Presses Universitaires de France (1998)
[12] Flusin, B.: La civilisation byzantine. Presses Universitaires de France (2006)

Structural Synthesis of the Ancient Chinese Zhuge Repeating Crossbow

Kuo-Hung Hsiao[1] and Hong-Sen Yan[2]

[1] Collections and Research Division, National Science and Technology Museum, Kaohsiung 80765, Taiwan
[2] Department of Mechanical Engineering, National Cheng Kung University, Tainan 70101, Taiwan

Abstract. This work synthesizes the structure of the Zhuge repeating crossbow, invented around 220 AD and depicted confusedly in ancient Chinese literature. Based on the analysis of the mechanism, the structural characteristics and design constraints of this device with uncertain members and joints are concluded. Then, according to the concepts of generalization and specialization subject to the concluded design constraints, eight feasible structures of mechanism that meet the technological standards of the subject's time period are systematically reconstructed.

1 Introduction

A crossbow is a bow mounted as a weapon on a stock with a trigger mechanism in it which holds the drawn bowstring. The crossbow was popular and widely used in ancient China before the Qin Dynasty (221-206 BC). The shooting process of the crossbow includes three steps: drawing the bowstring, placing the bolt and shooting the bolt. A repeating crossbow is a crossbow where the separate actions of three steps can be accomplished with a simple one-handed movement while keeping the crossbow stationary. The Records of the Three Kingdoms should be the first literature to mention the repeating crossbow named "Zhuge crossbow" and to describe the strategist Zhuge Liang (181-234 AD) of the Three Kingdoms Period (220-280 AD) as the inventor [1]. Figure 1 is a Zhuge crossbow in the book of Treatise on Armament in the Ming Dynasty (1368-1644 AD) [2]. The magazine contains a number of bolts on the top of the frame, and the device is worked by moving the input link forward and backward. One or several bolts can be shot at once in the oscillating motion of the input link. This allows a higher rate of fire than a normal crossbow. This repeating crossbow consists of the frame K_F (1), the bow K_{CB} (2), the bowstring K_T (3) and the input link K_I (4) at least. However, through the existing descriptions and drawings, the mobility of the magazine (the magazine is movable or fixed on the frame) is not certain. If the magazine is movable, it must draw the bowstring in the shooting process as the percussion link K_{PL} (5). If the magazine is fixed on the frame, the input link must draw the bowstring to shoot the bolt. Therefore, the number of members of Zhuge crossbow is uncertain as well as the type of some joints. It is hard to realize

T. Koetsier, M. Ceccarelli (Eds.): Explorations in the His. of Machines & Mech., HMMS 15, pp. 213–228.
springerlink.com

how the reciprocation of the input link transfers to draw the bowstring and shoot the bolt. This therefore causes a great difficulty to the readers for accessing and understanding the real technology from the original drawing.

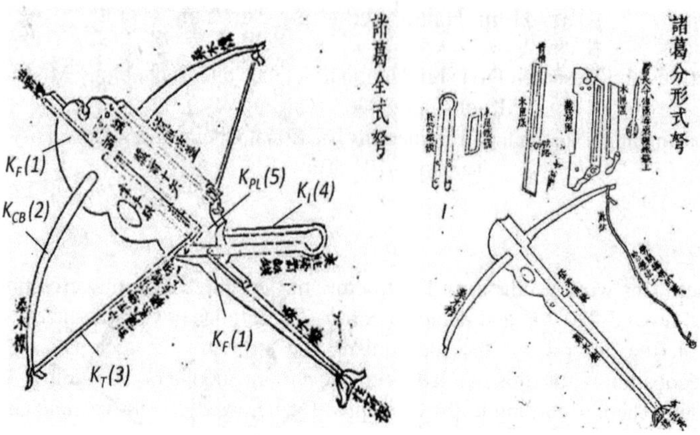

Fig. 1. A Zhuge crossbow [2]

The purpose of this work is to synthesize all feasible structures of Zhuge crossbow which described unclearly in ancient literatures. In what follows, the development of crossbow in ancient China is briefly introduced first. The representation of joints of ancient drawings is presented. Then, an approach for synthesizing the structures of mechanisms of ancient drawings is proposed. Finally, according to the mobility of the magazine, two types of Zhuge crossbow as examples are provided.

2 Development of the Crossbow

The ancient Chinese's crossbow with a trigger mechanism was developed in the later Spring Autumn Period (770-476 BC) and popular in the early Warring States Period (475-221 BC). The earliest one was found in the Tombs at Qufu, Shangdong Province and dated to the 6th century BC [3]. For two thousand years, it was the standard military issue of the ancient Chinese arsenal and changed warfare forever. The trigger mechanism was designed in such a way that it was able to store a large amount of energy within the bow when drawn, but was easily fired with little recoil when the input link was pulled. It also allowed the shooter to aim the target for precision shooting. Figure 2 shows a trigger mechanism which was excavated at Xian, Shanxi Province and belonged to the Western Han Dynasty (206 BC - 8 AD). The parts of the trigger mechanism were mass produced with precision and interchangeable. Sorted by the structure, the development of the crossbow can be divided into two stages in the Han Dynasty (206 BC - 220 AD). The trigger mechanism was made of bronze and consists of three working parts including the input link, the percussion link and the connecting link. The assembly was supported completely by the wooden stock of the crossbow in the early stage. In the Han Dynasty, the trigger mechanism had two improvements. One was the three working parts supported within a bronze casing sunk completely into a hollowed section of the wooden stock. Since the bronze casing allows a higher tension of shoot than the early type, the range of shoot was enlarged. The other was the trigger mechanism added the scale of the sight and the accuracy of shoot was raised. After the Han Dynasty, the structure of the crossbow and the trigger mechanism was similar and only increased the size.

Fig. 2. A trigger mechanism

With the accuracy improved, the next idea was the question of how to shoot more than one bolt. This led to the development of the repeating crossbow. The repeating crossbow accomplished the separate actions of drawing the bowstring, placing the bolt and shooting through the oscillation of the input link. As a result, it is possible to shoot at a faster rate compared to the normal crossbow. Based on the archaeological finds, the repeating crossbow first appeared around 400 BC in the tomb in Hubei Province [4]. The magazine contains twenty bolts on the top of the stock, and the device is worked by moving the input link forward and backward. Two bolts can be shot at once in the reciprocating motion of the input link. Since the tomb belonged to Chu State during the War States Period (722-481 BC), the device was named Chu State crossbow. However, there is no relevant historical documentation until the Three Kingdoms Periods. Since the range of Zhuge crossbow is limited, the head of bolt was often dipped in poison. This weapon became the standard crossbow used after the Song Dynasty (960-1219 AD). In the China-Japan war of 1894-1895, the repeating crossbow was still as common weapon among Qing Dynasty (1644-1911 AD) troops. The basic construction of this weapon has remained very much unchanged since its invention, making it one of the longest-lived mechanical weapons.

3 Representation of Joints

Through the study of ancient drawings, the joints can be classified into three types including joints that are easy to identify such as gear joints and wrapping joints, joints that are uncertain, and joints that are certain but are rare in modern kinematics of mechanisms. Since some special joints in ancient drawings can not be represented by modern schematic representations, a novel representation for such joints in ancient drawings is introduced here [5, 6].

The number of degrees of freedom is the number of independent parameters needed to specify the relative positions of the pairing elements of a joint. An unconstrained pairing element has six degrees of freedom including three translational and three rotational degrees of freedom. A joint is represented as:

$$J_{R_{xyz}}^{P_{xyz}}$$

in which the superscript P_{xyz} denotes that it can translate as a prismatic joint (J_P) in x, y and z axes; and the subscript R_{xyz} denotes that it can rotate as a revolute joint (J_R) in x, y and z axes. When a pairing element connects to another pairing element and forms a joint, a constraint is imposed and the motion of the original member is reduced by one or more degrees of freedom. Hence, a joint has a maximum of five degrees of freedom and a minimum of one degree of freedom. For example, joint J_{Rx} represents that a pairing element of a joint can rotate in x-axis with respect to the other pairing element. Joint J^{Pz} represents that a pairing element of a joint can translate in z-axis with respect to the other pairing element. Joint $J_{R_{yz}}^{P_x}$ represents that a pairing element of a joint not only translates in x-axis and also rotates in y and z axes with respect to the other pairing element.

Figure 3(a) shows a two-member mechanism with a joint from the book of Chinese Technology in the Seventeenth Century [7]. Since the drawing is not clear, link K_L is connected to frame K_F with an uncertain joint. Considering the type and the direction of motion of the link, the joint has two possible types. First, the link rotates in z-axis only, denoted as J_{Rz}. Second, the link rotates not only in z-axis but also translates in x-axis, denoted as J_{Rz}^{Px}. The x and y axes are defined as the horizontal and vertical directions in the drawing, respectively. The z-axis is determined by the right-hand rule.

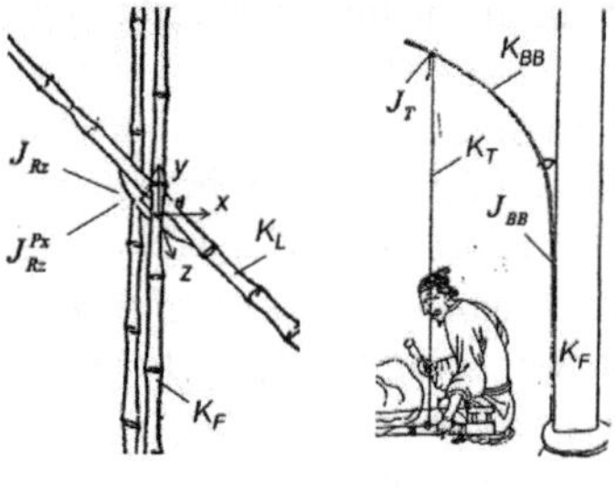

(a) (b)

Fig. 3. Special joints in ancient drawings [7]

Thread and bamboo often appear as members in ancient drawings especially in textile and agricultural devices. Figure 3(b) shows that bamboo K_{BB} is connected to frame K_F and thread K_T from the book of Chinese Technology in the Seventeenth Century [7]. The bamboo is firmly fixed to the frame and the thread ties the bamboo directly. The bamboo possesses elasticity and can go back to the original position after using. The motion of the bamboo can be regarded as a link connected to the frame with a revolute joint. The joint incident to the bamboo and the frame is defined as a bamboo joint, denoted as J_{BB}. For a bamboo joint in a planar mechanism, its characteristic is the same as a revolute joint (J_R). The number of degrees of freedom of the joint is 1. For a bamboo joint in a spatial mechanism, its characteristic is the same as a spherical joint (J_S). The number of degrees of freedom of the joint is 3. Using a thread tying a member to form a joint is rare in modern mechanisms but popular in ancient drawings. The joint incident to the thread and a member is defined as a thread joint, denoted as J_T. For a thread joint in a planar mechanism, its characteristic is the same as a revolute joint (J_R). The number of degrees of freedom of the joint is 1. For a thread joint in a spatial mechanism, its characteristic is the same as a spherical joint (J_S). The number of degrees of freedom of the joint is 3.

4 Design Approach

Figure 4 shows the process for synthesizing all possible structures of mechanisms in ancient drawings with uncertain members and joints [5, 6, 8, 9]. Each step of the process is explained as follows:

Step 1. Analysis of the structure of mechanism
By analyzing the structure of mechanism of a target device in the drawing, its structural characteristics are concluded including the possible number of members and the incidences among members and joints. Since most of indefinite parts in ancient drawings are the number of members and the structures of joints, the uncertain members and joints are identified and emphasized in this step.

Step 2. Generalized chains
The second step is to obtain or identify the atlas of generalized chains with the same numbers of members and joints subject to concluded structural characteristics based on the concept of number synthesis [10, 11]. Generalization is to transform the mechanism, that involves various types of members and joints, into a generalized chain with only generalized links and joints. A generalized link is a link with generalized joints; it can be a binary link, ternary link, quaternary link, ...etc. A generalized joint is a joint in general; it can be a revolute joint, prismatic joint, spherical joint, or others. When several members are connected by joints, they are said to form a chain. If every link in the chain is connected to at least two other links, the chain forms one or several closed loops and is called a closed chain. Graphically, a joint is symbolized by a small circle, and a member with i incident joints is symbolized by an i-side shaded polygon with small circles as vertices. Some of the important atlases of generalized chains have already been built in references [10, 11].

Step 3. Specialized chains
Specialization is to assign specific types of members and joints in the available atlas of generalized chains, subject to required design constraints to have atlas of

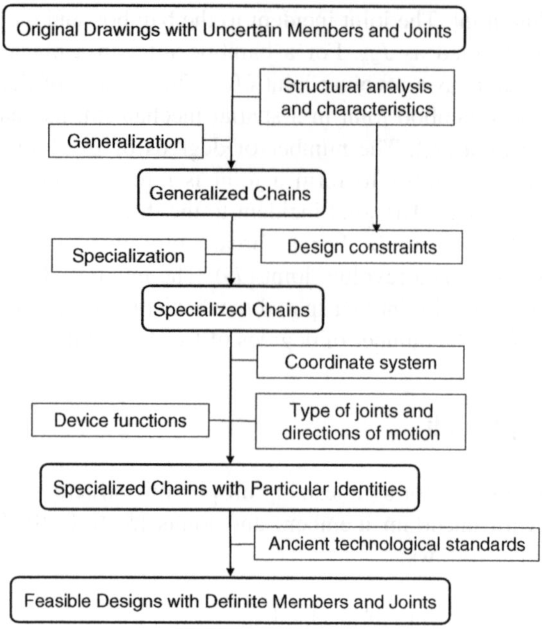

Fig. 4. Procedure for structural synthesis of drawings with uncertain members and joints

specialized chains [10, 11]. Design constraints are defined based on the concluded structural characteristics. A generalized chain after specialization is a specialized chain. Based on the process of specialization, all possible specialized chains can be identified after assigning the types of members and joints subject to the concluded design constraints for each generalized chain obtained in step 2.

Step 4. Specialized chains with particular identities
In order to represent the structure of mechanism, a right-hand rectangular Cartesian coordinate system is defined to describe each motion axis of the joints in the drawing. By studying the target device in the drawing, the function of the device is determined. In view of the device function, these uncertain joints can be represented by different types of joints to achieve the equivalent function. Considering the types and the motion directions of the uncertain joints, all possible types of the uncertain joints are generated. By assigning the possible types of the uncertain joints to the specialized chains obtained in step 3, specialized chains with particular identities are obtained.

Step 5. Feasible designs with definite members and joints
According to the motion and function requirements of the device in the drawing, all feasible designs that meet the technological standards of the subject's time period can be obtained from the specialized chains with particular identities.

5 Examples

Based on the mobility of the magazine, there are two types of Zhuge crossbows which are presented as examples to illustrate the proposed approach shown in Fig. 4.

Example 1. Zhuge crossbow with movable magazine

Through the reciprocation of the input link, Zhuge crossbow accomplishes the shooting process. If the magazine is movable, the device comprises five members including a frame, a bow, a bowstring, an input link and a percussion link with the magazine. The function of the percussion link with the magazine is to draw and release the bowstring to shoot the bolt. Since the motion of the bow and the bowstring is spatial, Zhuge crossbow with movable magazine is a spatial mechanism. From the descriptions and drawings in the existing literatures, the types of some joints are uncertain.

All feasible structures of Zhuge crossbow with movable magazine are synthesized through the following steps:

Step 1. The structural characteristics of the device are:

1. It is a cam mechanism with five-member (members 1-5).
2. The frame is a multiple link.
3. The bow (K_{CB}) is a binary link and connected to the frame (K_F) with a bamboo joint (J_{BB}).
4. The bowstring (K_T) is a binary link and connected to the bow (K_{CB}) and the percussion link with the magazine (K_{PL}) with thread joints (J_T), respectively.
5. The input link (K_I) is connected to the frame (K_F) with an uncertain joint (J_1).
6. The function of the percussion link with the magazine is to draw and release the bowstring through the oscillating motion of the input link. The relative motion between the frame and the percussion link is the combination of rolling and sliding, like a cam joint. Therefore, the percussion link with the magazine (K_{PL}) is a multiple link and connected to the frame (K_F) and the input link (K_I) with a cam joint (J_A) and an uncertain joint (J_2), respectively.

Step 2. Since this is a device with five-member, the corresponding generalized chains with five members are shown in Fig. 5, in which N is the number of links and J is the number of joints.

Step 3. There must be a pair of binary links as a bow and a bowstring. The pair of binary links must be connected to each other and to a multiple link as the frame. Therefore, only the generalized chain shown in Fig. 5(c) is qualified for the process of specialization. Now, all possible specialized chains are identified as follows:

Frame - link K_F

Since there must be a link as the frame K_F and a pair of binary links is connected to the frame, for the generalized chain shown in Fig. 5(c), the assignment of the frame generates one non-isomorphic result, as shown in Fig. 6(a). Two chains or mechanisms are said to be isomorphic if they have the same topological structures.

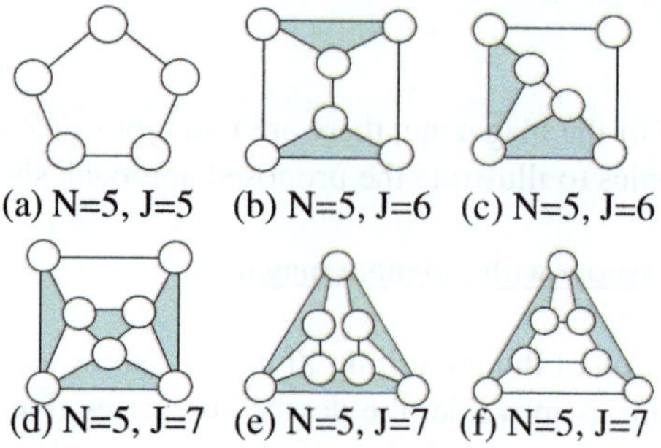

Fig. 5. Atlas of some generalized chains with five members

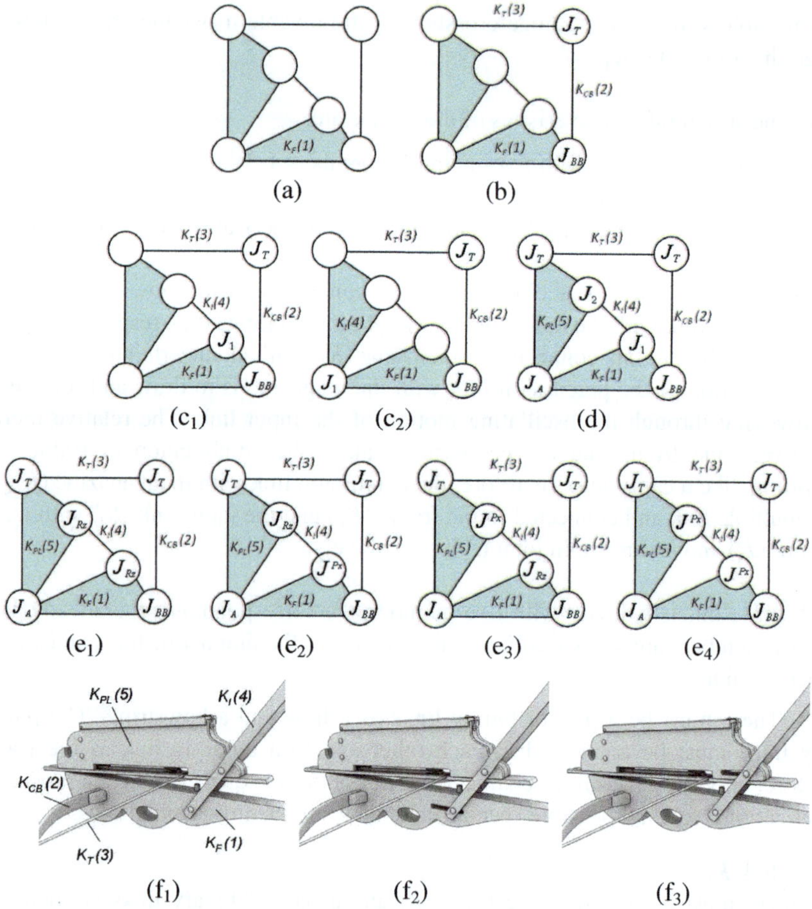

Fig. 6. Structural synthesis of Zhuge crossbow with movable magazine

Bow and bowstring - link K_{CB} and link K_T
Since there must be a pair of binary links as the bow (K_{CB}) and bowstring (K_T), and the bow is connected to the frame and the bowstring with a bamboo joint J_{BB} and a thread joint J_T, for the case shown in Fig. 6(a), the assignment of the bow and the bowstring generates one result, as shown in Fig. 6(b).
Input link - link K_I

Since there must be an input link (K_I) that is connected to the frame (K_F) with an uncertain joint (J_1), for the case shown in Fig. 6(b), the assignment of the input link generates two results, as shown in Figs. 6(c_1) and (c_2).

Percussion link with the magazine - link K_{PL}
Since the percussion link with the magazine (K_{PL}) must be a multiple link and connected to the frame (K_F) and input link (K_I) with a cam joint (J_A) and an uncertain joint (J_2), the percussion link with the magazine can be identified as follows:

1. For the case shown in Fig. 6(c_1), the assignment of the percussion link with the magazine (K_{PL}) generates one result, Fig. 6(d).
2. For the case shown in Fig. 6(c_2), since there is no the multiple link as the percussion link with the magazine (K_{PL}) that is connected to the frame, Fig. 6(c_2) is not qualified for the process of specialization.

Therefore, one specialized chain with identified frame K_F, the bow (K_{CB}), the bowstring (K_T), input link (K_I) and percussion link with the magazine (K_{PL}) is available as shown in Fig. 6(d).

Step 4. The coordinate system is defined as shown in Fig. 1. The function of the percussion link with the magazine is to draw and release the bowstring to shoot the bolt. The uncertain joints may have multiple types to achieve the equivalent function. Considering the uncertain joints J_1 and J_2, each of them has two possible types: the first one rotates in z-axis, denoted as J_{Rz}; the second translates in x-axis, denoted as J^{Px}. By assigning the possible types of the uncertain joints J_1 and J_2 (J_{Rz} and J^{Px}), to the specialized chains shown in Fig. 6(d), four specialized chains with particular identities as shown in Figs. 6(e_1)-(e_4) are obtained.

Step 5. Since a mechanism with double sliders has the drawback of the transmission and is rare in ancient China, it does not meet the technological feasibility of the subjects' time period. By removing the one shown in Fig. 6(e_4), three feasible specialized chains with particular identities are synthesized as shown in Figs. 6(e_1)-(e_3). Figures 6(f_1)-(f_3) show the corresponding 3D solid models of the feasible designs. Fig. 6(f_1) qualifies best to match the original drawing.

The number of degrees of freedom, F_P, of a planar mechanism with N_L members and N_{Ji} joints of type i is:

$$F_P = 3(N_L - 1) - \Sigma N_{Ji} C_{pi} \tag{1}$$

where C_{Pi} is the number of degrees of constraint of i-type joint.

The number of degrees of freedom, F_S, of a spatial mechanism with N_L members and N_{Ji} joints of type i is:

$$F_S = 6(N_L - 1) - \Sigma N_{Ji} C_{si} \quad (2)$$

where C_{Si} is the number of degrees of constraint of *i*-type joint.

For Fig. 6(f), the spatial five-member device, N_L=5, $N_{JRz} + N_{JPx}$=2, $C_{sJRz}=C_{sJPx}$=5, N_{JT}=2, C_{sJT}=3, N_{JBB}=1, C_{sJBB}=3, N_{JA}=1 and C_{sJA}=4. Based on Equation (2), the number of degrees of freedom, F_S, is :

$$\begin{aligned} F_S &= 6(N_L\text{-}1) - [(N_{JRz} + N_{JPx})\,5 + N_{JT}\,C_{sJT} + N_{JBB}\,C_{sJBB} + N_{JA}\,C_{sJA}] \\ &= 6(5 - 1) - [(2)(5) + (2)(3) + (1)(3) + (1)(4)] \\ &= 1 \end{aligned}$$

Therefore, the motion of the devices is constrained.

Example 2 Zhuge crossbow with fixed magazine

In the operating process, one hand holds the frame of Zhuge crossbow, the other hand pulls and pushs the input link to complete the shooting. If the magazine is fixed on the frame, the stability and the accuracy of the shooting can be improved. Since the motion of the bow and the bowstring is planar, Zhuge crossbow with fixed magazine is a planar mechanism. In this case, Zhuge crossbow with fixed magazine comprises four members including a frame, a bow, a bowstring and an input link at least. Since adding a percussion link to draw the bowstring is efficient and familiar, it may comprise five members. Therefore, for the magazine which is fixed on the frame, the number of members is four or five and the types of some joints are uncertain.

All feasible structures of Zhuge crossbow with fixed magazine are synthesized through the following steps:

Step 1. The structural characteristics of the device that are different from Zhuge crossbow with movable magazine described above are:

1. It is a mechanism with four-member (members 1-4) or five-member (members 1-5).
2. The bowstring (K_T) is a binary link and connected to the bow (K_{CB}) and the input link (K_I) or the percussion link (K_{PL}) with thread joints (J_T), respectively.
3. The input link (K_I) is connected to the frame (K_F) with an uncertain joint.
4. The percussion link (K_{PL}) is a multiple link and connected to the frame (K_F) and the input link (K_I) with uncertain joints, respectively.

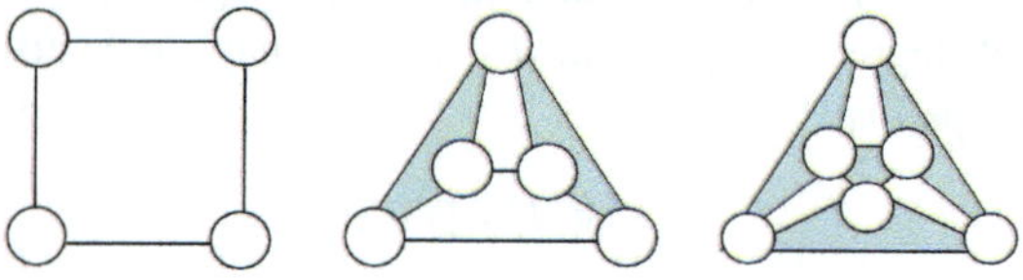

(a) N=4, J=4 (b) N=4, J=5 (c) N=4, J=6

Fig. 7. Atlas of some generalized chains with four members

Step 2. Since this is a device with four-member or five-member, the corresponding generalized chains with five and four members are shown in Figs. 5 and 7, respectively.

Step 3. For a device with four-member, there must be a pair of binary links as a bow and a bowstring and be connected to each other. For a device with five-member, the pair of binary links must be connected to each other and to a multiple link as the frame. Therefore, only two generalized chains shown in Figs. 5(c) and 7(a) are qualified for the process of specialization. All possible specialized chains are identified as follows:

Frame - link K_F

Since there must be a link as the frame K_F and a pair of binary links is connected to the frame, the frame can be identified as follows:

1. For the generalized chain shown in Fig. 7(a), the assignment of the frame (K_F) generates one non-isomorphic result, Fig. 8(a_1).
2. For the generalized chain shown in Fig. 5(c), the assignment of the frame (K_F) generates one non-isomorphic result, Fig. 8(a_2).

Therefore, two specialized chains with identified frame K_F are available as shown in Figs. 8(a_1)-(a_2).

Bow and bowstring - link K_{CB} and link K_T

Since there must be a pair of binary links as the bow (K_{CB}) and the bowstring (K_T), and the bow is connected to the frame and the bowstring with a bamboo joint J_{BB} and a thread joint J_T, the bow and the bowstring can be identified as follows:

1. For the case shown in Fig. 8(a_1), the assignment of the bow (K_{CB}) and the bowstring (K_T) generates one non-isomorphic result, Fig. 8(b_1).
2. For the case shown in Fig. 8(a_2), the assignment of the bow (K_{CB}) and the bowstring (K_T) generates one result, Fig. 8(b_2).

Therefore, two specialized chains with identified frame K_F, the bow (K_{CB}) and the bowstring (K_T) are available as shown in Figs. 8(b_1)-(b_2).

Input link - link K_I

Since there must be an input (K_I) that is connected to the frame (K_F) with an uncertain joint, the input link can be identified as follows:

1. For the case shown in Fig. 8(b_1), the assignment of the input (K_I) and the uncertain joint (J_3) generates one result, Fig. 8(c_1).
2. For the case shown in Fig. 8(b_2), the assignment of the input (K_I) and the uncertain joint (J_4) generates two results, Figs. 8(c_2) and (c_3).

Therefore, three specialized chains with identified frame K_F, the bow (K_{CB}), the bowstring (K_T) and input link (K_I) are available as shown in Figs. 8(c_1)-(c_3).

Percussion link - link K_{PL}

Since the percussion link (K_{PL}) must be a multiple link and connected to the frame (K_F) and input link (K_I) with uncertain joints (J_5 , J_6), the percussion link can be identified as follows:

1. For the generalized chain shown in Fig. 8(c_2), the assignment of the percussion (K_{PL}) generates one result, Fig. 8(d).
2. For the case shown in Fig. 8(c_3), since there is no the multiple link as the percussion link (K_{PL}) that is connected to the frame, Fig. 8(c_3) is not qualified for the process of specialization.

Therefore, one specialized chain with identified frame K_F, the bow (K_{CB}), the bowstring (K_T), input link (K_I) and percussion link (K_{PL}) are available as shown in Fig. 8(d).

Step 4. The coordinate system is defined as shown in Fig. 1. The function of the input link (in four members) and the percussion link (in five members) is to draw and release the bowstring to shoot the bolt. The uncertain joints may have multiple types to achieve the equivalent function. Considering the uncertain joint J_3, it has two possible types: the first one translates in x-axis, denoted as J^{Px}; the second rotates not only in z-axis but also translates in x-axis, denoted as J^{Px}_{Rz}. Considering the uncertain joints J_4 and J_6, each of them has two possible types: the first one rotates in z-axis, denoted as J_{Rz}; the second translates in x-axis, denoted as J^{Px}. Considering the uncertain joints J_5, it has one possible type: the percussion link K_{PL} rotates not only in z-axis but also translates in x-axis with respect to the frame K_F, denoted as J^{Px}_{Rz}. By assigning the possible types of the uncertain joints J_3(J^{Px} and J^{Px}_{Rz}), J_4(J_{Rz} and J^{Px}), J_5(J^{Px}_{Rz}) and J_6(J_{Rz} and J^{Px}), to the specialized chains shown in Figs. 8(c_1) and (d), six specialized chains with particular identities as shown in Figs. 8(e_1)-(e_6) are obtained.

Step 5. By removing the one shown in Fig. 8(e_6) with double sliders, five feasible specialized chains with particular identities are synthesized as shown in Figs. 8(e_1)-(e_5). Figures 8(f_1)-(f_5) show the corresponding 3D solid models of the feasible designs.

For Fig. 8(f_1), a planar four-member device, N_L=4, N_{JPx}=1, C_{pJPx}=2, N_{JT}=2, C_{pJT}=2, N_{JBB}=1 and C_{pJBB}=2. Based on Equation (1), the number of degrees of freedom, F_p, is :

$$
\begin{aligned}
F_p &= 3(N_L-1) - [N_{JPx}\,C_{pJPx} + N_{JT}\,C_{pJT} + N_{JBB}\,C_{pJBB}] \\
&= 3(4 - 1) - [(1)(2) + (2)(2) + (1)(2)] \\
&= 1
\end{aligned}
$$

Therefore, the motion of the device is constrained.

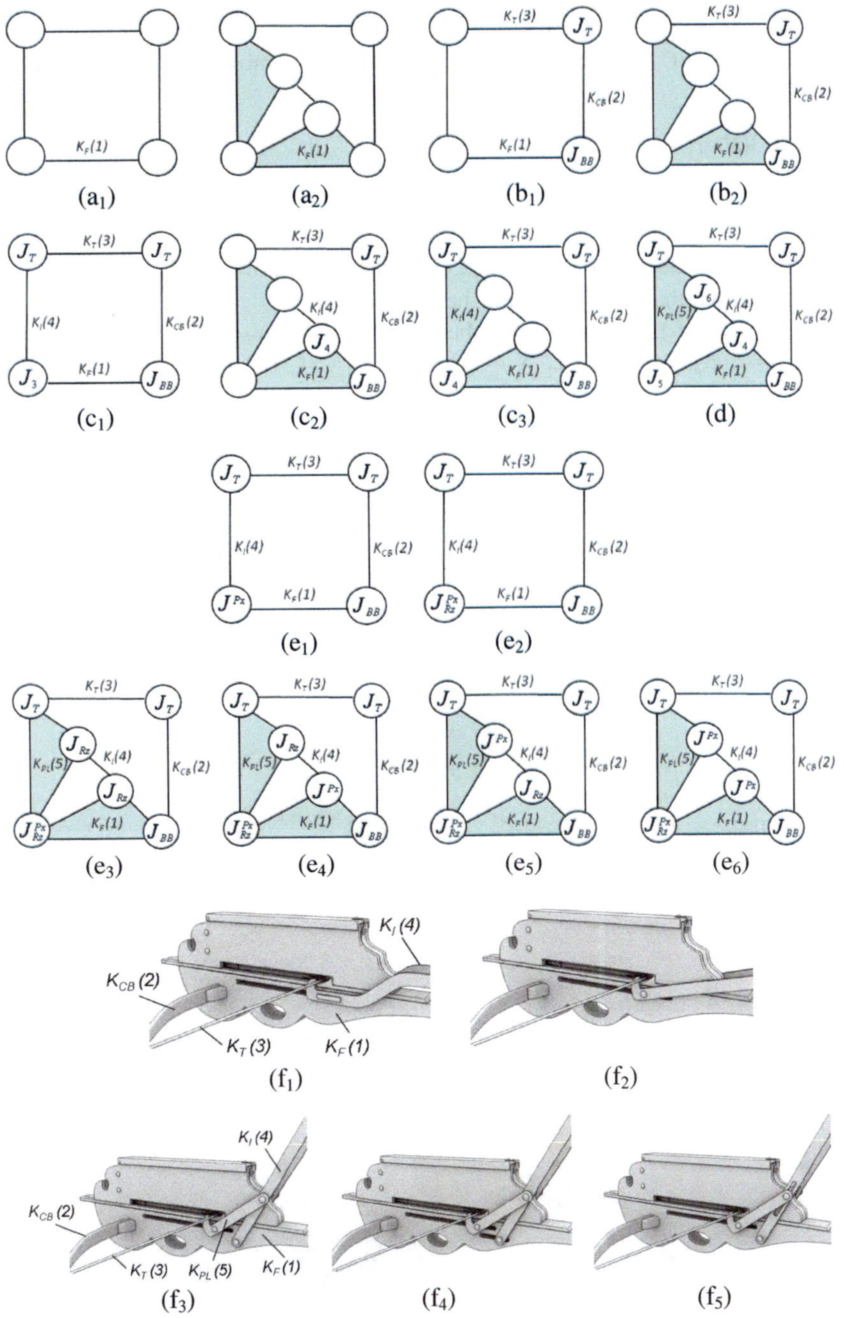

Fig. 8. Structural synthesis of Zhuge crossbow with fixed magazine

For Fig. 8(f_2), the planar four-member device, N_L=4, $N\,J_{Rz}^{Px}$=1, $C_p\,J_{Rz}^{Px}$=1, N_{JT}=2, C_{pJT}=2, N_{JBB}=1 and C_{pJBB}=2. Based on Equation (1), the number of degrees of freedom, F_p, is :

$$\begin{aligned} F_p &= 3(N_L-1) - [N\,J_{Rz}^{Px} C_p\,J_{Rz}^{Px} + N_{JT}\,C_{pJT} + N_{JBB}\,C_{pJBB}] \\ &= 3(4 - 1) - [(1)(1) + (2)(2) + (1)(2)] \\ &= 2 \end{aligned}$$

The motion of the device is constrained since the motion of input link 4 has an extra degree of freedom that does not affect the shooting.

For Figs. 8(f_3)-(f_5), planar five-member devices, N_L=5, $N_{JRz}+N_{JPx}$=2, C_{pJRz}=C_{pJPx}=2, N_{JT}=2, C_{pJT}=2, N_{JBB}=1, C_{pJBB}=2, $N\,J_{Rz}^{Px}$=1 and $C_p\,J_{Rz}^{Px}$=1. Based on Equation (1), the number of degrees of freedom, F_p, is :

$$\begin{aligned} F_p &= 3(N_L-1) - [(N_{JRz} + N_{JPx})\,2 + N_{JT}\,C_{pJT} + N_{JBB}\,C_{pJBB} + N\,J_{Rz}^{Px} C_p\,J_{Rz}^{Px}] \\ &= 3(5 - 1) - [(2)(2) + (2)(2) + (1)(2) + (1)(1)] \\ &= 1 \end{aligned}$$

Therefore, the motion of each device is constrained.

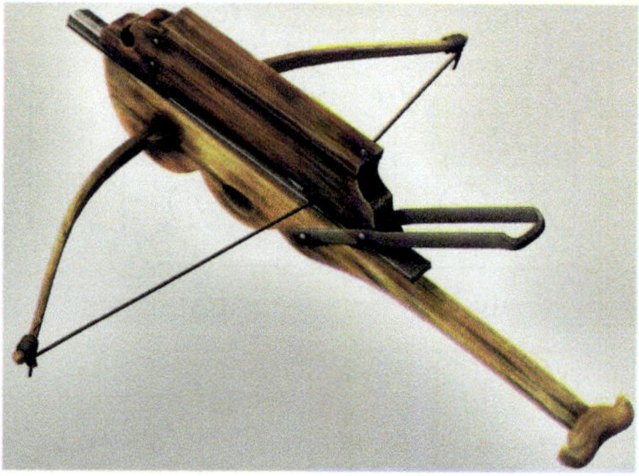

Fig. 9. Imitation of the original drawing of Zhuge repeating crossbow

Fig. 10. A prototype of reconstructed Zhuge crossbow

Figure 9 is the imitation of the original drawing to illustrate the incidences among members and joints of Zhuge crossbow. Fig. 10 is a prototype of reconstructed Zhuge crossbow based on Fig. 6(f_1).

6 Conclusions

Zhuge crossbow was one of the typical weapons in ancient China and described in many literatures. However, from the view point of modern mechanism and machine science, the structures of mechanisms in some ancient drawings are not clear, especially the number of members and the structure of joints. Through the study of available mechanisms and applying the concepts of generalization and specialization of mechanisms subject to concluded design constraints in the analysis process, all feasible specialized chains are generated. By assigning the possible joints to the feasible specialized chains, all structures of Zhuge crossbow are systematically synthesized. According to the mobility of the magazine, two examples are provided to illustrate the proposed approach. Via such an approach, the unclear drawings shown in ancient literatures can be reasonably reconstructed.

Acknowledgement. The authors are grateful to the National Science Council (Taipei, Taiwan) under Grant NSC 97-2221-E-006-070-MY3 for the financial support of this work.

References

[1] Chen, S.: History of the Three Kingdoms. Taiwan Commercial Press, Taipei (1968) (in Chinese)
[2] Mao, T.Y.: Treatise on Armament. Hainan Press, Hainan (2001) (in Chinese)
[3] Zhang, C.H., You, Z.H., Wu, Z.Z., Liu, Y.L.: History of Chinese Mechanical Engineering Invention. Tsinghua University Press, Beijing (2004) (in Chinese)
[4] Museum, J. (ed.): Important Archaeological Discoveries. Cultural Relics Press, Beijing (2009) (in Chinese)
[5] Hsiao, K.H., Yan, H.S.: Structural Identification of the Uncertain Joints in the Drawings of Tain Gong Kai Wu. Journal of the Chinese Society of Mechanical Engineers 31(5), 383–392 (2010)
[6] Hsiao, K.H., Chen, Y.H., Tsai, P.Y., Yan, H.S.: Structural Synthesis of Ancient Chinese Foot-operated Slanting Loom. Proceedings of the Institution of Mechanical Engineers, Part C, Journal of Mechanical Engineering Science 225, 2685–2699 (2011)
[7] Song, Y.X.: Chinese Technology in the Seventeen Century. Dover Publications, New York (1966) (in Chinese); trans. Sun E. Z. and Sun S. C.
[8] Yan, H.S., Hsiao, K.H.: Structural Synthesis of the Uncertain Joints in the Drawings of Tain Gong Kai Wu. Journal of Advanced Mechanical Design, Systems, and Manufacturing - Japan Society Mechanical Engineering 4(4), 773–784 (2010)

[9] Hsiao, K.H., Chen, Y.H., Yan, H.S.: Structural Synthesis of Ancient Chinese Foot-operated Silk-reeling Mechanism. Frontiers of Mechanical Engineering in China 5(3), 279–288 (2010)
[10] Yan, H.S.: Creative Design of Mechanical Devices. Springer, Singapore (1998)
[11] Yan, H.S.: Reconstruction Designs of Lost Ancient Chinese Machinery. Springer, Netherlands (2007)

Clocks and Dials with Automata: The Mosaic of Qasr El-Lebya

Helen Fragaki

UMR 7041, Archéologies et Sciences de l'Antiquité / Maison René Ginouvès –Nanterre / CNRS, Université de Paris I, Université de Paris Ouest Nanterre la Défense, Ministère de la culture

Abstract. The figure of Helios surmounting the Pharos of Alexandria on the mosaic of Qasr el-Lebya (539 A.D.) points its sword towards a semicircular object which has been convincingly recognized as a sundial. On the basis of the study of Alexandrian automata, it can be argued that this figure is a moving statue. In fact, clocks and dials, often associated with animated statues, were part of the ingenious inventions of Alexandrian engineers since the Ptolemaic period. One of the most famous ones was the water clock of Ktesibios which included automata indicating time by the same arm gesture as that of the Helios on the pavement of Qasr el-Lebya. During Late Antiquity, clocks connected to automatically moving figures seem to have been particularly popular. The clock of Gaza, described by Procopius (473-535 A.D.), included a statue of Helios, very similar to that of the mosaic, moving every hour and pointing its index finger to successively opening doors. Despite the absence of the personified Sun in Arabic time measuring devices, there is evidence of automata pointing to indicate time. Descriptions of moving statues on the Pharos by Arab authors may therefore refer to figures like the one depicted in this mosaic.

1 Introduction

Discovered by R. G. Goodchild in 1957, this mosaic decorates the floor of the basilica of Qasr el-Lebya, located 50 miles to the west of Cyrene[1]. Situated in the eastern part of the central aisle of the basilica, the pavement dates back to 539 A.D. and is composed of 50 square panels of 1.065 x 1.22 m. One of these panels depicts a lighthouse, securely identified by its form and by the inscription "Ο ΦΑΡΟϹ" referring to it as the Pharos of Alexandria (fig. 1). Indeed, the two-storeyed crenellated tower entered by an oblique staircase leading to an arch entrance is similar to the representations of the Alexandrian lighthouse on coins.

The upright male figure surmounting the building wears a sunray diadem which makes it possible to identify it as a statue of Helios, personifying the sun[2]. This figure is holding a sword in its right hand, pointing to a semi-circular object which was initially thought to be a mirror used as a telescope to detect boats sailing towards the harbour, but also as a weapon for setting fire to enemy fleets. Indeed, some Medieval Arab authors (Mas'ûdî, 915-56, Dimashqî, 1256-1327, Maqrîzî, 1364-1442 and Suyûtî, 1445-1505) mention a large glass or iron mirror still on the Pharos in their time. It is said to have been destroyed by the Byzantines who probably did not want

T. Koetsier, M. Ceccarelli (Eds.): Explorations in the His. of Machines & Mech., HMMS 15, pp. 229–250.
springerlink.com

their secret weapon to fall into the hands of their enemies. These accounts would furthermore imply that the position of the sword on the top of the so-called mirror had a specific meaning. A mechanical relationship between these two objects has therefore been suggested[3].

Fig. 1. Detail of the mosaic of Qasr el-Lebya, 539 A.D. Panel depicting the Pharos of Alexandria, 1.065 x 1.22 m. Photograph by Rupert Breitwieser / University of Salzburg.

However, according to a more recent interpretation, this semicircular object would sooner be a sundial of the *hemisphaerium* type. Sundials of this type determined time by the sunrays entering their concave interior through a perforation at a specific point on their upper convex surface. The sunrays acted as an hour hand indicating time by illuminating different parts of the interior surface of the dial according to the position of the sun. This most convincing interpretation led to the assumption that the sword of the statue of Helios symbolized the sunrays entering the cavity of the dial[4].

The figure of Helios has been included in the discussion about the various statues surmounting the Pharos in successive periods. It has been argued that a statue of the Graeco-Roman period was replaced during the Early Christian Age by a personification of the sun, symbolizing the light of faith and salvation[5]. Although the sun has certainly a Christian meaning within the iconographical context of the mosaic, this hypothesis has given rise to the objection that Helios is essentially a pagan figure. If such a statue really existed on the Pharos at the same time as on the pavement, it should have survived the edict of Theodosius, which strictly prohibited all pagan cults. The presence of the pagan figure on a public building as important as the Pharos during these times was thought to be improbable[6]. It could therefore be either a symbolic figure invented by the mosaists, or an older statue, maybe dating from the Greaco-Roman period.

2 The Construction of Alexandrian Dials and Automata

Regardless of how the statue is interpreted, the presence of a dial on the Pharos does not seem surprising, given the popularity of such devices in Alexandria. In fact, clocks and dials were part of the ingenious inventions of Alexandrian engineers who had been using the same methods to construct automata since the Ptolemaic period[7]. The most ancient water clock in this city is attributed to Ktesibios, who lived during the reign of Ptolemy II Philadelphus in the 3rd c. B.C. The works of this scholar are only indirectly known, through references in other ancient texts, especially those of Vitruvius. Ktesibios was interested in the properties of air pressure and managed to use the results of his research to construct different kinds of machines and weapons as well as musical instruments, jets, automata, amusing toys and water clocks[8].

Philon of Byzantium, who spent his life in Alexandria and Rhodes was at least one generation younger. He was the author of nine volumes on manufacturing techniques for various instruments and devices, some of which were war machines. The fifth volume, entitled *Pneumatica* and handed down through an Arab translation, seems to have been partly inspired by Ktesibios' works, to which frequent reference is made. On the basis of a study of compressed air and its effect on liquids, Philon's book describes a series of instruments and devices, many of which are automata and amusing objects that produce illusions[9].

Heron of Alexandria, who was active in the middle of the 1st c. A.D., was the latest representative of this engineering tradition[10]. He wrote theoretical as well as technical treatises[11], one of which dealt with water clocks[12]. Unfortunately, this work has been lost. One of his other works, entitled *Pneumatica*, has been handed down in about 100 manuscripts which mostly date back to the Renaissance and include drawings which accompany the descriptions of the machines. It assembles various devices which use air, steam and water pressure and produce movement or reactions through the use of siphons. In this book, which includes many animated statues, Heron goes back to Ktesibios' and Philon's works to which he adds his own inventions[13]. As the descriptions and drawings are clear and precise, the function of the devices is almost always obvious, and most of the automata can be easily reconstructed according to the author's instructions[14].

3 Ktesibios' Water Clock

The earliest Alexandrian clock, attributed to Ktesibios, has some resemblances with the one depicted on the mosaic of Qasr el-Lebya, and may have functioned in a similar way, although it does not include a sundial. It is known through a passage in Vitruvius, who describes a mechanism allowing the regular flow of water through an orifice, as a result of which a float is raised[15]. Rods set in motion by this float had serrated sides which engaged gears of cogwheels. The vertical movement of the rods also caused the rotation of the cogwheels which produced *"various kinds of movement; therein figures are moved, pillars are turned, stones or eggs are let fall* [16], *trumpets sound and other side-shows.*

Among these contrivances also, the hours are marked on a column or pilaster ; and these are indicated by a figure rising from the lowest part and using a pointer throughout the day"[17].

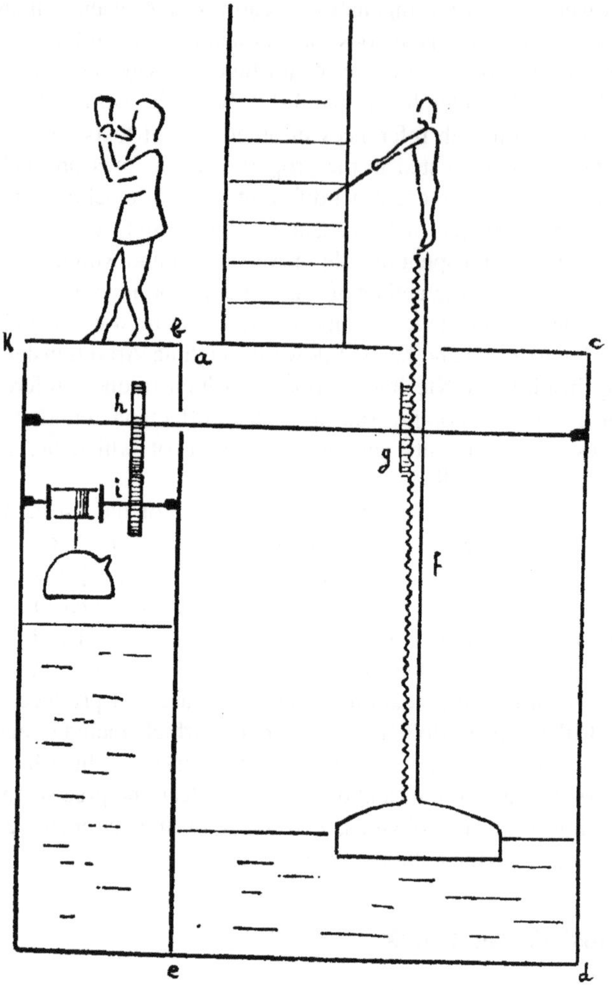

Fig. 2. Reconstruction of Ktesibios' water clock, in Meerwaldt, J. D. (1921), *Mnemosyne*, series 9, vol. 49, fig. 4. **f :** float and vertically moving rod with serrated sides lifting the figurine which indicates time, **h, i, g :** gears of cogwheels set in rotation by the rod **f** thanks to a horizontal shaft transmitting the rotation of the cogwheel **g** in contact with the float, **i :** cogwheel setting a horizontal drum in rotation. The *pnigeus* is suspended by a rope coiling itself round this drum, **h :** cogwheel with a smooth part acting as a sort of release as soon as it enters into contact with the gear **i** supporting the *pnigeus*.

Among the suggested reconstructions of the devices that produce the sound of trumpets, the most convincing one involves a *pnigeus* which is also used in water organs (fig. 2). It is actually a sort of bell with a slot through which air escapes and produces a whistling sound when the object is immersed in water. Suspended by a rope above the water contained in a vessel, the *pnigeus* gradually rises while the rope coils itself round a horizontal drum which is rotated by a cogwheel by means of a gear. This cogwheel is set in motion by the float which lifts the figurine indicating time thanks to a horizontal shaft transmitting the rotation of the cogwheel in contact with the float. The cogwheel corresponding to the *pnigeus* has a smooth part acting as a sort of release as soon as it enters into contact with the gear supporting the *pnigeus*. As the *pnigeus* is no longer supported by the engagement of gears, it falls into its container and produces a whistling sound, which sounds as if it were produced by the statuette of the trumpeter surmounting the device[18].

This sophisticated invention allows us to obtain a better understanding and reinterpretation of the sundial surmounting the Pharos on the mosaic of Qasr el-Lebya. A plausible link can be established between this sundial and the Alexandrian engineering tradition which, many centuries earlier, produced the elaborate water clock described by Vitruvius. Given that since the Ptolemaic period clocks invented in this city seem to have been traditionally associated with various automata[19], it is likely that the figure of Helios represents an automatically moving statue. This assumption is further supported by the similarity between the gesture of Helios and that of the figure indicating time on the water clock of Ktesibios. Indeed, both are upright and point to the main mechanism of the device with a sword or a stick so as to indicate the time. On the basis of this analogy, it can be argued that the sword moved as the hours passed by in the same way as the stick.

4 Late Antiquity Dials

Late Antiquity instruments measuring time provide further evidence for the incorporation of automata. Indeed, clocks with moving figures were particularly popular during this period, in which the mosaic of Qasr el-Lebya was also created[20]. In 507 A.D. Cassiodorus commissioned two very spectacular examples, a sundial and a water clock, for Gundibald, king of the Burgunds[21]. His own description of these devices is unfortunately rather vague, but mentions various astonishing automata[22], as well as a sunray acting as a pointer to show time[23], which is reminiscent of the sundial depicted on the pavement. The monumental clock of Gaza (fig. 3) described by Procopius (473-535 A.D.), which also dates back to the same period as this mosaic, undoubtedly provides the closest parallel to the one depicted on the top of the Pharos. Of course, it is more complex and sophisticated, as it forms an entire building and includes various automata, such as birds throwing spheres, the sounds of trumpets and cymbals as well as 12 doors that open in succession once an hour. But the figure of Helios dressed like a king, holding a celestial globe and moving every hour so as to point with its index finger to each of these opening doors, is almost identical with the one of the mosaic[24]. The resemblance between these two figures provides additional arguments in support of the interpretation that the Helios in the mosaic of Qasr el-Lebya is an automaton. It can be explained by the assumption that the constructors of the clock of Gaza were inspired by the works of Alexandrian engineers, especially those of Ktesibios and Heron, as demonstrated by subsequent similar clocks obviously following this scientific tradition[25]. Moreover, the presence of a statue of Helios on an important monument located in the centre of a Christian city like Gaza proves that the pagan personification of the sun was totally compatible with Christian architecture and iconography during Late Antiquity. It is therefore not surprising to find this figure on the Alexandrian Pharos after the edict of Theodosius against paganism.

Fig. 3. Reconstruction of the clock of Gaza described by Procopius, 473-535 A.D., in Diels, H. (1917), *Über die von Prokop beschriebene Kunstuhr von Gaza* (Abhandlungen der Königlich Preußischen Akademie der Wissenschaften, Philosophisch-historische Klasse 7), pl. II, Verlag der Königl. Akademie der Wissenschaften, Berlin

5 Helios on the Gaza Clock

Procopius' rather literary description of the Gaza clock does not deal with technical aspects, but the mechanism involving the figure of Helios can be deduced from Arabic treatises on similar devices which seem to be inspired by the works of earlier engineers. Indeed, Arab clock constructors often refer to their Alexandrian predecessors in their writings and recognize them as their main source of knowledge[26]. Two monumental water clocks described by Ibn al-Razzâz al-Jazarî in *The Book of Knowledge of Ingenious Mechanical Devices*, completed in 1204 or 1206, have many common features with the Gaza clock as Diels has already shown. The first example has 12 doors opening in succession, at one hour intervals, so as to reveal a figure, and also includes cymbalists, trumpeters as well as balls thrown from the beaks of falcons (fig. 4).

Fig. 4. The first clock illustrated in *The Book of Knowledge of Ingenious Mechanical Devices* by al –Jazarî, in Wiedemann, E. (1915) *Über die Uhren im Bereich der islamischen Kultur* (Nova Acta. Abhandlungen der kaiserlich leopoldinisch-carolinischen deutschen Akademie der Naturforscher 100), p. 63, fig.18, Druck von Ehrhardt Karras, Halle.

The devices are operated by a float that sinks at a steady rate into a cylindrical reservoir. A string attached to the top of the float passes through a pulley system setting a cart in motion (figs. 5)[27]. According to the author, the doors are opened as follows:

" *Twelve figures of bronze or wood are now made, each figure being of a size to fill one of the upper doors, and made according to the choice of the craftsmen. Each figure is suspended from his head by a ring above the top of the door. On the back of the figure is a bronze plate, which is wider than the door, and prevents the figure from coming out ; every figure is dealt with in like manner. Now one takes a wooden bar, its length greater than the width of the wall. This is erected behind the wall above the doors, its distance from the wall being equal to the height of one of the figures suspended in the doors. Opposite the centre of each door an iron hook is fixed in the bar, its bent end pointing in the direction towards which the Cart travels....At the bottom of each plate on the back of the figures a smooth iron peg is firmly fixed, its length ½ a finger. When each figure is lifted and the peg placed in the hook opposite the door of that figure, the head is hung above the top of the door, and the feet are held by the peg to the hook. The two door leaves are attached to each door by two light strings which are tied to the leaves and to the plate on the back of the figure. If the peg is lightly pushed from the hook in the direction towards which the Cart is travelling, the figure falls, opening the door leaves by its fall, and preventing them from closing again. He stands [there] as if he had come down from above*"[28].

The second clock described by al-Jazarî has a frieze projecting like a ledge from the face of the wall with 12 battlements along its edge. A man is standing at the edge of the ledge with his right hand stretched out and his index finger pointing towards the battlements so that he almost touches their points. At daybreak the figure is at the end of the frieze and moves smoothly until it reaches a point behind the first of the battlements one hour later[29]. Its movement is identical to that of the Helios of the Gaza clock which glides past the 12 doors to which it points as soon as they open. Musicians playing instruments are again included (fig. 6).

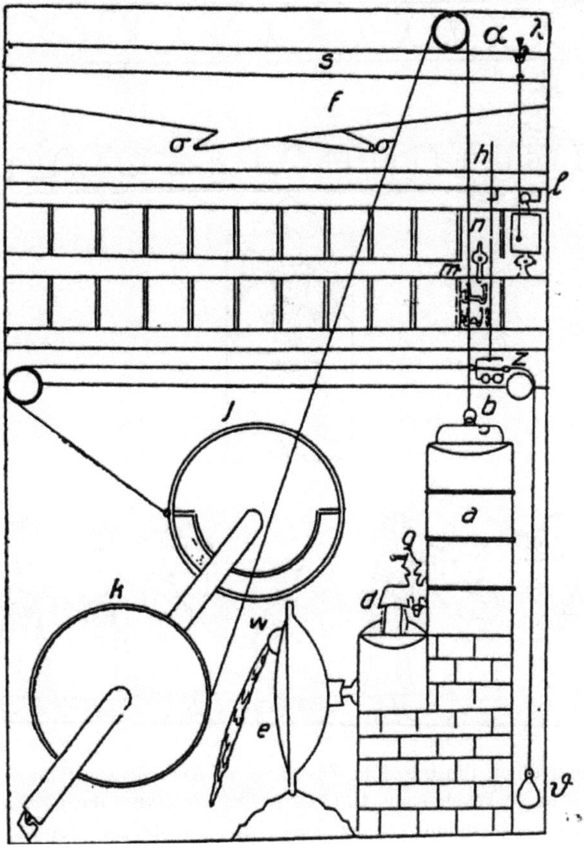

Fig. 5. The mechanism of the first clock as described by al-Jazarî in *The Book of Knowledge of Ingenious Mechanical Devices*, in Wiedemann, E. (1915) *Über die Uhren im Bereich der islamischen Kultur* (Nova Acta. Abhandlungen der kaiserlich leopoldinisch-carolinischen deutschen Akademie der Naturforscher 100), p. 82, fig.38, Druck von Ehrhardt Karras, Halle. **a** : reservoir **b** : float **g** : tap and its bent-down end, exactly over the centre of the float-chamber **d** : float's stopper entering the bent-down end of the tap **e** : regulator's mark erected on its footing **w** : onyx for the water outlet **z** : cart **l** : hook on which the back of the human figure is held by a peg at its base, h: rod pushing the peg off the hook, **θ** : weight behind the cart **j** : wooden disc completed with a ring **k** : balanced wooden disc on the other end of the axle

Fig. 6. The second clock illustrated in *The Book of Knowledge of Ingenious Mechanical Devices* by al –Jazarī, in Wiedemann, E. (1915) *Über die Uhren im Bereich der islamischen Kultur* (Nova Acta. Abhandlungen der kaiserlich leopoldinisch-carolinischen deutschen Akademie der Naturforscher 100), p. 99, fig.45, Druck von Ehrhardt Karras, Halle.

The figure standing in front of the battlements is attached to a cart which is set in motion by a float by means of pulleys as shown in fig. 7[30]. The basic principle is therefore similar to the one behind the Ktesibios' clepsydra, although the Arabic clocks, as well as the clock of Gaza, are more elaborate. In all the examples discussed so far figures pointing to indicate time are activated by a float, despite the differences in the ways of transmitting movement.

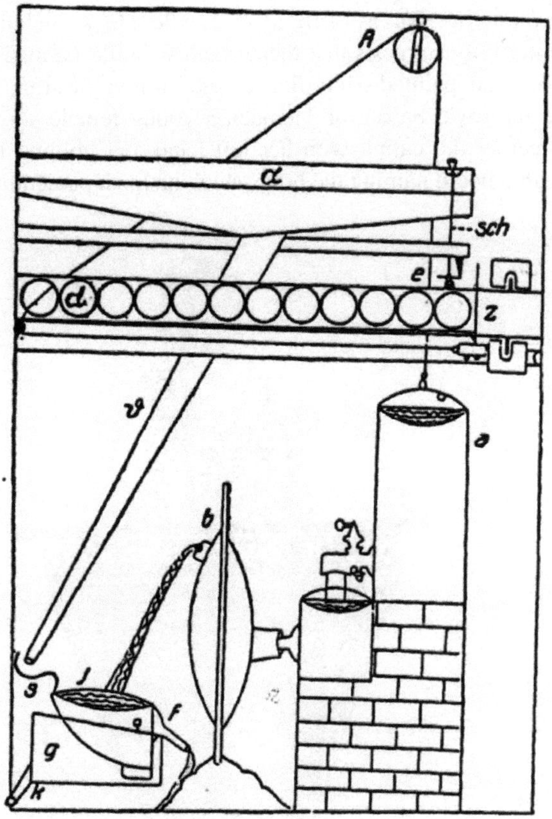

Fig. 7. The mechanism of the second clock as described by al-Jazarî in *The Book of Knowledge of Ingenious Mechanical Devices*, in Wiedemann, E. (1915) *Über die Uhren im Bereich der islamischen Kultur* (Nova Acta. Abhandlungen der kaiserlich leopoldinisch-carolinischen deutschen Akademie der Naturforscher 100), p. 106, fig.51, Druck von Ehrhardt Karras, Halle.

6 Figures Indicating Time on Arabic Clocks

As many monumental clocks of this period have close similarities with the Gaza clock, they certainly spring from the same engineering tradition. However, as personifications and allegoric figures are not common in Islamic art and culture, the absence of Helios as a statue with human features is not unexpected in Arabic time-measuring devices. On a clock offered to Charlemagne by Hârûn al Raschîd in 807 A.D., 12 windows opening in succession bring to mind the equivalent doors of the clock of Gaza, but according to Einhards account, they were thrown open by horsemen sallying forth at the end of each hour[31]. Also with 12 similar doors opening in succession, at one hour intervals, so as to reveal a figure, are the first water clock described by Ibn al-Razzâz al-Jazarî in *The Book of Knowledge of Ingenious Mechanical Devices* (fig. 4), as well as a candle clock included in this book. However, the writer only suggests that these figures should be made according to the craftsmen's choice and painted according to the current fashion[32]. On a 14th c. clock, located in the royal palace of Tlemcen a young female slave emerges out of similar doors, greeting the caliph with her left hand and holding in the right one a piece of paper with a poem naming the hour which had just passed by[33].

Fig. 8. The beaker water clock as illustrated in *The Book of Knowledge of Ingenious Mechanical Devices* by al-Jazarî, in Schulz, Ph. W. (1914), *Die persisch-islamische Miniaturmalerei*, pl. III, K. W. Hiersemann, Leipzig.

Therefore, moving figures that point to indicate time, although still present, do not represent the sun. On the beaker water clock, the design of which is described by al-Jazarî in the above mentioned book, a scribe sitting on the lid holds a pen which moves with a regular almost imperceptible circular motion from the beginning of the day, its tip passing over a graduated annulus carried by a raised section on the outer perimeter of the vessel, so as to successively reach the 15 divisions corresponding to each constant hour of the day (fig. 8)[34]. Despite the different identity and appearance of the figure, its function and gesture are identical to that of the Helios depicted on the mosaic of Qasr el-Lebya, the sword of the personified sun acting in the same way as the pen.

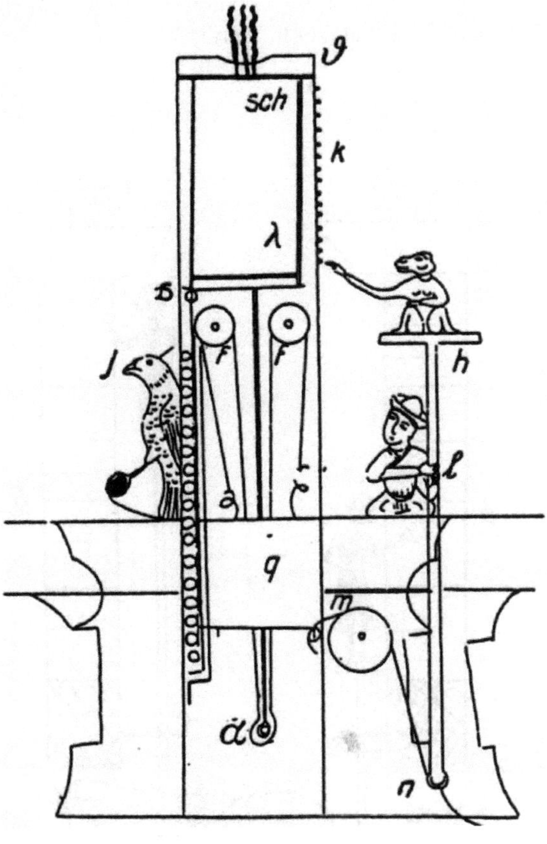

Fig. 9. The monkey candle clock as illustrated in *The Book of Knowledge of Ingenious Mechanical Devices* by al -Jazarî. Wiedemann, E. (1915) *Über die Uhren im Bereich der islamischen Kultur* (Nova Acta. Abhandlungen der kaiserlich leopoldinisch-carolinischen deutschen Akademie der Naturforscher 100), p. 160, fig.86, Druck von Ehrhardt Karras, Halle. **sch** : graduated candle **k** : sheath **λ** : disc **q** : weight connected to strings which are passed over the pulleys **ff a** : rod at the bottom of which the ends of the strings are tied **m** : pulley which is in a firm housing inside the candle-holder and its end is tied to the bottom of weight **q n** : lower end of the monkey's post to which is tied the string passing over the pulley **m** When the graduated candle **sch** is pushed down, the weight **q** rises from the bottom of the sheath **k**. As the candle burns away the weight sinks and the candle rises. The string connected to the bottom of the post pulls it upwards together with the monkey, whose finger moves from mark to mark until un hour has passed.

Both figures can be compared with the monkey appearing in a candle clock which is also described by al-Jazarî in the above mentioned book (fig. 9). Sitting on the top of a short post held on the shoulder of a kneeling man, this animal is clasping its knee to its breast with its left paw and outstretching the right one to point with its index finger to a sheath, containing a graduated candle, on which the numbers and the subdivisions of the hours are written. The monkey's keeper lifts the monkey up the post gradually until its finger reaches the successive hour-marks as time elapses[35]. Although this animal figure does not use a pointer, its movement and gesture have the same meaning as those of Helios and of the scribe.

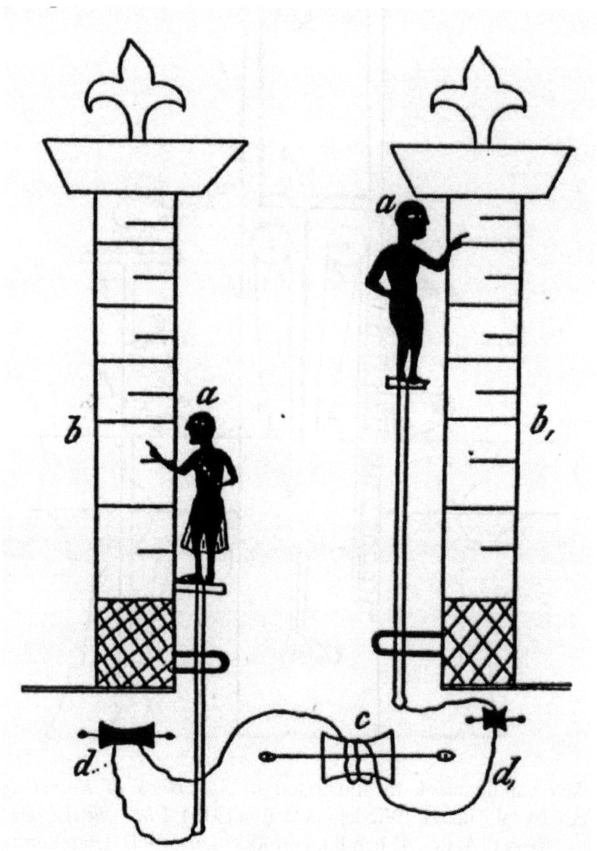

Fig. 10. Detail of the Arab manuscript attributed to Archimedes, 16th c., in Wiedemann, E. and Hauser, F. (1918), *Abhandlungen der kaiserlichen leopoldinisch-carolinischen deutschen Akademie der Naturforscher* 103.2, p. 188, fig. 7. This device functioned in a similar way to Ktesibios' water clock. The strings lifting or lowering the figures were wound onto drums set in rotation by a float through a pulley system.

On another water clock, which is described in an Arab text attributed to Archimedes and handed down in a 16th c. manuscript, but was probably inspired by Greek prototypes, an equivalent role is played by two male figures moving vertically as the hours pass by and pointing their fingers at a graduated column to show the time (fig. 10)[36]. It is therefore likely that this gesture, when it occurs in clocks and dials, is characteristic of automata indicating time.

7 The Sun in Arabic Clocks

Although it is no longer personified, the sun is still present in Arabic clocks as an hour-marker, represented as a disc. An impressive example of this kind of arrangement is given in *The Book on the Construction of Clocks and on their Use* by Riḍwân ibn al-Sâʿâtî, who relates how he restored the clock built by his father, Muḥammad b. ʻAlib. Rustam al-Khurâsânî al-Sâʿâtî, at the Jayrûn gate in Damascus around 1154 which had fallen into disrepair after his death (fig. 11)[37]. On this monumental device, the "hours of the sun" were represented by a large rotating wheel, around the perimeter of which each sign of the zodiac was painted in a circle. The sun was represented by a small disc set in one of these signs, to one of three positions, depending upon whether it was at the beginning, middle or end of the zodiacal period. A wooden screen concealed the lower half of the sun/zodiac wheel and its upper edge was the "horizon". At daybreak the "sun" rose to its correct sign and as the wheel rotated, turning 180 degrees by sunset, one sign appeared and another disappeared every hour[38].

A similar system was installed on the first clock described by al-Jazarî in *The Book of Knowledge of Ingenious Mechanical Devices* (fig. 4-5). As the author specifies, it recorded the passage of solar and unequal hours, that is to say, the hours of daylight and darkness were divided by 12 to give "hours" that vary in length from day to day and from daytime to night-time on a given day. This variation was indicated by the rotation of a zodiac circle crowning the clock. The movement of the circle was obtained thanks to a flow regulator (fig. 5) into which water running from a reservoir through a float-chamber entered. Below the zodiac circle bearing the 12 signs on its outer rim, only half of which was visible, the sun was represented by coloured discs of yellow glass, set in rotatable annuli so that they could be set in the correct position of the zodiac for a given day. At daybreak, the sun rose into its correct sign and the zodiac circle turned throughout the day, one sign rising as another set, until nightfall. Every hour one glass roundel became illuminated until at the end of the day all 12 were bright[39].

As the overall design of both of these clocks is very similar to that of the dial of Gaza, the disc or the lightened glass roundels seem to have replaced, during the Arabic period, the statue of Helios. These examples show that the movement of the sun, performed either by a human figure or by an image of the star in the sky was apparently a frequent means of materializing the course of time in clocks derived from the Alexandrian engineering tradition. It is therefore likely that the Helios of the mosaic of Qasr el-Lebya also executed an analogous movement.

Fig. 11. Reconstruction of the clock restored by Riḍwân ibn al-Sâ'âtî, in Hill, D. R. (1981), *Arabic Water-Clocks*, fig. 30, University of Aleppo, Institute for the History of Arabic Science.

8 Arab Authors

Although they appeared much later than the mosaic, the texts of some Medieval Arab authors tend to confirm the existence of mobile statues associated with time measuring instruments on the Alexandrian lighthouse. Of course, the descriptions contained in these texts are very confused and obviously fantastic, but they were probably inspired by reality[40]. Mas'ûdî (915-56), Dimashqî (1256-1327), Maqrîzî (1364-1442) and Suyûtî (1445-1505) refer to a statue permanently pointing its index finger in the direction of the sun and raising its hand at sunrise. They also mention another statue that produces pleasant, but different, shouts each hour, day and night. Moreover, according to a passage in the *Abridgment of Miracles* by Ibn Waçîf-Chah (± 1000), King Afrâous built a copper lighthouse, surmounted by a copper human head which shouted out every hour in the middle of his capital city. Another passage in this book mentions an effigy of the Sun turning with this star, facing towards the West in the evening and towards the East in the morning[41]. Although apparently distorting reality, mingling different existing devices and adding imaginary elements, it is plausible that these authors are vaguely referring to statues similar to the figure of the Helios depicted in the mosaic of Qasr el-Lebya.

9 Conclusion

This mosaic would therefore represent a sundial of the *hemisphaerium* type and a water clock to which the statue of the personified Sun would be attached, moving every hour so as to point with its sword to the sunray indicating time (fig.1). As the Medieval literary sources suggest, this movement may have been followed by a shout. The mechanism would be similar to the mechanism of Ktesibios' clepsydra (fig. 2) and its Arabic descendants (fig. 5, 7) and the gesture of the statue would have the same meaning as that of the various pointing figures included in this kind of device.

Despite their chronological discrepancies, these ingenious inventions probably had similar purposes and also had the same symbolical value. Indeed, they all combine time measuring instruments with automata, an association which seems to have its origins in Antiquity and continued through the Middle Ages up to the modern period, leading to the creation of Jack-o'-the-clocks, monumental dials found in cathedrals. This use of animated statues is not fortuitous, but seems to correspond to a concept which also has its roots in Antiquity. In fact, water clocks, sundials and other kinds of time-measuring instruments were considered to be projections of the celestial vault, reflecting the movement of the planets, the trajectory of the sun, the moon and the stars[42]. This is probably what the *hemisphaerium* sundial of the Qasr el-Lebya mosaic was meant to represent (fig. 1), as well as Arabic water clocks with the rotating zodiac circle accompanied by the sun and moon roundels (fig. 4-6). By adding artificially moving human or animal figures to these images of the heavens, such devices accomplished, or at least indirectly incorporated, all the movements existing in the universe. Perceived as "celestial automata", clocks and dials were completed by "biological automata" so as to compose a veritable *tour de force[43].*

It would therefore not be unexpected to find such a masterpiece on a public edifice as important and famous as the Alexandrian Pharos. As this kind of ingenious device was first invented by Alexandrian engineers it seems to be one of the city's hallmarks. The resemblance between the Helios as depicted in the mosaic of Qasr el-Lebya and different automata, especially the ones dating from Late Antiquity, corroborates this hypothesis, which is also supported by the evidence of Arab Medieval texts.

References

[1] Goodchild, R.G.: ILN 231, 1034–1035 (December 14, 1957); Ward-Perkins, J.B.: RACrist 34.1-4, 183–184, 190 (1958); Sichtermann, H.: AA 74, 345–348 (1959); Bianchi Bandinelli, R.: In: Dölger, F., Beck, H.G. (eds.) Diskussionsbeiträge zum XI. Int. Byzantinistenkongress, München 1958, pp. 58–60. C. H. Beck'sche Verlagsbuchhandlung, München (1961); Goodchild, R.G.: AntJ. 41, 218–219 (1961); Ward-Perkins, J.B.: In: Atti del VI Congresso Int. di Archeologia Cristiana. Ravenna 1962, pp. 654–657, Pontificio Istituto di Archeologia Cristiana, Città del Vaticano (1965); Brandenburg, H.: RömQschr 64, 112–113 (1969); Stucchi, S.: Architettura Cirenaica (Monog. di Archeologia Libica 9), pp. 401–402. "L'Erma" di Bretschneider, Roma (1975); Guarducci, M.: Atti Linc., Mem., 18, Ser. 8, 659–686 (1975); Giorgetti, D.: Atti Linc., Rend. 32, Ser. 8, 254–261 (1977); Alföldi-Rosenbaum, E., Ward-Perkins, J.B.: Justinianic Mosaic Pavements in Cyrenaican Churches (Monog. di Archeologia Libica 14), pp. 3–14, 33–37, 58–60. "L'Erma" di Bretschneider, Roma (1980); Quet, M.-H.: MÉFRA 96.2, 806, 813–820, 833 (1984); Empereur, J.-Y.: Le Phare d'Alexandrie, la merveille retrouvée, Paris, Gallimard, pp. 50–51(1998); McKenzie, J.: The Architecture of Alexandria and Egypt, c. 300 B.C. to A.D. 700, pp. 42, 252. Yale University Press, New Haven (2007)

[2] Goodchild, R.G.: AntJ. 41, 218–219 (1961); McKenzie: Op. Cit. (1), 42

[3] Goodchild, R.G.: AntJ. 41, 218–219 (1961); see also McKenzie: Loc. Cit. (1). For war machines invented by Alexandrian engineers, see von Staden, H.: In: Argoud, G., Guillaumin, J.-Y. (eds.) Sciences exactes et sciences appliquées à Alexandrie, Actes du colloque int. de Saint-Étienne 1996, pp. 149, 156–166. Publications de l'Université de Saint-Étienne, Saint-Étienne (1998); Argoud, G.: In: Argoud, G., Guillaumin, J.-Y. (eds.) Sciences exactes et sciences appliquées à Alexandrie, Actes du colloque int. de Saint-Étienne 1996, p. 128. Publications de l'Université de Saint-Étienne, Saint-Étienne (1998)

[4] Guarducci, M.: Atti. Linc., Mem. 18, Ser. 8, 669 (1975); Giorgetti, D.: Atti. Linc., Rend. 32, Ser. 8, 255–256 (1977); Quet, M.-H.: MÉFRA 96.2, 807 (1984); For the eventual use of Alexandrian technology for the construction of the horologium Augusti, see Fleury, P.: In: Argoud G., Guillaumin, J.-Y. (eds.) Sciences exactes et sciences appliquées à Alexandrie, Actes du colloque int. de Saint-Étienne 1996, pp. 109–110. Publications de l'Université de Saint-Étienne, Saint-Étienne (1998)

[5] Goodchild, R.G.: AntJ. 41, 218–219 (1961); Handler, S.: AJA 75, 60 (1971); Giorgetti, D.: Atti. Linc., Rend. 32, Ser. 8, 261 (1977); Auth, S.H.: JGS 25, 43–44 (1983); Empereur: Loc. Cit. (1); Fragaki, H.: Images antiques d'Alexandrie, *Ier* s. av. J.-C.-*VIII^éme^* s. apr. J.-C., pp. 8–9, 61–62. IFAO, Le Caire (2011)

[6] Empereur: Op. Cit. (1), 51

[7] Chapuis, A., Gélis, E.: Le monde des automates, étude historique et technique, pp. I.31–I.47. Éditions du griffon, Paris (1928); Sablière, J.: De l'automate à l'automatisation, pp. 18–30. Gauthier-Villars, Paris (1966); Hill, D.R.: A History of Engineering in Classical and Medieval Times, pp. 188, 201–202, 205–206, 231. Routledge, London (1996); id: Studies in Medieval Islamic Technology, pp. I.13–I.14, 17. Ashgate Variorum, Aldershot (1998); Fragaki, H.: Journal des savants, passim (January 2012)

[8] Vitr. 1.1.7; 7.(preface)14; 9.8.2-7; 10.7.4-5; Schmidt, W.: Herons von Alexandria Druckwerke und Automatentheater, pp. IX–XII, B. G. Teubner, Leipzig (1899); Chapuis, Gélis: Op. Cit. (7), I.32–I.33; Drachmann, A.G.: Ktesibios, Philon and Heron. A Study in Ancient Pneumatics. E. Munksgaard, Copenhagen (1948); Chapuis, A., Droz, E.: Les automates. Figures artificielles d'hommes et d'animaux. Histoire et technique, p. 33, 272, 405. Éditions du griffon, Neuchâtel (1949); Gille, B.: In: Singer, C., et al. (eds.) A History of Technology II, pp. 633–634. Clarendon Press, Oxford (1956); Drachmann, A.G.: The Mechanical Technology of Greek and Roman Antiquity, pp. 9–10. Munksgaard/University of Wisconsin/Hafner publ., Copenhagen/Madison/London (1963); Soubiran, J. (ed. and transl.): Vitruve, "De l'architecture" IX, pp. 271–287. Les Belles Lettres, Paris (1969); Hill, D.R. (transl. and annot.): The Book of Knowledge of Ingenious Mechanical Devices by Ibn al-Razzâz al-Jazarî, pp. 271–272, 277. D. Reidel, Dordrecht (1974); Gille, B. (dir.): Histoire des techniques, pp. 302, 329, 338–344, 347–348, 372, 396, 432, 1132, 1454, 1456–1457. Gallimard, Paris (1978); id: Les mécaniciens grecs. La naissance de la technologie, pp. 83–102, 104–105, 115, 118–119, 122, 132, 222. Éditions du seuil, Paris (1980); Hill, D.R.: Arabic Water-Clocks, pp. 7–9, 129. University of Aleppo, Institute for the History of Arabic Science, Aleppo (1981); id: Op. cit. (7), 7, 142–143, 200–201, 227–228 (1996); Argoud: Op. Cit. (3), 128–129; Fleury: Op. Cit. (4), 110–113, 128; Hill: Op. Cit. (7), I. 11, III. 189, 195, IX.12–IX.14 (1998); Tybjerg, K.: Studies in History and Philosophy of Science 34A.3, 443 (2003); ead.: In: Schürmann A. (ed.) Geschichte der Mathematik und der Naturwissenschaften, 3. Physik/Mechanik, pp. 208–213. Franz Steiner Verlag, Stuttgart (2005); ead.: In: Keyser, P.T., Irby-Massie, G.L. (eds.)The Encyclopedia of Ancient Natural Scientists. The Greek Tradition and its many Heirs, p. 496. Routledge, London (2008)

[9] Schmidt: Op. Cit. (8), 458–489; Wiedemann, E.: Über die Uhren im Bereich der islamischen Kultur (Nova Acta. Abhandlungen der kaiserlich leopoldinisch-carolinischen deutschen Akademie der Naturforscher 100), pp. 33, 35, 38, 42, 48–49, 58. Druck von Ehrhardt Karras, Halle (1915); Chapuis, Gélis: Op. Cit. (7), I.32–I.34, 37, 43, 59, 61–62, II.77–II.79, 82; Drachmann: Op. Cit. (8), especially pp. 41–126 (1948); Gille: Op. Cit. (8), 633, 636 (1956); Hill: Op. Cit. (8), 9–10, 12, 271, 273, 275–277 (1974); Gille: Op. Cit. (8), 341, 343–349 (1978); id: Op. Cit. (8), 85–87, 90–91, 93–97, 103–124, 129, 130–133, 138–141, 206–208 (1980); Hill: Op. Cit. (8), 12, 14, 16–17, 28, 34 (1981); von Hesberg, H.: MarbWPr, 53–54 (1987); Hill: Op. Cit. (7), 134, 148, 201–202, 205–206, 208, 212, 214–215, 231 (1996); Argoud: Op. Cit. (3), 128–129; Fleury: Op. Cit. (4), 112–113; Hill: Op. Cit. (7), I.12, 14, II.224, 227, III.192–III.193, IX.10–IX.11, 13–15, 17 (1998); Tybjerg: Op. Cit. (8), 208–213 (2005); ead.: Op. Cit. (8), 654 (2008)

[10] Schmidt: Op. Cit. (8), IX-XXV; Chapuis, Gélis: Op. Cit. (7), I.34–I.35; Drachmann: Op. Cit. (8), 74–76 (1948); Chapuis, Droz: Op. Cit. (8), 405; Drachmann, A.G.: Centaurus 1, 117–131 (1950-1951); id: Op. Cit. (8), 9–10, 12 (1963); Sablière: Op. Cit. (7), 20–21; Soubiran: Op. Cit. (8), LXX; Gille: Op. Cit. (8), 350 (1978); Fraser, P.M.: Ptolemaic Alexandria, vol. II, p. 616, Oxford University Press, Oxford (1982); von Hesberg, H.: MarbWPr, 54 (1987); Argoud, G.: In: Jacquinod B. (ed.)Actes du 28ème congrès APLAES. Saint-Étienne 1995, p. 29. Publications de l'Université de Saint-Étienne, Saint Étienne (1995); id: Op. Cit. (3), 127–128; Landels, J.G.: Engineering in the Ancient World, rev. edn., pp. 200–201. University of California Press, Berkeley (2000); Tybjerg: Op. Cit. (8), 444 (2003); ead.: Op. Cit. (8), 205 (2005); ead.: Op. Cit. (8), 384 (2008), See also Sidoli, N.: Centaurus 53, 55–61 (2011)

[11] Prou, V.: Les théâtres d'automates en Grèce au II*éme* s. avant l'ère chrétienne (Mémoires présentés par divers savants à l'Académie des inscriptions et Belles Lettres). Imprimerie Nationale, Paris (1881); Schmidt: Op. cit. (8); Schmidt, W., Nix, L.: Herons von Alexandria Mechanik und Katoptrik, pp. 300–373, 406–415. B. G. Teubner, Leipzig (1900); Chapuis, Gélis: Op. Cit. (7), I.36–I.47, 53, 71–72, 79, 191, 267, 335, II, 78, 80, 82; Gille: Op. Cit. (8), 631–632, 634–636, 638 (1956); Sablière: Op. Cit. (7), 18–30; Gille: Op. Cit. (8), 293, 306, 331, 337, 341, 343, 349–359, 532, 1132, 1441, 1447–1449, 1464 (1978); id: Op. Cit. (8), 94, 124–145, 198–199, 208, 218–219, 222 (1980); Argoud: Op. Cit. (10), 29–38; Hill: Op. Cit. (7), 2, 6, 8, 109, 120, 122, 142, 172, 188, 202, 206, 208, 212, 215 (1996); Argoud, G., Guillaumin, J.-Y. (éd. and tansl.): Héron d'Alexandrie, Les pneumatiques. Publications de l'Univ. de Saint-Étienne, Saint-Étienne (1997); Argoud: Op. Cit. (3), 129–145; Hill: Op. Cit. (7), III. 190–191, 196–197, IX. 11 (1998); Landels: Op. Cit. (10), 10, 26–31, 47, 58, 71–72, 75–81, 88–92, 99–105, 108–113, 118–121, 130–131, 192–194, 199–208; Tybjerg, K.: Op. Cit. (8), 444 (2003); ead.: Apeiron 37.4. In: Lang, P. (ed.) Re-inventions: Essays on Hellenistic and Early Roman Science, pp. 29–56 (2004); Eadem: Op. Cit. (8), 204–226 (2005); ead.: Op. Cit. (8), 384–387 (2008)

[12] Schmidt: Op. Cit. (8), 456–457; Wiedemann: Op. Cit. (9), 6, 19, 21, 35, 58; Chapuis, Gélis: Op. Cit. (7), I. 36, 191; Gille: Op. Cit. (8), 635 (1956); Sablière: Op. Cit. (7), 22; Hill: Op. Cit. (8), 10–12, 271, 273, 275–277 (1974); Gille: Op. Cit. (8), 124, 138 (1980); Hill: Op. Cit. (8), 12 (1981); Argoud, Guillaumin: Op. Cit. (11), 8, 11; Argoud: Op. Cit. (3), 128, 130; Hill: Op. Cit. (7), I. 12, 14, II. 224, 227, IX, 9–10 (1998); Landels: Op. Cit. (10), 208; Tybjerg: Op. Cit. (8), 208–209 (2005); Eadem: Op. Cit. (8), 386 (2008)

[13] Chapuis, Gélis: Op. Cit. (7), I. 37, 43; Drachmann: Op. Cit. (8), especially 41–126 (1948); Chapuis, Droz: Op. Cit. (8), 33–34; Fraser, P.M.: Ptolemaic Alexandria, vol. I, pp. 427–428. Oxford University Press, Oxford (1972); Gille: Op. Cit. (8), 353, 360–361 (1978); id: Op. Cit. (8), 93, 104, 122–123, 130–133 (1980); Fraser: Op. Cit. (10), 623–624; von Hesberg, H.: MarbWPr, 54–57 (1987); Carra de Vaux, B., Hill, D.R., Drachmann, A.G. (ed. and transl.): Héron d'Alexandrie (texte arabe de Qustā Ibn Lūqā), Les mécaniques ou l'élévateur de corps lourds, p. 10. Les Belles Lettres, Paris (1988); Argoud: Op. Cit. (10), 30; Argoud, Guillaumin: Op. Cit. (11), 12; Argoud: Op. Cit. (3), 131; Hill: Op. Cit. (7), III. 192 (1998); Tybjerg: Op. Cit. (8), 444–464 (2003); ead.: Op. Cit. (8), 206–213 (2005); ead.: Op. Cit. (8), 384, 387 (2008)

[14] Drachmann: Op. Cit. (8), 79, 160 (1948); Chapuis, Droz: Op. Cit. (8), 33–34; Drachmann: Op. Cit. (8), 19 (1963); Sablière: Op. Cit. (7), 22; Gille: Op. Cit. (8), 353 (1978); id: Op. Cit. (8), 131–132 (1980); Argoud, Guillaumin: Op. Cit. (11), 12; Argoud: Op. Cit. (3), 132, 137; Hill: Op. Cit. (7), III. 194 (1998); Tybjerg: Op. Cit. (11), 50 (2004)

[15] Vitr. 9.8.4-7

[16] A similar device, from which a sphere escaped and fell into the water to serve as an alarm clock is attributed to Aristotle by Diogenes Laertius (5.16); Soubiran: Op. Cit. (8), 278

[17] Vitr. 9.8.5; Diels, H.: Über die von Prokop beschriebene Kunstuhr von Gaza (Abhandlungen der königlich preussischen Akademie der Wissenschaften, Philosophisch-historische Klasse 7), p. 4, Verlag der königl. Akademie der Wissenschaften, Berlin (1917); Meerwaldt, J.D.: Mnemosyne, ser. 9, vol. 49, 409 (1921); Chapuis, Gélis: Op. Cit. (7), I. 33; Drachmann: Op. Cit. (8), 114 (1948); Chapuis, Droz: Op. Cit. (8), 33; de Solla Price, D.J.: Technology and Culture 5.1, 15 (1964); Soubiran: Op. Cit. (8), LXV–LXVI, 275–277; Gille: Op. Cit. (8), 97–100 (1980); Rice, E.E: The Grand Procession of Ptolemy Philadelphus, p. 65. Oxford University Press, Oxford (1983); Needham, J., et al.: Heavenly Clockwork. The Great Astronomical Clocks of Medieval China, 2nd edn., pp. 186–187. Cambridge University Press, Cambridge (1986). Compare with Philon (Pneumatica 36.40-42), where the movements of different figures are activated by floats, Carra de Vaux, B. (ed. and transl.) Philon de Byzance, Le livre des appareils pneumatiques et des machines hydrauliques, pp. 152–158, C. Klincksieck, Paris; Drachmann: Op. Cit. (8), 72 (1948)

[18] Meerwaldt, J.D.: Mnemosyne, series 9, vol. 49, 417–421 (1921); Soubiran: Op. Cit. (8), 279–281; Gille: Op. Cit. (8), 118 (1980); see also Hill: Op. Cit. (8), 8–9 (1981); id: Op. Cit. (7), 227–228 (1996); id: Op. Cit. (7), p. III. 189 who thinks that Vitruvius describes two different clocks invented by Ktesibios and therefore suggests another reconstruction of what he considers as the second one (1998)

[19] Soubiran: Op. Cit. (8), 278; Needham, et al.: Op. Cit. (17), 184–185

[20] Soubiran: Op. Cit. (8), 277

[21] Cassiod: Var. 1.45; Fridh (ed.): XXXVII–499 (1973); Diels: Op. Cit. (17), 16; Soubiran: Op. Cit. (8), 277; Argoud, Guillaumin: Op. Cit. (11), 16

[22] Cassiod: Var. 1.45.6

[23] Cassiod: Var. 1.45.8

[24] Diels: Op. Cit. (17), 7–17, 32–33; Chapuis, Gélis: Op. Cit. (7), I. 52; de Solla Price, D.J.: United States National Museum Bulletin 218, Contributions from the Museum of History and Technology, Paper 6, pp. 96–97 (1959); Goodchild, R.G.: AntJ. 41, 218–219 (1961); Soubiran: Op. Cit. (8), 277–278; Hill: Op. Cit. (8), 13 (1981); Needham, et al.: Op. Cit. (17), 97, 187, 211–212; Hill: Op. Cit. (7), 232 (1996); id: Op. Cit. (7), I.12–I.13, IX.16 (1998)

[25] See infra, pp. 235–239; see also de Solla Price, D.J.: In: Proceedings of the 10th Int. Congress of the History of Science. Ithaca 1962, pp. 599–602. Hermann, Paris (1964)
[26] Wiedemann: Op. Cit. (9), 6, 9, 12, 19, 21, 25, 32–35, 38, 42, 48–49, 54, 58; Diels: Op. Cit. (17), 3–13; Chapuis, Gélis: Op. Cit. (7), I.53; Weitzmann, K.: In: Miles G. C. (ed.) Archaeologia orientalia in memoriam Ernst Herzfeld, pp. 244–250, 264–266. Augustin Publisher, New York (1952); Gille: Op. Cit. (8), 635 (1956); Hill: Op. Cit. (8), 9–12, 271–277 (1974); id: Op. Cit. (8), 7–10, 12, 14, 16–17, 25–28, 34 (1981); id: Op. Cit. (7), 188, 201–202, 205–206, 230–231 (1996); id: Op. Cit. (7), I.11–I.12, 13–14, 17, II.224, 227, III, IX.9–IX.17 (1998)
[27] Wiedemann: Op. Cit. (9) 60–83; Hill: Op. Cit. (8), 17–32 (1974); id: Op. Cit. (8), 92–103 (1981); id: Op. Cit. (7), II.230–II.231 (1998). See also the facsimile edition of the book by Sezgin, F.: Compendium on the Theory and Practice of the Mechanical Arts by Ibn ar-Razzâz al-Jazarî. Institute for the History of Arabic-Islamic Science, Frankfurt am Main (2002)
[28] Wiedemann: Op. Cit. (9), 78–80; Hill: Op. Cit. (8), 28–29 (1974); See also Hill: Op. Cit. (8), 100–101 (1981)
[29] Wiedemann: Op. Cit. (9), 99; Hill: Op. Cit. (8), 42 (1974)
[30] Wiedemann: Op. Cit. (9), 101–102; Hill: Op. Cit. (8), 47 (1974); id.: Op. Cit. (8), 111–112 (1981)
[31] Einhard: Annales Regni Francorum 807, pp. 84–87 (1955); transl. Abel, v. Jasmund, Rau; see also Wiedemann: Op. Cit. (9), 36; Diels: Op. Cit. (17), 20; Chapuis, Gélis: Op. Cit. (7), I.62–I.63; Hill: Op. Cit. (8), 14 (1981)
[32] Wiedemann: Op. Cit. (9), 63, 78, 162; Coomaraswamy, A.K: The Treatise of al-Jazarî on Automata, pp. 5, 8, 11. The Museum of Fine Arts, Boston (1924); Hill: Op. Cit. (8), 18, 28, 92 (1974); See also supra, p. 236
[33] Wiedemann: Op. Cit. (9), 37–38; Chapuis, Gélis: Op. Cit. (7), I.66–I.67; Hill: Op. Cit. (7), II.233 (1998)
[34] Wiedemann: Op. Cit. (9), 134–141; Coomaraswamy: Op. Cit. (28), 12; Hill: Op. Cit. (8), 71–74 (1974); id: Op. Cit. (8), 112–113, 119 (1981); Sezgin, F.: Science et technique en Islam. III, pp. 103–105. Institut für Geschichte der arabisch-islamischen Wissenschaften, Frankfurt am Main (2004)
[35] Wiedemann: Op. Cit. (9), 160–162; Chapuis, Gélis: Op. Cit. (7), I.65; Hill: Op. Cit. (8), 90–91 (1974). See also Wiedemann, E.: Der Islam 13, 107–108 (1923), who associates the gesture of the monkey with those of the analogous figures indicating time in Ktesibios' and Archimedes' water clocks
[36] Carra de Vaux, B.: Journal asiatique sér. 8, vol. 17, 301–303 (1891); Wiedemann: Op. Cit. (9), 33; Wiedemann, E., Hauser, F.: Abhandlungen der kaiserlichen leopoldinisch-carolinischen deutschen Akademie der Naturforscher 103.2, 181–183, 188–189 (1918); see also Meerwaldt J.D.: Mnemosyne, série 9, vol. 49, 421–426 (1921); Hill: Op. Cit. (8), 25–26 (1981); Needham, et al.: Op. Cit. (17), 185, 191; Sezgin: Op. Cit. (34), 94–95
[37] Wiedemann: Op. Cit. (9), 167; Chapuis, Gélis: Op. Cit. (7), I.65–I.66; Hill: Op. Cit. (8), 69–70 (1981); id: Op. Cit. (7), 234 (1996); id: Op. Cit. (7), III. 203 (1998)
[38] Wiedemann: Op. Cit. (9), 171–177, 180, 182, 184, 186, 193, 202, 213, 223, 233–237, 241–244, 248–252, 255–263; Hill: Op. Cit. (7), 235 (1996); id: Op. Cit. (8), 71, 73–77 (1981)); Sezgin: Op. Cit. (34), 98–99

[39] Wiedemann: Op. Cit. (9), 60–64, 67–72, 75–78, 83, 88–98; Chapuis, Gélis: Op. Cit. (7), I.61–I.62; Coomaraswamy: Op. Cit. (28), 10–12; Hill: Op. Cit. (8), 18, 24–25, 27–28, 32, 35–41 (1974); id: Op. Cit. (8), 89–93, 101 (1981); id: Encyclopédie de l'Islam suppl. 5-6, new edn., p. 373, s.v. Hiyal (1982); Hill: Op. Cit. (7), 236 (1996); id: Op. Cit. (7), II.230–II.231, III.204 (1998); see also Wiedemann: Op. Cit. (9), 36–37, for a similar arrangement on a clock in Bagdad, described by many authors

[40] Goodchild, R.G.: AntJ. 41, 218–219 (1961); Mas'ûdî, de Meynard, B. (ed.): Les prairies d'or II, pp. 431–436, 439–440 (1863), de Courteille, P. (transl.); Thiersch, H.: Pharos, antike Islam und Occident, p. 40. Teubner, B.G. Leipzig (1909); Dimashqî, Mehren (ed. and transl.): Manuel de la cosmographie du Moyen Âge, p. 36 (1874); Thiersch: Op. Cit. 45–46; Maqrîzî, Bouriant (transl.): Description topographique et historique de l'Égypte. Mémoires de la mission archéologique française du Caire 17, p. 448 (1900); Thiersch: Op. Cit. 47

[41] Carra de Vaux, B. (transl.): Ibn Waçîf-Chah, L'abrégédes merveilles, p. 168, 194. Sindbad, Paris (1984)

[42] Vitr. 9.8.8-15; Carra de Vaux, B.: Journal asiatique sér. 8, vol. 17, 311–314 (1891); Diels: Op. Cit. (17), 7–10; Wiedemann, E., Hauser, F.: Abhandlungen der kaiserlichen leopoldinisch-carolinischen deutschen Akademie der Naturforscher 103.2, pp. 175–178 (1918); de Solla Price, D.J.: United States National Museum Bulletin 218, Contributions from the Museum of History and Technology, Paper 6, pp. 89–94 (1959); id: Technology and Culture 5.1, 13–14 (1964); Soubiran: Op. Cit. (8), 243, 299; Gille: Op. Cit. (8), 100 (1980); Needham, et al.: Op. Cit. (17), 64–65, 178, 185–186

[43] de Solla Price, D.J.: Technology and Culture 5.1, p. 15 (1964); Carra de Vaux, Hill, Drachmann: Op. Cit. (13), 19

The Drawings of Machines by the Great Architects of the Renaissance and Baroque Eras

Emanuela Chiavoni

Dep. of History, Drawing and Restoration of Architecture,
Faculty of Architecture, Sapienza University of Rome

Abstract. The purpose of this paper is to try and analyse some of the drawings of construction site machinery designed and illustrated by some of the greatest Italian architects of the Renaissance and Baroque eras, such as Filippo Brunelleschi, Francesco di Giorgio Martini and Domenico Fontana.

1 Introduction

The study aimed to analyse the aspect of representation of construction machinery, because, in the period under examination, some extremely interesting examples were designed and sometimes built for some of the most important construction sites of the era. Some elaborate drawings remain of some but, unfortunately, not all of these machines, which are of great interest, quality and beauty.

2 Filippo Brunelleschi's Drawings

The first protagonist examined was Filippo Brunelleschi (1377-1446), a very important figure in the Italian renaissance. A Florentine architect, engineer and sculptor, he worked considerably in the city of Florence, mainly producing works of architecture. He is known particularly for the ingenious construction of the dome of Santa Maria del Fiore cathedral in Florence, considered one of the masterpieces of architecture and engineering of the era. Brunelleschi laid the foundations for the architectural language of rebirth, which enabled creativity to be expressed in terms of space being enriched perspective, and also in the rigorous functionality given to space and clear proportions to the structures, for which he also drew inspiration from the classical world. This was the dawning of the new figure of the architect who, in addition to being involved in the technical and operational processes of construction, now also began dealing with aspects of design, intervening intellectually in mathematical, geometrical and historic issues. His architectural works were often characterised by clear geometries achieved by working on precise modularities and often using the Florentine yard, a unit of measurement used at that time in Florence, which provided multiple and sub-multiple units so that proportions could be given to individual elements and the building as a whole. It was during trips to Rome that the architect studied ancient architecture, leading to his understanding of mathematical laws and the proportions that

T. Koetsier, M. Ceccarelli (Eds.): Explorations in the His. of Machines & Mech., HMMS 15, pp. 251–259.
springerlink.com

governed them. Drawing has always formed the basis for all of the arts, a ritual that has repeated itself for centuries. The process of transcribing reality into graphic representation, and hence performed on location, carries an intrinsic energy and plays a fundamental and irreplaceable role in enabling an understanding of architectural works.

In 1418, a competition was held, again in Florence, for a model dome which would crown the immense structure of the church of Santa Maria del Fiore. One of the greatest problems was finding the expertise needed to erect frameworks to support, at a great height, the centrings on which the masonry structures would be built. For this enormous structure, Brunelleschi adopted a masonry system which he had studied from Roman monuments, but applied there in smaller dimensions and on hemispherical vaults. This procedure consisted of courses of herringbone brickwork which, as the bricks intersected, absorbed their own thrust, supported their own weight unaided with no supporting frames.

The work was completed in 1436 after many difficulties and disputes. In the same year, work began to erect the lantern, an important element needed to provide lighting in the church. The project was submitted in a competition again won by Filippo Brunelleschi a few years earlier. Although from the outside, the construction is still Gothic in appearance, the design was extremely new for the era and reflects the spirit of rationality which characterised the renaissance period. The lantern is the fulcrum and was designed with monumental forms reminiscent of classical architecture, while its eight buttresses extend the flow of the ribs, provide a limit to them and ensure the plasticity of the composition. One of the drawings presented reproduces the wooden structure placed on top of the dome to erect the lantern.

Many plans were devised and made by Brunelleschi for the innovative machinery to be built so that materials could be carried up, unfortunately few drawings demonstrating this work survive. It was thanks to the strategic organisation of the construction site and the availability of mechanisms and machinery capable of raising, lifting, turning and moving heavy weights that the building of the dome of Santa Maria del Fiore was possible.

The particular feature of these machines, which were innovative for their time, also attracted the attention of many XVth-century engineers including Taccola, Bonaccorso Ghiberti and Giuliano da Sangallo, each of whom has, in his own personal style, passed them down to us.

It is interesting to analyse the drawings of the machines created by Brunelleschi for various purposes on the construction site. A close look at them tells us that they are studies which the architect used to understand how to construct a useful mechanism for a certain operation. They are free-hand line drawings, probably done in ink and are accompanied by notes detailing the purpose of the study. Writing is used to complement the drawings, not just as a mnemonic note, but also because it is important that the accompanying text expresses the thoughts of the person who wrote it.

The method of representation often uses perspective, as in the examples shown, probably for a better understanding of the mechanism and also to facilitate comprehension of the individual parts and the relationship between them.

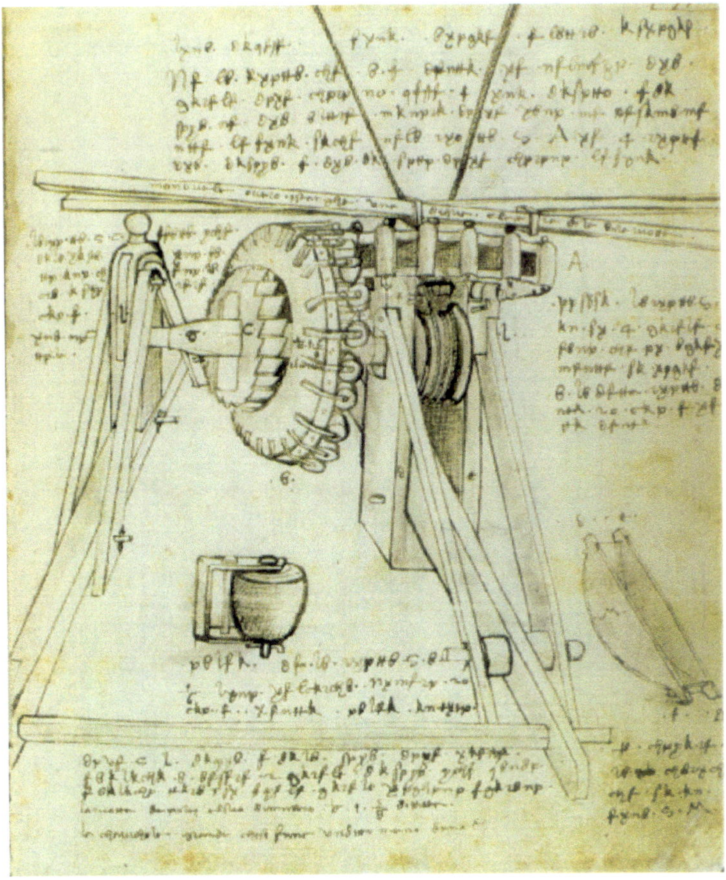

Fig. 1. Filippo Brunelleschi, c.a. 1420. Drawings for the design of the crane for the construction of the dome of Santa Maria del Fiore in Florence. "Zibaldone di Bonaccorso Ghiberti, fol 104 r., National Library, Florence

Sometimes, as in the case in fig. 1, enlargements of details are shown near the main drawing to explain an element which may not be clear or because it is hidden by other elements or rendered insignificant by its small size in the overall view.

These drawings are also characterised by extreme clarity and by lines that are very well controlled, typical of the era. Lines are carefully measured and show a lively representation, useful for explaining how the mechanism in the drawing works. The purpose of the drawings is to understand the object, with the aim of thoroughly explaining its mechanical capacities. They are apparently technical, yet highly refined in terms of graphical characteristics (Fig. 2a-b).

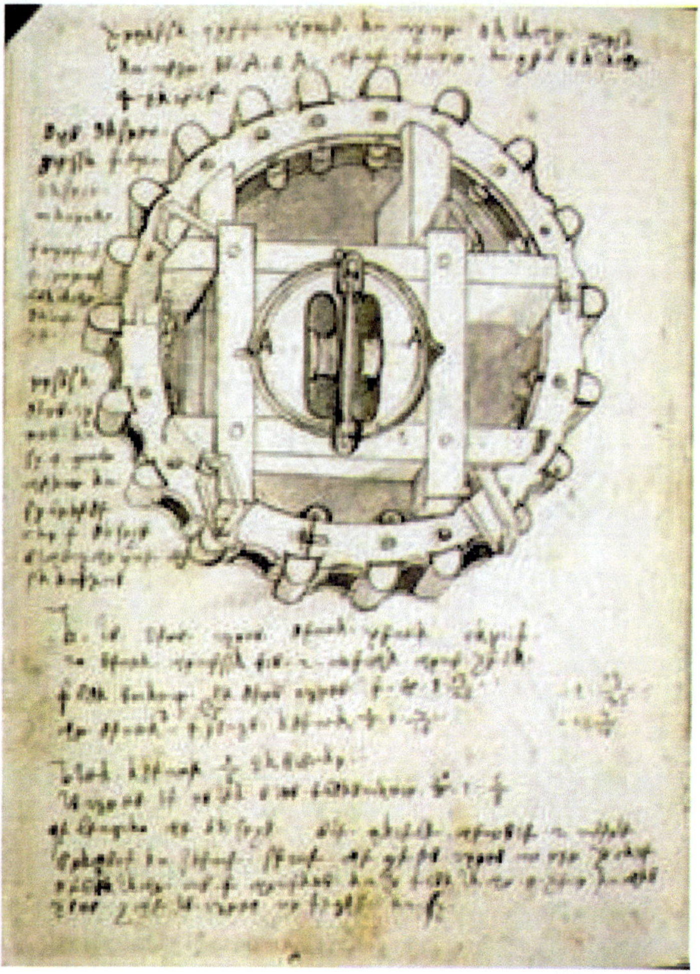

Fig. 2a. Filippo Brunelleschi. c. 1420 -1421 a- Drawings for the design of a wheel

Ceccarelli *et al* wrote: *During the Renaissance, technical drawing, in general and mechanical drawing.... saw the first intentional manifestations, not done for decorative or illustrative purposes ...the illustrations act as a mere support to the text. In the notes and workshop books drafted for the exclusive use of the authors, drawings prevail over text. Another interesting feature is the part concerning the authors of the drawings...the artists and the technicians. The artists displayed their skill in representing the machine in natural environments, using painting techniques; whereas the technicians sketched the objects and placed less emphasis on the surroundings, improving the information given on the object itself [1]*

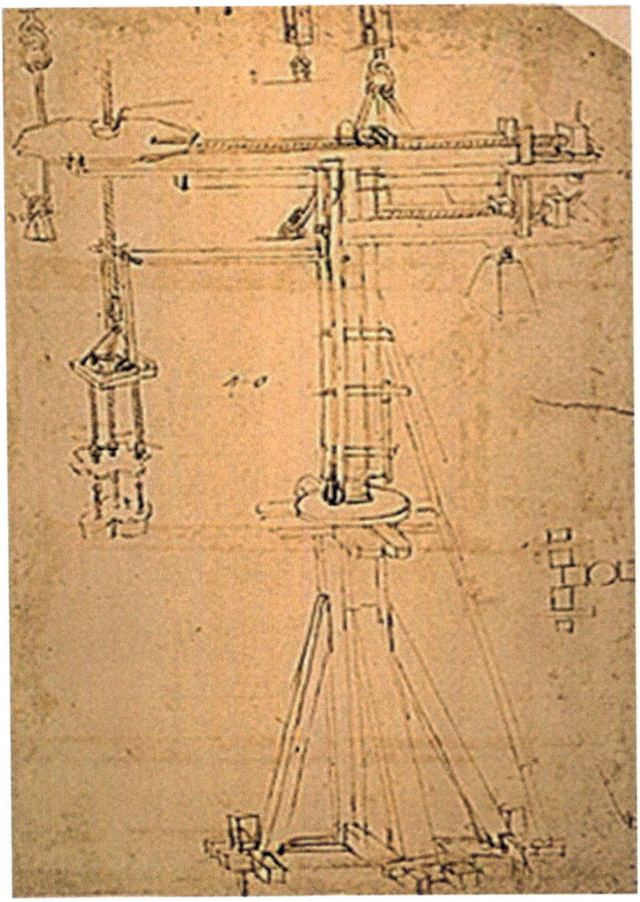

Fig. 2b. Drawings for the design of a revolving crane in "Zibaldone di Bonaccorso Ghiberti, National Library, Florence [2]

3 Francesco Di Giorgio Martini's Drawings

Another protagonist of the renaissance was Francesco di Giorgio Martini (1439 – 1501) a complex and eclectic figure: architect and architectural theorist, but also painter, engineer, sculptor and medallist. He was a scholar of architectural treatises; from the texts of Vitruvius to the manuscripts of Taccola (engineer, sculptor and fifteenth century writer, Mariano di Jacopo Detto Il Taccola, known for his treatises on technology such as *De ingeneis* and *De machinis* focusing on many machines typical of the era).

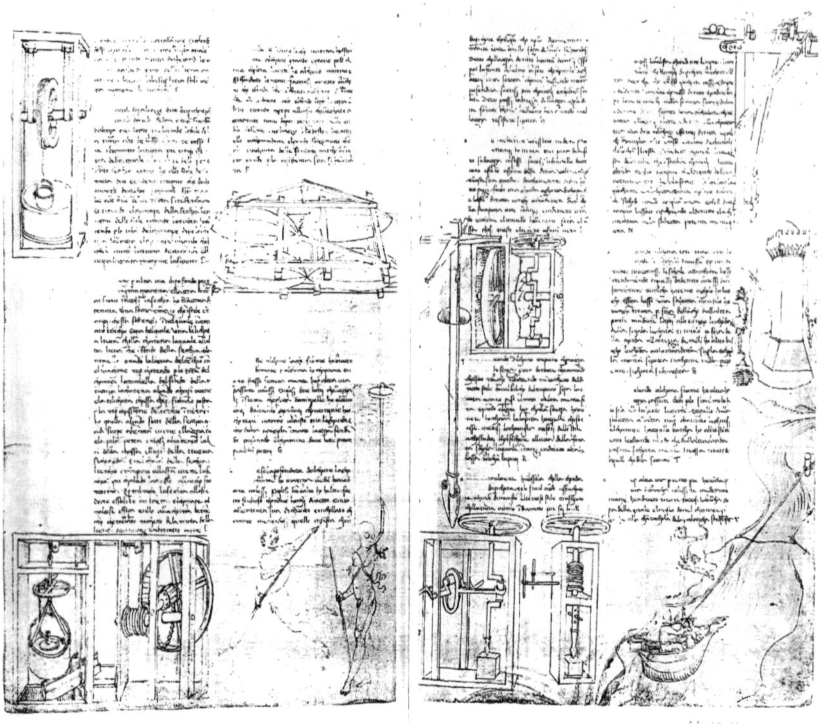

Fig. 3. Francesco di Giorgio, pages from his Treatise [4]

Francesco di Giorgio went on study trips particularly between Florence and Rome to observe the remains of classical architecture and study them in depth. He studied the works by carefully observing the architecture and then drawing it, usually free hand, directly on location. This practice, common to many artists over several eras, placed the discipline of drawing at the centre of training, mainly for the purposes of knowledge. The redrawing of the works of the great classical maestros is a fundamental operation for retracing the design of the work and to stimulate more modern design processes.

Francesco di Giorgio's architectural activity did not include many built works, but featured a large number of plans. However, important and well known were the collections of drawings of machines and military architecture which later made up the famous *Treatise of civil and military architecture,* written during his time in Urbino.

During this stay, his fame expanded and he became a reference figure for architectural design and renaissance engineering. This was because he was one of few to manage to hold the role of artist and technician simultaneously and was also capable of reading and judging the texts of Vitruvius, which were held in great esteem.

He found Urbino to be a welcoming city and a very stimulating environment in cultural terms. It is also likely that he met Leon Battista Alberti there. His research was directed towards the study of the ancient, the Latin language and Vitruvius. His "Treatise" was a continuous research project, collected over various manuscripts and several drafts. One of the most innovative elements of the treatise was the inclusion of drawings to clarify the text (Fig. 3) [3].

Of particular importance were his research and studies into the innovative principles of the art of fortification known as *"modern fortification"*, to which he dedicated many years of his life, together with his brothers Antonio and Giuliano da Sangallo.

4 Domenico Fontana's Drawings

Another representative of the end of the XVIth century was Domenico Fontana (1543–1607) a Swiss architect from Ticino who worked in Rome and Naples in the late Renaissance. At a young age, he went to study the architecture of the ancient and modern masters in Rome, where he met Felice Peretti who would later become Pope Sixtus V. The latter would go on to commission some works including the design of Villa Peretti on the Esquiline, unfortunately demolished for work on the underground railway.

In 1586, he studied how to raise the Vatican obelisk in Saint Peter's Square, using the complex systems and his solid knowledge of static problems. He wrote on this subject in the book "*On the transportation of the Vatican obelisk and the buildings of Sixtus V*" [5]. The height of the Vatican Obelisk in Saint Peter's Square, originally from Alexandria in Egypt, is around twenty-five metres and the work done to raise it sparked universal amazement at the time. Domenico Fontana also studied the raising of three other ancient Roman obelisks, one situated in Piazza del Popolo, the second in Piazza Santa Maria Maggiore and the third in Piazza San Giovanni in Laterano.

The figure 4 shows the precious and refined line drawing of the mechanism used to raise the Vatican obelisk. This representation indicates in the upper left the direction north, i.e., a reference to the cardinal points (north-south-east-west). This indicates a desire to contextualise the drawings and provide indications and references that were as precise as possible. The representation also indicates in the upper right of the drawing the graphic scale used and the unit of measurement. It also features the drawing of a compass which enriches the graphic composition and also calls attention to the fact that, thanks to this instrument, it is possible to show equal measurements. The representation is of the very moment of maximum effort to raise the obelisk and this is highlighted by the many men represented as they carry out the work. The composition is centred vertically on the sheet and is finished off with an elaborate frame to emphasise the importance of the drawing, as if it were a precious painting. The perspective chosen by the draughtsman highlights the wooden structure which acts as

Fig. 4. Domenico Fontana. Drawing of the raising of the obelisk in Saint Peter's Square in Rome [5]

scaffolding, built specially to lift the obelisk. The cables and ropes used to lift the obelisk are also present, and, although they come out of the composition, they are very evocative and useful for understanding the work that is being carried out and the human effort in progress. The representation of men is also important to enable a comparison of the proportions of the structure with the height of the human body. In fact, it is immediately evident that the mechanism is several metres tall. The upper part of the composition features several lines which give the effect of the sky through interrupted and discontinuous horizontal lines which fade into the background. The line drawing of the illustration was almost certainly done with Indian ink using a nib. Particularly evident are the inhomogeneous and, in some parts, faded lines, a sign of there being less ink on the nib and/or where the line begins with a full nib of ink and fades out when there is no more ink. Materials and some parts of the represented elements are enriched with textures and/or more or less homogeneous painted backgrounds, which help to delineate the drawing and define the various depths within the representation.

In addition to communicating their intrinsic content regarding the object or the mechanism or the process in question, all the drawings and representations of works, monuments, architecture also translate the methods of representation typical of the era in which they are produced and the style of the architect/artist that produced them. A close analysis of the images shows the clarity of the composition, the graphic penmanship as a form of art, methods of representation, individual or mixed graphic techniques, the graphic symbolism used. It is also possible to compare each of these parameters with all the others, or examine them against corresponding parameters of other artists and/or different historical eras. Even the layout has its own particular importance: even here there is a subjective choice characterising the style and always affecting the result of the final drawing. The choice of technique and instruments is also an important indicator of the style of an era.

References

[1] Ceccarelli, M., Cigola, M.: Evoluzione della rappresentazione grafica nel progetto dei meccanismi. In: Docci, M. (ed.): Atti del Convegno "Il Disegno di Progetto dalle origini a tutto il XVIII secolo. Storia e sviluppi", Roma Gangemi, pp. 427–433 (1997)

[2] Battisti, E.: Filippo Brunelleschi, Electa Milano (1976)

[3] Ceccarelli, M., Cigola, M.: Trends in the drawing of mechanisms since the early Middle Ages, in Journal of Mechanical Engineering Science. Proceedings of the Institution of Mechanical Engineers Part C 215, 269–289 (2001)

[4] Maltese, C. (ed.): Francesco di Giorgio Martini, tratatti di architettura, ingegneria e arte Militare, Milano Il Polifilo (1967)

[5] Fontana, D.: Della trasportatione dell'obelisco Vaticano et delle fabriche di nostro Signore Papa Sisto V, Roma (1590)

[6] Lefevre, W.: Picturing Machines 1400-1700, Trasformations: Studies in the History of Science and Technology. MIT Press (2004)

Cultural Heritages in Aterno Valley (Italy): Historical Watermills for Cereals Grinding

Romolo Continenza and Stefano Brusaporci

Architecture and Town Planning Department,
L'Aquila University

Abstract. Watermills were known and used from the Roman age; they diffused during medieval age and had a preponderance between working machines till the advent of steam engine. Fundamentally they can be classified as mills with horizontal wheel and mills with vertical wheel. Mills are complex structures, made by mechanisms, hydraulics works, buildings, and are witnesses of past cultures. The case study are the watermills of Aterno Valley (Italy), a territory between Gran Sasso and Sirente-Velino natural parks. The study involves different topics: history, hydrology, mechanics, architecture. Historical analysis and mill's surveying favour the comprehension of mill's documental values. They can be studied as indicators of modifications of technologies, society, economy, territory during centuries. Watermills are cultural heritages and we believe that only through the integration of all contributions deriving from different research fields will create a comprehensive and exhaustive description of the phenomena that have allowed the construction, the use and the actual neglected conditions of mills, and that can favour their conservation, protection and exploitation.

1 Introduction

The paper presents the results of a study on historical watermills for cereals grinding in Aterno valley (Italy). Mills are complex structures, made by mechanisms, hydraulics works, buildings, and are witnesses of past cultures. They can describe technological, social, economic and landscapes transformations that took place during centuries.

The territorial area is the Aterno valley, sited in the inner and mountain zone of Abruzzo region; this territory is characterised by deep modification phenomena that stratified and transformed the landscape, from italic and roman to medieval and modern age, at last with local industrialization processes.

The research method is based on the correlation between general historical studies and mill's surveying, to set local phenomena inside a larger experiences, and at the same time to understand their specific characteristics. The paper has the following articulation: the first section is about historical cereal mills; the second section focuses on history and surveying of Aterno valley's mills[1].

[1] In particular Stefano Brusaporci is the author of the section titled 'Historical cereals mills' and Romolo Continenza is the author of the section 'Aterno Valley watermills'.

T. Koetsier, M. Ceccarelli (Eds.): Explorations in the His. of Machines & Mech., HMMS 15, pp. 261–275.
springerlink.com

2 Historical Cereal Mills

Probably grinding had an agricultural origin in prehistoric age. The first and simpler grinding modality is the one with mortar and stone pestle (Fig. 1). The technique perfected during century, from cylinders to disks rolling on flat stones.

In roman age the so called 'conic mill' was invented, made by a lower conic stone fixed to the ground and a second concave one revolving on. This mill is also called 'hourglass mill' because the revolving stone has a sort of funnel on the top to collect cereals and let them enter into the mill, so it presents a doubled cone form. Romans called the superior stone *catillus*, and the inferior one *meta* (Fig. 2, 3, 4). In spite water techniques were known, the conic mill, moved by men or animals, was the most used because of the large availability of muscle energy, in particular supplied by slaves; we can estimate 5-6 rpm, producing about 0,10 pound of grain per minute.

Watermills diffused during medieval age thanks to several reasons: metallurgy development; the role of monastic orders that reorganised the economic system, with an integrated and differentiated management of large properties; urbanization from XI century that redesigned territories and increased water using for energy production and transportation.

Watermill realization was not homogeneous in European lands: according to territorial characteristics and urban settlement, mill's typologies corresponded to different economic and social organizations. A familiar economy required simple mill's techniques, with grinding made by men or animals, that is a low profitable process. Vice versa the urban settlement and economic development required the diffusion of water mills, that are more efficient and can serve large communities, but are more onerous because of their installation and maintenance. So water mills were a monopoly of local dominations and citizen were obliged to pay to grind cereals. Water mills configure as complex structures: buildings with several rooms, articulated mechanisms, hydraulics works, many workers; nevertheless there are no significant innovations from water mills used in ancient centuries. During centuries watermills became important as machines to produce movement, not only to grind cereals, but also to produce paper, to work metals, to weave, etc. However simpler grinding techniques were not abandoned inside familiar organization.

Main cereal grinding techniques are: simple compression, percussion, friction. Fundamentally water mills can be classified as mills with horizontal or with vertical wheel. Grinding wheels have smooth or grooved surfaces (Fig. 5).

Horizontal wheel mills have the grinder directly connected to hydraulic wheel. They need little volumes but rapid running water, conditions that characterises mountain territories. The water, taken from a river or a torrent by a proper canal, through an inclined pipe hits tangentially the mill's paddles and move the grinding stone. In this kind of mill the paddles have a spoon conformation, whereas in vertical wheels are planar. The low performance of horizontal wheel makes this kind of mill not suitable for commercial production, and was used to grind cereals by agrarian society based on self-consumption.

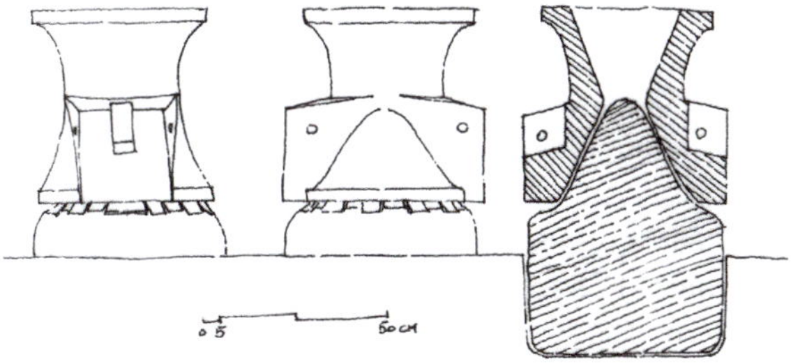

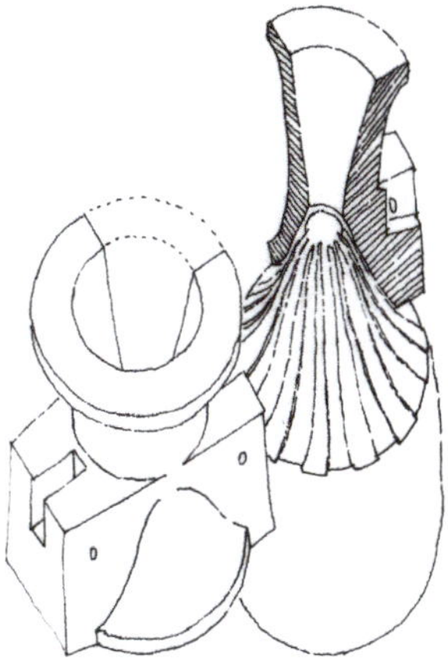

Fig. 1. The first and simpler grinding modality: mortar and stone pestle.

Fig. 2. So called ‘conic mill’ used in roman age, made by a lower stone fixed to the ground and a second concave one revolving on.

Fig. 3., 4. Sketches of ‘conic mill’: fronts, section and composition scheme.

Fig. 5. Grinding wheel with grooved surfaces.

Vertical wheel mills use plain rivers, with a constant flow and low water speed. The mill's paddles transmit the movement of the wheel and change the axe of the rotation from an horizontal one to a vertical one; besides it increases the grinding wheel rotation, according to the number of gear teeth[2].

The study of cereal mills can be favoured by the analysis of mills' drawings presented in historical treatises. The roman architect Vitruvius describes a water mill in his tractate *De Architectura*, probably written in 25 b.C. During centuries it was transcribed several times, reassumed and manipulated; then it was analysed, reorganised, printed and translated between XV and XVI centuries. The first effort to integrate the text of *De Architectura* with a systematic illustrations apparatus is presented in an anonymous manuscript realised between the end of XV and the beginning of XVI centuries, preserved in the Biblioteca Ariostea of Ferrara. Drawings are realised with the same inc used for writing. An image describes different kinds of *noria* that is a machine made by a vertical wheel used to rise water, and another drawing is about an articulated mechanism to elevate water and in second distance there is an image of a vertical wheel mill (Fig. 6). From the beginning of the XVI century many treatises were printed with illustrations to describe Vitruvius' texts: the first one was realised by Fra Giocondo from Verona in 1511; the first edition not in Latin, but in Italian, by Cesare Cesariano in 1521; than the famous edition by Daniele Barbaro with the illustrations of Andrea Palladio in 1556. Than between XVI and XVIII centuries innumerable works present watermills drawings. For example Jan van der Straet (1584) describes a mill that surmount a canal; he shows not only the mill but also the daily life of the village.

Otherwise Agostino Ramelli (1588) and Georg Andreas Böckler (1662) represent a mill not as building but like a model, to evidence mechanisms (Fig. 7, 8). Bernard Forest De Belidor (1737) produces numerous drawing of mills an its mechanisms: two of these present a cross and longitudinal section of a mill showing at the same time how it works and the spatial articulation of the rooms, the opposite of what Ramelli and Böckler do. In another representation De Belidor describes different vertical wheel's configurations, according if the water jet hits the wheel on the top or on the lower part; in particular there is a kind of mill with a spiral wheel (Fig. 9, 10).

The use of hydraulic energy for grinding is diffused in all the cultures, as the sketches of the Japanese painter Katsushika Hokusai (1815) report: in particular one drawing represents watermills modalities of functioning, with different mechanisms from Western ones, also with ropes (Fig. 11). Whereas another drawing describes a machine activated by water that work by percussion.

The graphics of each author reflects the culture and the history of his age. Anyway, both in Western and in Oriental countries, watermill had a preponderance between working machines till the advent of steam engine.

[2] About kinds, dimensions and power of historical mills see: *Dictionnaire technologique*, Ch. Thomine et Fortic, Paris, 1822.

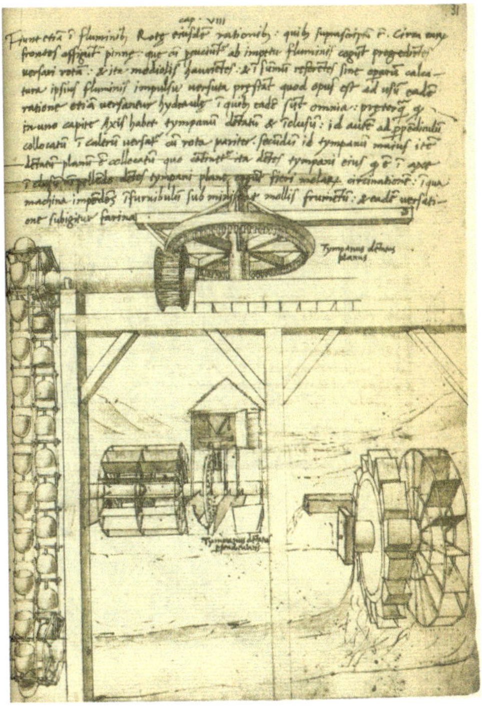

Fig. 6. Anonymous manuscript, end of XV - beginning of XVI century. Mechanism to elevate water and vertical wheel mill.

Fig. 7. Agostino Ramelli, *Le diverse et artificiose machine del capitano Agostino Ramelli dal ponte della Tresia ingegniero del Christianissimo Re di Francia et di Pollonia*, 1588.

Fig. 8. Georg Andreas Böckler, *Theatrum machinarum novum, exhibens aquarias, alatas, iumentarias, manuarias*, 1662.

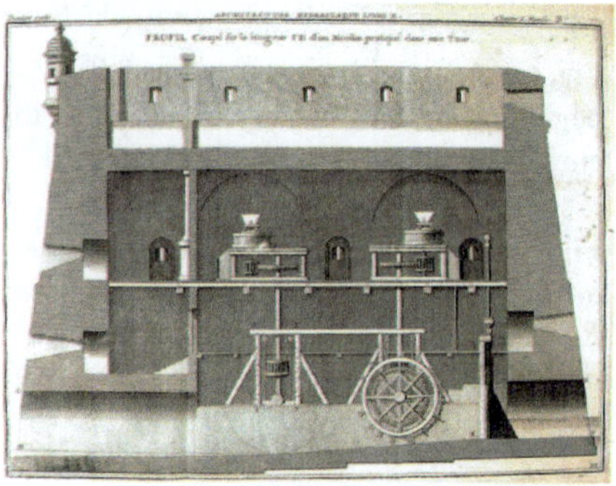

Fig. 9., 10. Bernard Forest De Belidor, *Architecture Hydraulique*, 1737.

Fig. 11. Katsushika Hokusai, *Manga*, 1815.

3 Aterno Valley Watermills

The origins

L'Aquila's basin is located below the Gran Sasso, in the centre of Italy, approximately at the height of 650-750 meters Above sea level. The basin is situated on the huge fracture originated from several tectonic movements which have given rise to the Appennini mountain range. This chain is made up of a long East-to-West directed valley and is marked by short hollows. The valley is constituted of two plateaus originated from lake basin leftovers, separated by a bottleneck and a high ground. It is right on this hill that the city of L'Aquila was built during the XIII cen. The Aterno river springs are located approximately 30 km westwards above the city, at one thousand metres height. The Aterno river passes across the entire valley collecting water during his path. At 400 metres height the Aterno flows into the Pescara and reaches the Adriatic Sea in Pescara town.

The mountain environment, characterised by rises and valleys, together with the rivers action, limited and interrupted because seasonal trends, have not allowed, throughout history, the use of hydraulic power. Therefore, there aren't archaeological evidences of the existence of hydraulic power devices or of flour production mills belonging to the Roman or High-Medieval period. Many proofs instead, confirm the Romans' ability in distributing and supplying water in the two wide roman settlements: the Amiternum (West) one and the Forcona (East) one[3].

Although first documents attesting the presence of hydraulic machines belonging to monasteries date back to first Middle Ages[4], earliest documents quotations date back to the XV-XVI centuries. The quotation of the so-called "*alla Molinella*" mill in Tempera close to the Vera river spring is dated 1482. In 1500 another mill is mentioned in the cadastral register of Camarda and, in the same year, is attested the payment of a census for the running of a mill in Tempera to S. Maria di Casanova abbey, by Giuseppe and Giovan Battista De Rosis. One may think that the introduction (or even the re-introduction) of technologies for the use of hydraulic power in the valley is monks' responsibility at the time of L'Aquila's institution. In fact there are, as previously said, documents attesting payments to the monastery. Moreover the Cistercians, skilful builders and experienced technicians, were permanently established in the area with "granches" and shelters located up in the mountains and down in valley bottoms.

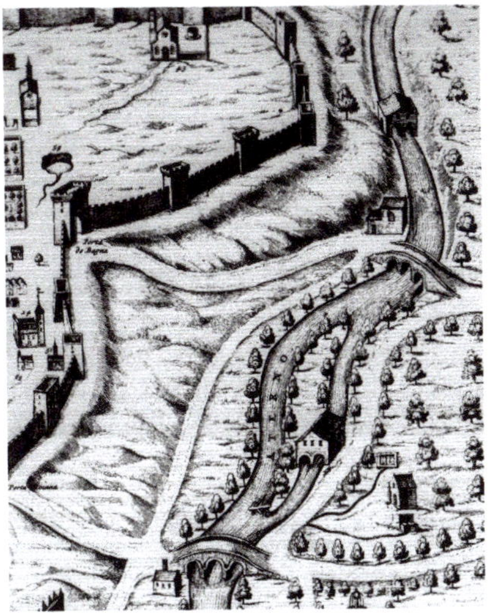

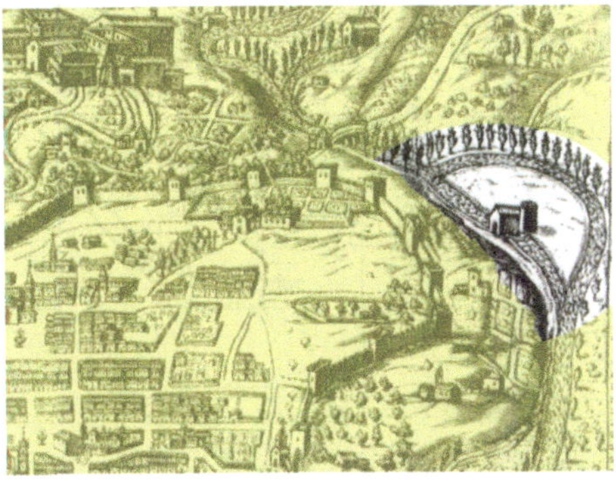

Fig. 12., 13 Mills located nearby the city walls of L'Aquila in the map drawn by Pico Fonticulano and engraved by G. Lauro (1600). In evidence Rosarolo mill.

---

[3] Great Roman stone blocks, dug in order to form half acqueduct pipe, emerged during the excavation of the Medieval tower of San Vittorino (close to the Amiternum area), integrated into the walls' pattern of the view. Eastbound, where lies the settlement of Civita di Bagno, ruins of a great thermal installation were found.

[4] In 962, upon the act of its constitution, the monastery of S. Bartolomeo in Carpineto in the county of Penne, Spoleto's Lombard dukedom, is given by the Penne count many estates and the permission of building mills and fulling mills.

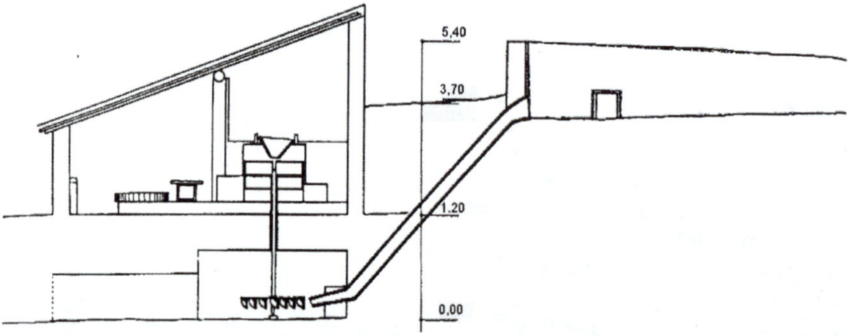

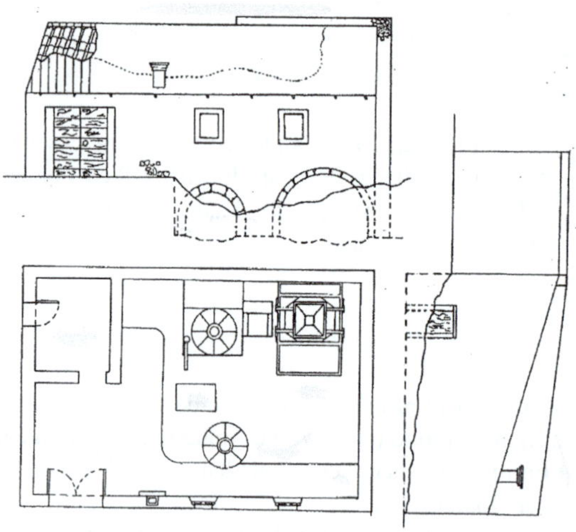

Fig. 14. The Rosarolo mill appears in a picture taken today.

Fig. 15., 16. Assergi mill's cross section, plan and fronts.

In the countryside, the activity of domestic hand made milling will persist for a long time in the area, especially in villages located at high heights, less populated and hard to reach during the cold winter. People's poverty conditions, the lack of resources and the distance from milling equipments, mainly located in valley bottoms, made it easier and more convenient the presence of this kind of techniques.

The hydraulic machine preminently used for the milling of wheat has been used for other purposes. The motive power could also be used during the process of cloth felting. This process consisted in the thermic and mechanical treatment of the wool cloth, after its previous under heat processing with soap, grease and some clays. Before the mechanic treatment the cloth was mashed, flattened and pressed in order to adjust its thickness and softness. In this last phase the hydraulic machine by means of a cam mechanism activated tilt-hammers replacing in that way the arduous work of men that, before, immersed in hot water until their knees, were trampling and pressing the cloth.

Economic and political aspects

Two legal conditions were necessary to ensure the establishment of this device which was very important for the time's societies: the water availability and the soil possession.

Water is, and has always been, a public natural resource and the right to use it was given by the king[5]. The administration of water was under the local landlord control. Therefore the landlord, according to feudal system, could authorize and earn the census paid for the use of this resource. The feudal baron, or the municipal power, were responsible for the control over mills industry and its products and could authorize the machinery installation. Moreover the landlord had the right to control water use, fees, water devices dimension and production timing. The baron, in addition, could also make use of the mill installation and devices for several days during the year. The mill, thanks to feudal privileges, guaranteed a remarkable income to the landlord – or to the municipality. This income was ensured also by the property ad administration of the public oven and of the oil presses. Since domestic milling techniques had been prohibited, people were forced to use these devices, being.

All these circumstances enable the mill and its hydraulics plants to play a relevant role in land organization and management. The canals, drawing water from the river and giving it back at a lower level, were essential for the enlargement of the irrigation network thus improving the productivity of soils. The mill could in the end find its own placement and role in the agricultural and in the town surrounding land thanks to its proximity to rivers, its position across the canal, its water spilling arches. Moreover the landscape was permanently modified by hydraulic devices and such conditions have resisted throughout history (see Fig. 13 and 14).

---

[5] "*Acqua e' una cosa de Realibus, e pero' non si possono edificare Molini, Valcheria ed altri*" ("*Water is a king's possession and no Mills can be built*") is possible to read in chapt. XXIV, art. 46 of the Guide to Universities of the Kingdom of Naples.

Old L'Aquila perspective plans show, around the edge of the city, where the river flows between two valleys, a number of mills whose presence was according to the population needs (Fig. 16).

At the beginning of 1800, in the wake of French Revolution changes, feudal privileges in the Kingdom of Naples were abolished: the institution of the Napoleonic government allowed private citizens to ask for the permission to build hydraulic machines[6]. This regulation permitted the spreading of hydraulic machines into the territory, triggering off competition phenomena, not just between owners but also between privates and already existing establishments. Previous structures tried to hamper in all possible ways the building of other private facilities. The mill network keeps expanding during the XIX Cen. until reaching high-altitude villages, where rivers and springs have, most of the times, torrential and seasonal speed. Only these structures have been longer operative. In fact, the introduction electric power (capable of delivering more intense power but also of ensuring the non dependence of manufacturing from seasonal fluctuations) and, afterwards, the introduction of new milling devices, such as the cylinder-shaped mill.

This machine had a superior productive capability compared to water mills one and was capable of joining automatically milling and other operations such as "sveccia-tura"- separation of the grain from its rigid wrapping -, flour sieving and flour separation from bran, producing a more refined and less expensive article.

The economic relevance of this business is, above all, attested by the controversial assessment, in 1869[7], of the milling tax. This taxation meant remarkable income for the treasury since it was affecting the most popular and cheap nourishments (bread and pasta). This imposition was rejected and criticized by lower classes who violently demonstrated their disappointment regarding the unsustainable political and normative situation. The law was abolished in 1884 since disorders and riots took place in many Italian cities.

At the beginning of the XX cen., although the mill is still playing an important role in the territory infrastructure network, the water mill starts to loose its importance. Only in villages situated far away from main transport links or in particular cases in which the producer wants to consume his own product, this kind of water mill maintains its former usefulness. During Fascism, self-sufficient ideologies have risen the interest in milling techniques improving its spread on the national territory. The so called "battle for wheat" undertaken by the fascist duce Mussolini in 1925 was developed in order to control the country's dependence on foreign supplies and in order to confirm the fascist topic of "back to the countryside" and the fascist economic plan based on traditional handmade production. The widespread impoverishment originating from the II World War and from the hard economic reconstruction process have permitted the use of mills until the 60s. In the following decades the use of mills has decreased dramatically. Today, in our case study basin, only 4 are still used by few committed customers and by several farmers. Lately new "old tastes rediscovery" trends are becoming popular and many people are appreciating this kind of flour

---

6 2nd of August 1806, 25 October 1806. Giuseppe Napoleone's laws.

7 The milling tax was imposed by Quintino Sella and came into effect the 1st January 1869.

which is obviously less refined but maintains all the vitamin contents that are usually lost during new milling techniques.

Technological, typological and morphological aspects

Centuries of concrete experience demonstrate that horizontal-turbine milling device is the one which best suits mountain environments.

Hydraulic devices. Mountain environment is characterized by moderate and uneven water flows, as well as by marked height differences. These differences allow the collection upstream of the proper amount of water required for the mill functioning (these plants are usually made up of one or more turbines and some auxiliary devices). Moreover they allow the setting-up of moderate declination paths for the water to approach the structure, the creation of an accumulation basin guaranteeing the constancy of the flow and an abrupt leap close to the turbine. In addition, is necessary to set up a channel driving the water back to the main stream. Some additional accessories may complete this device: the grids, for instance, help preventing heavy materials from wearing the turbines' blades; a gate system, necessary to adjust the intensity of the flow in the collecting channel but also between the basin and the turbine; the sluices, allowing the use of water for the purpose of irrigation when needed; additional bypass channels, whose purposes are to avoid dangerous overflows of the basin by letting away water excess, and to emptying and maintaining each segment of the structure operative.

Building elements. The space organization of a mill has a simple and regular shape. It has enough space to host all its turbines and other equipments, and to permit machinery maintenance. Close to the building, upwards, a water reservoir wide enough to avoid water shortage during worktime it is located. (Fig. 15) From this basin several penstocks directly reaching the turbines come out, one for each turbine. The pipes, usually made of masonry, direct the flow into the turbine compartment right in front of the paddles (Fig. 16). The small room containing the turbine has three walls made of solid refined masonry stones, a massive vault, characterized by a transmission motion axis located in its center. The fourth wall of this room is completely opened in order to let the water flow outside.

Machineries are located in a large room upstairs, which is made of masonry and covered with a roof laid on wooden trusses. Sometimes, usually in small buildings, a single pitch follows the slope in which the building is located. A small porch area works as a shelter in case of rain, to allow the loading and unloading operations. In several cases a small room adjacent to the wall where the entrance door is located was used for the storage of seeds and flour.

The imposition of strict health regulations, banning the presence of farming animals close to the mill and the fact that the mill was located far from villages, forced millers to settle in towns instead of living in the mill. Millers families preferred to live in urban environments in order to benefit of its facilities.

Simple stone frames decorate doors and windows complementing the wooden frames. Rarely the front door stone was decorated with the owner's arms or a date.

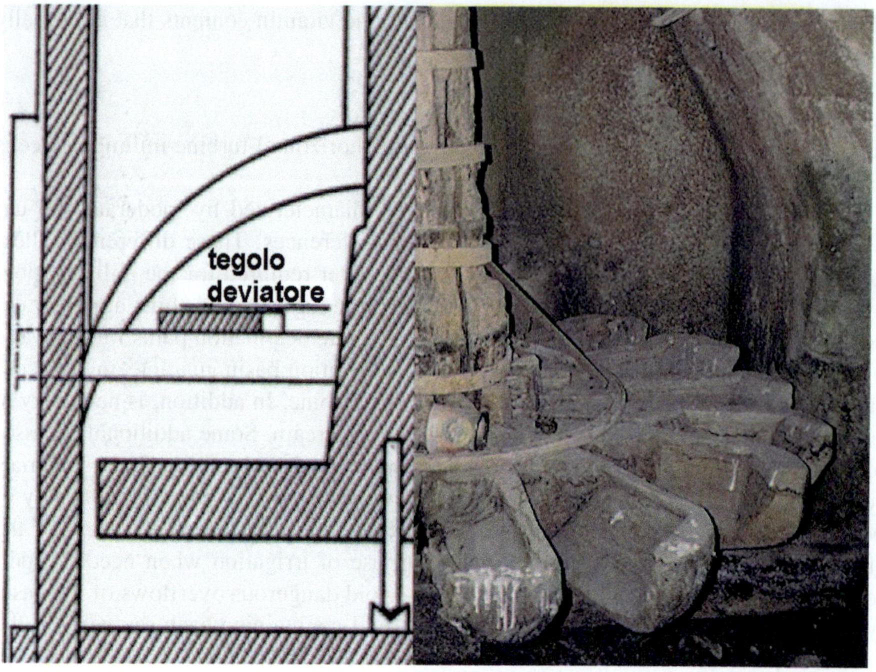

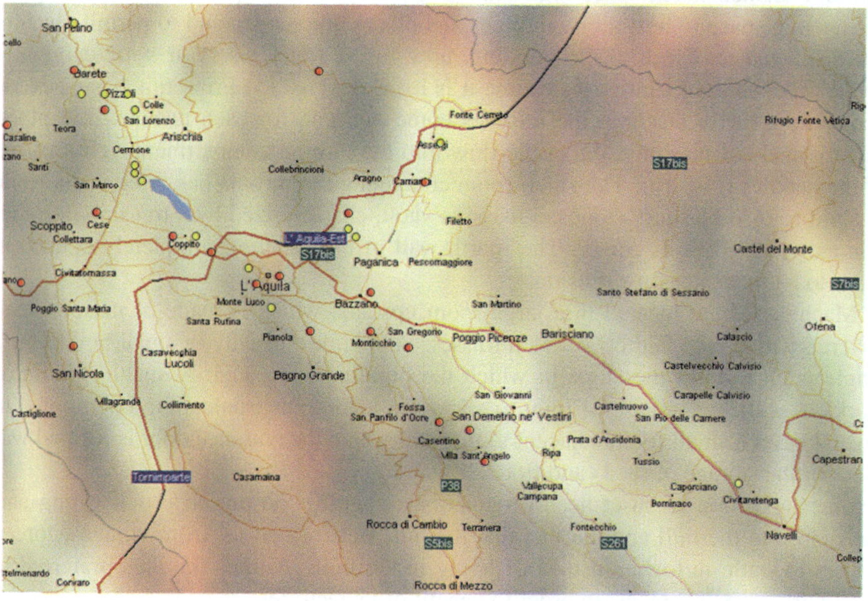

Fig. 17. Turbine's picture and section.
Fig. 18. Map showing the mills location in the area.

Mechanisms. During the survey campaign analogies between the installations came out: turbines, containment tanks, grain-channelling funnels, auxiliary mechanisms are substantially similar. These evidences led to the idea that all of these plants were produced by the same provider[8]. The only significant difference, the stone of the mills, coming from different and distant caves, suggested later maintenance substitutions.

Turbines, the most dynamic component, show remarkable hand-crafting abilities. Indeed, various different skills were involved in the construction process: the blades were carefully shaped with oak wood, while joints, bindings and pivot systems were forged with metal (turbine's original elements were probably rebuilt and refurbished several times). The axis of rotation, vertical and slightly conical, following the natural shape of the wood, is put in place by a cone/cup system (made in bronze or steel) on a wooden shelf-shaped brace. This one (which could be micrometerically adjusted by operating a lever and a flywheel/screw mechanism) allows to modify the distance between millstones, and then to choose the consistency of the product. The blades, from 6 to 12, are hand-crafted in wood and then plunged into the head of the rotating axis. This is reinforced with a double metal belt, in order to better resist to tangential stresses (Fig. 17).

The main device is located in the upper floor on a platform, centered on the turbines' axis. It is made up of a fixed millstone and a rotating one, of the same diameter (about 1 m.), linked to the moving axis by a plug. The lever mechanism, sustaining the axis, sets the distances between the millstone by micrometrical adjustments. A pyramidal funnel with adjustable outlet releases the grains into the mills. Then, because of their peculiar surface, milled flour is pushed out from the outer sides of the stones. It falls into a box with slanted bottom, collecting the product into sacks. A number of other mechanisms allows the control over the amount of water hitting the blades, the speed of the seeds falling between the milling stones, the distance between them and the outflow of milled product. The ability of the operator consists in mastering the control over the variables in order to obtain the desired final quality. Other instruments, mentioned before, are needed to complete the production process. The power produced by these machines can be evaluated using the following formula: $P = g \times Q \times H \times \mu$ (kW), where g: gravity acceleration; Q: pipe water delivery; H: hydraulic head (m); μ: machine efficiency. Using the values derived from mills' surveying ($Q = 0{,}1\ m^3/s$; $H = 4{,}0$ m; $\mu = 0{,}5$), the power is ~ 2 kW.

The mill survey campaign

The survey campaign started in 2007 and was suspended in 2009 as a consequence of the destructive seism occurred in L'Aquila. This interest originates from the observation of the cartography on a scale of 1:25.000 of the valley, produced by the Istituto Geografico Militare during the '50s. A careful examination of the Aterno river path and its affluents showed a great number of paddle turbines, indicating the presence of mills. This circumstance was in contradiction with our knowledge of this area. Willing to fill a gap in our knowledge we decided to undertake a study involving different

[8] In the Assergi mill, on the stone case, we found the carving "*FABB. FERRATA E VITATI BRESCIA*", the manufacturer's name.

topics: history, hydrology, mechanics, architecture. The examination of these facilities and devices has been performed through researches on their history in local and bibliographic archives and through the field research of milling vestiges. We believe that only through the integration of all contributions deriving from these different research fields will create a comprehensive and exhaustive description of the phenomena that have allowed the construction, the use and the actual neglected conditions of mills. Currently out of fifty-two mills shown on the cartography, many have disappeared, some are mere wall layouts and only in fifteen cases, mainly preserved installations, it has been possible to perform the survey of wall manifacture and machine ruins. Some mills are in very good conditions because of their transformation into restaurants, private association bases, B&B, etc. Only three are still operative. The map in the figure 18 shows the location of the still existing case study mills around the territory.

4 Conclusions

Historical analysis and mill's surveying favour the comprehension of mill's documental values. They can be studied as indicators of modifications of technologies, society, economy, territory during centuries. Thus watermills are cultural heritages and we have to favour their conservation, protection and exploitation. This research, still to be finished, permits the conservation of the memory of a phenomenon which has characterized the history of Aterno Valley. Moreover this study would like to arouse the attention of people in charge of natural parks which are controlling these territories (the National Park of Gran Sasso and Monti della Laga and the Park of Sirente Velino) in order to start the recovery of these former industrial activity vestiges. Finally, the mill is an excellent example of sustainable industrial business with a high level of integration with the surrounding territory. The mill is the effective evidence of a possible integration of man's activities with his living environment and an incisive testimony of an ancient device still bearing a product with great economic value. The mill might be the perfect testimony of those values and principles that a Park should communicate and promote.

References

Aliberti, G.: L'evoluzione delle tecniche molitorie in Italia nell'Ottocento. Giunti e Barbera, Firenze (1970)

Bloch, M.: Mélanges historiques. S.E.V.P.E.N., Paris (1963)

Centofanti, M., Brusaporci, S.: Topos/Antropos Le vie della transumanza e il sistema insediativo storico nel territorio tra il Gran Sasso ed il Sirente. In: Gambardella, C., Giovannini, M., Martusciello, S. (eds.) Le Vie dei Mercanti Rappresentare il Mediterraneo, La Scuola di Pitagora, Napoli, pp. 495–500 (2008)

Cigola, M., Ceccarelli, M.: Historical Development of Paper Mills and Their Machines in South Latium During the Nineteenth Century. In: Portilla, J.M., Ceccarelli, M. (eds.) History of Machines for Heritage and Engineerig Development, pp. 85–117. Springer, London (2011)

Continenza, R.: Best practice example. Case studies from inner Abruzzo. In: Rethinking Cultural Heitage, Experiences from Asia and Europe. Alinea Digitaldruck Gmbh, Dresden (2007)

Docci, M., Maestri, D.: Storia del rilevamento architettonico e urbano, Laterza, Bari (1993)

Duby, G.: Rural economy and country life in the medieval West. University of Soth Carolina Press, Columbia (1968)

Forbes, R.J.: Studies in ancient technology. E.J. Brill, Leiden (1965)

Capocaccia, A.A., Mondini, A., Forti, U.: Storia della tecnica. UTET, Torino (1980)

Grand, R., Delatouche, R.: L'agriculture au Moyen Age: de la fin de l'Empire Romain au 16 siecle. E. De Boccard, Paris (1950)

Le Goff, J.: Das Hochmittelalter, Fischer Bucherei, Frankfurt a. M. (1965)

Pirenne, H.: An Economic and Social History of Medieval Europe. Harcourt Brace, New York (1937)

Puato, U.: La presenza di mulini nei catasti antichi della Provincia di L'Aquila (secc. XVI, XVII e XVIII), BDASP, L'Aquila (1990)

Slicher van Bath, B.H.: Agrarian history of Western Europe. Arnold, London (1963)

Section IV

History of Technical Systems

The Front Dial of the Antikythera Mechanism

Michael T. Wright

Centre for the History of Science, Technology & Medicine,
Imperial College, London

Abstract. The Antikythera Mechanism, found in a shipwreck dating from the first century B.C., is the oldest geared mechanism known. Its general function was astronomical or calendrical, but the exact nature of the display on its front dial remains uncertain. We know that the places in the Zodiac of the Sun and Moon were shown, and the date. The author has previously argued that the places of the five planets known in antiquity must also have been shown. He devised mechanism to drive these indications, and demonstrated the viability of his reconstruction by building a working model. Reviewing his earlier publications, he clarifies obscure points and shows how recent study by others supports his arrangement of the display. Others, again, have proposed an interesting alternative to some parts of his restored mechanism for the planetary indications. It bears a close relationship to the device found in the original instrument for modelling the lunar anomaly, and may offer certain practical advantages. The author welcomes the new proposal, and suggests its further development. He concludes, however, that as yet there is no compelling reason to abandon any significant part of the reconstruction that he previously proposed.

1 Introduction

The Antikythera Mechanism [1], found in a shipwreck datable to the second quarter of the first century B.C. [2], is the earliest known geared instrument. Price gave the first correct account of its general arrangement, with front and back dial displays on two opposite faces of a wooden case, interconnected by internal mechanism worked by the rotation of an arbor projecting from one side [3]. Inscriptions on several surfaces had already indicated an astronomical context. Subsequently, making use of radiography, Price studied the instrument's internal mechanism with a view to identifying its precise function. In his reconstruction, the front dial showed the places of Sun and Moon in the Zodiac and the date, while the back showed the age of the Moon and other, less confidently identified, calendrical information. This instrument had no clear purpose, but he called it a "calendar computer" [4]. Price's reconstruction has now been superseded: partly by my own findings, described in a series of papers which are listed in two recent publications [5, 6]; and partly by those of the *Antikythera Mechanism Research Group* [7], who have published two papers [8, 9]. There is now general agreement over many features of the Antikythera Mechanism although, due to its poor state, minor points of uncertainty can probably never be resolved; but here I consider a major source of continuing debate, the reconstruction of its front dial.

T. Koetsier, M. Ceccarelli (Eds.): Explorations in the His. of Machines & Mech., HMMS 15, pp. 279–292.
springerlink.com

The Antikythera Mechanism, Reconstruction by M.T. Wright: left, front dial; right, back dial.

The two dials differ markedly in character. The back dial, which we now understand fairly fully, presents straightforward periodic information in two separate displays: principally a cycle of 235 months or 19 years (the calendrical cycle associated with Meton) above; and a cycle of 223 months (the cycle now called the Saros, indicating eclipse-possibilities) below. The arrangement is ingenious, elegant and efficient, particularly in its use of spiral scales to spread out these two long sequences of months over five and four turns respectively, making them easily legible and allowing the displays to yield further information; but the effect is simply that of sets of tables with moving pointers. The mechanical arrangements for driving these displays are essentially also simple. The user turns an input – a hand-knob in my reconstruction – which rotates a contrate wheel; this wheel engages a large spur wheel, one rotation of which represents one year, and the pointers on the back dial all derive motion from this wheel through straightforward compound trains of spur gearing embodying velocity-ratios reflecting the well-established period-relations noted above. I shall have occasion to refer to the large wheel again, and I follow Price in referring to it as wheel B1, the first wheel on axis B.

By contrast, although it is more poorly preserved, the front dial display was clearly more visually appealing. There were two large concentric graduated rings. The inner was divided into twelve parts bearing the names of the signs of the Zodiac, subdivided into degrees of longitude. The outer bore the names of the months of the year according to the Egyptian calendar, subdivided into days. Price suggested that two pointers showed the mean motions of the Moon and Sun, and that the Sun pointer also showed the date on the outer ring; or (for a reason that need not detain us) he envisaged the possibility of a third, separate date pointer.

I argue that the front dial originally carried further pointers, for the five planets known in antiquity. Price also thought of this possibility; or, more precisely, he suggested that an empty space behind the dial in his reconstruction might have contained gearing to drive such planetary indications [4]. It was characteristic of Price that he would drop into a paper any idea that he thought might possibly be relevant [10], but he appears not to have taken this one very seriously because he did not pursue it in any way. He passed over his seemingly prescient remark quickly and lightly, so that by the time I began developing my own reconstruction of the front dial display I had forgotten it; when I came to examine the Mechanism for myself I found Price's account so much at variance with what I saw that I chose to ignore the written word altogether and to look at the artefact with an "innocent eye".

As Price had conjectured and I later confirmed [11], wheel B1 and a central arbor on axis B passing coaxially through it were geared together so that if one rotation of B1 were to represent one year, one revolution of the central spindle would represent one sidereal or tropical month. ("Sidereal month" was Price's term. However, the dial includes a formal representation of the Zodiac and not a star-map; and the tropical year, not the sidereal year, had come to be regarded as a constant in Hellenistic astronomy [12]. Therefore the term "tropical month" is correct.) Significantly, it also emerged that Price had made a mistake in tracing this gear train. He believed that the two mobiles turned in opposite senses, and so he introduced a conjectural reversing arrangement. In reality, the two turn in the same sense, and pointers for the mean motions of the Moon and Sun could have been connected directly to the central arbor and to the wheel respectively [13].

By simplifying the arrangement behind the front dial, the removal of Price's reversing system made it easier to develop a new reconstruction based on hitherto unexplained features on and around wheel B1 which suggested strongly that epicyclic mechanism had been mounted on it. This in turn brought to mind the epicyclic theory of Hellenistic astronomy, a kinematic device for the compounding of two circular motions. The earliest trace of this idea is the interest shown in it by Apollonios of Perga (*fl. c.* 200 B.C.), reported by Ptolemy. In *The Almagest* [14], Ptolemy includes a theorem of Apollonios in which he explores the applicability of epicyclic theory to the apparent motion of the planets. Ptolemy follows Hipparchos (*fl. c.* 140 B.C.) when he discusses this same "Apollonian" epicyclic theory in connection with his treatment of the anomalous motion of the Sun and Moon. The construction of the Mechanism almost certainly postdates the work of Hipparchos, so the inclusion in it of a mechanised version of epicyclic theory to portray the motion of the Sun, Moon and planets is chronologically acceptable. This is the theoretical basis of my reconstruction of the front dial, as a concentric planetarium display of Moon, Sun and the five planets known in antiquity.

I first presented this scheme, illustrated by sketches and a crude cardboard model, in a conference paper read in 2001 (but publication of the proceedings was delayed) [15]. Early the next year I announced the completion of a working model [16], and later the same year I described both the model and some of the reasoning behind it more fully [17]. For reasons beyond my control all three papers were severely limited in length. Further papers were planned, but events took on a momentum of their own, with other aspects of the Antikythera Mechanism occupying what little time I could devote to the subject.

I felt no urgency. It seemed likely that this material would be of concern to only a small community composed mainly of those studying the history of instruments, of mechanism, and of astronomy. I supposed that anyone sufficiently interested to wish to discuss my work would apply to me for information and for the opportunity to inspect my model, which I regarded as a form of publication in itself. The main message, after all, had already been broadcast by Price: the Hellenistic world was capable of this degree of mechanical achievement; the detail was less important and, for those who worked in these fields, I thought much of it would be obvious.

That perception may have been mistaken. A relentless publicity machine has now brought the Antikythera Mechanism to the attention of a very much wider public, increasing the scope for misunderstanding and misrepresentation. The priority and value of my work has been contested. In revisiting the subject some ten years after having first presented my reconstruction, I show here that my contribution is still relevant and I welcome some newer developments.

Front dial of the Antikythera Mechanism, Reconstruction by M.T. Wright, second model, 2009.

2 The Planetarium Dial

I have explained my methodology, in using model-making both as an aid to developing my reconstruction and as a medium for subsequent communication [6]. The resort to model-making has however had an unintended consequence. As the model became well known, uncertainty grew both about it and about the reconstruction that it illustrates. Some suppose that I claim the model as a faithful representation of the original in every respect; others suppose that I do not intend my reconstruction to be taken as a serious essay in any way. Neither extreme view does justice to my work. I wrote about the design choices involved some ten years ago when the model was new [15, 16, 17], but it will be helpful to reiterate and emphasise some points before carrying the argument further forward.

From the outset this was a serious attempt to develop a reconstruction that would fit the extant fragments; was of the same character and class of work as the original; could account for hitherto unexplained features; and could endow the instrument – for which no-one had yet offered any satisfactory rationale – with a plausible function. These points, together with the further one that reconstruction as a planetarium connects the artefact to the literary evidence for planetaria in Graeco-Roman antiquity, have already been made [15, 16, 17]. Perhaps, however, I have not recorded in print how the evidence of an inscription on the instrument supports my reconstruction. Among the lettering that has long been legible to the naked eye, on a part that Price named the "Back Door", he records ΤΗΣ ΑΦΡΟΔΙΤΗ[Σ] (*tēs Aphroditēs*: of Venus); two lines further on is ΓΝΩΜΩ[Ν] (*gnōmōn*: pointer); and on the next line ΗΛΙ[Ο]Υ ΑΚΤΙΝ[?] ... ΗΛΙΟΝ (*hēliou aktin[?] ... hēlion*: Sun's ray ... Sun) [4]. The implication is that something was stated about Venus, and that Venus might be associated both with a pointer and with the Sun. There was a pointer for the Sun on the front dial, and it followed, as a strong possibility, that Venus was also displayed there. Price had earlier reported "terms ... that refer to stations or retrogradations of planets" [3]. These readings, implying strongly that the instrument was concerned with planetary phenomena, which (as we now know) can only have been displayed on the front dial, are now being gradually confirmed and extended, as we shall see.

Epicyclic theory was almost certainly widely known and discussed at the time that the Antikythera Mechanism was made, but it is proper to be cautious in considering just how freely the instrument's creator might have translated it into practical mechanism. Therefore I adopted a strict design criterion, introducing no subtlety into the process: as in the treatment attributed to Apollonios himself, the larger circular motion of the platform would always represent the Zodiacal motion of the body.

The general arrangement of the reconstruction followed directly from this point of principle. The epicyclic mechanism to be restored to wheel B1 – a platform rotating at the rate of the Mean Sun – was thereby limited to modelling the motion of the inferior planets, Mercury and Venus, and the anomalous motion of the Sun. Unfortunately, the ideal of devising a reconstruction accounting for all traces of structure on the wheel was – and remains – elusive. In my first model I was content to show that mechanism serving the three functions described could be fitted to the space, knowing that there were many ways in which an appropriate supporting structure could be devised that would make use of at least some of the fixings for which we have evidence. In a second model, I took more trouble to match my structure to the evidence, but still it is not all accounted for. A survey of this evidence is being prepared for publication. Identical sets of wheels, described and illustrated elsewhere, are fitted to each model [16, 17].

Reconstruction of the Antikythera Mechanism by M.T. Wright. The internal mechanism is shown removed from its case.

According to my design principle, mechanism for modelling the motion of the superior planets, Mars, Jupiter and Saturn, with their individual Zodiacal periods, called for three further epicyclic platforms. These could be interposed between the assembly on wheel B1 and the front dial, because the separation between those parts of the original instrument is indeterminate.

The wheel-work for the superior planets used in the first model has also been described and illustrated [16, 17]. In the second model, the same velocity-ratios and the same arrangements are used throughout, but the numbers of teeth are multiplied to make the wheels larger without making much change to the pitch, so that alterations to the design (whereby the central pipes and bosses are of larger diameter) are accommodated. I note in passing that the arrangements chosen have intrinsic advantages of interest to the designer of mechanism; but since this part of the scheme is conjectural I shall not discuss them here. It is however appropriate to say that the rather high number of wheels used represents no real objection; the trains run freely, and it is clear that the designer of the original mechanism was not afraid to use many wheels, some with high numbers of teeth.

The wheel-count was increased significantly by my decision to substitute compound gear trains for simple wheel-pairs in all the planetary models, in order to show that good approximations to the planetary periods could successfully be modelled. It is widely recognised that Babylonian planetary period-relations were available to the Hellenistic astronomer, and that these might have recommended themselves to the instrument-maker because they can be modelled using simple gear-pairs; but I thought it important to show that the designer might easily have attained a much higher accuracy in modelling the planets, comparable with that of the Metonic period-relation (determining the length of the month) which is already attested within the Mechanism. If compound trains realising good approximations would run, there could be no doubt that simple gear-pairs, having about the same velocity-ratio but representing coarser approximations (such as those derived from Babylonian period-relations), would run even more freely. The accuracies achieved have been described [17], but the justification for making this demonstration has not been fully argued. A discussion of the sort of refined periods that the Hellenistic designer might plausibly have aimed to approximate, and of the way in which he might have cast up approximations in designing compound gear trains, is the topic of a further paper now in preparation.

It was only some time after developing the basic reconstruction that I identified the boss of the Moon hand for the front dial, with its indication of the Moon's phase [18]. The addition of this central feature, together with a corresponding conjectural alteration to the Sun hand to provide a display of the day of the month, brought the front dial to its present state.

As well as having been described in print [6, 16, 17], a version of my model may be viewed as a computer-animation generated by Vicentini which may be accessed either directly from his own website [19] or through mine [20]; but note that the program "VCL Media Player" must be selected to play the animation. Currently it may also be seen, without the need to change default software, on the website of the Museo Galileo of Florence [21]; but there it is not seen in a full-screen view. This animation actually illustrates a second model, differing slightly from the first one; the few points of difference relevant to this paper are described here.

I devised my reconstruction before it was understood how the pin-and-slot modelled the lunar anomaly. So, intent on demonstrating just how complete a planetarium display the Hellenistic mechanic might easily have developed using only the epicyclic theory, I added to the first model a further small epicyclic system for the Moon. The discovery of the purpose of the pin-and-slot system made this conjectural addition redundant, whereupon I removed it. The certainty that lunar theory was included did, however, increase the probability that solar theory, for which I fitted conjectural epicyclic mechanism to both models, would have formed a part of the original scheme.

The identification of the pin-and-slot device and its purpose revealed greater subtlety of mechanical design than had been appreciated, and indicated that my adherence to a literal, direct realisation of epicyclic theory may have been unduly strict. Below I shall discuss the interesting suggestion of Carman *et al.*, that this device presents us with an alternative way of modelling the motion of the superior planets.

3 Other Conjectural Schemes

My restoration of planetary mechanism originated in the need to explain evidence seen in the original fragments, which is surely a *sine qua non* for any satisfactory reconstruction. However, two papers published by other authors at about the same time, exploring the idea of planetary displays from more abstract standpoints, should be mentioned.

Edmunds and Morgan [22] argued, plausibly, that hints found in the fragmentary inscriptions encouraged one to speculate about possible planetary indications. Basing their conjectural mechanisms on the epicylic theory, they outlined simple arrangements; but, having no first-hand knowledge of the instrument and apparently relying mainly on their reading of Price, they envisaged assemblies that simply could not be fitted to the original. Despite information and comments from myself, the authors thought it impossible to fit more than one epicyclic display on any one dial, and concluded that it would be "difficult" to incorporate more than two planets at once in the whole instrument.

Taking the ideas of Edmunds and Morgan as his starting point, Freeth [23] followed them in adopting directly epicyclic theory for the inferior planets. Unlike them, he thought both inferior planets might be fitted to the front dial, but only with "some difficulty". For the superior planets, Freeth developed several options. In the first, the epicyclic platform rotates – as in the case of an inferior planet – at the rate of the Mean Sun, but the epicycle is envisaged as a large internally-toothed gear-ring which runs around a smaller fixed central wheel. Freeth wisely expressed doubt as to whether the Hellenistic mechanic could have made this arrangement work, but he seemed unaware that it was, in fact, a realisation of the *eccentric* theory, understood by Hellenistic astronomers to yield the same effect as the epicyclic theory [14]. He then developed what was in effect the Apollonian epicyclic model, but again seemed unaware how close was the parallel between his scheme and conventional Hellenistic astronomy. In a first version he followed Edmunds and Morgan, applying torque to the platform, and driving the epicycle wheel by running it round a fixed central wheel. Freeth was right in suggesting that this system could work for Mars; but in advocating its use for Jupiter his judgement was more questionable because a significant step-up ratio was involved in driving the epicycle. For Saturn, with a much higher step-up ratio still, Freeth devised two further mechanisms, finally preferring one in which the epicycle was to be driven independently of the platform, using the same device that I had previously adopted.

Freeth's reasoning seemed to be mathematical, not directly related to Hellenistic astronomical theory, and arguably somewhat anachronistic; he acknowledged that his analysis "depend[ed] on a heliocentric knowledge of the Solar System". We are left unsure as to how readily the designer of the Mechanism might have arrived at some of these schemes.

It was however in the suggested placement of his mechanisms in the instrument that Freeth lost touch with reality altogether. Like Edmunds and Morgan, he indulged in a pure flight of fancy, envisaging modifications to the reconstruction of Price so that he might fit displays of the three superior planets (and one for the age of the Moon) to the back of the instrument.

In neither case had the authors actually examined the Mechanism, and their suggestions bore no very close relation to the surviving parts of the original. Neither paper need detain us further, except to remark that Freeth's first proposal for the superior planets – a platform rotating at the rate of the Mean Sun with an anomaly-generating device mounted on it – may be said to contain the germ of the idea more recently developed by Carman *et al.* [24] which I will discuss below.

4 The Antikythera Mechanism Research Group

The mechanical ensemble now known as the "pin-and-slot" lies within the extant parts of the Mechanism. It comprises two wheels lying face to face, running on a stationary stud that is stepped so that they turn about separate axes. One wheel has a pin planted in its face which projects to enter a radial slot in the other, so that if one of them were to rotate at a steady rate the angular velocity of the other would be modified by a cyclical variation. It was clear that the ensemble might be used to model an astronomical anomaly [25]. In their first publication the Antikythera Mechanism Research Group showed that the device was so arranged as to modify the motion of the Moon pointer on the front dial in accordance with a lunar theory like that of Hipparchos [8]. The conjectural epicyclic mechanism for the same purpose was therefore redundant, but it was the work of a single afternoon to remove it from the model and to effect other small alterations that made my provisional reconstruction conform to the Group's findings. None of this occasioned any outward change to the front dial display.

In the supplementary information appended to the same paper the Group presented provisional readings of the fragmentary inscriptions, extending those previously published. The "Back Door" inscription, mentioned above as including the words "of Venus ... pointer ... Sun's ray ... Sun", now seems even more telling. The words are now seen to lie within a text in which the word ΣΦΑΙΡΙΟΝ (*sphairion*: little sphere, globule) is repeated several times. The line immediately above the one naming the Sun has been expanded to read: ΓΝΩΜΩ ... ΚΕΙΤΑΙ ΧΡΥΣΟΥΝ ΣΦΑΙΡΙΟΝ. I translate this as: "... pointer ... there lies [or: "is placed"] a golden globule". In a later instrumental tradition a golden marker often denotes the Sun [26], and the mention of a golden globule, possibly in association with a pointer, just before the naming of the Sun, seems very significant. The repetition of the word "globule", before and after this short passage, suggests that "of Venus ... pointer ... there is a golden globule ... Sun's ray ... Sun" may be embedded within a longer descriptive list of markers for other celestial bodies besides Venus and the Sun. Such markers would have been useless unless individually distinguishable, and in the later tradition this was commonly achieved by making them of different materials or by painting or enamelling them in different colours. Thus the new reading seems to strengthen the evidence for planetary indications on the front dial along with those of the Sun and Moon, and even to give us a hint as to the appearance of the pointers: distinguished by raised globules or "buttons" of appropriate colours instead of, or in addition to, engraved names. The reading of the inscriptions is in the hands of epigraphic experts with direct access to the Group's images, whose work is still in progress, and I have hesitated to alter the pointers of my model until we know whether further details may emerge.

The Group's further publication in 2008 [9] led to some further small alterations, but only to the back dials.

5 Carman, Thorndike and Evans

Evans *et al.* [27] make the interesting observation that the surviving fragment of the Zodiac scale from the front dial appears to be divided unequally, in a way that is consistent with a deliberate intention that a single pointer should show both the date on the calendar scale (which is taken to be uniform) and the true position of the Sun on the Zodiac scale. They then suggest offsetting the axis of the Sun pointer and the dial so that the Moon pointer shows the Moon's longitude correctly (or more nearly so) on the unequally-divided ring. The scheme is practicable but appears untidy and unnecessarily complicated. Moreover, it seems possible to attribute the observed inequality to error in manufacture.

In the same paper, the authors outline an alternative to my concentric kinematic planetary display, suggesting that a set of five subsidiary dials might have been provided for the planets, each with a pointer indicating the phases of the planet's synodic cycle. The arrangement would certainly side-step problems that might be associated with the proposed increased complexity at the centre of the main dial, but the authors base their argument on the reported number of teeth on wheel B1, from which they derive the motion for working these indications. They also suggest an alternative explanation for the size of wheel B1 and the features on it, as a display of essentially decorative elements.

These ideas merit greater attention on another occasion; but I will now discuss the substance of a further paper by the same authors in which they offer alternative mechanism to drive pointers for the superior planets in a concentric display [24].

Freeth *et al.* [8] went to some lengths to show that the pin-and-slot ensemble can generate an output with a varying angular velocity identical to that of epicyclic theory. Carman *et al.* demonstrate the same point, but more clearly and more economically, as a prelude to showing how this ensemble might have been used to work planetary indications.

It is generally assumed that any kinematic planetary displays in the Antikythera Mechanism would have been based on a single-anomaly theory. That is, only Ptolemy's "second anomaly" – the anomaly with respect to the Sun which accounts for the spectacular episodes of retrogradation – would have been modelled. Carman *et al.* show that the arrangement for the lunar anomaly actually found in the Mechanism can be adapted to become just such a single-anomaly model for the superior planets, and they show in outline how it might have been fitted to the extant fragments.

According to the lunar theory associated with Hipparchos, the Moon's motion is described by a single anomaly with respect to a slowly-rotating line of apsides; so the anomalous motion must be compounded with the slow, uniform motion of the line of apsides itself. In the Mechanism this is done by planting the pin-and-slot ensemble, driven at the mean rate of rotation of the Moon and generating the anomaly, on a platform rotating at the rate of the line of apsides. The output motion – comprising two rotations superimposed one on the other – is taken out through the centre of the platform, and transmitted through to the dial pointer.

In the single-anomaly model of a superior planet, the anomaly is symmetrical about the moment of opposition with the Mean Sun; that is, it is symmetrical about the line through the Mean Sun and the Earth. So the motion of a superior planet may be generated in the same way as is that of the Moon, by superimposing the anomaly on the rotation of the Mean Sun. Wheel B1 rotates at this rate, and Carman *et al.* suggest that three pin-and-slot devices, one for each of the three superior planets, could have been mounted either on wheel B1 itself or on further structure carried on it. There would have been a fixed central wheel for each planet, around which the appropriate pin-wheel would run. The three pin-wheels would, therefore, have been set at different heights on wheel B1. The numbers of teeth in the fixed wheel and in the pin-wheel would have been chosen in each case so that the pin-wheel, and with it the slot-wheel, rotated relative to the Mean Sun Wheel with a frequency approximating the planet's synodic period. The slot-wheels would also have been set at different heights so that each could deliver its output – comprising the anomaly superimposed on the motion of the Mean Sun – to the appropriate pointer through a central wheel with the same number of teeth as itself.

Without having tried it out, one cannot be quite confident that this system is really practicable, especially as applied to Mars. Modelling the very large anomaly for Mars calls for a large offset between the axis of the pin-wheel and that of the slot-wheel. The pin must therefore approach close to the centre of the slot-wheel once every revolution, when the pressure between the pin and the side of the slot, and the reactions at the pivots, will be high. Carman *et al.* appreciate this potential difficulty and claim to have satisfied themselves that it can be overcome. They report having modelled the system for Mars and having tried it in practice, apparently as an independent device. However, they give no details of their trial – scale, materials, parameters &c. – and there is room to doubt how well the device would work, and how long it would remain workable, as part of a complete planetarium instrument. For example, friction between the concentric pipes carrying the hands would place a load on the pin-and-slot mechanism which might, if not carefully managed, lead to its rapid wear; and a quite minor accident could cause pointers to foul one another and give rise to sudden heavy loads.

Carman *et al.* restrict themselves to designs based on approximate period-relations drawn from Babylonian astronomy which can be realised by the use of simple gear-pairs. Better approximations to the planetary periods for the superior planets, such as those used in my models, could be obtained by driving the pin-and-slot ensembles through compound trains. The pin-and-slot device produces a symmetrical effect, so it may run in either direction on wheel B1, and the number of axes in the train driving it may be odd or even; but obviously the final output must run the right way, and so the overall number of epicyclic axes for each planet must be odd. Such compound trains would be less economical in the number of wheels used, and the arrangement would almost certainly be less compact, than the simple gear-pairs envisaged by Carman *et al.*; but the idea is interesting because pin-and-slot devices worked by compound trains could achieve results as good as those of my models while taking less space and using fewer wheels than the gearing that I have suggested. Correspondingly, the epicyclic systems that Carman *et al.* envisage for the inferior planets could also be driven using compound trains, just as they are in my models.

Carman *et al.* appear to urge, as a point in favour of their scheme, the consistency that it offers in using for each superior planet a mechanical ensemble closely comparable with that found modelling the lunar anomaly in the Mechanism itself; but they introduce inconsistency by falling back on the use of the Apollonian epicyclic system for the inferior planets. It is however quite straightforward, in principle, to adopt the pin-and-slot ensemble for each of the inferior planets too. (The same device could also be used for the solar anomaly, but Carman *et al.* adhere to the main idea of their earlier paper in suggesting that the solar anomaly was accommodated by having an unequally-divided Zodiac scale.) Here the stepped stud of the pin-and-slot ensemble might remain stationary, fixed to the frame. Motion from wheel B1, or from any other mobile turning at the rate of rotation of the Sun, would be transmitted to the pin-wheel through gearing with the correct velocity ratio to ensure that the anomaly was generated with the right periodicity. Gearing with the inverse velocity ratio would then return the output to the central axis, so that the resultant motion would have the required mean Zodiacal period of one year. Whoever was sufficiently acute to have devised the known mechanism for the lunar anomaly, and perhaps the similar mechanism that Carman *et al.* propose for the superior planets, should surely have had no difficulty in imagining this application to the inferior planets also.

Therefore, if it is thought to be a point in favour of any reconstruction that the several anomalies are modelled by similar mechanisms, then the pin-and-slot scheme that I have just outlined for the inferior planets (and perhaps for the solar anomaly) must be examined in connection with that of Carman *et al.* for the superior planets. The most obvious way to mount the non-rotating mechanism would be to plant it on the frame plate, or on a bridge or similar structure spanning the other wheels planted directly on the plate. Either way, this conjectural mechanism for the inferior planets would most naturally have lain behind wheel B1 (now carrying only the mechanism for the superior planets), but there is no room for anything of that sort there.

Perhaps there were such non-rotating pin-and-slot ensembles for Mercury and Venus, arranged in some less obvious way. If, on the other hand, we must accept the use of epicyclic mechanism for these planets, the case for using any different mechanism for the superior planets is weakened. The use of the pin-and-slot in modelling the lunar anomaly would then become the "odd one out": a point that might fit with an idea – first conceived to account for different oddities, some of which still remain unexplained – that the back part of the Antikythera Mechanism may have been partly or wholly rebuilt at some time during its working life [28].

6 Conclusion

In my reconstruction of the front dial of the Antikythera Mechanism, a satisfactory (and satisfying) planetarium display, with concentric pointers for date, Sun, Moon and the five known planets, is plausibly fitted to the instrument. This central focal feature provides the instrument with a rational purpose, and it is in accord with the evidence for an interest in planetary phenomena indicated by the inscriptions on the instrument. The back-dial displays, very different in character, make more sense as supplementaries to a main, planetarium display than as principal functions in their own right; they are like the smaller tools on the back of a pocket-knife. Fitted with a planetary display, the instrument accords with literary evidence that attests the existence of planetaria; whereas other reconstructions present us with rather complicated devices of very little real purpose, which find no mention in the literary tradition.

The mechanical arrangement that I devised for driving the planetary display accounts for more of the artefactual evidence than does any other reconstruction to date. It is made up of elements which are all based on precedents found within the original fragments, and all are made in the same style, using the same techniques, as the original parts. I have shown that this arrangement works well.

It is important to distinguish between the general principle of the reconstruction and those parts of the particular designs described and modelled that are no more than conjectural. The latter may be changed without vitiating the reconstruction as a whole. In particular, where choice presented itself I elaborated the reconstruction as far as I thought the designer of the original could possibly have desired. In this way I erred – if anything – on the safe side in making my demonstration that the scheme was practicable. I was cautious in another respect, in choosing to implement only the simple epicyclic theory associated with Apollonios, and only in the most straightforward way. This led inevitably to the addition of a significant number of wheels in order to add the superior planets to my concentric display. I went further in showing how, using compound gear trains, all the planetary periods could be approximated very closely. This was intended to demonstrate that the accuracy attainable by the Hellenistic mechanic knew no practical bounds, and that cruder periods, such as those derived from Babylonian period-relations and capable of being realised using simple gear-pairs, could certainly be modelled with ease. In the atmosphere that obtained when I devised my reconstruction, of considerable doubt as to what the Hellenistic mechanician might have been able to achieve, this demonstration was important.

Our subsequent understanding of the working of the pin-and-slot mechanism, as modelling the lunar anomaly in place of my conjectural epicyclic mechanism designed for the same purpose, has led to thoughts about its wider application. Carman *et al.* have shown how it could have been used to drive concentric pointers for the superior planets in a particularly economical way. I point out here that it could also have been applied to modelling the motion of the inferior planets and the solar anomaly. Such changes to the planetary mechanism might allow us to economise the number of gear-wheels used, or to make the wooden case of the Mechanism a little slimmer, but they would not necessarily make any difference to either the appearance or the function of my front dial display.

The extended reading of inscriptions on the instrument suggests a small cosmetic enhancement, with the addition of "globules" of differing materials or colours to distinguish the several pointers on the front dial instead of – or as well as – engraved names; but at the same time it seems to srengthen the evidence supporting my conjectural restoration of a full planetarium display with pointers for Moon, Sun, five planets and the date. So far I see no reason to make any significant changes to this reconstruction. As for the mechanism driving the planetary pointers, the arrangements that I have put forward represent just one set of possibilities among many. It is interesting to consider other suggestions, but there is as yet no compelling reason to reject my arrangement in favour of another.

References

[1] National Archaeological Museum, Athens, inventory number X.15087
[2] Yalouris, N.: In: Descœudres, J.-P. (ed.) Eumousia, Meditarch, Sydney, p. 135 (1990)
[3] Price, D.J. de S.: Scientific American 200(6), 60–67 (1959)
[4] Price, D.J. de S.: Transactions of the American Philosophical Society 64(7) (1974)
[5] Wright, M.T.: Interdisciplinary Science Reviews 32(1), 27–43 (2007)
[6] Wright, M.T.: In: Staubermann, K. (ed.) Reconstructions, pp. 1–20. NMSE Publishing Ltd., Edinburgh (2011)
[7] `http://www.antikythera-mechanism.gr/`
[8] Freeth, T., et al.: Nature 444, 587–591 (2006)
[9] Freeth, T., et al.: Nature 454, 614–617 (2008)
[10] Simms, D.L.: Private communication
[11] Wright, M.T.: Bulletin of the Scientific Instrument Society (85), 2–7 (2005)
[12] Pedersen, O.: A Survey of the Almagest. Odense University Press (1974)
[13] Wright, M.T., Bromley, A.G.: In: *Αρχαία Ελληνική Τεχνολογία* (Ancient Greek Technology), Proceedings of the 1st International Conference under that title, EMAET, Athens, pp. 19–25 (1997)
[14] Toomer, G.J.: Ptolemy's Almagest. Duckworth, London (1984)
[15] Wright, M.T.: In: Paipetis, S.A. (ed.) Extraordinary Machines and Structures in Antiquity Proceedings of a Conference under that title, pp. 81–94. Peri Technon, Patras (2003)
[16] Wright, M.T.: Horological Journal 144(5), 169–173 (2002); 144(6), 193 (2002)
[17] Wright, M.T.: In: Η *Αρχαία Ελλάδα και ο Σύγχρονος Κόσμος* (Ancient Greece and the Modern World) Proceedings of a Conference under that title, pp. 86–97. University of Patras (2003)
[18] Wright, M.T.: Antiquarian Horology 29(3), 319–329 (2006)
[19] `http://www.mogi-vice.com/Antikythera/A-W-M.zip`
[20] `http://www.mtwright.co.uk`
[21] `http://brunelleschi.imss.fi.it/galileopalazzostrozzi/multimedia/TheAntikytheraMechanism.html`
[22] Edmunds, M., Morgan, P.: Astronomy & Geophysics 41, 6.10–6.17 (2000)
[23] Freeth, T.: Mediterranean Archaeology & Archaeometry 2(2), 45–58 (2002)
[24] Carman, C.C., Thorndike, A., Evans, J.: Journal for the History of Astronomy 43, 24 (2012) (seen in pre-print)
[25] Wright, M.T.: Antiquarian Horology 29(1), 51–63 (2005)
[26] Field, J.V., Wright, M.T.: Early Gearing. The Science Museum, London (1985)
[27] Evans, J., Carman, C.C., Thorndike, A.S.: Journal for the History of Astronomy 41, 1–39 (2010)
[28] Wright, M.T.: In: *Αρχαία Ελληνική Τεχνολογία* (Ancient Greek Technology) Proceedings of the 2nd International Conference under that title, TEE, Athens (2006)

The Historical-Technological Study of Monumental Tower Clocks: The Town Hall Clock of Alcalá la Real, Jaén (Spain)

López-García Rafael[1], Dávila-Rufián Iván[2], and Dorado-Vicente Rubén[1]

[1] Department of Mechanical Engineering, University of Jaén (Spain)
rlgarcia@ujaen.es
[2] Student of Technical Industrial Mechanical Engineering, University of Jaén (Spain)

Abstract. This article is a contribution to industrial archaeology, the science dealing with the study of past industrial remains. The paper offers a historic technological study of the mechanical clock in the tower of the Town Hall of Alcalá la Real in the province of Jaén (Spain). It is part of a project, which encompassed the compilation of historic documentation, fieldwork, the description of the functioning and the graphic reconstruction of the clock's machinery. After a brief introduction about the history of clocks and the description of the working of monumental bell tower clocks we proceed to the details of the town hall clock. We describe how we constructed on the basis of fieldwork by means of infographic techniques a 3D model that gives a virtual simulation of the working of this mechanical device.

1 Introduction

The origin of the term Industrial Archaeology (IA) goes back to the XX century. Pioneers of this discipline were the Baron of Verneilh who proposed in 1876 the study of French forges from the XVI to the XVIII century, and the Portuguese archaeologist and historian Sousa Viterbo who, in the article called "Portuguese Industrial Archaeology", aimed at mills that then were on the verge of disappearing [1].

One of the first IA definitions was [2]: "a scientific discipline whose aim is the discovering, cataloguing and study of physical remains of past industrial activities, to know through them significant aspects of working conditions, technical processes and productive processes".

Industrial Archaeology sprung linked to the raise of interest for physical remains from the Industrial Revolution in England at the beginning of the 60s, and it is there where the term was first used by authors such as Dudley and Rix. Among the first carried out tasks was to begin the inventory of Industrial Revolution remains with the drawing-up of Industrial Revolution Index with the aim to know and preserve what still remains.

Keneth Hudson considers IA a practical history, which studies industrial remains from the past. On the other hand, Angus Buchanan does consider IA as archaeology, because it requires fieldwork, archaeologist techniques, and expects the research, analysis, register and preservation of the studied objects. For him, industrial

T. Koetsier, M. Ceccarelli (Eds.): Explorations in the His. of Machines & Mech., HMMS 15, pp. 293–307.
springerlink.com

archaeology is a practical activity of research, documentation and classification of industrial monuments, and it is also "an historic study which allows to evaluate the meaning of these monuments in the social-history and technology context" [1].

Nowadays the Industrial Archaeology studies the remains of technology to understand the Industrial past. It must be supported with a research methodology and technology tools, as well as the use of information technologies (IT) [3].

This work, apply Industrial Archaeology methods to understand the mechanical clock situated in the tower of the Town Hall of Alcalá la Real, carrying out a historic and technological study of the machinery by means of fieldwork and currently available technological tools and analyse its influence in society, indicating time in life, and society organization, time synchronization, etc.

To attain the previous goal the following tasks have been completed:

- Recompilation of historic and technological information of clocks to analyse the evolution of its technological advances and types throughout the times.
- Description of the generic working of clocks, their main elements and functions.
- Description of belfry clocks performance. Our main goal is the belfry clock of the Town Hall of Alcalá la Real, Jaén (Spain).
- Fieldwork: measure of each clock's elements, as well as an inventory of photographs and videos, which complete the graphic documentation of the clock. The fieldwork provides information to create and analyse a virtual model by infographic tools [4].
- 2D and 3D graphic reconstruction of the machinery. We make plans, which complete the graphic description and the working from the only existing original plan of the clock.
- By means infographic techniques, 3D modelling texture and the carrying-out the simulation of different mechanisms of the clock. The videos obtained from the previous simulations show the clock cinematic and performance.

Section 2 shows brief historic information about clocks, whereas section 3 is devoted to describe the clock tower of Alcalá la Real. We detailed our fieldwork and the 3D model reconstruction in Sections 4 and 5. Finally the main conclusions are drawn in Section 6.

2 CLOCKS

A brief explanation of clocks evolution is included in the first half of this section. Because the proposed study is devoted to tower clocks, the second part of this section describes the technological characteristics of mechanical clocks.

Historic evolution

The first clock known is the Gnomon clock, a vertical style stuck in a quadrant or sphere, which shows the hours guided by the sun movements. The first Gnomon was probably the Chaldaeans, around 2234 B.C. [5].

The Egyptian solar clock appears around 600 B.C. It consists on a flat stone with a stone head facing the rising of the sun. As it rises, due to its celestial orbit, it shows the hours, more briefly as it reaches the vertical. From midday it moves towards east and starts showing inversely afternoon hours. The only inconvenience is that you have to wind it or move it twice a day.

Sundials do not work at nights or cloudy days. Other systems like sand clocks, fire clocks (based on the speed of fuel consumption), and water clocks solve the previous drawback.

The first fire clocks were in ancient China around the 5th century A.D. It consisted of using tarred ropes with equidistant knots. When it started burning from one end, one knot after another started to consume, in the calculated space of time of one hour.

Sand clocks are dated around year 25, which consists of two bottles joined by the neck, in which there is a narrow space, which allows the sand to slide from one bottle to another at a short pace. After the upper container poured all its sand the clock had to be turned the other way round.

The water clock (also called clepsidra by the Greeks which means "water thief") determines the time by the speed the water spilled from one container to another. Water level indicated precisely the time. To prolong the interval of the measured time sometimes several similar recipients were made [6]. Temperature disturbs the measures indicated by these clocks. They also showed deviations due to water impurity and corrosion [7].

Around the X century appeared mechanical clocks powered by weights. Weights pulled a small wheel, subject to controlled pressure, which at the same time pulled a hand on a quadrant, indicating the time. Huygens introduced the pendulum, whose mission was to regulate the wheel's movement in a uniform way. A few years later appeared clocks powered by spiral springs and regulated with a steering wheel and a weaker spiral spring [6].

Soon a reduction of the size was demanded. These clocks became private instruments with which people did not depend on bells or belfry. Tools and materials limited the size reduction. New technologies opened the door to portable pieces powered by an interior motor. For it, a spiral spring was adopted, which meant having to reduce the train and improve its articulation, in a way that the required weight was relatively very small. The introduction of the spring motor, with snail wheel, made the real movable domestic clock possible and, finally, thanks to the advances in reducing sizes, the portable chronometer, which we know as pocket or waist watch (Fig. 1).

Quartz (electronics) and atomic watches are based on nowadays technologies. Electronic watches use a phenomenon known as piezoelectricity. This phenomenon produces mechanical vibrations, which are used to count time intervals. The first quartz clocks were made in the USA around 1928 [7].

Atomic clocks use an atomic resonance frequency as a frequency standard for their timekeeping element [8]. Agencies of national norms keep exactitude of 10 seconds per day [9].

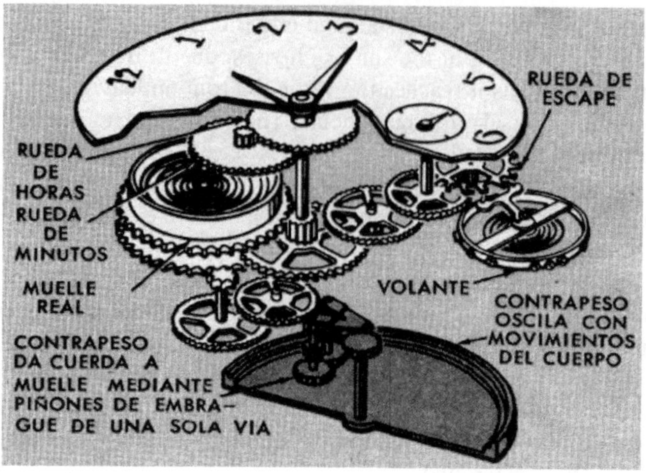

Fig. 1. Scheme of a waist/pocket mechanical clock

Tower Clock Components

A clock can be analysed from its pieces and its different functions [6]. A mechanical clock is divided in six main parts:

- Impulse motor (Fig. 2). It maintains the clock movement. The wheel train and the escapement device link the impulse motor and the oscillator. The impulse can take place in different ways: with weights, royal spring, with rope, with electricity, etc.

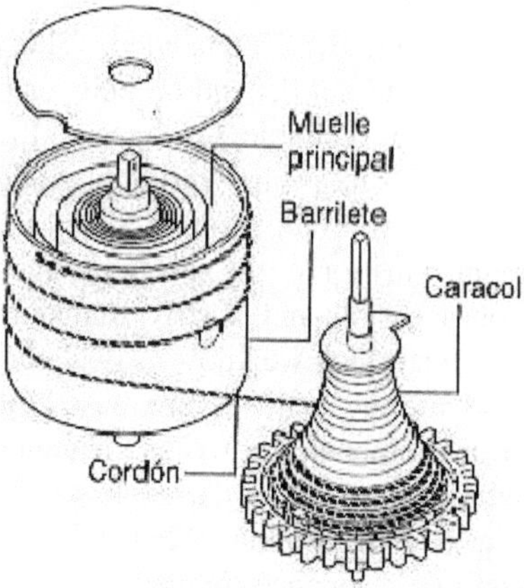

Fig. 2. Spiral spring

- Wheel train (Fig. 4). Gear train, which transmit the energy from the impulse motor to the escapement device. It is also connected to the hour system.
- Escapement device (Fig. 3). It moves the oscillator system with a constant frequency.
- Oscillator. The periodic movement needed to measure time is realised by a pendulum or a wheel driven by the escapement device.
- Remontoire. It runs the timekeeping mechanics and provides a secondary source of power.
- Hour system. Thanks to this device we can read the hour indicated by the clock.

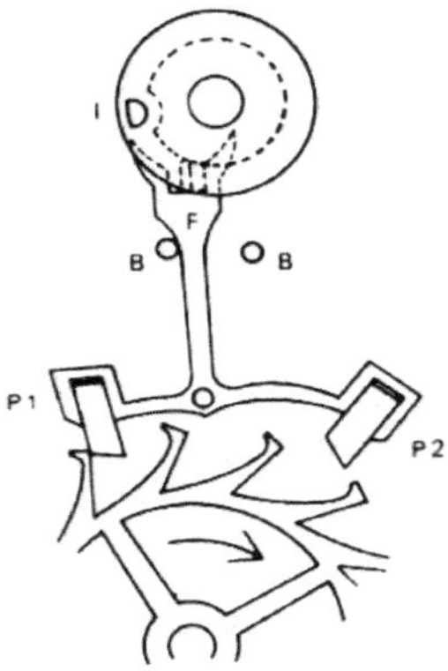

Fig. 3. Escapement wheel and ancora

According to the previous clock components, the following list enumerate the main features of tower clocks:

- Impulse motor: by weights
- Wheel Train: solid, composed of five or six wheels (Fig. 4)
- Escapement: variable
- Oscillator: by pendulum
- Hour system: composed by spheres and needles
- Remontoire: by lever arm key

The solidness of all the set of wheels is great: it is made to resist heavy weights and friction for hundred of years. It is not unusual that fifty years past before a clockmaker intervenes in them [6].

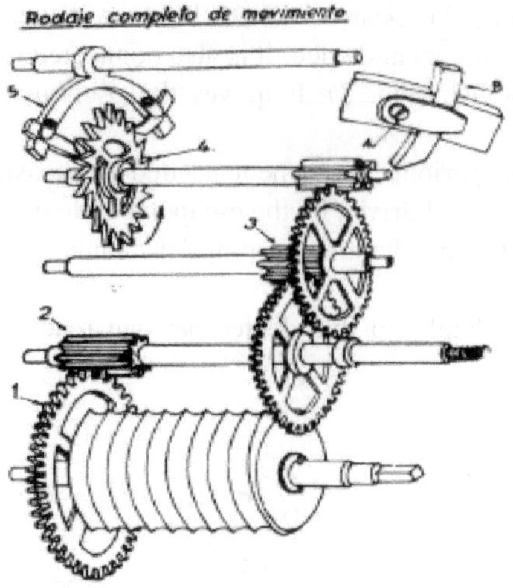

Fig. 4. Wheel train of a tower clock

3 Clock Tower of the Town Hall of Alcalá la Real

Brief biography of Fernando de Tapia [10]

The clock of the Town Hall of Alcalá la Real was made by Fernando de Tapia (Alcalá la Real, 1749-1841). Fernando de Tapia y Castilla was the Mayor of Torre de Fuente Álamo and Perpetual Town Councillor of Alcalá la Real. Throughout his life he was dedicated to public issues and his work as a clockmaker gave him greatest fame. Born in Alcalá la Real in 1749 and at the age of 29 he displayed his first clock to king Carlos III who appointed him as Manager of the Royal College of Clockmaking of Málaga. At the age of 35 he displayed several watches, probably pocket, to the King and Prince of Asturias (future Carlos IV).

At the age of 37, Fernando de Tapia started making the clock of the Town Hall of Alcalá la Real and he finished it in twelve years. He published two books related to watchmaking: "*Conocimientos prácticos en el Honesto y Delicado Ejercicio de la Reloxería*" partially conserved, and "*Nuevo método sobre el modo de poner la luna en toda clase de reloges de ruedas con exactitud, y sin el herror que hasta ahora se ha observado*" (Fig. 5) written in 1810 [11], in which he gives the solution to the problem of the lunar cycle duration related to the inaccuracy of the 59 tooth wheel for two cycles [12]. At the age of 65 he took two watches to King Fernando VII. He died in 1841.

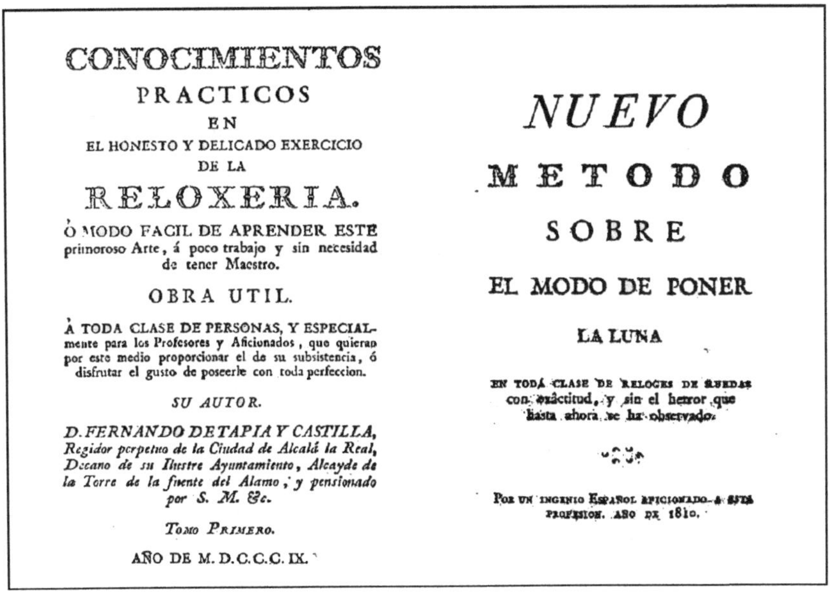

CONOCIMIENTOS
PRACTICOS
EN
EL HONESTO Y DELICADO EXERCICIO
DE LA
RELOXERIA.
Ó MODO FACIL DE APRENDER ESTE
primoroso Arte, á poco trabajo y sin necesidad
de tener Maestro.
OBRA UTIL.
À TODA CLASE DE PERSONAS, Y ESPECIALmente para los Profesores y Aficionados, que quieran por este medio proporcionar el de su subsistencia, ó disfrutar el gusto de poseerle con toda perfeccion.
SU AUTOR.
D. FERNANDO DE TAPIA Y CASTILLA,
Regidor perpetuo de la Ciudad de Alcalá la Real, Decano de su Ilustre Ayuntamiento, Alcayde de la Torre de la fuente del Alamo, y pensionado por S. M. &c.
Tomo Primero.
AÑO DE M. D. C. C. C. IX.

NUEVO
METODO
SOBRE
EL MODO DE PONER
LA LUNA
EN TODA CLASE DE RELOGES DE RUEDAS
con exâctitud, y sin el herror que
hasta ahora se ha observado.
POR UN INGENIO ESPAÑOL AFICIONADO A ESTA
PROFESION. AÑO DE 1810.

Fig. 5. Front page of Fernando de Tapia's book

Description of the Machinery
The clock is in the tower of the Town Hall of Alcalá la Real, in the province of Jaén (Spain), in a closed small-sized room with ventilation, on the third floor of the tower (Fig. 6). Over this room is the campanile, which lodges different bells.

Fig. 6. Hall of Alcalá la Real and current view of the clock

This clock has the following features:

- Motor drive. By weights. The weights are joined to the mechanisms with pulleys and proportionating the movement thanks to the gravitational force, as they are suspended from a rope, which is winded to the drum of the corresponding axis. There are four weights in total, one for each one of the mechanisms.
- Wheel Train. It is made of four gear trains that transmit the movement. Each train has the right number of gears and teeth in function to the desired movement.
- Escapement. It is a Catherine wheel with ratchets (Fig. 7 and Fig. 14). The ratchets stop the Catherine wheel to obtain a desired movement. There are two ratchets that alternatively lock-liberate the Catherine wheel. The ratchets are integral to two different axes that rotate in inverse direction.
- Oscillator: It is a pendulum joined to the axis that moves one of the ratchets.
- Remontoire (Fig. 8). Each mechanism has a squared section bar (Fig. 8.a) in which there is an integral toothed wheel (Fig. 8.b), which fits together with another wheel (Fig. 8.c) integral to the axis where the rope (Fig. 8.d), joined to the weight, is winded in such a way that while turning around this bar with a lever it winds up the clock as the weight raises. It is winded once a day.
- Hour system. The hour is marked by the top sphere, which has a hand for hours and another hand for minutes. The lower external sphere marks the lunar phases.

Fig. 7. Ratchets.

Fig. 8. Remontoire

The tower clock of Alcalá la Real carries out four functions nowadays: tells the time in upper external sphere, bells quarters and hours, and marks the phases where the moon is in the lower external sphere. The mechanisms are laid out in a structure, in cubic form, which is divided into four cubes in which the four mechanisms are fitted in (Fig. 9). To carry out these functions, the clock has three differentiated mechanisms, joined by a set of rotating axes with different levers (Fig. 11 shows this transmission system):

Fig. 9. Clock structure and mechanisms

Timing mechanism

It is in charge of providing the hands of the sphere the necessary movement to mark the time. It consists of a gear train with a drum (in the lower axis) on which a rope is winded up in such a way that due to the gravitational force, a weight joined to the rope, transmits a circular movement. The escape is located in this mechanism, which starts delaying the movement of the wheels by means of ratchets, which stop periodically the Catherine wheel. Without this organ, the movement of the wheel would accelerate until the weight falls freely to the ground.

The parallel and alienated gears train have four axes, which transmits the movement from the lower axis (which receives the movement of the weight) to the top axis, where the Catherine wheel is.

The remontoire engages with the timing mechanism at the bottom axis end. The weight ascends turning the square section handle bar. At the other end is a wheel that transmits motion to the clock needles.

Quarter chime

This mechanism provides the movement to sound the bell of the quarters-hour. A wheel rotates and a set of ledges on its circumference hits a lever that rings a bell (Fig. 10). The motion duration determines the number of bell strokes, four to one. The mechanism has a gear train and aligned parallel axes of five and nine gears.

Hour chime

This mechanism provides the movement to sound the bell of hours. The system is similar to the previous mechanism. It has a gear train and aligned parallel axis of five and eight gears. This mechanism is in motion only when it has to strike the bells. The rest of the time is stopped.

Fig. 10. Levers to ring the bells

Moon phases

This mechanism provides the movement to display the moon phase. The hour system transmits the movement through the end of its axis. It meshes with a perpendicular wheel, which is outside the set and part of the mechanism of the moon. A gear train joins the aforementioned mechanism to a big wheel that indicates the moon phase. This wheel makes one turn every eighty-five thousand, fifty minutes, corresponding to two moons.

The fourth mechanism, which is not working, was used to give the curfew.

Fig. 11. Transmission axes.

4 Fieldwork

In order to preserve and understand the Town Hall clock of Alcalá la Real, we will create a Computer aided model. This model must be an accurate geometrical reproduction to guarantee an adequate kinematic simulation.

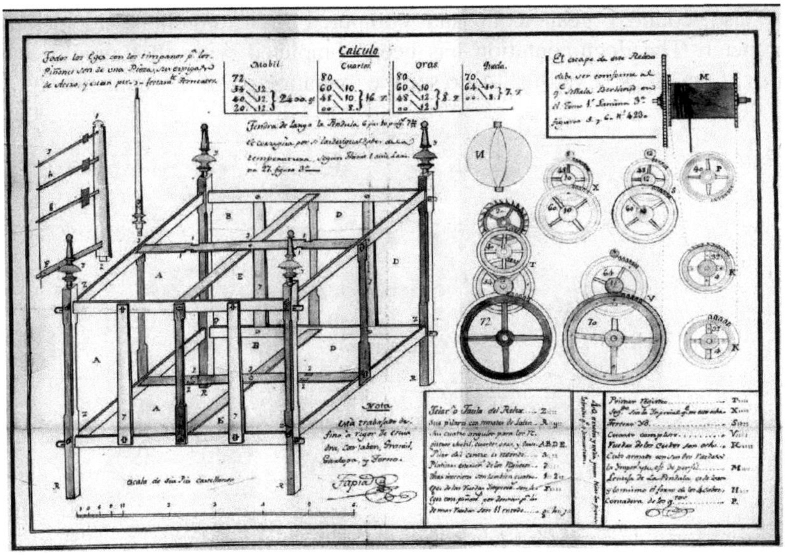

Fig. 12. Original clock plan

To implement the previous model we need information: dimensional and technological. We searched and compiled the available information about the Town Hall clock of Alcalá la Real. Hall files and several publications showed the costs, dates or the people involved in the construction. However, few data were found regarding its technology, operation or mechanical properties. The most valuable document is the original clock plan that is preserved in the town Library (Fig. 12).

Because of the lack of documentation, we measured the clock parts. We portrayed the measures in hand sketches as the one showed in Fig. 13.

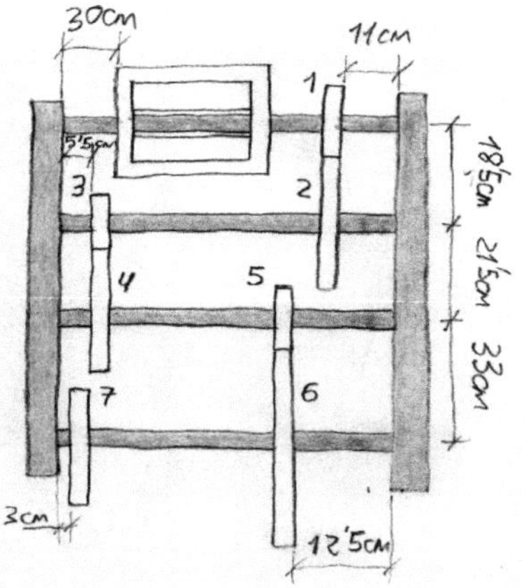

Fig. 13. Example of clock sketch

We also counted gear teeth and computed their angular velocities using a chronometer. The documentation has been completed with photographs and video captures of the clock (like Fig. 14) to see the operation of all mechanisms.

Fig. 14. Escapement (Catherine wheel)

5 Clock 3D Model

We used SolidWorks to model the clock. SolidWorks is an easy-to-use CAD commercial software. It helps to model 3D mechanical systems from their individual components. Each component can be built by operations like extrusion, revolution and cutting. SolidWorks can animate the system and make static and dynamic analyses.

These are the main steps in defining the 3D clock model:

- **Step 1.** Parts, gears, levers, bumpers and other parts unassembled are created in this step (Fig. 15).

Fig. 15. Example of modelled components

- **Step 2.** Gear trains and the transmission mechanism are introduced in this step (Fig. 16)

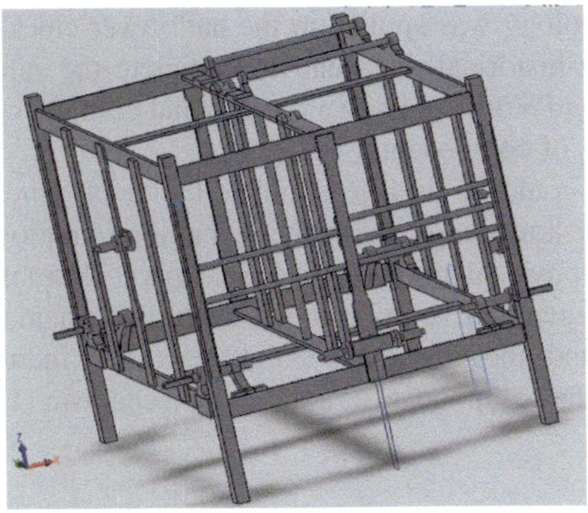

Fig. 16. Modelled structure

- **Step 3.** Once all the pieces are modelled, we proceed to assembly the mechanical system (Fig. 17). Note that we have 4 mechanisms and the transmission (Section 3). These mechanisms are indicated on the Figure 17.
- **Step 4.** Final assembly of the previous mechanisms (Fig. 17)

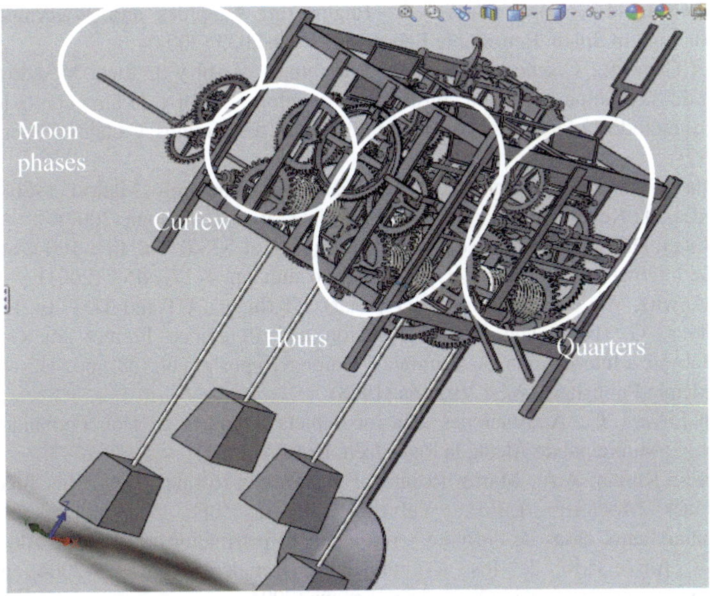

Fig. 17. Final clock assembly

6 Results and Conclusions

This work describes a methodology to preserve and to document ancient mechanical devices. The methodology was applied to the hall tower clock of Alcalá la Real. We compile the existing historical data, and to complete the information, we model a virtual clock with SolidWorks (a known commercial computer aided design software) from actual measures of components.

This CAD model could be used to simulate the clock kinematic and dynamic, in order to improve our knowledge about the technology employed in this mechanical system. Known the physical properties of each part, CosmosMotion, a SolidWork module can provide quantitative information about the kinematics (position, velocity and acceleration), and the dynamics (joint reactions, inertial forces and power requirements) of all the components of a moving mechanism [13].

References

[1] Casado Galván, I.: Introducción a la arqueología industrial: origen de la disciplina y metodología. IES Alhambra, Granada (2009)

[2] Hudson, K.: Industrial Archaeology. Thames, London (1964)

[3] Rojas-Sola, J.I., López-García, R.: Arqueología Industrial en la Sierra Sur de Jaén: Aplicación a los molinos hidráulicos. La memoria del trabajo y propuesta para su recuperación. In: Anales de Ingeniería Mecánica, Año 15, pp. 1653–1662. León, España (December 2004)

[4] Rojas-Sola, J.I., López-García, R.: Engineering Graphics and Watermills: Ancient technology in Spain. Renewable Energy 32, 2019–2033 (2007)

[5] Soler Gayá, R.: Diseño y construcción de relojes de sol y de luna. Métodos gráficos y analíticos. Prontuario para la construcción de relojes de sol y de luna con la justificación de los métodos y fórmulas. Colegio de Ingenieros de Caminos, Canales y Puertos, Madrid (1997)

[6] Belda González, P.G.: Manual de relojería. Editorial Paraninfo, Madrid (1986)

[7] Landes, D.: Revolución en el tiempo. Editorial Crítica, Barcelona (2007)

[8] Sullivan, D.B.: Time and frequency measurement at NIST: The first 100 years. In: 2001 IEEE International Frequency Control Symposium, pp. 4–17. NIST (2001)

[9] Markowitz, W.: Comparisons of ET (solar), ET (lunar), UT and TDT. In: Babcock, A., Wilkins, G. (eds.) The Earth´s Rotation and Reference Frames for Geodesy and Geophysics, International Astronomical Union Symposia, vol. (28), pp. 413–418. Kluwer Academic Publishers, West Virginia (1988)

[10] Juan Lovera, C.: Aportaciones a la incompleta biografía de Don Fernando de Tapia. Archivo Municipal de Alcalá la Real, Jaén, España (2006)

[11] Heredia Rufián, A.A., Martín Rosales, F.B., Murcia Rosales, D.C., Rus Jiménez, L.D., Sanjuán Monforte, L.E.E., Valverde Atienza, J.F., Valverde Atienza, R.G.: Ayuntamiento, casas de enfrente y plaza. In: El patrimonio artístico de Alcalá la Real. XXV Aniversario de los Ayuntamientos Democráticos 1979-2004, pp. 16–19. Ayuntamiento de Alcalá la Real, Jaén, España (2004)

[12] Montañés, L.: Don Fernando de Tapia y Castilla. Una insólita y tenaz vocación relojera, que floreció en Alcalá a caballo entre dos siglos. Revista Iberjoya, Enero, 60–73 (1981)
[13] Dorado, R., López-García, R., Torres-Jiménez, E., Díaz Garrido, F.: Teaching Turbochargers via Computer Aided Design Software. Int. J. Engng. Ed. 26(1), 59–67 (2010)

Improbable Mechanics: A Short History of Fake Automata

Nadia Ambrosetti

Dipartimento di Informatica e Comunicazione
Università degli Studi di Milano

Abstract. A fake automaton is a pseudo-mechanical device that appears running by itself, while someone, generally hidden into or near the device itself, makes it work thanks to repeated inputs. This study aims to investigate their history in the long period ranging from the Antiquity to the Modern Age, by analyzing literary and historical sources.

1 Some Preliminary Remarks

Giving the definition of the Greek word "automaton" can be considered a quite easy task nowadays, since such topic is well studied in computer science, in automation, and in robotics. We must however remark that only starting from the Hellenism the word assumed the meaning of mechanical device, which performs, after a user's input, a finite number of default actions, typically, but not necessarily in a periodic sequence. Before, it was largely used to refer to any event that happened spontaneously, that is without any external input. Automatic devices were generically referred to as statues, or objects capable of performing some actions.

Such a consideration shows entirely the difficulty of any study about automata in Antiquity, since it is very hard to locate passages on the topic in the huge corpus of ancient literature without the help of semantic fields of lexical references; in addition, sometimes the description of their behaviour is so vague that the question whether we are dealing with an automaton or not, remains. For example, a commonplace in ancient Greek and Latin literature is to describe an image, a statue or a bas-relief highlighting the vibrancy of the representation of movement: the image, says the author, seems to move, and the scene appears to take place before the eyes of the beholder. The ambiguity of some of these descriptions has led distinguished scholars [1] to the hypothesis that those objects could be examples of static automata, being similar to the small theatre described by Heron in his *Automatica* [2].

A fake automaton, on the contrary, is a pseudo-mechanical device that appears running by itself, while someone, generally hidden into or near the device itself, makes it work thanks to repeated inputs. Designers probably created fake automata in order to meet too demanding requirements, they were otherwise unable to satisfy, due to the lack of skills, stuff, time, or funds.

T. Koetsier, M. Ceccarelli (Eds.): Explorations in the His. of Machines & Mech., HMMS 15, pp. 309–322.
springerlink.com

2 Fake Automata in History

Egyptian civilization

The ancient Egyptian civilization, thanks to its wide technological knowledge, has always evoked in the collective imagination the idea that they could actually build automated devices, useful, for example, to keep out predators from the treasures buried in the graves, or to contribute to the complex religious or civil liturgy with what we would now call special effects. Unfortunately no positive evidence is available at the moment. On the contrary, archaeologists found that talking (or, better, giving out sounds that could be interpreted as words uttered by a divine voice) statues of gods were widely used to impress the faithful; they worked by means of a trumpet hidden in hollows leading down from the mouth. In the Thirties of the 20th century, the Egyptologist Loukianoff [3] analyzed a statue housed at the Egyptian Museum in Cairo (catalogue #66143), the bust of Re-Harmakhis from Lower Egypt: such artefact is a large white limestone bust of the god, and shows an orifice where the priests could speak in order to amplify and alter the sounds of their voices, simulating the god's tone of voice. Since the god symbolized not only the rising sun, but also the repository of the deepest wisdom, faithful were likely to consult him in order to be advised or warned. Such statues didn't move or act anyway, but this is not surprising and is indeed perfectly consistent, if we consider that hieratic immobility was considered by the ancients the most appropriate attitude of a deity.

The Greek and the Roman World

Despite the large amount of finds from the ancient Greek and Roman world, we lack any physical evidence of automata, whether they were actual or fake. The main reason is probably the perishability of the materials used to build them. Vice versa many interesting mythological and historical sources have survived.

Daedalus. Descendant of the Athenian royal line of Cecrops, Daedalus, which literally means artful, was a mythical skilled craftsman and artisan; the origin of his legend probably dates back to the 5th century BC [4-5].

Daedalus would have been the first who carved statues of the gods [6]; their beauty was such as to induce those who had seen them, to believe that statues could not only move, but also see or speak. The mythical sculptor is then mentioned as one of Socrates' ancestors in Plato's Euthyphrones (11ξ), where Socrates refers to his works as to moving statues. Since we are obviously not dealing with actual automata, some remarks are due. The separation of the legs surely refers to a precise moment in the history of Greek sculpture, the archaic phase, in which the figures were represented standing with their legs near one another, in total immobility; such hieratic attitude was also reinforced by the almost total absence of any form of glance, which would explain why in two texts appears the reference to the opening of the eyes. Daedalus would therefore represent the personification of an important transition in Greek sculpture, when statues legs began to be separated, and eyes were vividly carved. Secondly, the reference to the alleged ability to speak is probably derived from contamination with stories related to sacred statuary. Finally, their being tied up is probably hyperbolic and due to the high realism of these statues, so similar to humans that they would like to move. This does not exclude, however, that some sculptors (whether or not Daedalus) simulated some sudden movement, which would require binding the statue, until the moment it was expected to move.

Demetrius of Phalerum. The Greek historian Polybius (200-118 BC) refers a quite curious episode in his *Histories*, XII, 13, 10-11. The Athenian orator Demetrius of Phalerum (350-280 BC) had been criticized by his political opponent Demochares (355-275 BC), for his entrance into the theatre followed by a mechanical slobbering snail and by monkeys, during the Great Dionysia held in 308. The description is very brief: the snail is moving of itself and is followed by monkeys. According to Demochares (quoted by Polybius), it would have been an allegory of the slowness and stupidity of the Athenians, who had yielded to others the honour of defending Greece, and were tamely submissive to the king of Macedonia, Cassander (350-297 BC). Even if we assume that Demochares was speaking allegorically, the reference to a mechanical contrivance, which doesn't appear as an oddity neither to Polybius, nor to Demochares, is interesting. We have no further detail on the monkeys, so we can't conclude that they were automated puppets; it is more likely to think that they were humans dressed as monkeys, since it would have been extremely risky to leave some monkeys free to move around the theatre. As for the snail, we can exclude that it was an animal, because the remark on the spontaneity of the movement would have been unnecessary; so it remains to investigate which energy source could move it, what size it could be to be visible to everyone in the theatre and also to contain a reservoir for slime, and moreover how it could move on the steps of the theatre, or at least overcome obstacles in the way.

If we consider the overall context, it appears that the snail could hardly be an actual automaton. In fact, the object could not be small, because it would have been invisible in the theatre and it could neither move on the steps of the cavea, nor overcome obstacles. In addition, Demetrius' show would have been extremely slow and therefore boring, or at least not as impressive, as he expected. Furthermore, the snail could not be moved by the action of hot air or of steam, because it would imply the presence of a boiler in it, with noteworthy drawbacks from both the practical and the scenographic point of view (need for wood and for a man to keep the fire, risk of fire, emission of smoke and steam). Some scholars [7] tried to reconstruct the snail operation, by assuming that it was large enough to contain some men who moved in a wheel (similar to the one currently used as a pastime for hamsters), pushing the whole device.

Canachus of Sycion. In a passage by Pliny the Elder (Natural History, XXXIV, 75), the Roman historian refers to a brazen statue of Apollo, carved in the 6th century BC by the Greek sculptor Canachus of Sicyon, for the temple of Miletus. The statue Canachus executed was a nude Apollo, which was known as the Philesian (kissing): it was at Didymi, near Miletus, and was made of bronze from Ægina. The sculptor had also added a stag, so nicely poised on its hoofs, that we must admit that a thread passed beneath. One fore-foots, too, and the alternate hind-foot are made so as to grip the base, the socket being so indented on either side, as to admit of the figure being thrown at pleasure upon alternate feet. The statue would have had in its right hand a stag, that by means of a rope could seem to walk, or, better, to move its legs. The absence of any reference to power sources lets us infer that the rope was pulled by humans.

Fig. 1. A drawing of a bas-relief of the theater of Miletus, showing the Apollo by Canachus

Philon of Byzantium. He was one of the most celebrated engineers of the ancient world, living in the Alexandrian era. Nothing is however known about his life and works, apart from some minor news [8]. He made studies about automated theatres and about *Pneumatics*, as we can argue from some extant fragments of a work about mechanics. The book about *Automatics* is lost, on the contrary the part dealing with *Pneumatics* survived in an Arabic translation [9], so we can have a general idea about his works. Philon's devices generally require that someone activates them secretly, if possible, to allow viewers to enjoy the spectacle of the operation. This input must generally be given, before device operation begins. In one case, #59, according to Carra de Vaux numeration, a repeated input is required.

Device #59 is probably one of the most complex ones, and involves mythology: a young shepherd and a dragon. It should be placed in a temple, near running water from a cave or a steep rise, as Philon recommends in order to have an unlimited quantity of available running water. The builder should also provide a system to drain water surplus. As we can see in Fig. 2, water flows from the top, and falls into the basin, unless the shepherd turns away; in this case, water remains on the device surface, filling a small basin in front of the dragon. The plot of the scene is quite easy: until the shepherd is watching the water source, the thirsty dragon hisses; whenever the shepherd turns away, the dragon finally drinks. Water flow is controlled by a user, who contributes to the operation of the device while it is working, and alters the system state, preventing the water to escape through the pipes hidden in the dragon. In this device the human figure works mainly as a cork, while the theriomorphic one as a pipe.

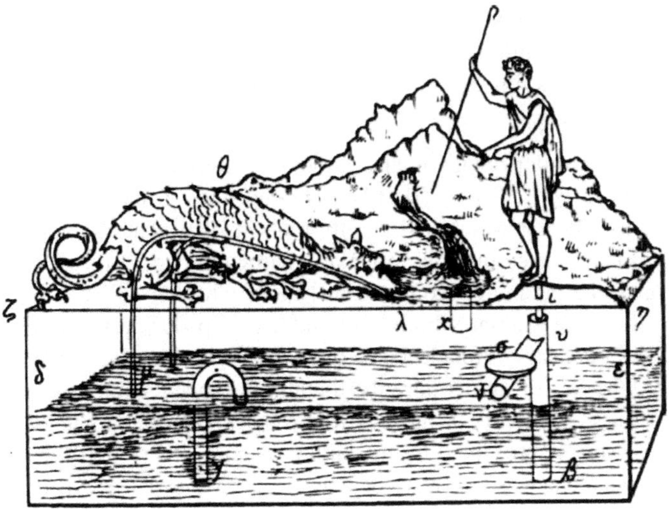

Fig. 2. Philon's device #59

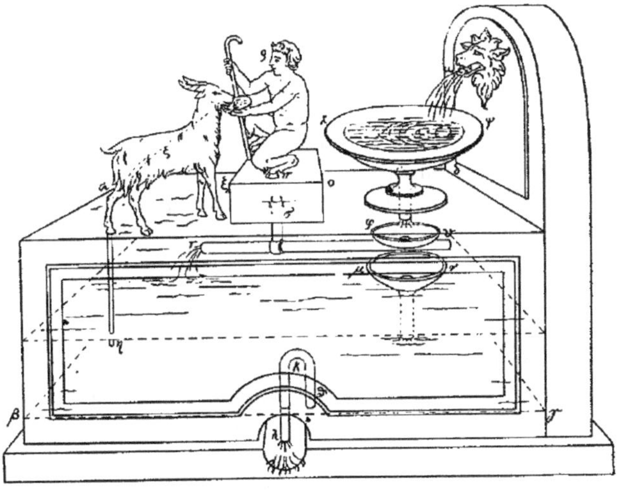

Fig. 3. Heron's device #30, in Pneumatics, I

Heron of Alexandria. He was another famous Greek engineer, who probably lived in the first century AD in Alexandria. Heron's *Pneumatics* is the oldest work, conserved in original Greek manuscripts, describing amazing applications of pneumatic and hydraulic control principles, operating and moving without human involvement. In the first book of Heron's *Pneumatics* [2], device # 30 is a variation of Philon's #59; the narrative scheme is quite different, due to the presence of a turning Paniscos, drawing water from the source, and watering the animal, which may drink whenever a liquid is being presented to it. A user's input is still necessary to make the god turn from time to time. Heron, however, implements in another device (I, 16) a different variation of Philon's #59, where the turning of the guardian (an owl), is not triggered by a user's input, but is actually automatic, working by means of a counterweights system.

Puppets. A very similar device is also quoted by the Greek historian Appianus (95-ca. 165), in his *Roman History* (II, 147). He describes a macabre scene taken from Caesar's funeral, during which, as secretly requested by Antony, a self-moving wax statue of Caesar was used in order to make it show the twenty-three bleeding wounds and to instigate the people of Rome against those who had killed Caesar: the statue, that was hidden near Caesar's body, got suddenly up during the funeral and began moving around the hearse. It had to be a quite impressive scene, as the Roman people immediately burned the curia where the dictator had been slain, at least according to Appianus. Source reliability is not so high, especially due to its uniqueness; the anecdote is, however, presented by the author as so ordinary, to give the impression that such a spectacle was not rare, at least at the time of the historian, i.e. under Trajan's reign. We can be confirmed about the absence of automated movement thanks to Petronius' quotation of the so-called *neurospastes* (1st century AD): they were a kind of puppets that were lowered from the ceiling during banquets, to entertain or, more probably, to scare guests. Caesar's wax statue was likely one of them, activated by some hidden servants.

Medieval China and Tatars

The only Western work that traces the major issues of history of science and technology in China is Needham's *Science and Civilization in China* [10]; in this book, the author faces also the question of automata in ancient China, quoting some instances in the second volume. Some fifteen years ago, Chen restudied the subject, focusing exactly on the oldest works by Chinese scholars about science, and included automata [11].

A Brazen Music Band. According to Needham, this description is taken from the work *Miscellaneous Records of the Western Capital*, that in 1st century AD collects anecdotes about the Han dynasty. The author describes a brazen music band, probably dating back to the end of the 3rd century BC, since he refers that it was found in the treasury of Qin Shihuangdi, the emperor of the terracotta army, who died in 210 [11]. Its working, shown in Fig. 4 as a UML sequence diagram, depends strictly on continuous inputs, since the activator must work for the whole show, by blowing into a tube, and pulling a rope. Obviously by increasing the number of musicians, and consequently of activators, the performance would have been more various, therefore the anonymous used a 12-element band. We can also consider the band as a set of parallel couples of activator-device systems, all working together in the same place.

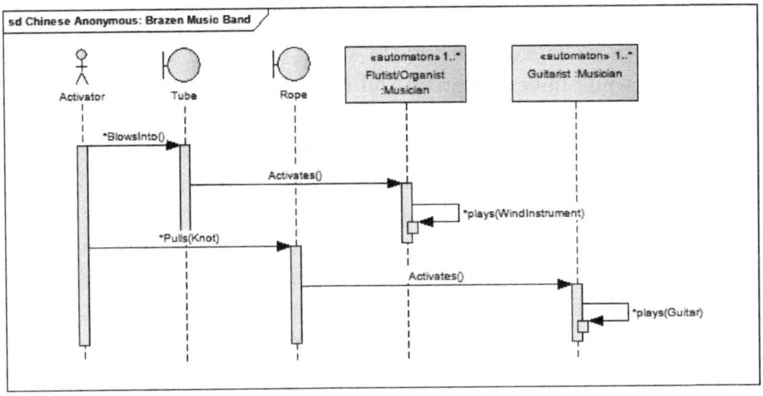

Fig. 4. working of the Chinese Brazen Music Band

Travel Account. Chinese works are however not our only sources about automata in the Eastern lands, as in the 13th-14th centuries, the European friars William of Rubruck, and Odoric from Pordenone made detailed descriptions of automated, or apparently automated systems in their travel accounts, though referred to the Tartars. The Flemish Franciscan William of Rubruck (c. 1220 - c. 1293) in 1253, on king Louis of France's orders, left from Constantinople on a missionary expedition to convert the Tartars. His account can be considered one of the most important works of medieval geography [12]. In his journey he visited the Khan's palace at Karakorum, where he witnessed the famous silver tree by master William the Parisian, that is the French goldsmith Guillaume Boucher, taken away from Belgrad by the Mongols, and brought to their capital, according to Olschki [13].

Fig. 5. The Karakorum Tree

"In the entry of this great palace, [...] master William the Parisian had made a great silver tree, and at its roots are four lions of silver, each with a conduit through it, and all belching forth white milk of mares. And four conduits are led inside the tree to its tops, which are bent downward, and on each of these is also a gilded serpent, whose tail twines round the tree. And from one of these pipes flows wine, from another cara cosmos, or clarified mare's milk, from another bal, a drink made with honey, and from another rice mead, [...]; and for each liquor there is a special silver bowl at the foot of the tree to receive it. Between these four conduits in the top, he made an angel holding a trumpet, and underneath the tree he made a vault in which a man can be hid. And pipes go up through the heart of the tree to the angel. In the first place he made bellows, but they did not give enough wind. [...] When then drink is wanted, the head butler cries to the angel to blow his trumpet. Then he who is concealed in the vault, hearing this blows with all his might in the pipe leading to the angel, and the angel places the trumpet to his mouth, and blows the trumpet right loudly. Then the servants who are in the cellar, hearing this, pour the different liquors into the proper conduits, and the conduits lead them down into the bowls prepared for that, and then the butlers draw it and carry it to the palace to the men and women." [13]

The tree recalls Byzantine and Arabic courtly golden trees, due to the precious metals it is made of; any action is however performed by servants, playing precise roles in order to get a striking effect: pouring drinks for an unsuspecting audience. The tree working is completely wanting of automation, since automata are only the terminal parts of pipes. Boucher tried to imitate the automata described in some texts, or requested by the king; since he was more skilled as a goldsmith than as an hydraulic engineer, he only preserved the external wonderful appearance.

Medieval and Renaissance Europe

During the Middle Ages and later during Renaissance, we see a plain separation among the sources: on the one hand, we have literary works that describe automata created by magicians or sorcerers (including a pope, Gerbert of Aurillac, or theologians, like Albertus Magnus). In these cases, the imagination of the authors has been satisfied to devise a wide variety of devices: talking heads, as automatic pilots for ships, navigators for paladins who have lost their way toward the goal of their heroic enterprise, automatic chessboards, recognition devices, and many more. They all actually work thanks to magic. On the other hand, we have technical sources, that describe real, though simple, automata, mainly operated by counterweights; they were used during religious ceremonies (like the angel by Villard de Honnecourt) or to entertain aristocratic people in their castles and villas.

Hesdin. One of the most interesting examples of automata implementation in the Middle Ages is the whole castle of Hesdin, in Artois. According to our first quotations, in the castle or in its huge park, the counts of Artois, and later, Philip the Good or the dukes of Burgundy, built and/or maintained a large number of automata, at least until 1470, when we have the last quotation, made by the English writer and diplomat William Caxton (1412-1490), fascinated by the place [14]. This castle [15] was an actual *hortus deliciarum* (garden of delights), where noblemen could rest. Nowadays nothing survives but written sources.

Probably the most extraordinary automated system was the *Salle des Distractions*, a large room made available to guests, whose operation was probably the following: the victims entered the gallery, where they were forced to face one after another a group of automata, that criticized them (a statue dressed as a hermit); wet them (floor, generic human figures, fountain, guardian); covered them with flour (a roof, a generic human figure); made them fall (bridge); ordered to leave the gallery as soon as possible, only to send the victims to face another automaton (a bugler). The Lord enjoyed from outside the spectacle of his terrified guests, prisoners in a sort of Medieval candid camera. Although engineers of the Burgundian court had certainly the skills to build automata, it would have been extremely difficult to manage the whole system automatically and dynamically, once the visitor had entered; therefore it is implausible to assume that the entire *Salle* system was automatic. It is more likely that the servants of the Duke controlled the activation of the different devices to make them react properly and timely to the intentional, or unintentional, inputs of the terrified victims, whose behaviour was mainly unpredictable.

Florence. Another complex system, according to Giorgio Vasari, is alleged to the renown Florentine architect Filippo Brunelleschi (1377-1446); it deals with *ingegni*, that is mechanical systems for public feasts from the second quarter of the 15th century [16-17]. Art historians have attributed to him, or possibly to il Cecca, a master of his circle, the authorship of the theatre for the representation of the Annunciation in Florence in 1439, during the council of the Eastern and Western church, in the SS. Annunziata or in S. Marco; and the other ones for S. Maria del Carmine or S. Felice in Piazza, even in Florence. The description of the ceremony by the bishop Abraham of Suzdal [18] reports that it was needed the presence of "persone non vedute" (hidden people) in order to manage some devices, probably for safety reasons (the angels that were raised were not statues, but people).

Fig. 6. A modern reconstruction of the Ingegno dell'Annunciazione

Pratolino. The architect Bernardo Buontalenti (1531-1608) worked for almost all his life at the Medicis' court in Florence [19] and began experimenting in the creation of gardens featuring fountains and caves, habitually populated by automata:

the most famous is the villa of Pratolino. Built as a gift from Francesco I de' Medici (1541-1587) to his mistress and later wife, Bianca Cappello (1548-1587) from Venice, the villa was a permanent exhibition of technical skills, in order to stress the primacy of Florence [20]. Today very little remains of the villa, but its appearance can be reconstructed thanks to the descriptions by De Vieri [21], written in 1587, and by an anonymous in Ms. 2312 236, housed at the Biblioteca Riccardiana in Florence; to a report made by the Modenese draughtsman and painter Giovanni Guerra (1544-1618), sent by the pope to visit Pratolino in 1598; and, as to the Southern part, to a painted lunette by Giusto Utens, now housed at the Museo di Firenze com'era.

The set of automata in the garden of the villa has not remained constant over the centuries and provided descriptions show that this very evolution, probably due both to the high cost of maintenance and to the need to continue to impress even usual visitors. The plumbers (*fontanieri*), who are mentioned in the descriptions as always present in every corner of the park, were responsible for ensuring maintenance, but probably they were also the activators of the automata. Otherwise their ubiquitous presence in the park would have not made sense. Again, as for the castle of Hesdin, besides verifying that automata worked properly, they probably took care to start them at the right time, so that visitors of the park could enjoy the show, and brought systems back to their initial state, once the guests left.

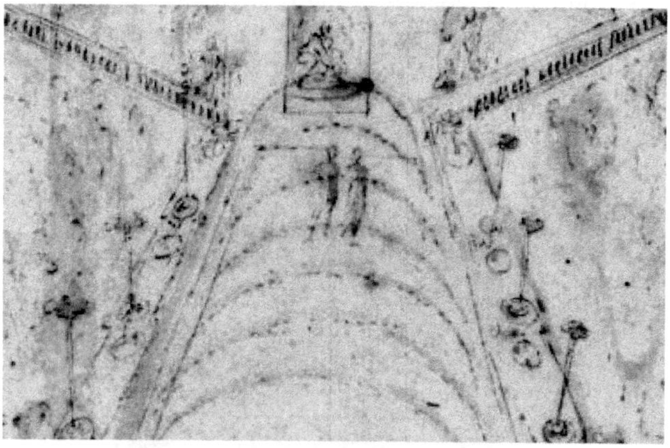

Fig. 7. The "viale delle fontane"

Renaissance Engineers. During the 15th and 16th centuries, the surviving fragments of the works by Heron were found, copied and translated into Latin; they spread widely in Europe and contributed to the development of mechanics related to automata, together with the precision mechanics, already in full development from the late Middle Ages. These are probably the reasons why we find, in later centuries, a large number of actual automata: the emergence of expertise in many artisans and technicians did decrease the need for fake devices. Just to mention a name among the many skilled technicians, we can cite Giannello Torriano (ca. 1500-1585), who worked at the courts of the emperor Charles V and of his son Philip II, building automata that could play and dance [22] or pray [23].

Fig. 8. A 16th-century praying automaton

Modern Times

The court of Vienna at the time of the Empress Maria Theresa (1717-1780) is the setting of one of the episodes in this historical journey. She would have asked the Hungarian Baron Wolfgang von Kempelen (1734-1804) to design and build an automaton that could entertain the court [24]. So in 1769 he made a chess-player, called the Turk, that was exhibited in European courts and in America by the engineer himself and later by others until 1840. Even some of the most famous people of that age could watch the automaton and play chess: the automaton even suggested them some moves, and it defeated some extraordinary adversaries, like Frederick the Great, Benjamin Franklin, Napoleon, or Charles Babbage. [25-28]. It answered questions made by the audience, by indicating letters on a board, and could also perform some more tricks, thanks to a man hidden in the pedestal. Many intellectuals of different backgrounds studied the Turk and decided it was a hoax: the engineer Karl Gottlieb von Windisch (1725-1893) spoke of a couple of forces (vis motrix and vis directrix), representing the two components of the device, the visible and invisible one. The writer and journalist E.A. Poe on the contrary wrote of the Turk as of a fake, because, in his opinion, a mechanical device could only perform a finite set of deterministic actions (like a calculator) and therefore could not determine the next chess move among the hundreds possible moves [29-30].

The Turk was probably the last known specimen of a fake automaton created for the sole purpose of surprising the audience for the exceptional technical and technological knowledge. The era when the Turk was designed and built, is the First Industrial Revolution, when technicians could make even complex machines, capable to ease human effort mainly in repetitive tasks. During the 19th century, the scientific and technological development that would lead to the Second Industrial Revolution, would allow to design and produce devices for widespread use. Audience familiarity with such devices would lead to decrease the need for fake automata.

Fig. 9. The hidden mechanism of the Turk

3 Conclusions

From this brief historical survey of the topic the presence of some constants emerges.

A first interesting issue pertains to control mechanisms. In fact, the main problem the designers of fake automata had to face, was the coordinated management of the subsystems composing the automaton. This was in terms both of reaction time on inputs and in the order of actions required to obtain a specific result. The Turk deserves however a slightly different remark: von Kempelen did not have to deal with the problem of coordinating the movement of the automated arm with the position on the chessboard or with the move of the opponent, but he rather had to accept that the reproduction of human logical procedures was still far from being reached.

Then, we can focus on the two main contexts in which such devices have been used: the religious space and the social space. The religious space is connected with the idea of a trick, done by the priests at the expense of the faithful, sometimes a little too prone to be deceived. In this case, deception is functional to keep alive in the faithful the impression of being in touch with the supernatural. It was therefore essential that the deception was based upon an experience completely different from the ordinary ones: a statue giving relevant answers to questions posed by the consultant, is a typical example. Ancient and Medieval technology were absolutely unable to generate similar devices. The second context is the social one: automata are openly presented as mechanical devices, whose technology already exists and it is also used in other contexts of everyday life (hydraulics, ballistics, and so on). Their aim is thus to impress an audience by applying serious knowledge to a play in an unpredictable way. In this context, the existence of a designer capable of building the device is necessarily implied. In the case of fake automata, however, the skills of the designer are not enough to get the very results and this forces them to introduce further human inputs, in order to make automata perform required (possibly wonderful) actions.

As a matter of fact, along centuries mechanics has always been in high esteem, and consequently expectations from it were also great, as shown in this passage by the

Roman philosopher Cassiodorus: "What a wonderful art is Mechanics! The mechanician, if we may say so, is almost Nature's comrade, opening her secrets, changing her manifestations, sporting with miracles, feigning so beautifully, that what we know to be an illusion is accepted by us as truth." [31]

References

[1] Russo, L.: La rivoluzione dimenticata. Il pensiero scientifico greco e la scienza moderna, Feltrinelli, Milano (1997)
[2] Schmidt, W., et al.: Heronis opera quae supersunt omnia. Teubner, Leipzig (1899)
[3] Loukianoff, G.: Une statue parlante, ou Oracle du dieu Re-Harmakhis. In: Annales du service des Antiquités de l'Egypte, vol. XXXVI, p. 187 (1936)
[4] Morris, S.P.: Daidalos and the Origins of Greek Art. Princeton University Press, Princeton (1992)
[5] Muller-Dufeu, M.: La sculpture grecque. Sources épigraphiques et littéraires. Ecole des Beaux-Arts, Paris (2002)
[6] Marcinkowski, A., Wilgaux, J.: Automates et créatures artificielles d'Héphaïstos: entre science et fiction. Techniques & Culture 43-44, 2 (2004)
[7] Rehm, A.: Antike Automobile. Philologus 46, 317 (1937)
[8] Drachmann, A.G.: Philo of Byzantium. In: Coulston Gillespie, C. (ed.) Dictionary of Scientific Biography, vol. X, pp. 586–589. Charles Scribner's Sons, New York (1974)
[9] Carra de Vaux, B.: Le livre des appareils pneumatiques et des machines hydrauliques de Philon de Byzance d'après les versions arabes d'Oxford et de Constantinople. Académie des inscriptions et des Belles Lettres: Notices et Extraits des mss. de la Bibliothèque Nationale, Paris 38(1), 27 (1903)
[10] Needham, J.: Science and Civilization in China. Cambridge University press, Cambridge (1965)
[11] Chen, C.-Y.: Early Chinese Work in Natural Science: a Re-examination of the Physics of Motion, Acoustics, Astronomy and Scientific Thoughts. Hong Kong University Press, Hong Kong (1996)
[12] Chiesa, P.: Testo e tradizione dell'Itinerarium di Guglielmo di Rubruck. Filologia Mediolatina 15, 133 (2008)
[13] Olschki, L.: Guillaume Boucher: A French Artist at the Court of the Khans. The John Hopkins press, Baltimore (1946)
[14] Franke, B.: Gesellschaftsspiele mit Automaten - "Merveilles" in Hesdin. Marburger Jahrbuch für Kunstwissenschaft (1977)
[15] Brunet, M.: Le Parc d'attractions des ducs de Bourgogne a Hesdin. Gazette des Beaux-Arts 78, 331 (1971)
[16] Zorzi, L.: Il teatro e la città: saggi sulla scena italiana. Einaudi, Torino (1977)
[17] Ventrone, P.: Una visione miracolosa e indicibile: nuove considerazioni sulle feste di quartiere. In: Garbero Zorzi, E., Sperenzi, M. (eds.) Teatro e spettacolo nella Firenze dei Medici. Modelli dei luoghi teatrali. Catalogo della mostra (Firenze, Palazzo Medici Riccardi, 1 aprile - 9 settembre 2001), pp. 39–51. Olschki, Firenze (2001)
[18] Todarello, N.L.: Le arti della scena, Latorre editore, Novi Ligure (2006)
[19] Botto, I.M.: Buontalenti, Bernardo. In: Ghisalberti, A. (ed.) Dizionario Biografico degli Italiani, vol. 15, pp. 280–284. Istituto della Enciclopedia Italiana, Roma (1972)
[20] Battisti, E.: Per un'iconologia degli automi. In: Battisti, E. (ed.) L'antirinascimento, pp. 220–253. Feltrinelli, Milano (1962)

[21] de Vieri, F.: Discorsi delle meravigliose opere di Pratolino, et d'Amore. Giorgio Marescotti, Firenze (1587)
[22] Bedini, S.A.: The Role of Automata in the History of Technology. Technology and Culture 5(1), 24 (1964)
[23] King, E.: Perpetual Devotion: a Sixteenth-Century Machine. In: Riskin, J. (ed.) Genesis Redux: Essays in the History and Philosophy of Artificial Life, pp. 263–292. University of Chicago press, Chicago (2007)
[24] Walker, G.: Anatomy of the Chess Automaton. Fraser's Magazine 19, 725 (1839)
[25] Sussman, M.: Performing the Intelligent Machine: Deception and Enchantment in the Life of the Automaton Chess Player. In: Bell, J. (ed.) Puppets, Masks, and Performing Objects, pp. 81–96. MIT Press, Cambridge (2001)
[26] Standage, T.: The Turk: the life and times of the famous eighteenth-century chess-playing Machine. Walker Publishing Company, New York (2002)
[27] Carroll, C.M.: The Great Chess Automaton. Dover Publications, New York (1975)
[28] Riskin, J.: The Defecating Duck, or, the Ambiguous Origins of Artificial Life. Critical Inquiry 29(4), 599 (2003)
[29] Evans, H.R.: Edgar Allan Poe and Baron von Kempelen's Chess-Playing Automaton. International Brotherhood of Magicians, Kenton (1939)
[30] Poe, E.A.: Maelzel's Chess-Player. Southern Literary Messenger 2, 318 (1836)
[31] Hodgkin, T. (ed.): The Letters of Cassiodorus. H. Frowde, London (1886)

Centrifugal Governors: The Story of a Mistake

Max Sakharov and Valentin Tarabarin

Bauman Moscow State Technical University, Moscow, Russia
{max.sfn90,vtarabarin}@gmail.com

Abstract. In this paper we study the centrifugal governors from the BMSTU collection in Moscow. The brief description of the governors, their distinctive features and working principles are given. The article deals with the main problems of the governors and their solutions suggested in Maxwell's, Vyshnegradsky's and Chebyshev's fundamental works. Some wrong directions in the development of governor design were identified. Inoperative mechanisms, some of which can be found in the BMSTU collection, were the result of the described dead-end directions. The article also provides information about the mathematical simulation of the operation of the governors.

1 Introduction

Creating a machine, it is important to provide the required uniformity of output shaft's motion. Uneven movement is associated with change in the process of resistance forces and moving forces during the operation. These changes are caused by the periodic character of working process or external reasons. For example, in the pump the load can be significantly lower during the stage of absorption, than during the stage of injection. In the internal combustion engine the force accelerates the engine during the combustion stroke and slows it during the intake, exhaust and compression strokes. These processes result in uneven rotation, which varies with the frequency of the working cycle. Flywheels are commonly used to prevent such irregularity.

System of automatic control is required to control the speed of an output shaft for long periods of time. When we speak about steam engines, this system is a centrifugal governor. The steam engine had to provide a uniform rotation of the output shaft, which machine tools were connected to. The task of the governor is to maintain, within a narrow range around given value, the average angular velocity of steam engine when load changes, for example when machine tools and pumps are turned on/off or when the pressure varies in the boiler. The governor automatically changes the steam flow in order to restore the equilibrium between the moving and loading moments [9].

Governors built on Watt's governor principle were successfully used in industry until the middle of the XIX century. However, later the difficulties in setting up of governors, the cases of unstable operation and oscillations appeared. Why did it

T. Koetsier, M. Ceccarelli (Eds.): Explorations in the His. of Machines & Mech., HMMS 15, pp. 323–336.
springerlink.com

happen? Nowadays, after studying Maxwell's and Vyshnegradsky's publications on control theory [10, 11], it's easy to answer this question.

The first steam engines were low-power engines with small shaft speed. They had huge flywheels and lightweight governor elements. As the quality of movable joints was low, governor elements moved with significant friction. The strict requirements to uniformity of rotation were absent. Later it will be shown that all these aspects together ensured the steam engine with the centrifugal governor stable operation. At the end of the XIX century powerful and high-speed steam engines were created. The growth of the shaft speed has made it possible to reduce the size of flywheels. The precision of regulation became important; hence the insensitiveness and the irregularity of governor had to be decreased. Increasing of steam engines power required augmentation of desired effort for governor elements handling and this resulted in gaining of load block weight.

Several improvements were introduced to increase the sensitivity of the governor, such as more accurate manufacturing of the mechanism's parts and better lubrication of movable joints. There were also many attempts to reduce the irregularity of the governor, modifying Watt's governor original design. Engineers believed that these measures would improve the regulation, but the process went vice versa – governors lost stability. Maxwell's paper [10, 11], where he urged to create astatic governors (i.e. governors with zero coefficient of irregularity) exacerbated the issue. Solutions of many problems were found in Vyshnegradsky's works [11]. He proved that an astatic governor is inoperative and a static governor must be equipped with a damper.

2 Classification and Basic Parameters. Static Governors

There are eight models of centrifugal governors in the BMSTU collection of mechanisms. These models of various designers and manufacturers clearly illustrate different stages of development in centrifugal governor's theory. First of all, let's consider the common classification of the governors. They are divided into three groups, such as static, astatic and pseudoastatic [17, 18].

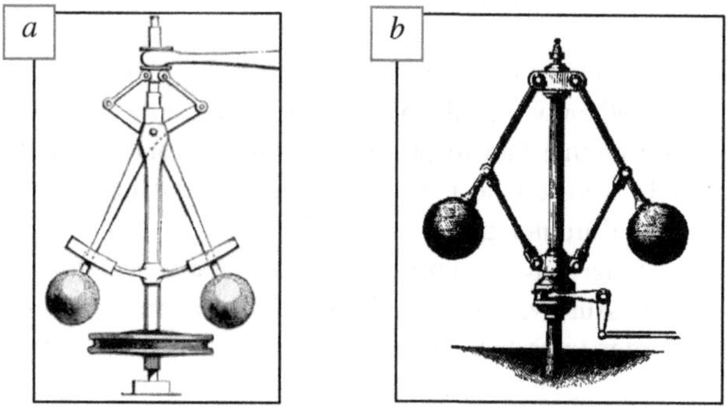

Fig. 1. a - Watt's centrifugal governor, b – modified Watt's centrifugal governor.

Static governors are governors, where each value of the shaft speed corresponds to a certain position of its sleeve, and, consequently, a certain position of throttle valve, which controls the steam flow. Watt's governor is static one. A steam engine equipped with the Watt's mechanism can't run at any speed with given load. It means that every value of the load corresponds to certain value of the speed.

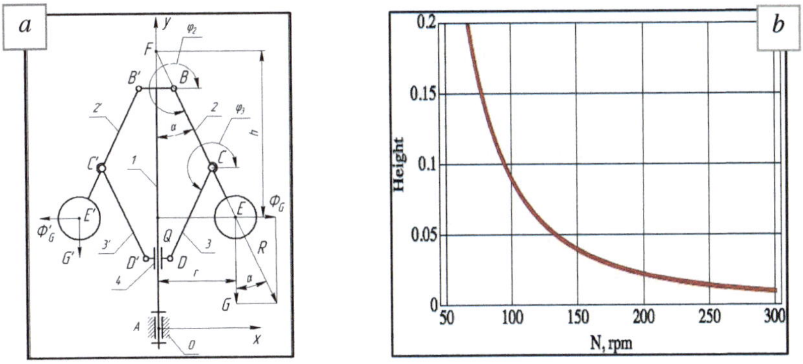

Fig. 2. a - modified Watt's centrifugal governor's force diagram, b – its theoretical characteristic.

In figure 1a the original design of Watt's governor is shown. However, the modified more compact Watt's governor was used in practice instead of the original design (Fig. 1b). Let's consider the modified mechanism. In figure 2a: 1 - the shaft of the governor 2 and 2' - levers with loads, 3 and 3'- levers, connected to the sleeve, and 4 – the sleeve. The shaft 1 is connected to the shaft of the engine and rotates with speed ω_1. Balls are in equilibrium under the inertia force Φ_G and the gravity G when the speed is constant. With the speed increasing, the equilibrium is disturbed because the inertia force Φ_G varies and the weight remains unchanged. Governor is symmetrical, so it suffices to consider a force diagram only for one half of the mechanism. Friction in the joints, the weight of the levers and the sleeve are neglected [15, 17]. The loads will be in equilibrium, if the projections of forces Φ_G and G onto an axis are equal (Eq. 1).

$$\frac{m\omega^2 \cdot r}{\sin\alpha} = \frac{m \cdot g}{\cos\alpha}, \tag{1}$$

from which it comes out

$$\omega^2 = \frac{g}{h} = \frac{g}{l_{EF} \cdot \cos\alpha}, \tag{2}$$

here m – the mass of the load, r – the distance between the load's center of gravity and rotation axis, $h = l_{EF} \cdot \cos\alpha$ – the subnormal (the projection of the normal to the trajectory from the point F onto y-axis, it's shown in figure 4a). In equation 2 l_{EF} and

g are constants, each value of the speed ω corresponds to the value of the angle α and subnormal, the growth of the speed ω increases the angle α and decreases the length of subnormal. Equation 3 defines the known relation between angular speed and rotation frequency.

$$\omega = \frac{2\pi}{60} \cdot N, \tag{3}$$

here N is the number of rotations per minute. The graph of the relation between position of the sleeve h and rotation frequency N is shown in figure 2b. This graph shows that the governor is not suitable for powerful steam engines. The curve of the relation is almost horizontal at speed values after 250 rpm that means there isn't any sleeve movement; hence there is no regulation. At the same time speed continues to grow, it can lead to accidents. Despite these shortcomings modified Watt's governor was widely used and successfully worked in low-powered steam engines.

To understand the common problems of the governors, as well as design features of various mechanisms, we will use the concepts of the irregularity and the insensitiveness of the governor. We assume that the shaft of the engine rotates with a constant speed ω_0, and the governor is in equilibrium. Let the speed of the engine has increased. Resistance forces and friction forces must be overcome when governor moves into new position. The sleeve will begin to rise only when the centrifugal force can balance the weight of the loads and resistance forces. Let's assume that this will happen at the speed of $\omega_1 > \omega_0$. Similarly, the sleeve will begin to descend only when the speed is reduced to a certain value of $\omega_2 < \omega_0$. When the speed varies within range from ω_2 to ω_1 the governor keeps its configuration of the levers corresponding to the speed ω_0. Therefore, the governor is insensitive within this speed range. To quantify the insensitivity of the governor, coefficient of the insensitiveness ε is used (Eq. 4); its optimum value is withhin range from 0.03 to 0.05 [17, 18].

$$\varepsilon = \frac{\omega_1 - \omega_2}{\omega_0}. \tag{4}$$

The governor is designed to maintain an average speed of engine shaft within a certain range of its change. Top position of the sleeve corresponds to the maximum speed ω_T; bottom position does to the minimum speed ω_B. Governor will be in equilibrium at all values of the speed from ω_B to ω_T. However, the equilibrium speed will be different from the controlled value ω_0. The difference between new equilibrium speed and ω_0 is an error of regulation. The coefficient of irregularity is the ratio of maximum change in speed to its mean (Eq. 5).

$$\Delta = \frac{\omega_T - \omega_B}{\omega_0}. \tag{5}$$

For astatic governor $\omega_B = \omega_T$, that means Δ equals zero. According to this, the less coefficient of irregularity is, the closer the governor is to astatism. The optimum value of Δ is 0.06, i.e. the speed fluctuation is six percent [17, 18]. Astatic governors are also called isochronous. Such name is explained by the fact that the revolution time of governor around its axis is the same for any position of the loads; the governor has only one equilibrium speed [9].

The first problems arose when the powerful steam engines appeared. Heavy sleeve was added to the Watt's governor, for proper work with high-speed steam engines. That governor was patented in 1858 by American engineer Charles T. Porter of New York (Fig. 3). As one can see in Fig. 3a that loads are at the intersection of the upper and lower levers [2, 12, 18].

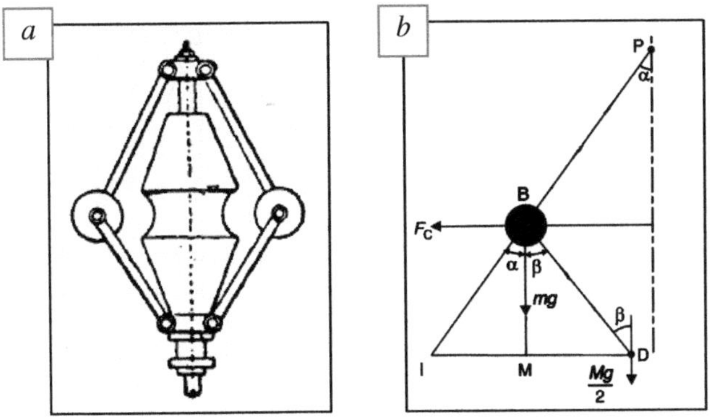

Fig. 3. a - Porter's centrifugal governor, b – its force diagram.

Porter significantly reduced the weight of the balls, thus it increased the sensitivity of the governor. Heavy sleeve was at the governor axis of rotation and its moment of inertia did not affect sensitivity. The weight of the sleeve adjusted the value of the equilibrium speed. Let's define the equilibrium speed of the governor. We take moments about the point I (Eq. 6).

$$F_C \cdot BM - mg \cdot IM - \frac{Mg}{2} \cdot ID = 0, \tag{6}$$

here m – the mass of the load, M – the mass of the sleeve, F_C – the inertia force. Let's express the length of the moment arms in terms of tangent and introduce the ratio K (Eq. 7,8)

$$F_C = mg \cdot \tan\alpha + \frac{Mg}{2} \cdot (\tan\alpha + \tan\beta), \tag{7}$$

$$K = \frac{\tan\beta}{\tan\alpha}, \tag{8}$$

from which it comes out

$$\omega^2 = \frac{g}{h} \cdot \left(1 + \frac{M}{2 \cdot m} \cdot (1 + K)\right). \tag{9}$$

Equation 9 defines the equilibrium speed of Porter's governor.

3 Astatic Governors

As mentioned above, by the middle of XIX century static governors constructed according to Watt's type of governor no longer managed with the task.

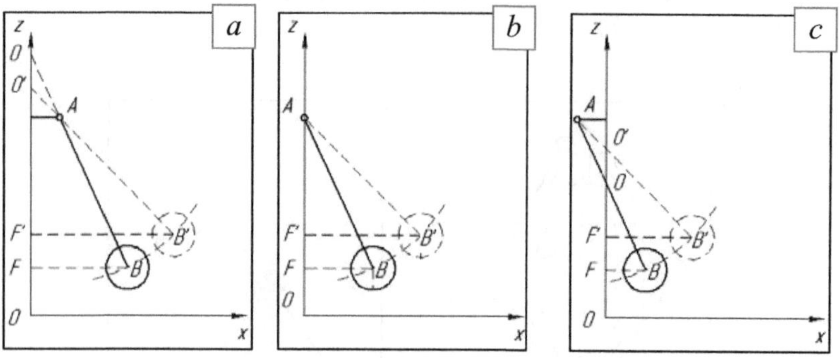

Fig. 4. Lever's layout.

The requirement for more precise regulation, arisen with the appearance of powerful steam engines, has made engineers invent new methods for reducing the insensitiveness and the irregularity. Before early research papers on the control theory appeared, irregularity and high insensitiveness had been considered to be the key problems of the governors. Watt's governor had a significant value of the coefficient of irregularity. It was attractive to reduce it in order to create more precise governor. That's why engineers created many new designs of governors, which were different from Watt's one. The simplest way to decrease the irregularity is the crisscross mounting of the upper levers (Fig 4). Equation 2 defines the equilibrium speed of Watt's governor, where h - subnormal (in our case, the distance between projection of the gravity center of the ball onto rotation axis and the point where the lever intersects with this axis - segment OF). The more h changes, the more the speed varies. Hence, the way of suspension of balls showed in figure 4c is the best and can minimize the speed fluctuations. But this is not enough to get totally astatic governor.

The main way to achieve isochronism was the creation of the governor with a parabolic trajectory of loads. Let's consider the equation of a parabola (Eq. 10)

$$x^2 = 2 \cdot p \cdot y, \tag{10}$$

here p – the subnormal of a parabola, which is constant at any point of the curve, i.e. at any position of the balls. The value of the subnormal determines value of the equilibrium speed. Equation 11 defines the equilibrium speed of the governor if the loads move along parabolic trajectory.

$$\omega^2 = \frac{g}{p}. \tag{11}$$

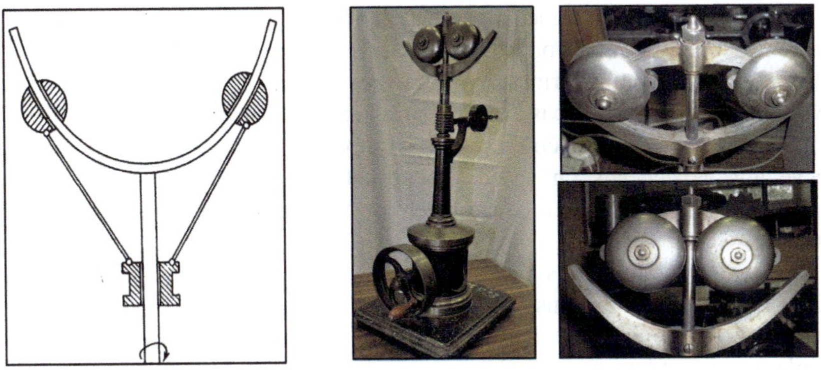

Fig. 5. Garnett's centrifugal governor and its sketch.

Since g and p are constants in this equation, then at any point of parabolic trajectory the only one equilibrium speed ω is possible.

There are several different designs of parabolic governors. Let's note Garnett's governor, which is kept in the BMSTU collection (Fig. 5). Here the loads are attached to the rigid guide frame, which provides a parabolic trajectory of their movement [4, 17].

Let's consider a couple of unusual governors, which were created to overcome the shortcomings of Watt's governor. The first governor was designed by German engineer F. Redtenbacher [14, 16]. The student of Moscow Craft School (MCS) P. Ivanov manufactured the model of this governor in 1862 [5]. Redtenbacher offered a completely different design, where the balls were replaced with a massive ring (Fig. 6). There is a groove in the upper part of the ring at the arc of approximately 160 degrees; this lightens one side of the ring. The ring is fixed so that it controls the movement of the sleeve through the levers, which, in turn, affects the position of throttle valve. Redtenbaher supposed to create a sensitive governor without an error of regulation. But in his works, he wrote that current model did not realize this idea [16].

Fig. 6. Redtenbacher's centrifugal governor annd its force diagram.

Next governor was designed by F. Jenkin. Here the heavy sleeve is designed in the form of a cover, the inner part of which serves as the guidepath for divergent loads (Fig. 7). The operating principle of the governor is that the centrifugal element revolves around the main axis and is always kept at a constant angle, as the loads slide on the inner surface of the cover, which is set freely on this axis. The pressure on the sleeve always depends on speed. The disadvantage of the governor is that the equilibrium speed depends to some extent on the coefficient of friction between two surfaces.

Many different governors had been described and created until Chebyshev showed that it was possible to construct governor with the coefficient of irregularity any arbitrarily close to zero just considering Watt's original governor design by means of simple fracture of levers.

Fig. 7. Jenkins's centrifugal governor and its force diagram.

As a starting point Chebyshev took the governor with bended and intersecting levers [1, 3]. Using the principle of virtual displacements, he got an expression, which defined the squared ratio of the equilibrium speed ω to the variable speed ω_0 (Eq. 12).

$$\frac{\omega^2}{\omega_0^2} = \frac{\left[1+\frac{\cos(\varphi+\alpha)}{\sqrt{m^2-\sin^2(\varphi+\alpha)}}\right]\cdot\sin(\varphi+\alpha)-A\cdot\sin(\psi-\varphi-\alpha)}{B\cdot\sin 2(\psi-\varphi-\alpha)}, \quad (12)$$

here ψ – the angle of lever's inflection, φ – the angle between levers and axis at equilibrium speed ω_0, when speed ω changes, the angle will change by a certain value α, m- the length of the upper levers. Coefficients A and B were introduced to simplify the expression (Eq. 13).

$$A = \frac{2r}{P}, \; B = \frac{\omega_0^2 \cdot r^2}{P \cdot g}, \quad (13)$$

here P – the weight of the sleeve taken as a unit relative to the weight of the load, r – the length of the levers from the point of attachment to the axis to the center of the loads, the length of the levers from the point of attachment to the upper arms to the levers' inflection is also taken as a unit. It's obvious that if $\alpha = 0$ the ratio of angular speeds must be equal to one. Astatic governor has only one equilibrium speed at any possible position of the sleeve and, hence, at any possible value of angle α, i.e. equation 11 must be equal to one. Using a power series expansion and equating to zero all coefficients of the first five powers we can get a system of equations, by solving which we determine the optimal values of parameters A, B, m, φ, ψ. This set of parameters allows to reduce the speed fluctuations down to 0.1%.

Depending on the desired equilibrium speed, levers' radiuses of curvature were determined. Let's note that in contrast to conventional methods of designing centrifugal governors, Chebyshev didn't neglect the mass of levers. He designed levers of complex shape with variable width and radius of curvature. For example, in figure 10 the governor, where the regulating mass is concentrated not only in the loads themselves but also in the heavy parts of levers, is shown. Chebyshev's design governors were manufactured in IMTS (Imperial Moscow Technical School) workshops (Fig. 8, 9).

In practice, astatic governors had serious problems. While operating, the governor performed, in fact, oscillatory motion, and skipped all the time the equilibrium speed, that was totally unacceptable. What's more, there was also a problem of governor stability. Working near the resonance frequencies, the governor "went crazy" - the amplitude of the sleeve increased indefinitely. Nineteenth century engineers called this unstable behavior "hunting" and devoted much effort to improve the design of the governors. For example, the French physicist J. B. L. Foucault believed that instability could be avoided by decreasing backlash in the gears, sag in driving belts, etc. [2]. In 1871, English engineer, J. Head wrote that the solution of this problem was the usage of air damper [6].

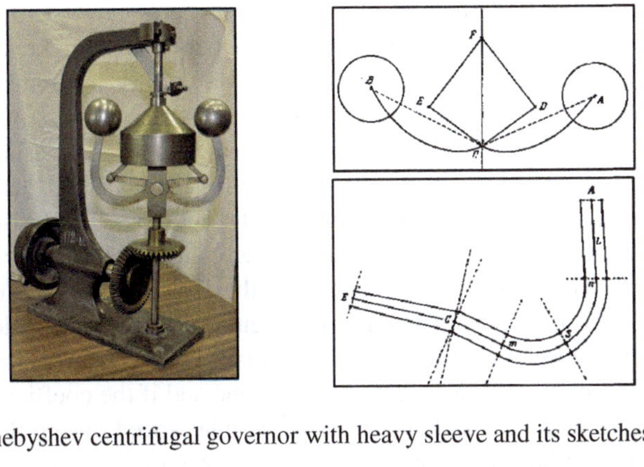

Fig. 8. Chebyshev centrifugal governor with heavy sleeve and its sketches.

Two fundamental works on the control theory of that time will be considered to realize stability and generation of oscillations problems in details. In Maxwell's article [10, 11] (published in 1868) he attempted to formulate the basic principles of different governor types. Maxwell tried to draw attention of engineers to astatic governors phenomenon. They, as Maxwell, thought were capable of more precise regulating, than static ones.

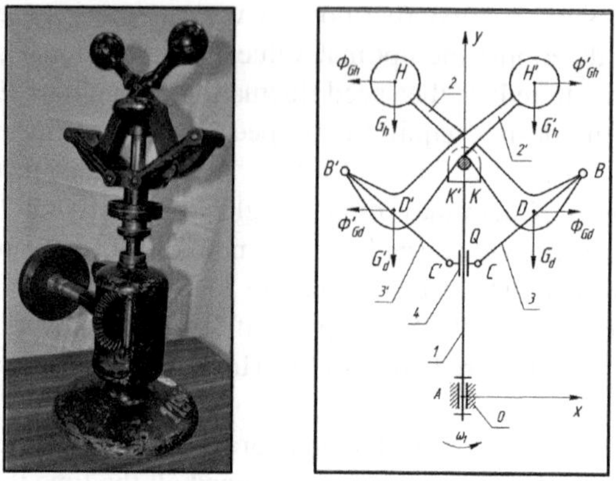

Fig. 9. Chebyshev's centrifugal governor with curved heavy levers and its force diagram.

This explains lots of different designs of astatic governors. Being involved in use of governors in astronomical instruments he didn't realize what parameters, in industry, defined the stability of governors in industry and what the relative values of certain constants in the equations of motion for steam engines were. Maxwell determined the range of parameters within which the stable operation of astatic governor was possible. But he didn't notice that this range of parameters had no practical significance for the ordinary steam engine. So this paper didn't conatain practical advice on governor design.

The origine of control theory of machines in industrial practice began in 1876 when Vyshnegradsky published his article [11]. Only by comparing the works of Maxwell and Vyshnegradsky, we can realize Vyshnegradsky's impact. He solved the equations of motion for Watt's governor and proved that astatic governors were unsuitable, essentially using the same equations that Maxwell had used to make the conclusions about the suitability of astatic governors. Vyshnegradsky got the mathematical condition of absence of self-actuation of the governor, which allowed him to make two basic theses. The first thesis is that the governor can't operate normally with a zero coefficient of irregularity, that is astatic governors aren't suitable. The closer governors to the astatism, the less is the range of parameters within which the transition process is monotone and if the coefficient of irregularity is small enough, then the oscillations of the angular speed during the transition process are inevitable. While operating, astatic governor will perform an oscillatory motion. This Vyshnegradsky's thesis is in conflict with Maxwell's intention to use astatic governors. The second thesis is that any governor, except astatic one, in order to operate properly should have a special device generating a viscous friction (damper) [13]. Astatic governors aren't suitable even with these devices.

After Vyshnegradsky's works engineers started to understand how any changes of design parameters affect the stability of the governor. Whereas before these works no one of the engineers could have guaranteed that that governor created by him would operate properly, because of complete misunderstanding of how design features affected the stability.

4 Pseudoastatic Governors

As a result, in actual practice successful governors were those with abidance of balance between value of the coefficient of irregularity, required for stable operation of governor and for small speed fluctuations. These governors were called pseudoastatic governors.

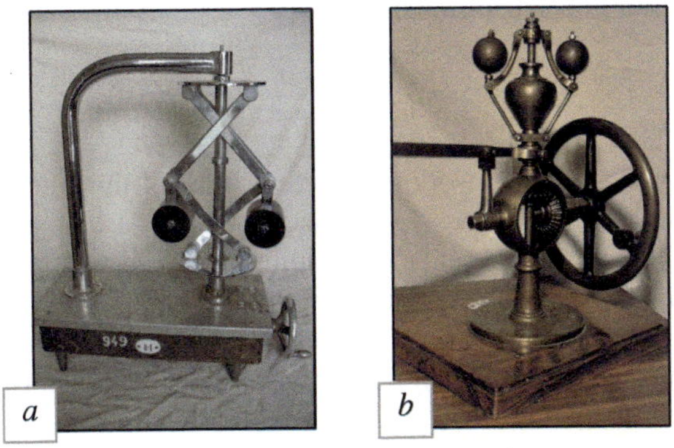

Fig. 10. a - Kley's centrifugal governor, b - Proell's centrifugal governor.

Let's consider Kley's model of governor from the BMSTU collection (Fig 10a). Contrary to certain opinions, this governor is pseudoastatic [7]. Since the exact definition of pseudoastatic governors doesn't exist then, for a rough estimate we assume that pseudoastatic governor has the coefficient of irregularity $\Delta \leq 0.04$ [15, 17]. The crisscrossed levers decrease the coefficient of irregularity. However, that design requires high height of the construction, so the shortcoming of this model is its large size.

Proll's governor (Fig. 10b) is a successful modification of Kley's governor, providing the same small speed variations at twice smaller sizes. The upper lever arms are bent, in order to allow the free movement of the loads; the sleeve is heavy and has a pear-shaped form. Let's notice that unlike the classic design, here the loads are attached to the lower levers and directed upward [4, 8]. A few designs of Proll's governor are known to exist (Fig.11a,b). One of them is kept in the BMSTU collection. Let's determine the equilibrium speed of Proll's governor rotation and plot a graph of governor's working characteristic.

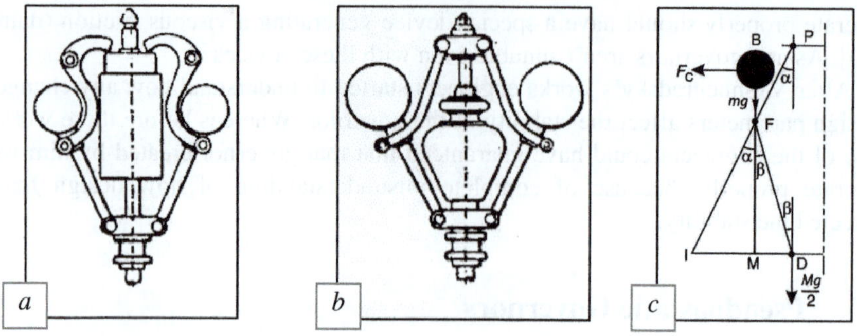

Fig. 11. a, b – sketches of different Proll's governors, c – its force diagram.

The expression for equilibrium speed (Eq. 14) differs from equation 9 by the ratio of the height of the point where lower and upper levers are intersected (segment FM) to the height of the loads' attachment (segment BM). This ratio is variable, thereby reducing the equilibrium speed fluctuations.

$$\omega^2 = \frac{g}{h} \cdot \left(1 + \frac{M}{2 \cdot m} \cdot (1 + K)\right) \frac{FM}{BM} \,. \tag{14}$$

The graph in figure 12 shows working characteristics of three governor types: static, astatic and pseudoastatic. Obviously, the characteristic of astatic governor is a vertical line that corresponds to the absence of equilibrium speed fluctuations. The curve of smaller slope corresponds to pseudoastatic governor that allows slight variations of the equilibrium speed. The curve of the bigger slope describes the work of static governor, which is characterized by the largest equilibrium speed fluctuations.

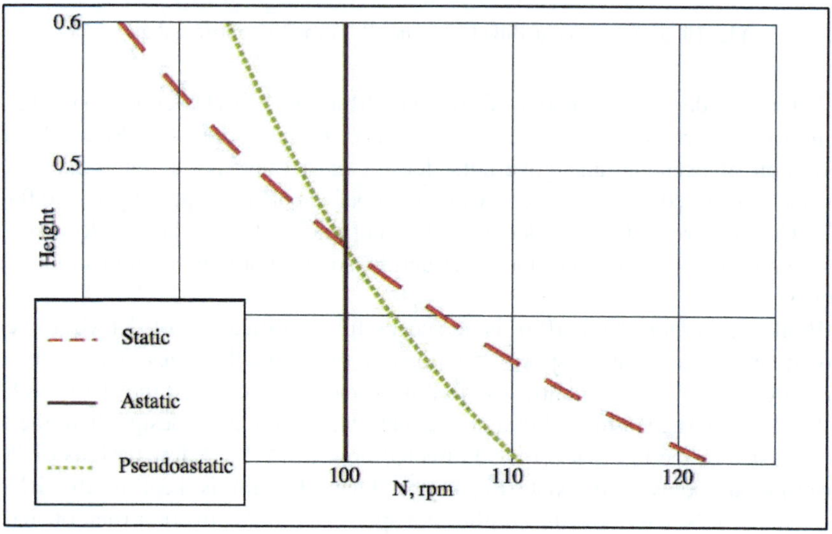

Fig. 12. Theoretical characteristics of static, astatic and pseudoastatic governors.

In the end, let's notice that the search of ideal governor has not been stopped. After pseudoastatic gravity-controlled governors engineers have started creating pseudoastatic spring-controlled governors. In these governors springs were used to react against the centrifugal force. They could be designed to operate at high speeds and were comparatively smaller in size, because there was no need in heavy sleeve and loads. Their equilibrium speed could be adjusted by changing the initial spring compression. These governors were very suitable for internal combustion engines. Among spring controlled centrifugal governors Hartnell's and Hartung's ones were the most commonly used.

5 Conclusions

Attribution is one of the main tasks of the history of technology; it answers such important questions as time of mechanism creation and inventorship. An important part of this work is to analyze causes that have led to creating and development of mechanisms. Lots of different governor designs [7] show that previous models haven't met the increasing demands. Attempt to eliminate the shortcomings of static governors with the help of astatic governors was a mistake. New models of astatic governors were unstable and failed to operate properly. However, astatic governors helped to identify the reasons of unstable behavior. They became the foundation for the development of pseudoastatic governors, which proved to be efficient in conditions of high-speed steam engines. Theoretical researches of centrifugal governors laid the foundations of the control theory. This article also provides information on mathematical simulation of basic governor types. These studies have confirmed the advantage of pseudoastatic governors over others. Simulation was based on fundamental works on the theory of governors, gathered in the lectures of IMTS professors [15, 17, 18]. Centrifugal governors are milestone in the history of technology. The governors from the BMSTU collection were officially recognized as monuments of science and technology. Even nowadays they are widely used in engineering education. For example, German firm GUNT (www.gunt.de) produces equipment for laboratory works. This equipment allows future engineers to determine working characteristics of basic governor types experimentally.

Acknowledgements. The authors thank an anonymous referee for his/her valuable remarks and comments on the paper.

References

1. Artobolevsky, I.I., Levitsky, N.I.: Mechanisms of P.L. Chebyshev. A Collection of scientific Heritage of P.L. Chebyshev. USSR Academy of Science (1945)
2. Bennett, S.: A History of Control Engineering 1800–1930. Institution of Electrical Engineers, London and Peter Peregrinus Ltd., Stevenage, UK (1979)
3. Chebyshev, P.L.: Complete Set of Works. USSR Academy of Science (1946)

4. Gardner, D., Hiscox, M.E.: Mechanical Movements. Powers and Devices, The Norman W. Henley Publishing Co., NY (1907)
5. Golovin, A., Tarabarin, V.: Russian Models from the Mechanisms Collection of Bauman University. Springer, Heidelberg (2008)
6. Head, J.: On a simple construction of steam engine governor having a close approximation to perfect action. Proceedings of the Institution of Mechanical Engineers 22, 213–228 (1871)
7. Header, H.: A Handbook on the Steam Engine. Crosby Lockwood and Son, London (1902)
8. Hopkins, A.A., Bond, A.R.: Scientific American Reference Book. MUNN & Co., NY (1914)
9. Manychkin, N., Sakharov, M., Tarabarin, V.: The Models of Centrifugal Governors in the Collection of BMSTU. In: Proceedings of EUCOMES 2010, pp. 511–518. Springer, Heidelberg (2010)
10. Maxwell, J.C.: On Governors. Proceedings of the Royal Society of London 16, 270–283 (1868)
11. Maxwell, J.C., Vyshnegradsky, I.A., Stodola, A.: Theory of automatic regulation. USSR Academy of Science (1949)
12. Mayr, O.: The Origins of Feedback Control, English edn. MIT Press, Cambridge (1970)
13. Merkin, D.R.: Introduction to the Theory of Stability of Motion. Science (1976)
14. Moon, F.C.: The Machines of Leonardo da Vinci and Franz Reuleaux. Springer (2007)
15. Orlov, F.E.: Steam Engines. Lections of prof. Orlov. IMTS (1890)
16. Redtenbacher, F.: Die Bewegungs-Mechanismen. F. Bassermann, Heidelberg (1866)
17. Zernov, D.S.: The Building of Steam Engines. Lections of prof. Zernov. IMTS (1895)
18. Zernov, D.S.: Theory of Steam Engines. IMTS (1896)

The Anahuac Propeller: A Century after

J. Romero-N[1], C. López-C[1], A. Colín-R[2], and M. Arroyo-C[1]

[1] Fac. de Ingeniería, Univ. Aut. de Querétaro
[2] Mechatronics Division, Instituto Tecnológico de Huichapan, México

Abstract. During the Mexican Revolution, engineer Juan Guillermo Villasana-López developed a special aviation propeller that he called *Anáhuac*, aimed at coping with the efficiency problems that contemporary propellers had on the Mexican plateau due to its high altitude. Such particular model included a radial exit border and a constant chord width at its middle length. In spite of its successful use in many Mexican airplanes, this technological development eventually vanished from the universal aeronautics literature. This work aims at re-assessing such contribution of Mexico to worldwide aviation, and further assessing its potential use in other technology areas. A comprehensive critical review of literature was carried out, as well as the building and testing of a scale model of such propeller, which was manufactured taking into account Villasana's original patent.

1 Introduction

At one hundred years of the beginning of aviation in Mexico, reviewing the standpoint, both, technologically and historically, of the remarkable proposed advances at that epoch by Mexican engineers is a must under the perspective of the recent trends of the aviation industry in Querétaro, México. Moreover, this is also an acknowledgement to the impact that the first social revolution had on specific aspects of the national technological development.

It is well known that propeller airplanes operation is essentially based upon the air displacement and the reaction caused by the effect of acceleration being received by the propeller. Thus, the force on the propeller disk is given by

$$F = 0.5A\rho(v_s^2 - v_0^2) \tag{1}$$

where A is the projected area of the runner, ρ is the air density, and v_s and v_0 are the air speeds before and after the propeller disk, respectively. The thrust produces a speed enough for lifting by the airflow around the wings [1]. Now, taking into account that the air density varies from 1.225 kg/m^3 at sea level to 0.996 kg/m^3 at the Mexican plateau height [2], a loss of about 18 % of power thrust occurs. Thus, when the engines were used at those altitudes, there was a loss of power. This was the fundamental problem addressed by the engineer Juan Guillermo Villasana-López. Indeed, he designed the "*Anáhuac* propeller" to compensate the loss of power thrust due to the low-density air.

T. Koetsier, M. Ceccarelli (Eds.): Explorations in the His. of Machines & Mech., HMMS 15, pp. 337–348.
springerlink.com

In this work, an exhaustive critical review of literature is presented on the aviation technological developments materialized at the down of nineteen century, aiming at identifying potential Mexican technological contributions to aviation universal history.

On the other hand, experimental tests on a scaled down *Anahuac* propeller are reported in order to prove its enhanced thrust characteristics, and to assess its efficiency when used as a wind turbine.

2 The Anahuac Propeller in the Literature

While Villasana-López installed and field-tested its new model of propeller in the autumn of 1915, he ran preliminary tests since the summer of 1910, just few weeks before the commencement of the Mexican Revolution, which occurred on September 15th. And the patent was issued on April 2, 1918 [3], shortly after the official ending of such armed movement. A photograph of the only one illustration included on the patent document is shown in Fig. 1. For that figure, the description (loose translation) is as follows:

> *"the exit or posterior edge of the propeller "a", is a straight line normal to the shaft axis that constitutes a sort of generatrice line. The anterior border "b" is an equidistant curve along all of its points, with respect to the straight line of edge "a" thus creating a constant width bar for maximum efficiency of the vane".*

The propeller design described and registered by Villasana-López is original in many aspects. Actually, when compared with other contemporary designs, the edges of the *Anahuac* propeller make the difference, as these are straight and radial.

Castillo-Tenorio [4] describes the several phases of the Mexican aeronautics, from the aerostatic era that ended at 1907, to the theoretical investigations performed by Julio Fuentes, Villasana-López (1908), the brothers Juan Pablo and Eduardo Aldasoro-Suárez (1909), and Miguel Lebrija- Urtegui (1910). According to this author, previous to its *Anáhuac* propeller, Villasana-López designed and tested a planner specifically to work at the altitudes of central Mexico. The conceptual design of his propeller would have resulted from discussions with an American classmate, in June 1910. Castillo-Tenorio, however, acknowledges some confusion with the dates. In that respect another historian, José Villela-Gómez, points out that the design and manufacturing was in 1915 [5]. Thus, it seems that several *Anáhuac* propellers had existed, the first one around 1911. According to Villela-Gómez, the first propeller was manufactured on September of 1915 at the workshops of the School of Arts and Works, ending up the manufacture on October 7th of that year. After a second test flight on October 12, a record flying height was reached for this continent: 19700 ft. above the sea level. As a result of this success, Rovirosa [6] describes that the *Anáhuac* propeller was considered the "queen of the Mexican skies", until the wooden propellers were substituted by the metallic propellers.

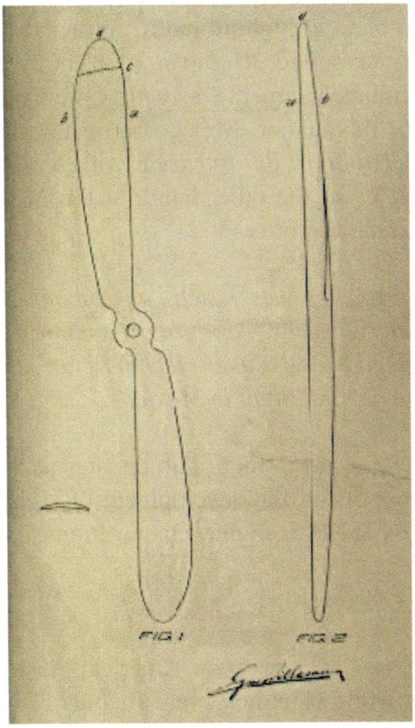

Fig. 1. Photograph of the only one illustration on the patent document by Villasana-Lopez [3].

The precise location of the flat area on which Villazana-López made his flying practices is in the State of Hidalgo, and is called *Llanuras de Venta Prieta*.

González [7] stresses the importance of aviation during wartime, by recalling phrases such as: "aviation will save democracies". In this military context, the *Anáhuac* propeller was especially appreciated due to the logistic impact that such transportation mode had on the Mexican revolution. For example, Taylor Hansen [8] recalls the use of airplanes by General Pascual Orozco in 1912, to carry mail and materials across the Mexico-USA border. According this author, the propeller was used in mono- and bi- planes, from 1915 to 1920. Romero-Salinas [9] reports that Villasana-López flew his first planner when he was very young and that in 1909, at the age of 18, he participated in the foundation and organization of the *Sociedad*

Impulsora de la Aviación (Society for the Promotion of Aviation). Regarding the *Anáhuac* propeller, it is further suggested by this author that it was already in public use by 1912. According to another author (Vargas [10]), Villasana-López got his baccalaureate in Aeronautics in 1918, but he designed the propeller in 1915, to further affirm that this propeller was used in almost every airplane owned by the Mexican Air Force, from 1915-1920. Paz-Fernández [11] acknowledges the *Anáhuac* propeller as a solution to the thrust problem at high altitudes, further revealing that some of these propellers were actually exported to Japan. As an interesting fact, this author identifies *Quetzalcoatl* (plumed serpent), as a first pre-Columbian concept of flying beings. In the opinion of this author, Mexico aeronautic industry consequent decay was the result of the *Tratados de Bucareli* with the United States ("Bucareli Treaties"), signed in 1919. On the other hand, at the Mexican Air Force webpage [12], the following paragraph can be read:

> "*the Anáhuac propeller was tested on October 26 (1915) with successful results; it was designed by Cap. 2/o. Mechanic Juan-Guillermo Villasana-López, who worked in 1910 as a mechanic for Alberto Braniff, to whom he helped to realize his first flight in Mexico*".

About the problem for taking-off at high-altitude sites such as Mexico City, Lazarín Miranda [13] cites another effort: the development of higher octane fuel, so that the traditional planes could fly at Mexican plateau elevations.

An invention consistently disregarded

Simmons and Withington [14] are two American historians that analyzed the development of aviation worldwide; however, the only reference made to Mexico in their book is anecdotic: recalling that celebrated aviator Charles Lindbergh met his fiancée in Mexico. British author E. Charles Vivian [15] wrote one of the classic books on aviation history. He included in his publication some Latin-American contributions for the development of aviation in general. While from the design viewpoint, the author cites propeller designs made by Brazilian Santos Dumont; from airplane operation the author alludes to the flying feats of Peruvian Georges Chávez. The author further weighs on Wright brothers´ contribution to aviation by acknowledging Stringfellow´s work in developing the first motorized plane in 1848. As far as Mexico is concerned, the author asserts that the only two civilizations in the history of mankind that lack flying beings in their mythological background were the Aztecs and the Chinese. Americans authors T. L. Edward and J. S. Crenshaw [16] published a text dealing with technical details of aeronautic developments, noting that in 1916 the 1st American Air Fleet tried to assist General Pershing in capturing the Mexican rebel General Francisco Villa, but that their actions were frustrated due to the low-output of their airplanes at Mexico's high altitude. In his book, Argentinian Santaló Sors [17] acknowledges contributions to aviation from pioneering Argentinian pilots, and highlights the impact of several technological developments on the worldwide aviation. Juan M. Aguilar, a Spaniard author, published his text on aviation history, acknowledging some European inventions for improving airplanes [18]. In particular, this author describes achievements of Spaniard engineers: Juan de Sierva with his helicopter, and Ramón Franco with his hydro-airplane. In this text, reference to Mexico is limited to describing Charles Lindberg's visit to Mexico in 1927. So, it seems that authors tend to report domestic outputs, instead of objectively recognize developments from abroad [19]. Nevertheless, on the Spanish version of an international European magazine issued in Mexico in 2011 [20], some Mexican contributions to aviation are cited; in particular, the design and fabrication of the *Anáhuac* propeller, that represented "a technological advance due to its structural strength, lightness, and lifting regime". Unfortunately, in [21] there is a curious suggestion about the design originality, highlighting the name of Francisco Santarini over Villasana-López, and noting that according to a military communication of the US embassy dated on December 3, 1918, the Mexican aviator Horacio Ríos had shown to Villasana-López a book on aviation, from which the description and drawings had been obtained for the *Anáhuac* propeller. While it is evident that Villasana-López was influenced by contemporary propeller designs, suggested piracy act results hard to believe as low-thrust issues due to low air density were prevalent at that time, and unsolved until Anáhuac propeller was invented.

3 Experimental Tests of a Scale Down Model of the Anahuac Propeller

A scale-down model of the *Anáhuac* propeller was built for laboratory tests, based upon the drawing included in the patent and by direct observation of an existent full size sample of the propeller. The experimental setup for testing the model is shown in Fig. 2, involving the propeller-driven motion of a vehicle within a wind tunnel. Measuring the distance run and the elapsing time, the accumulated kinetic energy E_k on the vehicle of mass m reaching the speed v is

$$E_k = \frac{1}{2} m v^2 \tag{2}$$

If the propeller works now as a wind turbine, the kinetic energy E_{kr} due to the rotation is:

$$E_{kr} = \frac{1}{2} I \omega^2 \tag{3}$$

where I is the mass polar moment of inertia of the propeller and ω its angular speed. Fig. 3 illustrates the corresponding experimental setup.

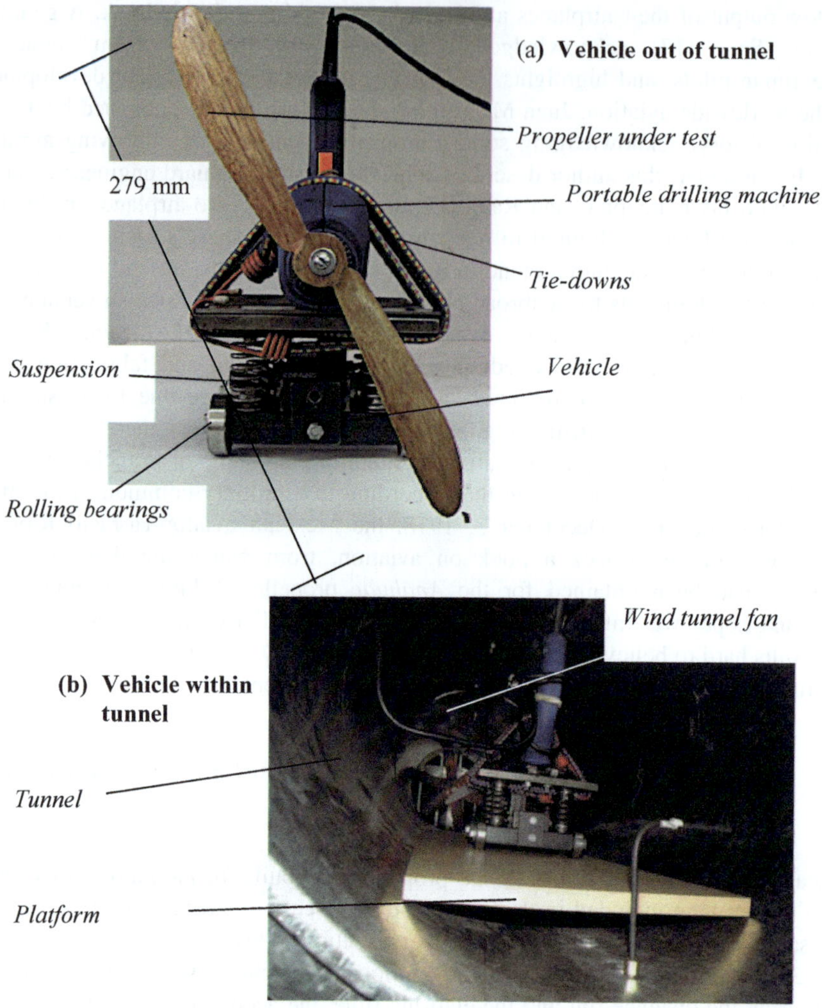

Fig. 2. Experimental setup for thrust testing.

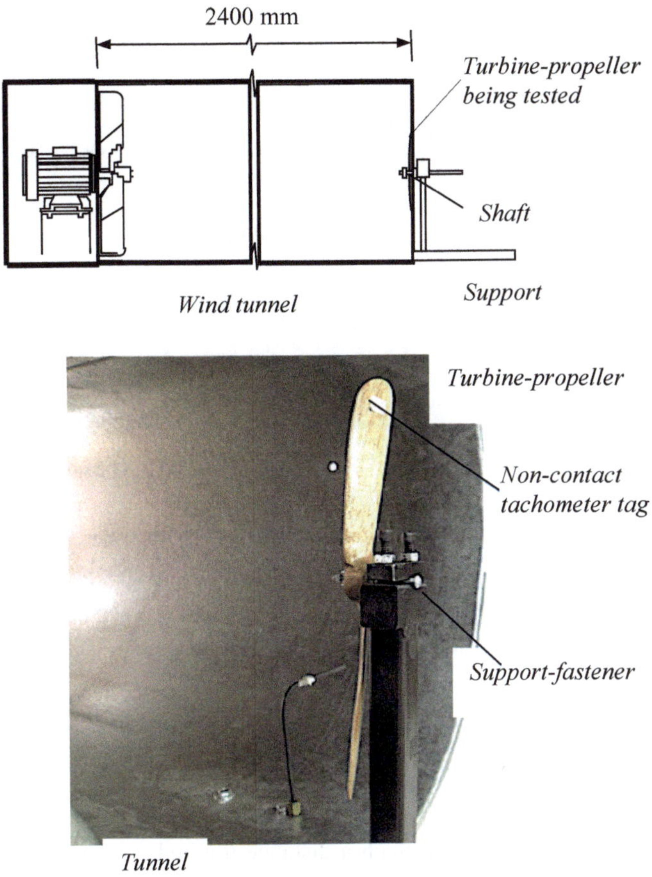

Fig. 3. Experimental setup for measuring the speed of the *Anáhuac* propeller working as a turbine.

4 Fabrication and Test of the Anahuac Down-Scale Model

The scale model was manufactured by Mr. Carlos García-Zepeda, owner of the workshop called *GAMA*, using drawings provided by the first author of this paper. For comparison purposes, two commercial propellers were selected for testing, namely, the JZ_11x10 made of wood and the APC_C2_11x9 made of plastic, with pitches of 10 and 9 inches, respectively. Masses of the *Anáhuac,* the JZ_11x10, and the APC_C2_11x9 propellers were 0.015, 0.020 and 0.040 kg, respectively. The three models are shown in Fig. 4.

(a) *Anáhuac* propeller.

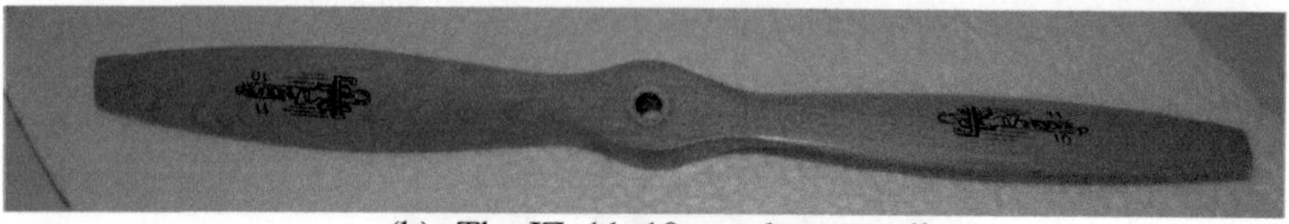

(b) The JZ_11x10 wooden propeller.

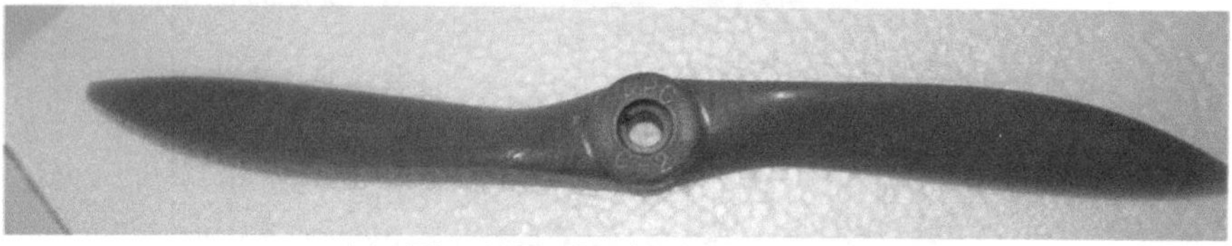

(c) The APC_C2_11x9 plastic propeller.

Fig. 4. Photos of the three propellers used for comparison.

For thrust evaluation testing, and after some preliminary trials, the running distance for the vehicle was set to 0.2 m. Such a distance implied a running time compatible with the available instruments´ resolution (digital chronometer). Results of different test replicas are shown in Fig. 5, where it can be seen that consistently, the vehicle equipped with the *Anahuac* propeller ran the selected displacement in less time that with the other two propellers, with the slowest run being that with the APC_C2_11x9.

The amperage in the portable drilling device power cable was measured as a complementary indicator, but no difference in readings was detected with any of the propellers (1.5 A). Angular speed of the drilling device was 3120 rpm. The normalized values of the kinetic energy imparted to the 8-kg vehicle by the thrust of the different propellers are shown in Fig. 6. These plots reveal that the *Anahuac* propeller was 2.5 times more efficient than the APC_C2_11x9, and almost 67% more efficient than the JZ_11x10.

For assessing the energy conversion ability of the different propellers when working as wind turbines, ten measurements were taken of the angular speed attained by the different propellers after being exposed to a steady 55 kph air flow within the

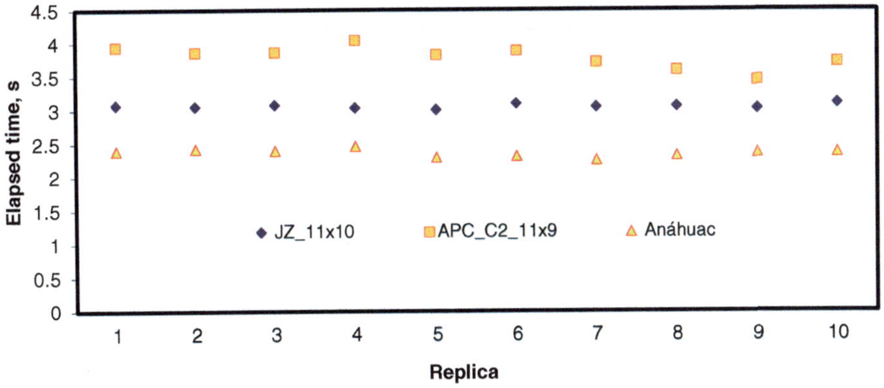

Fig. 5. Elapsed time for the vehicle running a 0.2 m length inside the tunnel with different propellers.

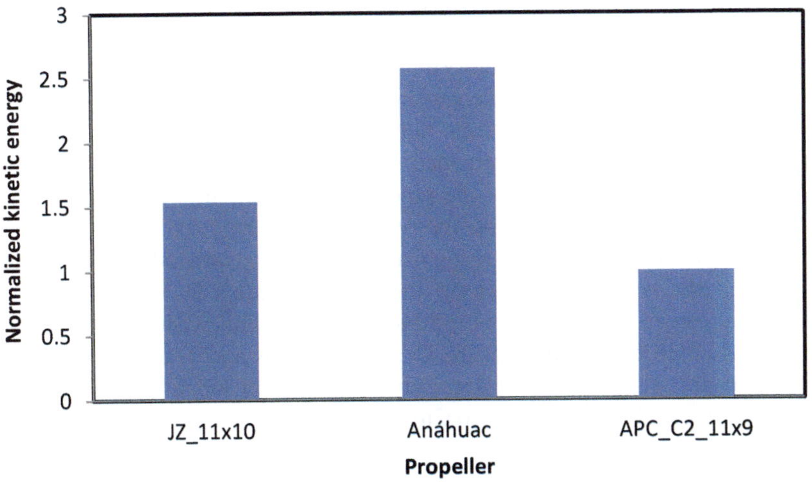

Fig. 6. Average normalized amounts of energy imparted to the vehicle during the thrust test.

wind tunnel. While such air speed was measured previously with no restriction in the tunnel [22][23], angular speed of propellers was measured through a non-contact tachometer. Results of angular speed measurements for the three propellers are shown in Fig. 7 that reveal that the APC_C2_11x9 was the most efficient propeller to transform air´s energy, followed by the JZ_11x10, and in the last place the *Anáhuac* propeller. Consequently, the first one was 21% more efficient than the second one, and 32 % more efficient than the *Anáhuac*.

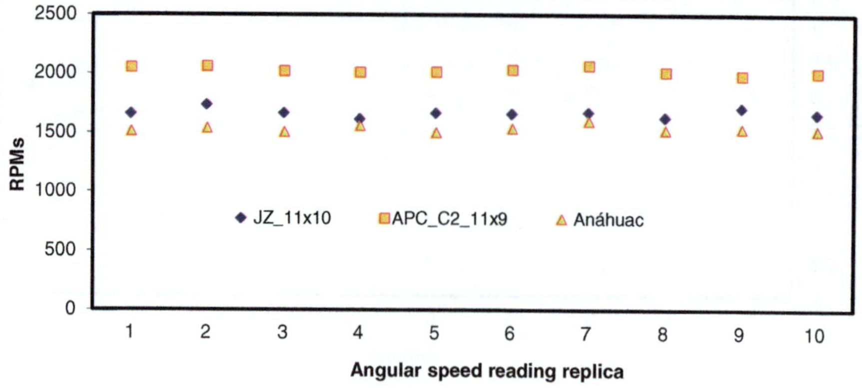

Fig. 7. Measured angular speed for the propellers when subject to 55 kph air flow within the wind tunnel.

5 Biographical Data of Engineer Villasana-López

The life of engineer Juan Guillermo Villasana-López is marked by his avid interest in aviation, nationalism, and entrepreneurship. He was born on February 10, 1891, in Pachuca, State of Hidalgo, where his father somehow transferred carpenter skills to him. Later on, these skills provided him with the necessary knowledge to conceptualize in a precise manner the process of fabrication of the wooden propeller that he called *Anáhuac* to remind the Valley where he lived. As a youngster he designed his own kites and as a recognized engineer, he was asked by the Mexican president in the thirthies to fabricate the first Mexican car. While he started his professional career as a member of the Mexican Army, he eventually moved to civil life as head of different offices at the Communications and Transports Ministry. He received several distinctions and decorations, even from Japan and El Salvador. He died on February 23, 1959 in Mexico City.

6 Conclusions and Future Work

It can be considered a significant achievement for Mexico having been able to assemble airplanes of domestic design a century ago, which is greatly attributed to the on-going revolutionary movement. In this context, the *Anáhuac* propeller was the most important artefact for such development, further influencing international aviation and solving existing airplane limitations for taking-off and flying at high elevations. Such propeller was the result of the ingenuity of a Mexican engineer: Juan-Guillermo Vissasana-López, who at a very young age demonstrated great intuition for developing flying objects. A currently evaluation of the *Anáhuac* propeller has proved the advantages of its design for providing thrust, whereas its capabilities to transform the air flow into rotational energy were slightly less than other models used in aero-modelling. However, given the complex aerodynamic features involved, it seems necessary to carry out more testing for characterizing the response of the propeller to wind circulation. Finally, it seems that there exists a trend not to recognize achievements not given within the traditional industrialized world.

Acknowledgements. For this research, the wind tunnel built by Ibert Zárate-Pena, Diego A. Rivera- Anaya and César Hurtado-Mora, students at the School of Engineering (UAQ, San Juan del Río Campus) was used. The authors further appreciate the help from Ibert Zárate-Pena for manufacturing the shaft for testing the APC_C2_11x9 propeller.

References

[1] NASA (2010), Propeller thrust, Glenn Research Center. National Aeronautics and Space Administration 10 (February 2010), http://www.nasa.gov
[2] WASP, WASP–The Wind Atlas Analysis and Application Program (2010), http://www.wasp.dk/Support/FAQ/WebHelp/AirDensityTable.html
[3] Villasana-López, J.G.: Nueva Hélice, Mexican Patent 17219, code 20089 (1918)
[4] Castillo-Tenorio, I.: Origins of aeronautics at IPN, manufacturing of the prototype of the Anáhuac propeller. El Cronista, Nueva Época, Año 11(41), 19–22 (2009) (in Spanish)
[5] Villela-Gómez, J.G.: Pioneers of Mexican aviation, p. 59. Colofón, México D.F. (1964) (in Spanish)
[6] Salazar-Rovirosa, A.: World and Mexican aviation history, p. 144. Ed. Económicas, México D.F. (1970) (in Spanish)
[7] González, E.: Chronicles of the aviation in Sinaloa. Dirección de Investigación y Fomento de la Cultura Regional, Méx. (1994) (in Spanish)
[8] Taylor Hansen, L.D.: Origins of the Mexican Air Force, 1913-1915. Revista HMex, LVI 1, 175–230 (2006) (in Spanish)
[9] Romero Salinas, G., et al.: Aviation in the City of Pachuca. Revista Enlace - Expresión de la Sociedad Civil (7), 7–11 (2004) (in Spanish)
[10] Vargas, C.M.: Design and construction of Mexican Aeroplanes 1909-1939. Cemanáhuac – Boletín Informativo, Año XIV VI, (23), 14–15 (2005) (in Spanish)
[11] Paz Fernández, J.M.: Non-linear dynamic structures through energy relaxing methods. MSc Thesis. ESIME-IPN, Méx. (2006) (in Spanish)
[12] SDN-FAM, Mexican Air Force, Evolution (February 1, 2010) (in Spanish), http://www.sedena.gob.mx/index.php?id=97
[13] Lazarín Miranda, F.: Air transport in Mexico and its economic impacts. In: Proceedings, Second National Congress on Economic History, Symposium 13, School of Economics, UNAM (2004) (in Spanish)
[14] Simmons, D., Withington, T.: History of Aviation. Parragon Books Ltd., London (2007) (in Spanish)
[15] Vivian, E.C.: A history of aeronautics. Nabu Press, London (1920)
[16] Edward, T.L., Crenshaw, J.S.: Aviation album. Popular Mechanics, USA (1953)
[17] Santaló Sors, L.: History of aeronautics. Espasa–Calpe, Buenos Aires, Argentina (1946) (in Spanish)
[18] Aguilar, J.M.: Aviation. From Icarus to spacecraft. In: Encyclopedia: The world and Man, Ed. Bruguera, Barcelona. (1962) (in Spanish)

[19] WPX, 1915 in aviation (December 15, 2009) (2010) http://en.wikipedia.org/w/index.php?title=1915_in_aviation&action=edit&section=2
[20] Alva Merino, J.: Mexican Air Force. Avion Revue Internacional (123), 36–43 (2010) (in Spanish)
[21] Hagedorn, D.: Conquerors of the sky: A history of aviation in Latin America. In: Smithsonian National Air and Space Museum, Washington, D.C., University Press, Florida (2010)
[22] Zárate Pena, I.: Design and construction of a scaled-down wind tunnel for general purpose. Thesis, Fac. de Ingeniería. UAQ, México (2010) (in Spanish)
[23] Rivera Anaya, D.A.: Instrumentation and control of a wind tunnel. Thesis. Fac. Ingeniería, UAQ, México (2010) (in Spanish)

The Rotary Aero Engine from 1908 to 1918

Giuseppe Genchi and Francesco Sorge

Dipartimento di Ingegneria Chimica, Gestionale, Informatica,
Meccanica – Università degli Studi di Palermo

Abstract. The rotary aero engine is a special type of air-cooled radial engine, where the cylinders are arranged like the spokes of a wheel and turn around the crankshaft. The propeller is connected to the cylinders, while the crankshaft is fixed to the frame. The rotary aero engine, developed in 1908, set new standards of power and light weight within the aircraft industry. It was adopted by many pioneer aviators and widely used to set records of endurance, speed and height. Many aero engine manufacturers produced different models and variants of this type of engine, which was extensively used until the end of the First World War. The latest evolution of the rotary engine was the counter-rotary arrangement, which was devised and designed by the Siemens-Halske company. The distinctive feature of this type of engine was that the engine body (with cylinders and propeller) rotated in one direction while the crankshaft rotated in the opposite one. This result was obtained by using a bevel gear mechanism. However, rotaries were quickly and definitively replaced in 1918 by new kinds of conventional engine, which were developed in the same period by other manufacturers. The main features of rotary and counter-rotary aero engine and the performance limits that caused their decline will be described in this paper. The rotary engine will be compared with the conventional one in terms of power output, specific consumption, weight and inertia loads transferred to the frame.

1 Development of the Rotary Aero Engine

The power units used in the early aircrafts at the beginning of the XX Century were strictly derived from water-cooled automobile engines or air-cooled motorcycle engines. The water-cooled types were penalized by their heavy weights while the air-cooled were prone to suffer for overheating. The aero engines started their development as a consequence of the increasing specific power demand of aircraft motors. In the early aeronautical era, various kinds of engines were proposed for this purpose by many manufacturers and inventors. Many of these prototypes were of unusual and odd design but were not produced in series or did not have commercial success because of their poor performances. The rotary engine was one of the solutions experimented with success at the beginning of XX Century, a period very reach in innovations and scientific progress. Before its aeronautic application, the rotary principle was already known. It was used in various applications, such as the Millet's motorcycle in 1888 (Fig. 1), which however did not reach success and large

T. Koetsier, M. Ceccarelli (Eds.): Explorations in the His. of Machines & Mech., HMMS 15, pp. 349–362.
springerlink.com

diffusion. The rotary aero engine was first introduced into the aircraft industry by the Seguin brothers. They started to develop it by 1906 and then produced their first model, the Gnome 50 HP "Omega", in 1908. This engine is usually considered as the archetypal of the aeronautical rotaries. The 50 HP Gnome, thanks to its peculiar features, granted better performance than its contemporary conventional engines, setting new standards of power and light weight in the aircraft industry. It was adopted by many pioneer aviators and widely used to set records of endurance, speed and height. The "Société des Moteurs Gnome" of the Seguin brothers, was the first and most important rotary aero engine manufacturer of the early rotary era (until about 1912); their engines were also produced under license in many countries. From 1910 others manufacturers, like Le Rhone, Clerget, Bentley, Oberursel and Siemens-Halske, started to design and produce a large number of models and variants. Rotaries were extensively used until the end of the First World War: their success and technical development were as quick as their decline.

Fig. 1. 1888 Millet's motorcycle with 5-cylinder rotary engine into rear wheel.

Actually, rotaries were quickly and definitively replaced in 1918 by new kinds of conventional engine with higher performances, which were produced in the same period by other manufacturers. However the rotary engine gave a significant contribution to the development of the aviation industry.

2 Rotary Aero Engine Features

The rotary aero engine is a special type of air-cooled radial engine, in which the cylinders are arranged like the spokes of a wheel and turn around the crankshaft. The propeller is connected to the cylinders, while the crankshaft is fixed to the frame. In a generic radial engine, both of the rotary and conventional type, a master rod is usually mounted on the crankshaft while a number of slave rods are connected to the master rod. In this arrangement all the rods lie on the same plane with clear advantages with regard the inertia load balancing and the engine length. The success of the aero rotary engine was mainly due to its good cooling capability and its construction lighter than the contemporary aero engines. The first water-cooled aero engines were limited by their heavy weight, whereas the lighter conventional air-cooled ones suffered from a bad or insufficient cooling capability, due to the low aircrafts speed and the scarce cooling efficiency of the cylinder shape. These factors represented an important limit to the performance and reliability of the first aero engines. Studies of those days indicated that a conventional radial engine would have to be given a forward speed of 50 km/h to be cooled to the same degree as a comparable rotary engine running on a test stand.

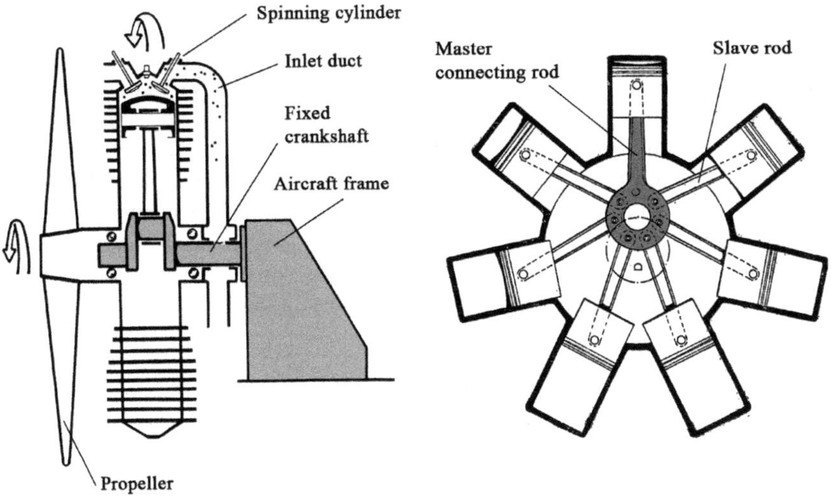

Fig. 2. Typical rotary engine features.

The efficient cooling of the rotaries resulted from the combination of the wind milling and the propeller slipstream. This also allowed the rotary engines to be realized with a construction lighter than the conventional ones. Rotaries were usually designed with extremely fine section in most components, operating close to the limits of the permissible thermal distortion. At the beginning of the aviation era, the excellent cooling of the engine and the lightweight construction made the rotary engine superior to his contemporaries. However, the rotary design imposed some problems about the admission of the petrol-air mixture to the turning cylinders. In this type of engine, the mixture is drawn through the hollow crankshaft and the carburettor is mounted on the outer end of it. The crankcase is thus utilized as distributor chamber from which the mixture reaches the cylinders in different ways, depending on the particular design. This represents one of the most relevant differences among the various rotary aero engines.

In the early Gnome engines the mixture passes from the crankcase to the combustion chambers through automatic valves in the piston crown. These are of the spring-loaded type like the "atmospheric" valves used in the early motor car engines. When the cylinder pressure falls during the intake stroke, the reduced pressure opens the inlet valve and the mixture is drawn in; as the compression builds up, the inlet valve closes automatically. This type of inlet valve reduces the volumetric efficiency of the engine: it opens later and closes sooner than the proper valve timing. The exhaust valve is of the actuated type and mounted on the cylinder head.

In 1913, the Seguin brothers introduced a new design of rotary, the Monosoupape, in which the inlet automatic valve of the piston crown was eliminated. This type of engine has a single overhead actuated valve, with an unusual timing, that is used both to draw fresh air into cylinder and to discharge the exhaust gases. A rich mixture is transferred from the crankcase to the combustion chambers by a row of ports on the bottom of the cylinders. These ports are similar to the ones of a two stroke engine and are uncovered by the piston at the bottom of its stroke. However the Monosoupape, like most rotary engines, operates along a four stroke Otto cycle following the conventional phase sequence. The rich mixture drawn into the cylinder mixes with the fresh air which is already there to form a proper combustible mixture. The engines with this odd feeding system were not less efficient than the contemporary rotaries with conventional design. The Monosoupape engines in fact were used by the Allied air forces throughout the First World War. This kind of rotary was introduced in order to replace the crown inlet valve, which frequently gave some trouble caused by a dangerous backfire on the crankcase. The new design of the Gnome engine exhibited fewer parts, reduced weight, better reliability and about the same power.

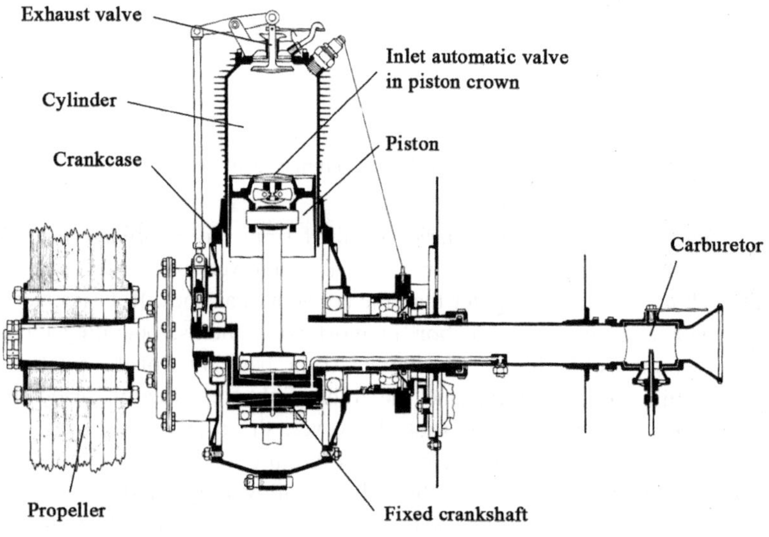

Fig. 3. Rotary engine Gnome Omega (1908)

The rotary engines produced by other manufacturers, such as Le Rhone, Clerget, Oberursel and Bentley, usually had a more conventional valve system in which the inlet manifolds (one per cylinder) were fitted radially out of the crankcase and reached each cylinder head. The mixture admission and the exhaust gas discharge were controlled by pushrod actuated overhead valves, usually two per cylinder.

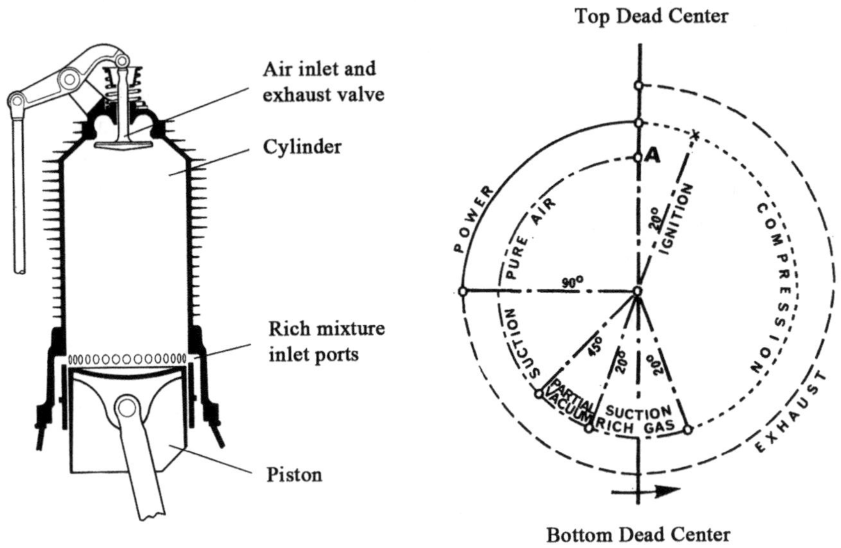

Fig. 4. Gnome Monosoupape cylinder and its engine cycle features.

3 The Counter-Rotary Aero Engine

The latest evolution of the rotary engine was the counter-rotary type, first and only designed in 1914 by Siemens-Halske, a brand of Siemens AG. The distinctive feature of this type of engine was that the crankcase (including cylinders and propeller) rotated in the one direction while the crankshaft rotated in the opposite one. This effect was obtained by using a bevel gear mechanism with speed ratio -1.

The best counter-rotary engine was the Siemens-Halske Sh.III, which featured the considerable power output of 200 HP with a power-to-weight ratio of about 1 HP/kg. This engine, which was produced in considerable quantities (about 1200 exemplars until 1918), was fitted to different aircrafts including the Siemens-Schuckert D.III and D.IV. These aircrafts featured an exceptional rate of climb thanks to the performance of their Sh.III engine, and have been considered as the best German fighters produced during the First World War.

The counter-rotary design was devised in order to reduce the propeller speed with the purpose of improving its efficiency without reducing the power output. Improvements in performance to this point in time had been achieved by increasing engine speed, with systems immediately preceding the counter-rotary design, featuring maximum speeds of about 1400 rpm. Nevertheless, this caused some problem regarding the efficiency of the fixed-pitch wooden propellers in use in that period. A reduction gear mechanism had been introduced before the propeller in the contemporary conventional engines but it was very difficult to arrange a similar mechanism in the rotaries. In a Sh.III engine, the fine counter-rotary design allows both the propeller to run efficiently in its usual range of 900÷1000 rpm, and the engine to produce a power output corresponding to a speed of about 2000 rpm of the simple rotary, i.e. twice that of the propeller. In fact, other engine parameters being equal, the power output is dependent on the frequency of the thermodynamic cycle: in a counter-rotary, thanks to the bevel gear mechanism, this corresponds to the sum of the absolute values of both the crankcase speed (cylinders plus propeller) and that of the crankshaft. The reduction of the cylinder speed gave two other advantages: a significant decrease of the windage losses from the spinning cylinders and a moderate reduction of the engine gyroscopic effect for the benefit of the aircraft manoeuvrability, due also to the counter rotating masses of the crankshaft together with its balance weights and the connecting rods. Unlike the early rotary engines, the 200 HP power output of the Sh.III ensured an aircraft speed sufficient for a correct cooling of the cylinders, which ran at "only" 1000 rpm, rather than at the usual 1300÷1400 rpm of the contemporary rotaries.

On the other hand, the Siemens-Halske Sh.III, such as all the other German rotaries, suffered for the use of poor quality lubricant oil. In that period, the castor oil was the best lubricant available, clearly superior to the contemporary mineral oils, and it was widely used for this reason on the rotaries, whose fine construction was extremely sensitive to the lubricant quality. In Germany, there were insufficient castor oil supplies and so the use of mineral oils limited the operating life of the early Sh.III engines to about ten hours before the pistons began to seize.

Sh.IIIa of the Historic Museum of Engines and Mecanisms of the University of Palermo

Sh.III of the Science Museum of London.

Fig. 5. Counter-rotary engine Siemens Sh.IIIa, with detailed views of bevel gears.

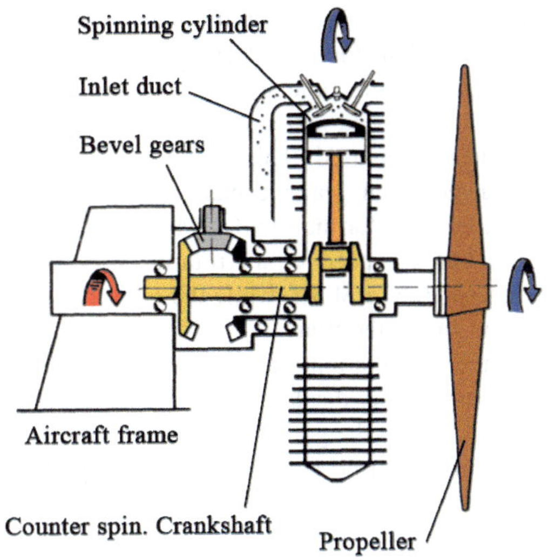

Fig. 6. Counter-rotary engine Sh.IIIa features.

4 The Limits of the Rotary Aero Engine

The peculiar features that granted the rotary engine success were the same that caused its decline. The cylinders' spinning gave a superior cooling effect and permitted the engine lighter construction than other conventional air-cooled or water-cooled earlier engines. Nonetheless, during the short life of this kind of engine, it was experienced that the peculiar characteristics themselves of the design implied significant limits to some of the main factors on which the engine performance depends: volumetric efficiency, engine speed and displacement.

In a rotary engine, the mixture passes through the hollow crankshaft, emerges into the crankcase and at least arrives into the combustion chambers through long inlet ducts, or else transfer ports as in the Gnome Monosoupape engines. The intake path is very long and extremely tortuous and this implies remarkable pressure drops with consequent significant reduction of the volumetric efficiency, that is of the power output. By comparison, the more advanced conventional engines realized near the end of the War are equipped with simple and direct inlet ducts. Therefore, the use of odd inlet valve system, such as the automatic valves on the piston crown or the transfer ports for the intake of rich mixtures, implies a certain reduction of the number of parts and of the weight but causes a further decrease in the volumetric efficiency with respect to the conventional valve system operation. In Table I, which reports the specific power, the displacement and the brake specific fuel consumption of several engines, the negative effect of the poor volumetric efficiency and of the significant inlet pressure drop typical of the rotary design appears evident with respect to the conventional engine. In synthesis, the poor volumetric efficiency was a peculiar trouble of the rotary design which implied an irremediable disadvantage for the improvement of its performance.

The mechanical stressing of the materials due to the centrifugal loads did not represent an important problem for the rotary engine, mainly because the maximum engine speed was limited at the time by the standard use of fixed-pitch wooden propellers. As the crankcase speed increased (together with the propeller speed), the airscrew thrust efficiency decreased. This trouble was solved on the conventional engine by mounting a reduction gear mechanism before the propeller, whereas the only effective solution adopted for the rotaries was the counter-rotary design. For this reason the maximum engine speed of all rotaries was about 1400 rpm with the exception of the Siemens Halske Sh.III, while the other types could rotate up to 2000 rpm on approaching the end of the War.

The mentioned technical limitations led many manufacturers to improve the engine performance by increasing the displacement. However this solution implied to increase the sizes, the mass and the gyroscopic effect, with negative consequences on the aircraft manoeuvrability. Furthermore the bore increase of an air-cooled thin-wall cylinder implied some thermal distortion trouble: it was caused by the cooling flow rate differences between the front and the rear part of each cylinder.

Table 1. Performance comparison among rotary and conventional engines.

Engine Type Year	Displ. [litres]	Weight [kg]	Power [HP] (rpm)	Brake specific fuel cons. [g/(HP*h)]	BMEP [bar]	Power to displ. ratio [HP/litres]	Power to weight ratio [HP/kg]
Gnome Omega* 7-cylinder rotary 1908	7.98	75	50 (1200)	388	ca. 4	6.3	0.67
Clerget 11 EB 11-cylinder rotary 1916	23.60	230	200 (1300)	300	5.65	8.5	0.87
Bentley BR2 9-cylinder rotary 1918	24.92	226	230 (1300)	286	6.34	9.2	1.02
Siemens Sh.IIIa 11-cyl. count.-rot. 1918	18.60	198	200 (1000**)	270	5.52	10.8	1.01
Daimler D.IVa 6 in-line cyl.*** 1916-1918	21.70	405	260 (1450)	230	ca. 8.0	12.0	0.64
Hispano-Suiza 8FB 8-cyl. V 90° *** 1917-1918	18.48	275	300 (1800)	230	ca. 8.0	16.2	1.09

*: First aero rotary engine. **: Propeller speed. ***: Water-cooled engine.

Finally, the power output of the high-performance later rotaries implied a significant increase of the aircraft speed, so that the cooling effect of the spinning cylinders began to be considered as just negative due to its related windage losses. On the other hand, the cooling troubles of the early air-cooled aero engines was also faced in other ways, e.g. by improving the cylinder shape or by using materials with high heat transfer coefficients, such as aluminium alloys.

The conventional aero engines (specially of the water cooled types) that were developed in the same period by important manufacturers such as Daimler, BMW, Rolls Royce, Salmson, Hispano-Suiza and Fiat, did not suffer for the rotary limitations and so bridged the early performance gap. Table 1 gives a synthetic comparative analysis among some important rotaries and some other contemporary conventional engines. The most significant parameters of the aero engine performances are taken into consideration: the brake specific fuel consumption, which is an engine efficiency indicator, the brake mean effective pressure (BMEP), which is proportional to the torque-to-displacement ratio, the power-to-displacement ratio and the power-to-weight ratio.

5 Rotary Engine Dynamics

The rotating crankcase arrangement turns out to be beneficial to the overall engine dynamics and this aspect may have probably influenced its temporary success. On the one hand, apart from the potentiality of halving the angular speed for fixed power level, the counter-rotary architecture offers the advantage of improving the aircraft manoeuvrability due to a sort of compensation of the angular momentum between the contra-rotating parts. On the other hand, both rotating-crankcase schemes may be shown to produce a non-negligible lowering of the inertia forces acting on the frame. Such forces can be generally scaled by the product of three factors, the motor mass, the square of the angular speed and some reference motor length. Therefore, for geometrically similar engines, the inertia force amplitude is proportional to $N^2V^{4/3}$, where N is the rev number per unit time and V the volume, whilst the inertia torque amplitude is proportional to $N^2V^{5/3}$. Wishing to compare the radial engines on the basis of the same power and the same sizes, the angular speed of the counter-rotating engine must be one half of the other two and its inertia forces and moments must be multiplied by a factor $1/2^2$ for the comparison.

The inertia force system of the counter-rotary engine is different from the fixed-crank and the conventional fixed-crankcase arrangements. As the outer cylinder body is perfectly balanced and symmetric with respect to the axis of rotation, the counter-rotary dynamical behaviour would seem to be similar to the rotating crank configuration, the main differences arising from the double frequency of the force reciprocation with respect to the output shaft. As a matter of fact, comparing the three engine types in the absence of counterweights, the fixed crank arrangement would generally exhibit better balancing conditions because of the suppression of the predominating centrifugal effect of the crank rotation, whereas the reciprocating forces arising in the rods and the pistons balance each other approximately due to the star shaped engine structure. Nevertheless, the centrifugal forces associated with the crank rotation may be efficiently balanced by mounting suitable counterweights on the crankshaft back, which restore the rotating crank competitiveness. Actually the counter rotary engine ShIII is equipped with two proper balancing masses on the crankshaft.

Figure 7 shows a scheme of the rods and Table II summarizes the inertia force system, which was derived by conventional procedures (e. g. see [6]). The introduction of the auxiliary dummy coefficients d_c, d_i and d_o permits applying the compact formulation of Table II to all radial engines. The rod obliquity angles φ vary during the cycle and are to be calculated by proper closure equations of the triangle OA_0B_0 and the quadrilateral OA_0AB, as shown in Fig. 7.

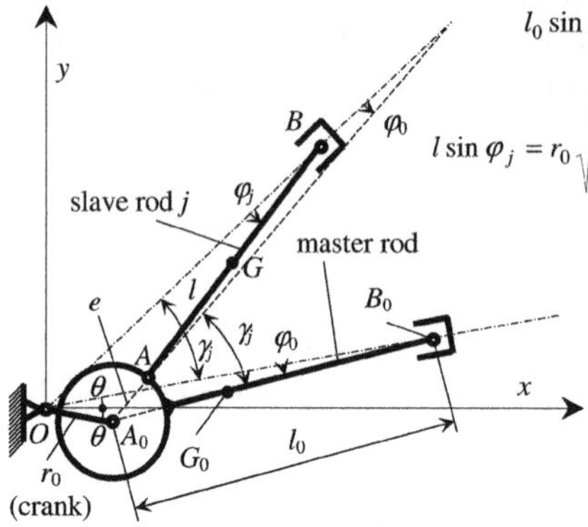

$$l_0 \sin\varphi_0 = r_0 \sin(d_c\theta)$$

$$l \sin\varphi_j = r_0\sqrt{1-2\frac{e}{l_0}\cos\gamma_j+\left(\frac{e}{l_0}\right)^2}\,\sin\left[d_c\theta+\arctan\left(\frac{\sin\gamma_j}{\cos\gamma_j-\frac{e}{l_0}}\right)\right]$$

counter-rotary	$d_c = 2$
fixed crank	$d_c = 1$
fixed crankcase	$d_c = 1$

Fig. 7. Geometric scheme of master rod and generic slave rod ($\gamma_j = 2\pi j/n$)

The moments of the inertia forces with respect to the engine axis through O are resisted in part by the crankcase and in part by the crank. The fraction loading the crankcase can be calculated once the mutual forces exchanged between the pistons and the cylinders orthogonally to their contact walls are found. This can be done for the slave rods (j = 1 to 10) by simply imposing the rotational equilibrium with respect to the crankpins A. With regard to the master rod, after calculating the mutual forces between slave and master rods through the crankpins A by the translational equilibrium of each slave rod, the rotational equilibrium around A_0 yields the reaction force exchanged between the master rod and the master cylinder. Once the inertia torque on the cylinder body is calculated, the torque fraction loading the crank is then obtainable by difference. In the engine configuration with just one rotating part, either the crank or the crankcase, the described calculation gives the inertia torque on the frame and the residual torque fraction applied to the propeller shaft, which merges in practice with the driving torque but does not affect its value on the average. In the counter-rotary engine on the contrary, the crankcase torque fraction is transferred to the propeller directly, while the crank torque is transferred to the bevel gearing set. As the crank and crankcase speeds are equal and opposite, the latter torque fraction is balanced by an equal reaction moment applied by the opposite bevel gearwheel, which is connected in turn to the output shaft. Therefore, the moment $2(M_{O,\mathrm{tot.}} - M_{O,\mathrm{crankcase}})$ is applied to the aircraft frame, while the sum $M_{O,\mathrm{crankcase}} + (M_{O,\mathrm{crankcase}} - M_{O,\mathrm{tot.}})$ merges with the driving torque applied to the propeller, without affecting its average value of course during one complete cycle.

Figure 8 shows the diagrams of the inertia forces and moments for the three examined configurations. They were traced in the hypothesis of perfect balancing of the "centrifugal" forces by means of suitable counterweights, i. e. by putting $d_i = 0$ identically in the formulae of Table 2. Moreover, as an equivalence criterion between the counter rotary and the simply rotary configurations, it is assumed that the angular speed of the latter ones is twice the former, in the hypothesis of equal sizes and equal power of the three engines. As the inertia forces and moments are scaled by $m_s l_0 \omega^2$ and $m_s l_0^2 \omega^2$ respectively, where ω refers to the counter-rotary engine, a factor 4 arises in the scaling process of the two other engines.

Table 2. Inertia force system

$$\frac{F_{x,j}}{\omega^2} = (m_{r,j} + m_p)\left\{ r_0 d_i \cos\theta + e\,\mathrm{sgn}(j)\left[\left(d_o + \frac{\dot{\varphi}_0}{\omega}\right)^2 \cos(\theta d_o + \varphi_0 + \gamma_j) + \left(\frac{\ddot{\varphi}_0}{\omega}\right)^2 \sin(\theta d_o + \varphi_0 + \gamma_j)\right]\right\} +$$
$$(m_{r,j} l_{G,j} + m_p l_j)\left[\left(d_o + \frac{\dot{\varphi}_j}{\omega}\right)^2 \cos(\theta d_o + \varphi_j + \gamma_j) + \left(\frac{\ddot{\varphi}_j}{\omega}\right)^2 \sin(\theta d_o + \varphi_j + \gamma_j)\right]$$

$$\frac{F_{y,j}}{\omega^2} = (m_{r,j} + m_p)\left\{ - r_0 d_i \sin\theta + e\,\mathrm{sgn}(j)\left[\left(d_o + \frac{\dot{\varphi}_0}{\omega}\right)^2 \sin(\theta d_o + \varphi_0 + \gamma_j) - \left(\frac{\ddot{\varphi}_0}{\omega}\right)^2 \cos(\theta d_o + \varphi_0 + \gamma_j)\right]\right\} +$$
$$(m_{r,j} l_{G,j} + m_p l_j)\left[\left(d_o + \frac{\dot{\varphi}_j}{\omega^2}\right)^2 \sin(\theta d_o + \varphi_j + \gamma_j) - \left(\frac{\ddot{\varphi}_j}{\omega}\right)^2 \cos(\theta d_o + \varphi_j + \gamma_j)\right]$$

$$M_{O,j} = -I_{G,j}\ddot{\varphi}_j - F_{(r)x,j}\, y_{G,j} + F_{(r)y,j}\, x_{G,j} - F_{(p)x,j}\, y_{B,j} + F_{(p)y,j}\, {}_j x_{B,j}$$

Notation:

$\omega = \dot{\theta}$, angular speed of propeller shaft (dots indicate time derivatives);

$\gamma_j = 2\pi j/n$, angle between generic rod j and master rod ($j = 0 \ldots n-1$, $j = 0$ refers to master rod and n is the number of cylinder, e. g. $n = 11$ for Sh.III);

m_p, piston mass;

$m_{r,j}$, $I_{G,j}$, l_j and $l_{G,j}$, masses, barycentric moments of inertia, total length and barycentric distances from crankpins of rods ($m_{r,j} = m_r$, $I_{G,j} = I_G$, $l_j = l$ and $l_{G,j} = l_G$ for $j > 0$; $m_{r,j} = m_{r0}$, $I_{G,j} = I_{G0}$, $l_j = l_0$ and $l_{G,j} = l_{G0}$ for $j = 0$);

subscripts (r) and (p) refer to the rods and the pistons separately;

$d_i = \mathrm{sgn}|\omega_{\mathrm{crank}}|$ and $d_o = \mathrm{sgn}|\omega_{\mathrm{crankcase}}|$ are unitary dummy coefficients that vanish for fixed inner crank and fixed outer crankcase respectively.

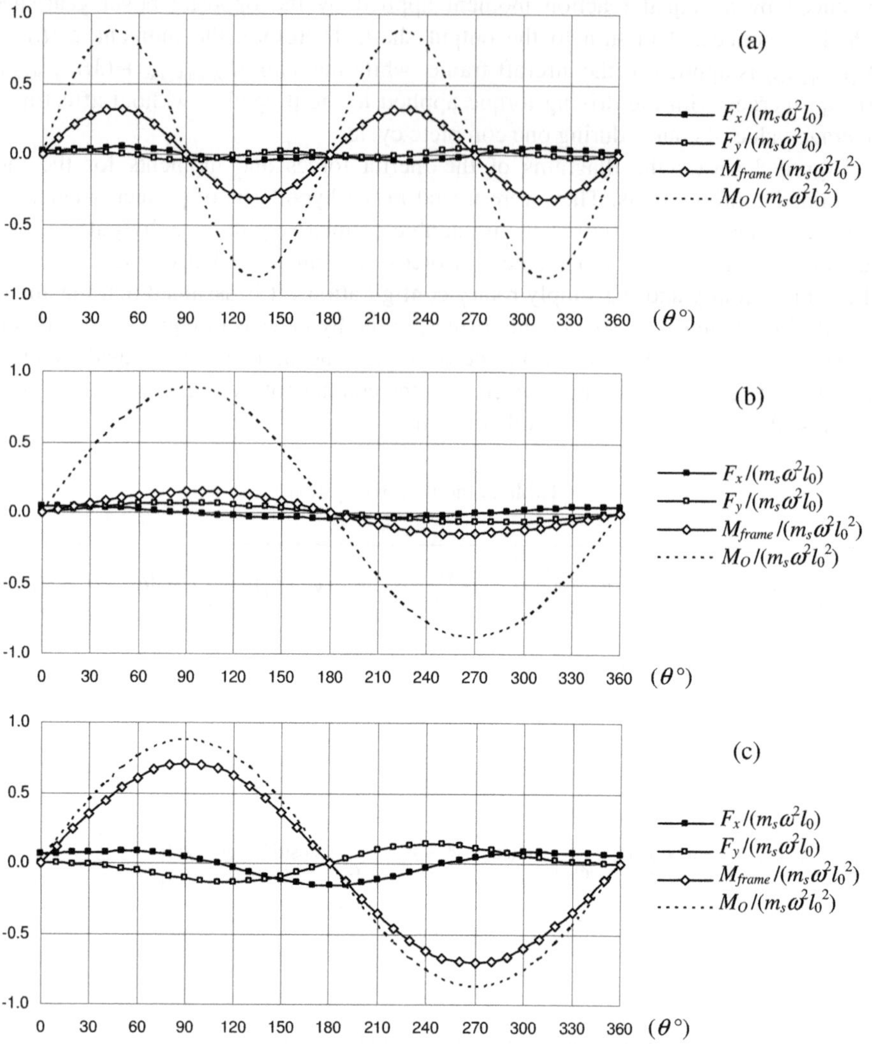

Fig. 8. Inertia forces and moments. (a) counter-rotary; (b) fixed-crank; (c) fixed-crankcase. Equal motor sizes, ω = counter-rotary engine speed ($\omega_{(b)} = \omega_{(c)} = 2\omega$). Massive and geometrical characteristics assumed by previous figures and engine plans.

The doubling of the inertia reciprocation feature is clearly observable in the counter-rotating case, together with a common characteristic of the three arrangements: the x component of the resultant inertia force is symmetrical with respect to the configurations $\theta = 0$ and $\theta = \pi$ (crank and master rod aligned), while the y component and the moments are anti-symmetric. Numerical tests with unbalanced "centrifugal" forces may be found to yield much higher levels of F_x and F_y for the contra-rotating and fixed-crankcase arrangements, due to the nearly stellar configuration of the reciprocating inertia forces and the parallelism of the centrifugal ones. Nonetheless, considering properly counter-weighted cranks, the fixed-crank and counter-rotary configurations appear to produce the lowest average levels of the inertia forces, which implies smoother vibratory conditions and may have been perhaps one of the reasons why these solutions had a rather larger success during the early decades of the 20th century. The inertia torque on the frame is lowest in the fixed crank arrangement and highest in the fixed crankcase one, the counter-rotating case taking place in an intermediate position. In short, the rotating cylinder configuration was no doubt advantageous with regard to the dynamics of the engine, whose balancing requirements did not certainly impose any development limitation.

6 Conclusion

The rotary aero engine set new standards of power and light weight in the field of aircraft industry from its dawn in 1908. During the ten years of its development it gave an important contribution to the early aeronautic industry. It was adopted by many pioneer aviators and widely used to set records of endurance, speed and height. The peculiar features granting the early rotary engine success were the high power output and the good cooling effect on the engine cylinders. Moreover, its vibratory behavior appeared quite smooth. But these properties were obtained thanks to the particular rotary design that implied, as experienced during its development, significant limits on the improvement of the engine performance. The major limitations were the low volumetric efficiency and the difficulties to enhance power by increasing displacement and speed engine. The conventional aero engines (specially the water-cooled types) were developed in the same period by important manufacturers such as Daimler, Rolls Royce and Hispano-Suiza, and from 1918 these engines were able to provide equal or better performances than rotaries. Furthermore these ones, unlike the rotaries, showed considerable improvement possibilities, which were realized in the following years. To conclude, it is possible to say that from the end of the First World War, rotaries were quickly and definitively replaced by the conventional engines which, after evolving in the famous high-power piston engines of the Second World War, were widely substituted in a very short time in the aeronautic industry by the turbojets about 30 years later.

References

[1] Filippi, F.: Dall'elica al getto, Associazione Industriali Metallurgici Affini. Edizioni EDA, Torino (I) (1983)
[2] Giger, H.: Kolben-Flugmotoren. Motorbuch Verlag, Stuttgart (D) (1986)
[3] Nahum, A.: The rotary aero engine. HMSO Books, London (1987)
[4] Gunston, B.: Development of piston aero engines. Patrick Stephens Limited, Haynes Publishing, Sparkford (2006)
[5] Von Gersdorff, K., Schubert, H., Hebert, S.: Flugmotoren und Strahltriebwerke. Bernard und Graefe Verlag, Bonn (D) (2007)
[6] Biezeno, C.B., Grammel, R.: Engineering Dynamics. Blackie & Sons Lt., London (1954)

Section V

History of Technical Systems

The Historical Development of Catrasys, a Cable System

Marco Ceccarelli

LARM: Laboratory of Robotics and Mechatronics,
University of Cassino, Via G. Di Biasio 43, 03043 Cassino, Italy

Abstract. CaTraSys (Cassino Tracking System) is a cable system with parallel kinematic architecture that has been developed for monitoring and measuring large displacements both in position and orientation. It is portable, mechanically robust, and user-friendly operable. It was conceived in the early 1993 at LARM in Cassino and since then it has been continuously improved in operation and design as well as it has been applied in different applications up to the today solution. The paper illustrates historical evolution of CaTraSys as an example in keeping historical track of modern systems since their early stages of development.

1 Introduction

The measure and monitoring of large displacements are problems for an efficient operation of multi-body mechanical systems and at the most for performance evaluation of robots, [1-3].

Since a displacement can be treated as a sequence of positions, instruments can be used from the field of measure of distances. Thus, several measuring instruments can be available in principle for such a task with moving multi-body systems when they can refer to measuring principles that are based on triangulation or trilateration with fundamentals established already in the past by using the geometry of triangles, as outlined in [3].

However, in the last decades economical obligations and suitable capacities for an easy implementation on modern systems with large operation ranges and with needs of reduced encumbrance have stimulated developments for alternative new systems and/or feasible adaptation of existing (traditional) devices. Thus, besides instruments that are based on or are themselves theodolites, [4], lot of efforts have been spent in developing instruments that are based on modern vision techniques, like laser systems [5, 6], or video capture systems, like in [7].

Special reduced attention has been addressed to cable systems that nevertheless since the beginning have been appreciated for mechanical robustness and user-friendly oriented operation. Early attempts with cable systems in robotic measurements are dated in 1997, [8] and 1999, [9] from other laboratories later than at LARM where the problem was attached since 1993, [10]. Just later other cable parallel systems were developed like for example those reported in [11-16] for different applications and design purposes.

T. Koetsier, M. Ceccarelli (Eds.): Explorations in the His. of Machines & Mech., HMMS 15, pp. 365–379.
springerlink.com

CaTraSys has been also of direct inspiration for deign of similar systems, like for example those indicated in [17, 18]. It has been also applied with adjustments in other research activities like for example in the Waseda locomotor design, [19].

It is believed interesting to present in this paper how a novel system like CaTraSys has been developed to fulfill practical needs for other research and design activities and how it has been then extended to other applications, even with other design solutions.

2 Catrasys (Cassino Tracking System)

CaTraSys (Cassino Tracking System) is composed of a mechanical part, an electronics/informatics interface unit, and a software package, Fig.1.

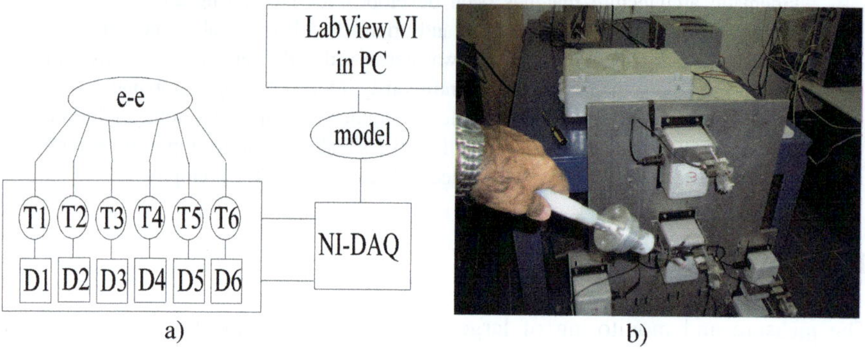

Fig. 1. CaTraSys design: a) a scheme; b) mechanical part.

The mechanical part consists of a fixed base, which has been named as Trilateral Sensing Platform TSP, Fig. 2 a), and a moving platform, which has been named as end-effector for CaTraSys, Fig. 2 b). The two platforms in the current architecture are connected by six cables, whose proper tension is maintained by pulleys and spiral springs that are fixed on the base and it is monitored by suitable force sensor units. The moving platform operates as a coupling device since it connects the cables of the transducers to the extremity of a movable system. This allows the cables to track a system while it moves with the moving platform attached to it. Signals from cables transducers and force sensor units are fed through an amplified connector to the electronic interface unit, which consists of a Personal Computer for data analysis with LabView hardware and software.

Suitable reference points are used to determine the position and orientation of the moving platform, which is the end-effector for CaTraSys whose parallel architecture can be used in several configurations but preferable with 3-2-1 solution. The term 3-2-1 is a technical term from Robotics indicating the number of cables that are attached in each of the three attachment points that are used to guide the platform motion.

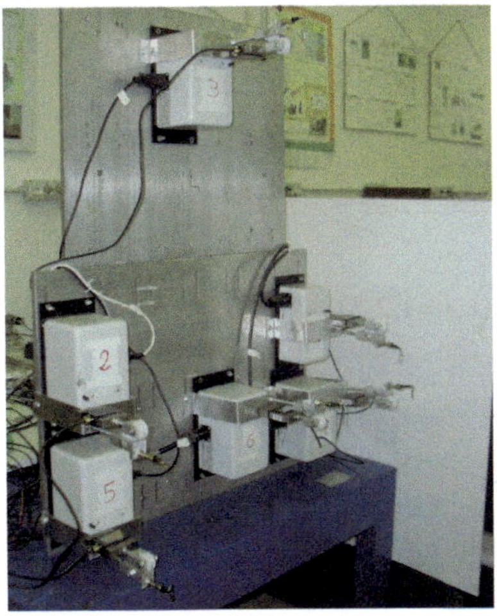

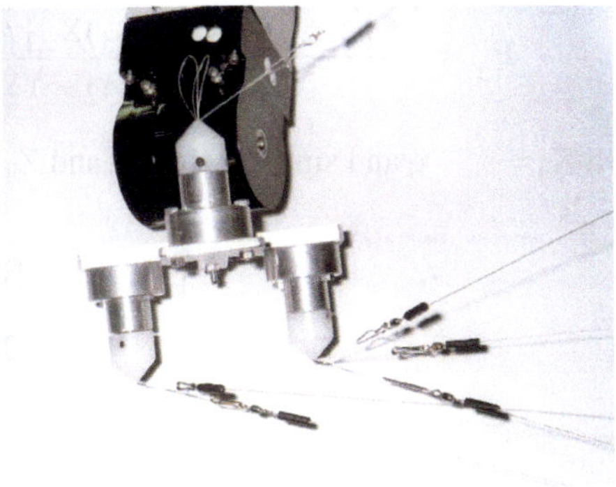

a) b)

Fig. 2. Mechanical components of CaTraSys: a) the Trilateral Sensing Platform with sensors; b) the end-effector coupling platform.

Trilateration is used for evaluation and computation of measured position and orientation of the moving end-effector platform as based on the scheme in Fig.3. By referring to a position of a general point H, when the distances u_1, u_2 and u_3 are measured, the coordinates x, y, z of H can be solved by the system of equations

$$u_i^2 = (x - x_i)^2 + (y - y_i)^2 + (z - z_i)^2 \quad \text{for } i = 1,2,3 \tag{1}$$

which describe the trilateration geometry of Fig.3 as a function of coordinates x_i, y_i, z_i of centers O_i of cable transducers.

Eqs.(1) can be solved to obtain

$$\begin{aligned} x &= H_x\left(-B_2 - \sqrt{B_2^2 - 4B_1B_3}\right)/2B_1 + E_x \\ y &= H_y\left(-B_2 - \sqrt{B_2^2 - 4B_1B_3}\right)/2B_1 + E_y \\ z &= \left(-B_2 - \sqrt{B_2^2 - 4B_1B_3}\right)/2B_1 \end{aligned} \tag{2}$$

where

$$A_i = u_i^2 - x_i^2 - y_i^2 - z_i^2 \qquad B_1 = H_x^2 + H_y^2 + 1$$

$$B_2 = 2H_x(E_x - x_3) + 2H_y(E_y - y_3) - 2z_3$$

$$B_3 = E_x^2 + E_y^2 + A_3^2 - 2E_x x_3 - 2E_y y_3$$

$$H_x = -(Z_{21} + H_y Y_{21})/X_{21} \qquad H_y = \frac{Z_{21} - Z_{32}X_{21}/X_{32}}{Y_{32}X_{21}/X_{32} - Y_{21}} \tag{3}$$
$$E_x = (A_1 + A_2 - E_y Y_{21})/X_{21}$$
$$E_y = \frac{-A_1 + A_2 + (A_2 - A_3)X_{21}/X_{32} - A_3}{Y_{32}X_{21}/X_{32} - Y_{21}}$$

with $X_{ij} = x_i - x_j$ and similarly for Y_{ij} and Z_{ij} (i, j =1,2,3).

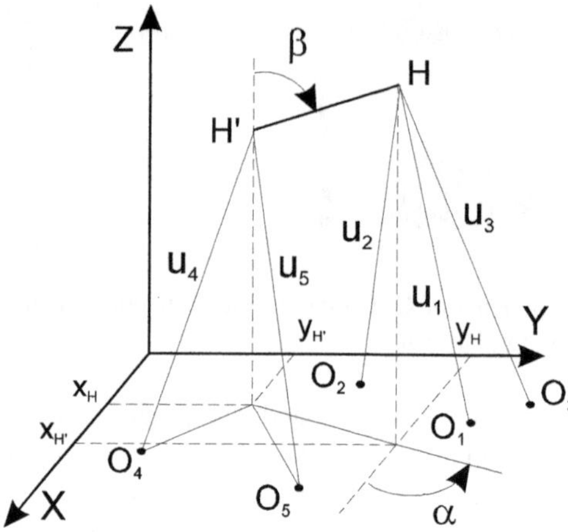

Fig. 3. A trilateration scheme for position and orientation measure of points in the CaTraSys end-effector.

Once x, y, z coordinates of H are computed by using the cable measures u_1, u_2 in Eqs.(2), position of H' can be computed by using two more measures u_4 and u_5 since the third one can be assumed as the given distance HH'. Thus also the orientation of the direction HH' can be evaluated by means of the angle α and β in Fig.3 though the expressions

$$\alpha = \tan^{-1}\left|(y_H - y_{H'})/(x_H - x_{H'})\right|$$
$$\beta = \tan^{-1}\left|\frac{(z_H - z_{H'})}{\sqrt{(y_H - y_{H'})^2 + (x_H - x_{H'})^2}}\right| \tag{4}$$

A further operation scheme is reported in Fig. 4 a) for the end-effect with full capability to measure position and orientation of the moving platform by using the 3-2-1 parallel architectures with three attachment points H, F, Q. All the possibility orientation angles in Fig. 4 b) can be computed by

$$\alpha = \tan^{-1}\left(\frac{y_H - y_F}{x_H - x_F}\right) \quad \beta = \tan^{-1}\left(\frac{z_H - z_F}{\sqrt{(y_H - y_F)^2 + (x_H - x_F)^2}}\right)$$

$$\psi = \tan^{-1}\left(\frac{z_Q - z_S}{\sqrt{(y_Q - y_S)^2 + (x_Q - x_S)^2}}\right) \qquad (5)$$

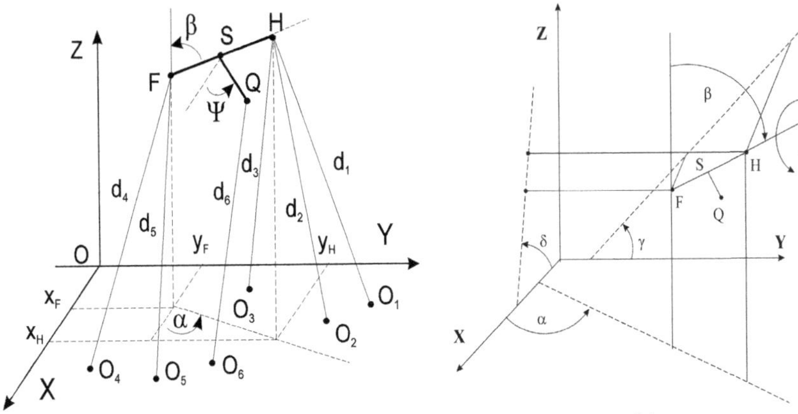

a) b)

Fig. 4. A scheme 3-2-1 parallel architecture for position and orientation measure of the CaTraSys end-effector: a) layout; b) angle identification.

In addition, Fig. 5 illustrates the tetrahedrons that can be identified within the operation scheme of Fig. 4 a). Their degeneracy to planar figures or even to lines can be monitored during the operation with the aim to avoid not only singularities but mainly to ensure configurations with stable computation capabilities. In particular, such an analysis once supported by experimental results brought to a CaTraSys layout with transducers located on STP on the vertices of two triangles apposite oriented.

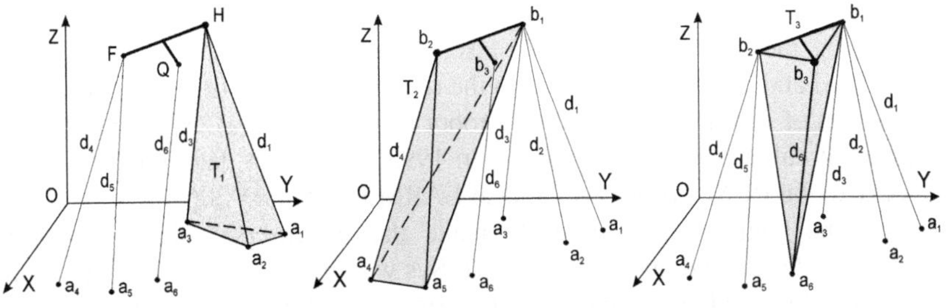

Fig. 5. Tetrahedrons avoid singular configurations to ensure robust efficient output.

The advantage of a robust mechanical design suggested a use of CaTraSys in applications with well defined mechanical features, like analysis of robot workspace and human arm actions. Limits were experienced and still characterizes CaTraSys operation as regarding with the bulky appearance and design of the system due also to the motions of the cables, even with risky aspects. Easy operation with user-oriented characteristics has been tested with practices by bachelor students and by performing lab tests in different applications with quick reprogramming and re-adaptation.

3 A History of Catrasys

The history of CaTraSys development is summarized in the chronicle of Table 1 with indication of main events. More information and full list of publications are reported in LARM webpage http://webuser.unicas.it/weblarm/larmindex.htm.

Table 1. Chronicle of main CaTraSys developments

Year	Event	Figure & Reference
1993	Design and construction of first prototype	Figs 6, 7; [10, 20]
1998	Full design and second prototype	Fig. 8; [21,22]
1998	New design for the end-effector	Fig.9, [21, 24]
2000	Application for analysis of human arm motions	Fig.10, [23-25]
2002	Error and singularity analysis	[26-28]
2006	Application for tracking human walking	Fig.11, [30,31]
2008	Application for rehabilitation	[29]
2008	Tension sensor and readable output	Fig.12, [30]
2009	New end-effector design for rehabilitation	[32]

Main developments are illustrated in Figs. 6 to 15. In particular, the historical evolution has been characterized by the following aspects.

Since 1990 workspace analysis has been an important research line at LARM both for robot analysis and design. While developing models and algorithms for numerical computations, since the beginning an interest arose for an experimental validation of procedures and results of the computer-oriented approaches. Thus, it was attempted to use existing measuring systems like laser trackers but they were experienced very complicated and costly as compared to the required tasks. A first attempt to determine the position of a workspace point of a robot manipulator was experienced by using 3 commercial cable position transducers in 1993 as a direct determination of position workspace of a PUMA robot available at LARM Fig. 6, [10, 20].

Since the beginning, in order to make the system portable and user-friendly oriented the design and operation of CaTraSys was conceived as based on commercial transducers installed on a portable platform, an end-effector coupling device with easy attachments, Fig. 6 b), and on virtual instruments (VI) of LabView, Fig. 7.

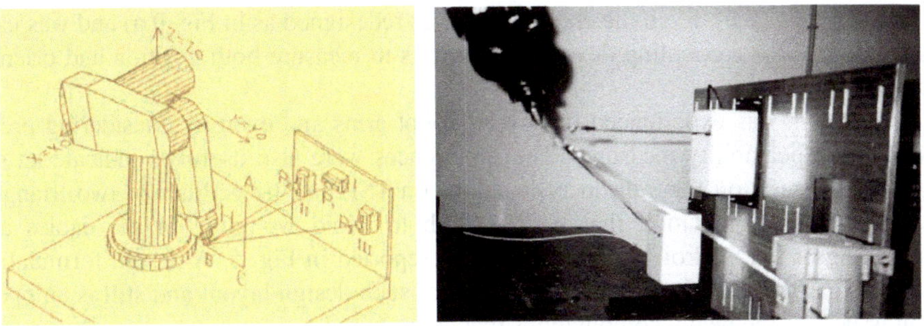

Fig. 6. CATRASYS Trilateral Sensing Platform prototype in 1993, [10]: a) CAD scheme; b) mechanical design.

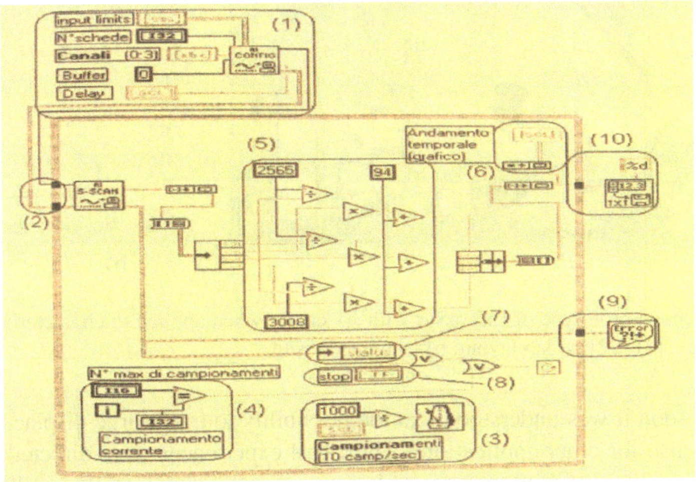

Fig. 7. Diagram of LabView Virtual Instrument for first prototype in Fig. 6, [10].

In particular, the end-effector was designed with features of an easy installation on robot wrist flange and having mobility of the attachments that will avoid wrapping of cables coming from transducers. The design was simulated as in Fig. 6 a) and then applied with direct experimental results both to determine workspace area and to identify D-H parameters. The cable transducers were chosen as measuring sources with suitable characteristics of mechanical design so that they made possible to design the sensing Platform TSP and end-effector, as well as to program an on-line measuring activity via VI, with developments made mainly by students for their thesis works. The modeling of the operation required a special attention for a proper formulation that brought to Eqs. (1) – (5), for a suitable implementation in VI programs.

The efficiency of the first CaTraSys encouraged to enlarge its capability for a full determination of a rigid body pose and thus 6 cable transducers were used for a final layout of CaTraSys in 1998, Fig. 8, [21, 22]. The TSP was enlarged, Fig. 8 a) and the workspace determination was elaborated in algorithms to track directly the workspace boundary, Fig. 8 b). Even the end-effector was redesigned as in Fig. 9 a) and was used as a module for a coupling device that permits to measure both position and orientation, Fig. 9 b).

CaTraSys was experienced on several robot arms and even by considering precision and repeatability measures. The experiences were also useful to indicate an optimal configuration for transducer locations on TSP on the vertices of two triangles with opposite orientation that permits both to avoid wrapping of the cables and shrinking of tetrahedrons to singularities as reported in Fig. 5. A design formulation was also attempted to get a better rationale of such design layout and still is an opening issues for the heavy computations that are required.

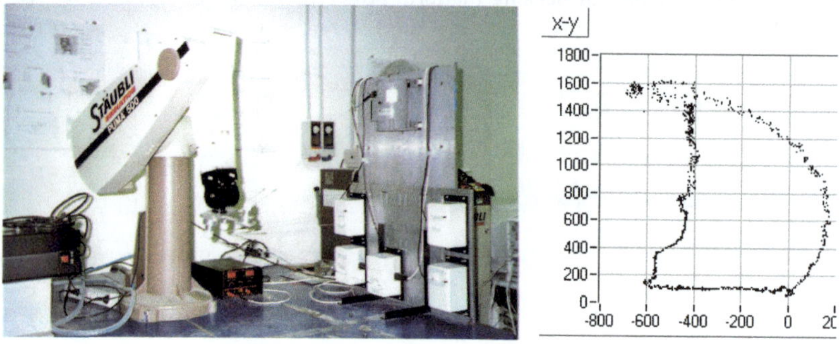

a) b)

Fig. 8. Second prototype of CaTraSys with six cables when applied to characterization of robot workspace, [21, 22]: a) lab layout; b) computed result.

Very soon it was understood that the possibility to track large displacements could be used also for other applications. Thus, first experiences were directed to the experimental analysis of motion range of human arm with successfully results as reported in Fig. 10, [23-25]. Features of user-orientated operation suggested the system to be useful for diagnosis procedures in determining human arm movements with great details, at a proper resolution levels. This application was tested with improvements in the elaboration of the results and with tests of users from different frames, even with the help of students within Erasmus programs.

The progress of operation capability and its success required a better characterization of CaTraSys performance mainly in terms of its precision. Thus, specific attention was addressed to model and evaluate error and singularity features, [26,28]. A maximum error of 3 mm was identified for a displacement of 2 m, [27] and dangerous and non dangerous singularities were characterized experimentally, [28].

After human arms, human legs were considered for their experimental characterization. Human walking was attached with few adaptations since 2006, Fig. 11 with indication of usefulness both for diagnosis and rehabilitation therapies, [29]. The

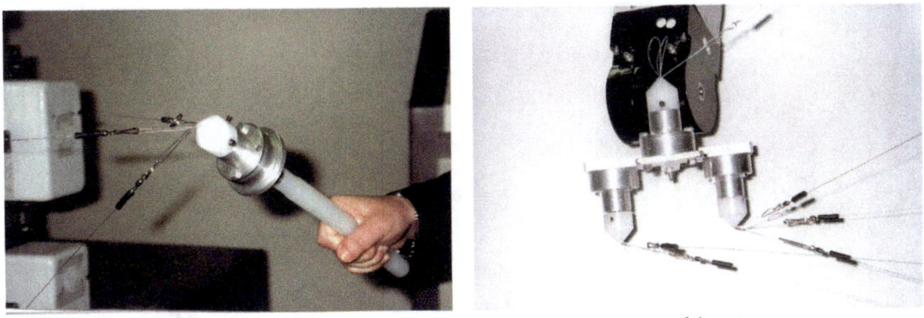

a) b)

Fig. 9. Design of End-effector for CaTraSys in 1998: a) a single module; b) full platform, [21].

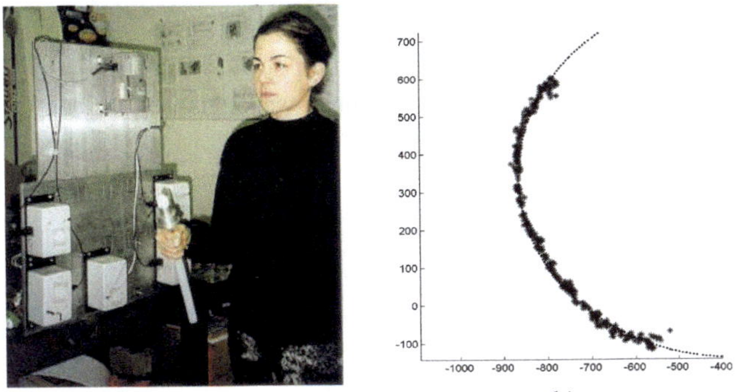

a) b)

Fig. 10. CaTraSys application to analysis of motion range of human arms in 2001, [23]: a) lab test; b) computed output.

operation was improved with a tapis-roulant Fig. 11 b) to give a continuous test operation, [30,31] that attracted the interest also of hospital physiotherapists. Unfortunately, doctor reluctance and medical protocols have not yet permitted a practical application of the system in hospital frames.

Portability and user-friendly operation were improved at design levels with also indications of necessary future work for a better patient safety and acceptance mainly through a better interface attachment as end-effector, [32]. The output of CaTraSys was also improved for a better online readability and to give both kinematics and force data, Fig. 12. Force sensors were installed on the transducers to measure the cable tension so that the force of the user can be measured during the motion as related to spring reaction from transducer tensioning systems. Curves like those in Fig. 12 were recognized of a great help to characterize human walking and its defects at a numerical level not possible with the current Physiotherapy practice as a first attention of doctors pointed out in 2009.

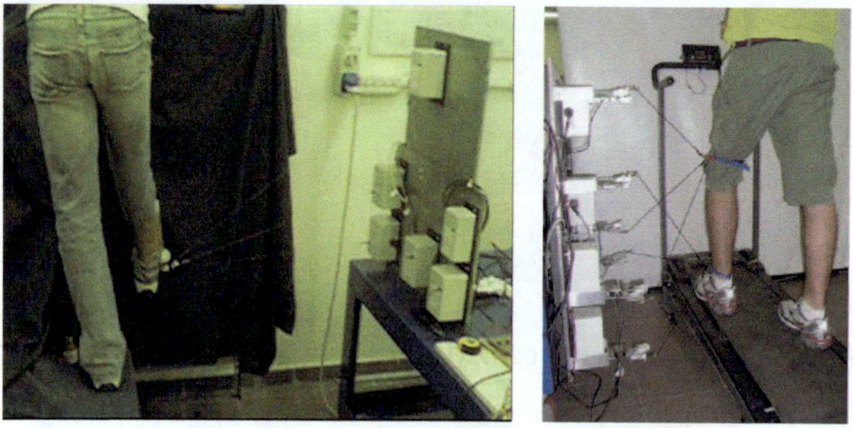

a) b)

Fig. 11. CaTraSys application to analysis of human walking: a) in 2006, [29]; b) in 2008, [30, 31].

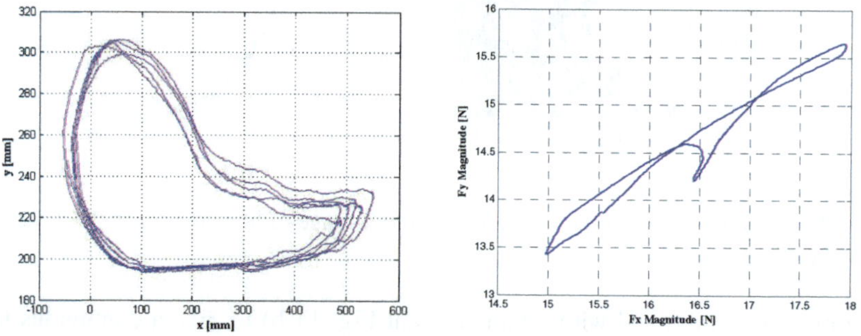

a) b)

Fig. 12. Outputs from CATRASYS implementation for human walking analysis in 2009, [30]: a) ankle point trajectory; b) leg force action.

Nevertheless, CaTraSys was continuously used to measure robot workspace and the above improvements give the possibility of better measures and new characterizations. It was also used to measure the workspace of parallel manipulators with tests on CaPaMan (Cassino Parallel Manipulator) prototypes that are available at LARM.

The success of CaTraSys design and operation suggested other designs as based on its functionality with kinematics of cable-based parallel manipulators. A first extension was Milli-CaTraSys that was developed by using LVDTs for more accurate measures. With a suitable additional force measuring feature it gives the possibility to evaluate stiffness behavior of any system, [33], Fig. 13. In particular, it was tested on CaPaMan 2bis also to implement a novel formulation for stiffness matrix determination with few tests, [33].

A relevant extrapolation of CaTraSys design can be considered the conceive of CaLoWi (Cassino Low-Cost Wire System) as an active version of CaTraSys. CaTraSys gets the input motion from a user through the end-effector and CaLoWi moves from input motion by motors to guide the end-effector to desired motion.

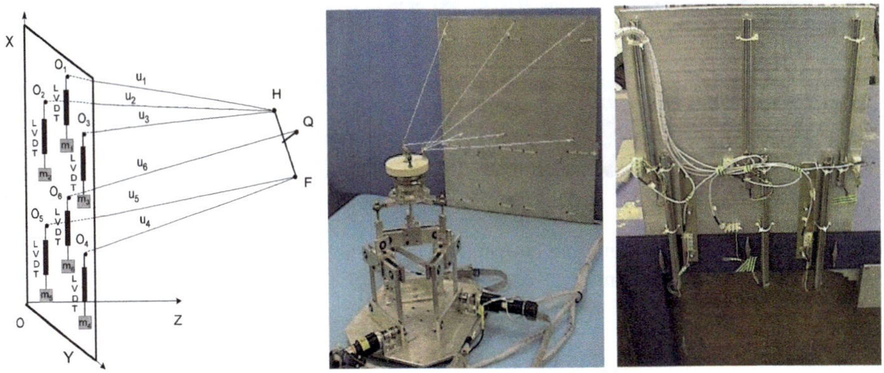

a) b)

Fig. 13. A scheme of Milli-CaTraSys in 2007, [24]: a) with the reference frame, LVDTs and masses m_i as applied to the cables u_i (i=1,…,6); b) lab layout.

The original design of CaLoWi was started in 2003 and after formulating its operation kinematics a feasible a prototype was built in 2004, Fig. 14, [34, 35]. CaLoWi was tested as a rescue operating system at the same time of similar designs proposed in Japan, [15]. Just after, it was combined with CaTraSys as active counterpart for rehabilitation guided exercises, Fig. 15, [36]. This included also a better path-planning that nevertheless was approached later in more general way as in [37]. In 2007 a first novel use of CaLoWi was proposed for hospital environments for bed transfer in [38].

In 2009 an extension of cable use with features both from CaTraSys and CaLoWi was conceived in designing a novel leg for robot humanoid as in Fig. 16, [30].

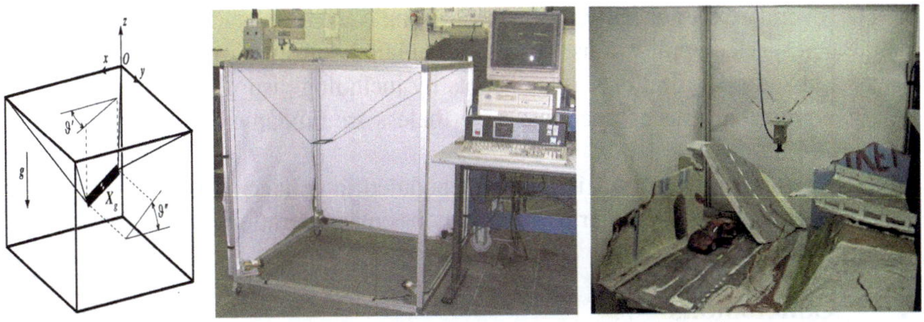

a) b) c)

Fig. 14. CaLoWi design and prototype in 2004, [34]: a) scheme; b) mechanical design; c) an application.

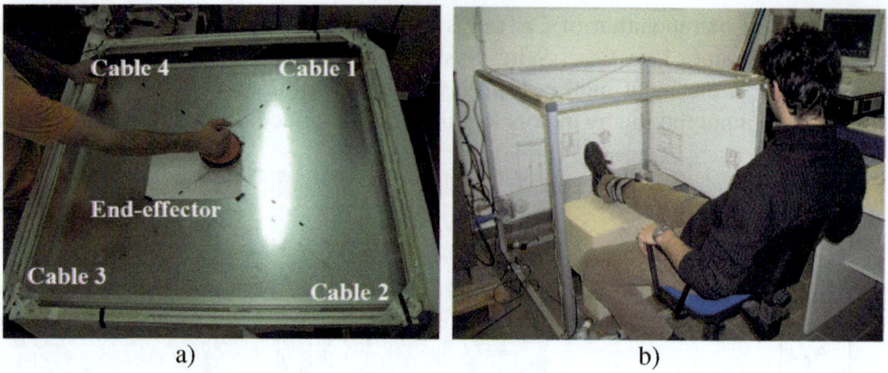

Fig. 15. CaLoWi layout for guiding human limb motion in 2008, [36]: a) in writing with human arm; b) in articulating leg joints.

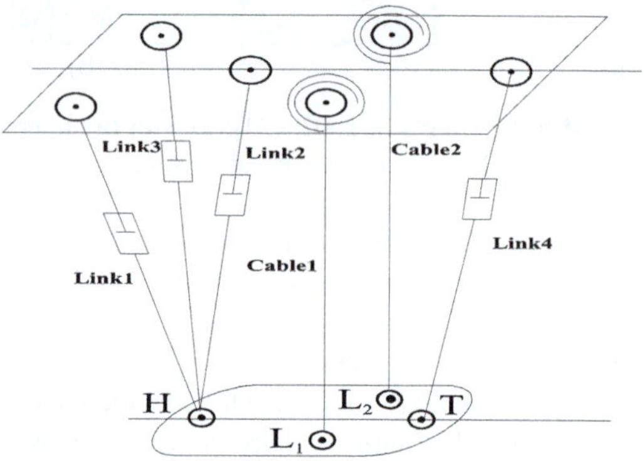

Fig. 16. A new leg design for humanoid robots by using cable actuation and control as proposed n 2009, [39].

The overall experience with CaTraSys and CaLoWi as directed to the above-mentioned service applications has also stimulated definition of a general frame both for design and operation of service robots by including also no technical aspects like safety, comfort, and acceptance that nevertheless are fundamental for success of those devices, [40].

Future work on CaTraSys is mainly scheduled in enhancing those aspects that affect i design features for better user-friendly oriented operation.

4 Conclusions

The paper presents the brief history of development of CaTraSys, a cable system for measuring large displacements in its mains aspects as occurred at LARM in Cassino. The system has been developed since 1993 with step-by-step improvements by lab needs also with considerable contributions of students for their thesis works and today is well used for several applications. The success of CaTraSys can be recognized in having been of inspiration for other similar systems and also in teaching and research activities within the many international collaborations that have widespread successful experience results. It is believed to be an interesting case of system development memory of technical evolutions.

Acknowledgments. The research work reported here was partially made possible by Grant n.250485 CIP-ICT-PSP2009.2.3 for ThinkMotion project by the European Community.

References

[1] ANSI-RIA, Report R15.05-1 Point-to-Point and Static Performance Characteristics Evaluation (1990)
[2] ISO, ISO 9283 - Manipulating Industrial Robots: Performance Criteria and Related Testing Methods (1990)
[3] Van Brussel, H.: Evaluation and Testing of Robot. Annals of the CIRP 39(2), 657–664 (1990)
[4] Driels, M.R., Pathre, U.S.: Vision-Based Automatic Theodolite for Robot Calibration. IEEE Trans. Robotics & Automation, 351–360 (1991)
[5] ASL Automatic Systems Laboratories Ltd., Lasertrace- Operator's Handbook, Bradville (1996)
[6] Gilby, J.H., Parker, G.A.: Laser Tracking System to Measure Robot Arm Performance. Sensor Review 2-4, 180–184 (1982)
[7] Vicon Motion System, Vicon Nexus 1.7 (2011), http://www.vicon.com (visited October 2011)
[8] McMaster, R.S.: Calibration Pus System: A Low Cost. In: High Performance, Dynamic Testing Device for Tracking the Tool Point of Industrial Robots, UK, pp. 1–6 (1997), http://www.rosl.com (visited in December 1997)
[9] Jeong, J.W., Kim, S.H., Kwak, Y.K.: Kinematics and Workspace Analysis of a Parallel Wire Mechanism for Measuring a Robot Pose. Mechanism and Machine Theory 34, 825–841 (1999)
[10] Gabriele, E.: Numerical and Experimental determination of a robot workspace, Master Thesis. LARM at University of Cassino (1994) (in Italian)
[11] Kamamura, S., Kino, H., Won, C.: High-speed Manipulation by Using Parallel Wire-drive Robots. Robotica 18(1), 13–21 (2000)
[12] Kossowski, C., Notash, L.: CAT4 (Cable Actuated Truss—4 Degrees of Freedom): A Novel 4 DOF Cable Actuated Parallel Manipulator. Journal of Robotic Systems 19(12), 605–615 (2002)
[13] Lytle, A.M., Saidi, K.S., Bostelman, R.V., Stone, W.C., Scott, N.A.: Adapting a teleoperated device for autonomous control using three-dimensional positioning sensors: Experiences with the NIST roboCrane. Automation in Construction 13(1), 101–118 (2004)
[14] Williams II, R.L., Albus, J.S., Bostelman, R.V.: 3D Cable Based Cartesian Metrology System. Journal of Robotic Systems 21(5), 237–257 (2004)
[15] Hiller, M., Fang, S., Mielczarek, S., Verhoeven, R., Franitza, D.: Design, Analysis and Realization of Tendon-Based Parallel Manipulators. Mechanism and Machine Theory 40, 429–445 (2005)

[16] Krut, S., Ramdani, N., Gouttefarde, M., Company, O., Pierrot, F.: A parallel cable-driven crane for SCARA-motions. In: Proc. of the ASME Int. Design Engineering Technical Conferences & Computers and Information in Engineering Conference, IDETC/CIE 2008, Paper DETC2008-49969 (2008)

[17] Carvalho, J.C.M., Oliveira Jr., A.A., Werlang, R.C., Hatayama, C.H.: An Experimental System for Measuring Roboth Path. In: Proc. of II Congresso Nacional de Engenhaia Mecânica, João Pessoa (2002)

[18] Merlet, J.-P., Daney, D.: A new design for wire-driven parallel robot. In: Proc. of 2nd Int. Congress, Design and Modelling of Mechanical Systems, Monastir, March 19-21 (2007)

[19] Sugahara, Y., Carbone, G., Hashimoto, K., Ceccarelli, M., Lim, H.O., Takanishi, A.: Experimental Stiffness Measurement for WL-16RII Biped Walking Vehicle during Walking Operation. Fuji International Journal of Robotics and Mechatronics 19(3), 272–280 (2007)

[20] Ceccarelli, M., Gabriele, E.: Determining Primary and Secondary Workspaces of Industrial Robots. In: Proc. of 4th Int. Workshop on Robotics in Alpe-Adria Region, Portsach, vol. II, pp. 259–262 (1995)

[21] Toti, M.E.: Design and Experimentation of a Measuring System for Evaluation of Kinematic performance of Robots, Master Thesis. LARM at University of Cassino (1998) (in Italian)

[22] Ceccarelli, M., Toti, M.E., Ottaviano, E.: CATRASYS (Cassino Tracking System): A New Measuring System for Workspace Evaluation of Robots. In: Proc. of 8th International Workshop on Robotics in Alpe-Adria-Danube Region, RAAD 1999, Munich, pp. 19–24 (1999)

[23] Sbardella, F.: Optimization and Experimental Validation of the Measuring System CaTraSys (Cassino Tracking System), Master Thesis. LARM at University of Cassino (2002) (in Italian)

[24] Ottaviano, E., Lanni, C., Ceccarelli, M.: Experimental Determination of Workspace Characteristics of Human Arms. In: Proc. of 9th Int. Conference on Control and Automation, MED 2001, paper n.017, Dubrovnik (2001)

[25] Ottaviano, E., Ceccarelli, M., Sbardella, F., Thomas, F.: Experimental Determination of Kinematic Parameters and Workspace of Human Arms. In: Proc. of 11th International Workshop on Robotics in Alpe-Adria-Danube Region, RAAD 2002, Balatonfured, pp. 271–276 (2002)

[26] Thomas, F., Ottaviano, E., Ros, L., Ceccarelli, M.: Uncertainity Model and Singularities of 3-2-1 Wire-Based Tracking System. In: Advances in Robot Kinematics, pp. 107–116. Kluwer, Dordrecht (2002)

[27] Ottaviano, E., Ceccarelli, M., Toti, M., Avila Carrasco, C.: CaTraSys (Cassino Traking System): A Wire System for Experimental Evaluation of Robot Workspace. Fuji International Journal of Robotics and Mechatronics 14(1), 78–87 (2002)

[28] Ottaviano, E., Ceccarelli, M.: Numerical and experimental characterization of singularities of a six-wire parallel architecture. International Journal Robotica 25, 315–324 (2007)

[29] Ceccarelli, M., Ottaviano, E.: Cable-Based Parallel Manipulators of Rehabilitation Purposes of Human Limbs. In: Proc. of 2nd International Colloquium Collaborative Research Centre 52, Braunschweig, pp. 53–67 (2008)

[30] Grande, S.: Operation and Optimization of the cable System CaTraSys for a characterization of human walking, Master Thesis. LARM at University of Cassino (2010) (in Italian)

[31] Ottaviano, E., Ceccarelli, M., Palmucci, F.: An application of CaTraSys, a cable-based parallel measuring system for an experimental characterization of human walking. International Journal Robotica 28, 119–133 (2009)
[32] Ceccarelli, M., Romdhane, L.: Design issues for human-machine platform interface in cable-based parallel manipulators for physiotherapy applications. Journal of Zhejiang University Science A 11(4), 231–239 (2010)
[33] Carbone, G., Ceccarelli, M.: A Procedure for Experimental Evaluation of Cartesian Stiffness Matrix. In: Proc. of 15th CISM-IFToMM Symposium on Robot Design, Dynamics and Control, ROMANSY 2004, Montreal, paper Rom04-24 (2004)
[34] Ottaviano, E., Ceccarelli, M., Paone, A., Carbone, G.: A Low-Cost Easy-Operation 4-Cable Driven Parallel Manipulator. In: Proc. of IEEE Int. Conference on Robotics and Automation, ICRA 2005, Barcellona, pp. 4019–4024 (2005)
[35] Paone: Design and Experimental Validation of a Cable Parallel Manipulator, Master Thesis. LARM at University of Cassino (2004) (in Italian)
[36] Ceccarelli, M.: Problems and experiences in using parallel manipulators for medical applications. In: Proc. of International Workshop ROBOMED 2010: New Trends in the Development of Parallel Robots for Medical Applications, Cluj-Napoca, paper no.5 (2010)
[37] Lahouar, S., Ottaviano, E., Zeghoul, S., Romdhane, L., Ceccarelli, M.: Collision free path-planning for cable-driven parallel robots. Journal Robotics and Autonomous Systems 57, 1083–1093 (2009)
[38] Ottaviano, E., Ceccarelli, M., De Ciantis, M.: A 4-4 Cable-Based Parallel Manipulator for an Application in Hospital Environment. In: Proc. of 15th Mediterranean Conference on Control and Automation, MED 2007, Athens Paper T30-011 (2007)
[39] Ceccarelli, M., Carbone, G.: A Study of Feasibility for a Leg Design with Parallel Mechanism Architecture. In: Proc. of IEEE/ASME Conference on Advanced Intelligent Mechatronics, AIM 2009, Singapore, paper no. 131 (2009)
[40] Ceccarelli, M.: Problems and issues for service robots in new applications. International Journal of Social Robotics 3(3), 299–312 (2011)

A Contribution on the History of Ropeways

K. Hoffmann[1] and Nenad Zrnić[2]

[1] Vienna University of Technology, Faculty of Mechanical Engineering,
Institute for Engineering Design and Logistic Engineering,
A-1060 Wien, Getreidemarkt 9, Austria
hoffmann@ikl.tuwien.ac.at
[2] University of Belgrade, Faculty of Mechanical Engineering,
Department for Material Handling and Design Engineering,
11000 Belgrade, Kraljice Marije 16, Serbia
nzrnic@mas.bg.ac.rs

Abstract. The history of ropeways is directly related to the development and manufacturing of ropes and cables. Therefore the invention of the steel cable in 1834 by Albert increased the speed of the further development of new ropeway systems. Initially the development was focused on ropeways for transportation of goods, later interest turned to ropeways for passenger transport.

The first part of this contribution gives a brief overview of the historical development of rope-making, while the second part deals with the various steps of development of ropeway systems from the ancient times to the period between the two world wars.

1 Introduction

If we speak about the development of ropeways in the course of time it is necessary to report first on the manufacturing of ropes. The rope as an element to transmit movement and forces was probably known in prehistoric times. They were used mainly for hunting, carrying, lifting, and climbing purposes. Ropes were originally made by hand using organic material, such as fibres of date palms, flax, grass, papyrus, leather, hemp and others.

In the course of thousands of years only a few improvements in manufacturing of ropes are recorded. It was mainly the invention of the steel cable by the German mining official Albert in 1834 that a period of vigorous development of production and application of ropes and cables began [1]. Considering this very important invention, it is obvious to divide up the history of ropes as well as the history of ropeways in the two periods:

a) From prehistoric times up to 1834 (invention of Albert)
b) After 1834

T. Koetsier, M. Ceccarelli (Eds.): Explorations in the His. of Machines & Mech., HMMS 15, pp. 381–394.
springerlink.com

2 Ropes from Prehistoric Times up to 1834

Nothing is known about the beginning of rope making, but there seems to be evidences of ropes being made as far back as 20,000 BC. Figure 1 shows an individual - a tailed man? - which is climbing along a rope. This picture was found in a cave in eastern Spain [2].

Fig. 1. Climbing man, cave art, eastern Spain [2]

A very early indication of taking advantage of the use of ropes comes from early Egyptian findings. Egyptian rope dates back to 4000 to 3500 B.C. and was generally made of water reed fibres. Other Egyptian ropes were made from the fibres of date palms, flax, grass, papyrus, leather, or camel hair. All these early ropes were twisted by hand or braided.

A very interesting picture of early Egyptian rope making comes from the Tomb of Ti (see Fig. 2), which was built during the fifth dynasty of the Old Kingdom about 2400 B.C. [3]. The figure shows the twisting process requiring at least two workers standing at opposite ends of the cordage. One of the workers is imparting a twist on the strands by fastening short pieces of chord with weights attached, and the other is combining the strands by twisting them together in the opposite direction.

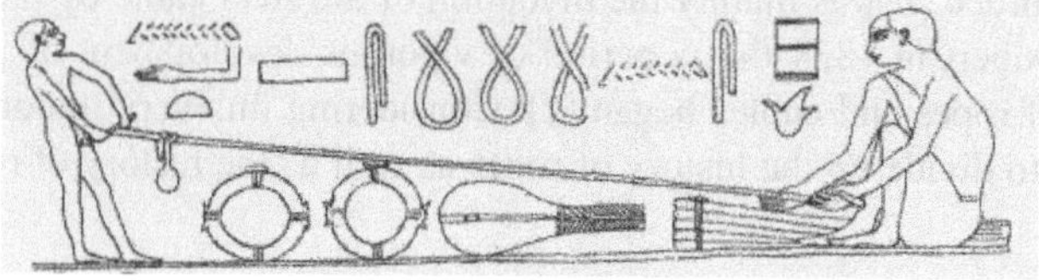

Fig. 2. Tomb of Ti, [3]

Another example for early Egyptian rope making shows Fig. 3. Simple hand tools were used to improve the hand twisting process. In this method, a stick with a rock tied to it would be swung around, to assist in the twisting of the strand. This method, called a "Spinner", worked well for producing shorter ropes.

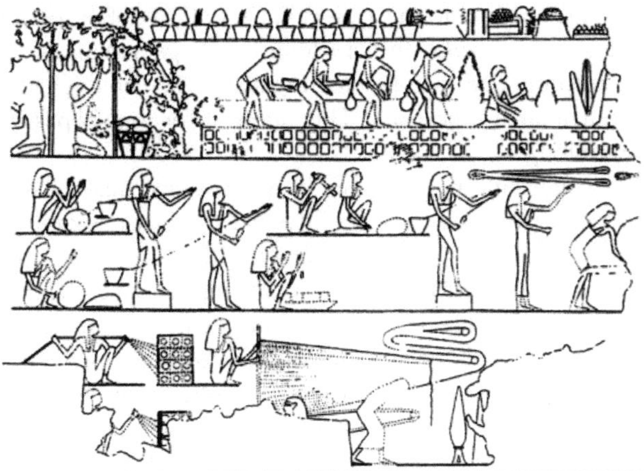

Fig. 3. Ancient Egyptian artwork depicting tools for rope making [4]

The use of such ropes pulled by thousands of slaves allowed the Egyptians to move the heavy stones required to build the pyramids.

During the early times not only in Egypt but also in Mesopotamia the use of ropes was well known. Figure 4 shows the use of ropes for the transportation of a heavy

Fig. 4. Transportation of a statue in Ninive [5]

statue in combination with simple technical tools such as levers and rollers. This Figure is a copy of H.A. Layard from a relief found in Ninive [5].

In China ropes made of hemp (cannabis sativa) were in use as early as 8500 years ago. The knowledge of hemp rope making was spread along the Silk Road finally to Europe.

In the Middle Ages and up to the 18th centuries all over Europe ropes were manufactured in so-called *Ropewalks*. In very long buildings strands of the full length of the rope were spread out and then twisted together. The cable length was thus set by the length of the available rope walk. This allowed for long ropes to be made up to 100 meters or even longer.

With the increasing building activities in Renaissance times and the growing importance of European maritime trade the demand for ropes augmented. The guild of rope makers was a highly respected one at that time. In many cities the names of streets or districts refer to their trade; there is for instance the "Reeperbahn" in Hamburg, the Seilerstätte in Vienna, or the "Lijnbaansgracht" in Amsterdam.

A very interesting example of an extensive use of ropes is the transportation and erection of the obelisk in front of the St. Peters Cathedral in Rome in the year 1586, where Domenicus Fontana used more than 45 ropes with a diameter of approximately 80mm and a length of 220m to carry out this project. After successful erection of the obelisk Fontana described all technical details in a book [6]. Figure 5 shows a painting of this event.

Fig. 5. Erection of the obelisk in Rome in the year 1586 [2]

During the course of the Industrial Revolution it became more and more necessary to improve the ropes to higher strength and increase the length of it. There was also a great need to develop more efficient means of transport. It was obvious to use steel for the manufacture of ropes. Only with the invention of the steel cable by the German mining official Albert in 1834 did a period of lively development of the various cable-drawn transport systems begin.

3 Ropeways before 1834

A ropeway is a system for the transportation of passengers or goods whereby the passengers or goods are conveyed in different types of cars. The runway is made of ropes (one or two) or rails and concrete carriageways respectively. The hauling function is carried out by means of one or more ropes. Speaking about aerial ropeways we have to distinguish between monocable aerial ropeways and bicable aerial ropeways. The difference is simply the number of rope systems. While with a monocable ropeway one or two ropes assume the carrying and hauling function (carry-hauling cable), with a bicable aerial ropeway the cars are carried by one or more ropes (track ropes) and moved by a further rope system (haul rope).

The principle of a cable-drawn transport system was known even in ancient times. But there are only a few historic documents from these times. As early Chinese historical drawings demonstrate, this principle was already in use at that time for passenger and material transport. Figure 6 shows the earliest known picture of passenger transportation from 250 B.C., South China [1]. The use of cables in China was e.g. implemented in a water-lifting device called “Gao zhuan tong che” (chain conveyor water machine – machine similar to sloping bucket elevator) [7].

Fig. 6. Brush drawing of a Chinese aerial ropeway dated 250 B.C. [1]

Centuries later in a Japanese historical epic the "Taiheiki", written in the late 14th century a ropeway was mentioned. This document relates how a Japanese emperor escaped via ropeway over a valley when he was surrounded by enemy forces.

In the Middle Ages a sort of a monocable aerial ropeway was depicted in a book by Johannes Hartlieb (Austrian National Library, Vienna, Ref. Nr. 3069). It shows a basket being conveyed in a kind of a monocable ropeway to a fortified castle by a hand operated winch (Fig. 7).

Fig. 7. Drawing of a monocable aerial ropeway from A.D. 1411 [1]

Another sketch by Marianus Jacobus is showing a primitive bicable ropeway dated back to A.D. 1440 (Figure 8). A pitcher is being transported over a ravine driven by an ox (National Library, Paris).

Fig. 8. Drawing of a primitive bicable aerial ropeway from A.D. 1440 [8]

Fausto Veranzio of Venice was a scientist working at the court of the Roman-German Emperor King Rudolf II in Prague. He collaborated with the famous astronomer Tycho de Brahe and Johannes Kepler. Veranzio illustrated a refined bicable passenger ropeway in the "Machinae novae", published in 1615 (Figure 9). It consisted of a wooden box in which the passengers rode, travelling on pulleys over a fixed rope. The vehicle was propelled over the river by the occupants pulling themselves along by means of this slack loop.

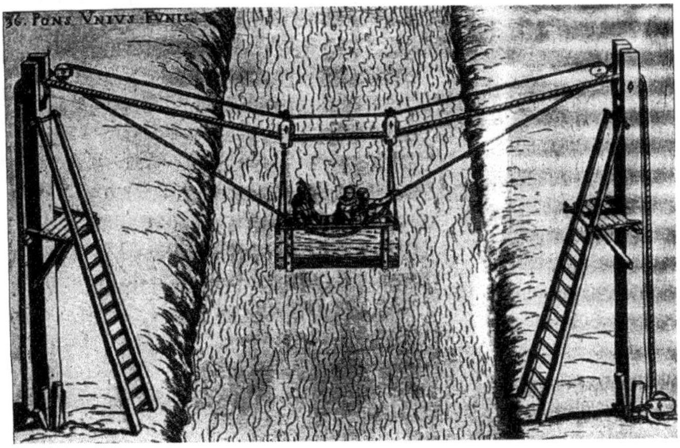

Fig. 9. Sketch of a bicable aerial ropeway by Fausto Veranzio [12]

At the start of the modern age, ropeways were also used for faster transport of goods and building materials to construction sites. The first efficiently working monocable ropeway was constructed by Adam Wybe, a Dutchman, who was a major contributor to the construction of Gdansk/Danzig (Polland). Figure 10 shows this ropeway from 1644 with securely fastened baskets which were used to transport material from the mountain to the city during the construction of fortifications of Gdańsk.

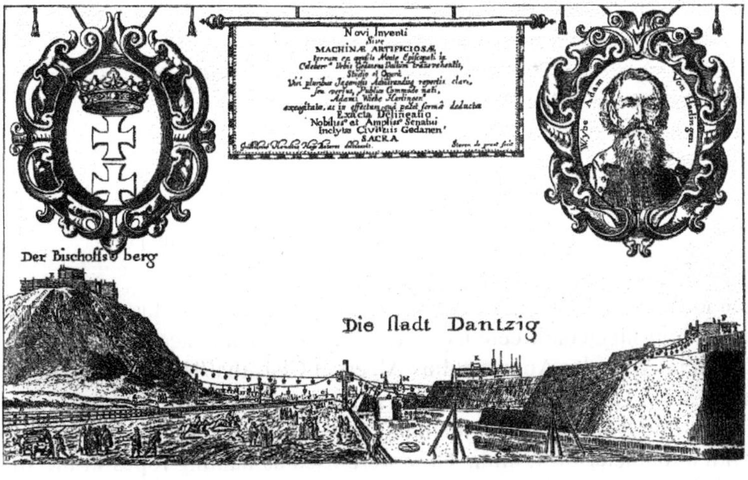

Fig. 10. Monocable aerial ropeway built by Adam Wybe in 1644 [9]

From the middle of the 17th century until the invention of the steel cable in 1843, no really further progress was made. The technology had reached its maximum potential at the time, limited by the strength and length of available ropes. With these restrictions it is astonishing that in the year 1804 the most powerful funicular of this time was brought into operation by the Austrian engineer Joseph Gainschnigg. It was in Badgastein, Austria were this 1.4 km long funicular was used both for transportation of ore from the goldmine down and for transportation of the miners up to the entrance of the mine. The rope was made of hemp with a diameter of about 60 mm and driven by a water wheel with a diameter of 15 m. This funicular was in operation more than 60 years without a change.

However, more than 200 years before the engineer Buonaiuto Lorini from Florence [10] designed a very similar funicular for transportation of building material up to a fortress (Figure 11).

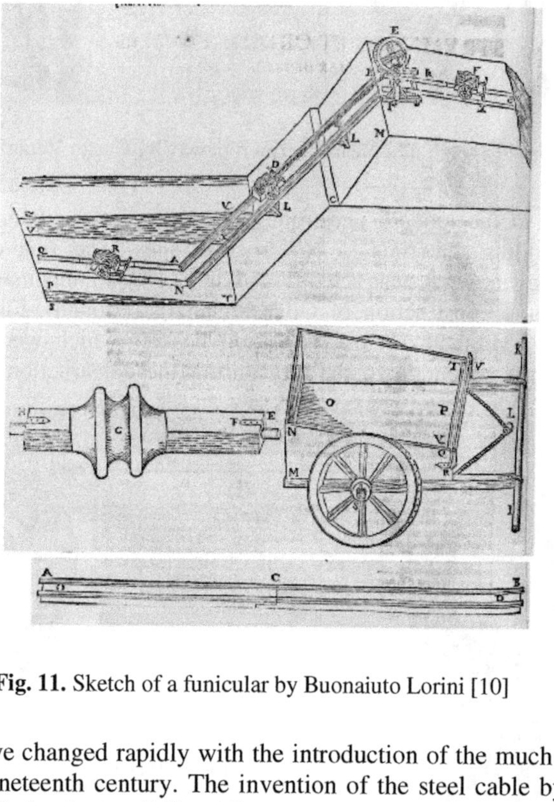

Fig. 11. Sketch of a funicular by Buonaiuto Lorini [10]

All these ideas have changed rapidly with the introduction of the much stronger wire rope in the mid-nineteenth century. The invention of the steel cable by the German mining official Wilhelm August Julius Albert at Clausthal/Harz, Germany in 1834 led to the most successful period of all kinds of ropeways in history [11]. Albert produced for the first time two ropes with a length of 605 consisting of 3 strands, each of four wires with a diameter of 3.5mm. They were used as hauling ropes in the silver mine in Clausthal.

4 Ropeways after 1834

During the course of the Industrial Revolution, not only the invention of the steel cable by the German mining official Albert in 1834 heralded a period of lively development of the various cable-drawn transport systems but also new power sources, first steam engines and later on electric motors.

To speak about the development of the various types of cable-drawn systems which have evolved over the decades it is necessary to classify the more recent systems according to the type of track (carrier) used. Figure 12 gives an overview of the various systems, which are of course in use with very diverse frequency. The development of funiculars and aerial ropeways will subsequently be described in more details.

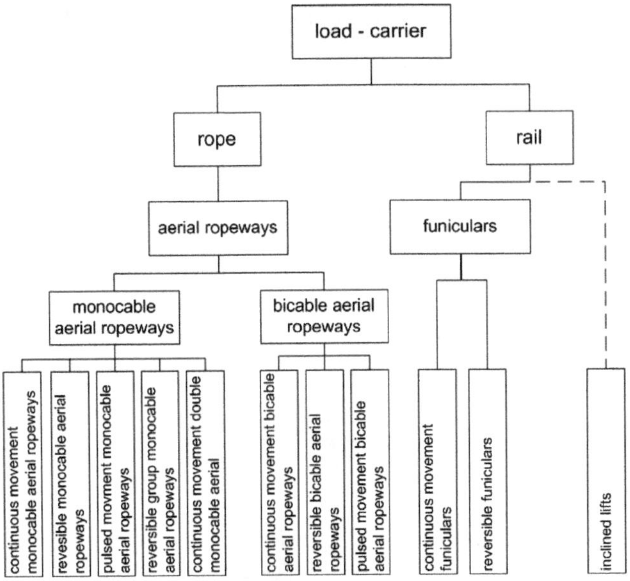

Fig. 12. Classification of ropeways [14]

Funiculars

Initially, after the introduction of steel cable the development of cable-drawn systems was mainly focused on funicular railways. These are railways on which one ore more wagons run on the rail track, hauled by the hauling rope which is wound round a driving drum. Such funiculars used as slope lifts have long been in operation in industry. Beside the funicular of Badgastein described before it was reported that Stephenson in 1825 adopted a rope for transporting a steam engine up a slope of about 1:50 [8].

The next step of development was to adopt the funiculars for passenger transport. The first funicular for public transport was put into operation in Lyon in 1862. It was driven by a steam engine.

On the occasion of the World Exhibition in 1873 a funicular of the same type was constructed at Leopoldsberg, Vienna (see Fig. 13). Unfortunately the funicular was demolished 3 years later because of financial reasons [13].

Fig. 13. Funicular at Leopoldsberg, Vienna [13]

It was A.S. Hallidie the mechanical genius born in London 1836 who made many patented inventions for suspension bridges and as his most important work he designed the world-famous Cable Car in San Francisco (Fig. 14). In 1871 he completed plans by which street cars could be propelled by underground cables and finally on 1st of September 1873 the cable car was successfully put into operation. The drive unit for the hauling cable was a steam engine in the powerhouse.

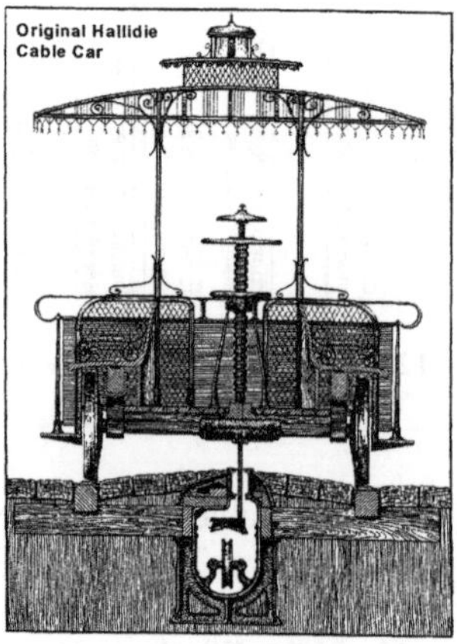

Fig. 14. Cable Car, San Francisco, 1873 [1]

The further development of funiculars during the 20th century can be divided into two groups. Both types have been developed parallel into a very high standard. These two types are:

- Reversible funiculars
- Continuous movement funiculars

Figure 15 shows the basic layout of a reversible funicular railway in shuttle operation. Two cars are linked together by a haul rope and this moves them along the route. The cars travel on tracks using steel rollers on rails. The drive unit for the system is housed usually in the upper station. Funiculars are normally designed as single-track systems. For this reason a passing point is provided in the middle of the route.

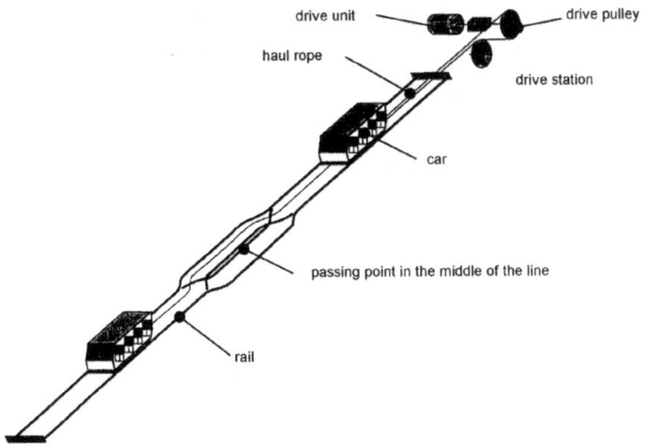

Fig. 15. Layout of a reversible funicular [14]

In contrast to shuttle operation, the continuous movement funiculars are working in circular operation, requiring two completely separate tracks. The cars are coupled to the haulage rope at specific distances by normal releasable clamping devices. Placement of stations along the line is possible without technical limits, according to local service requirements. This ropeway system is particularly suitable in urban areas as means of transportation for small and medium sized cities. It is also known as automated people mover, or APM system. However, the development of APM-systems began only in the late 1960s as the interest in environment friendly transport systems grew and is subject of a different study.

Aerial Ropeways

While initially after the invention of the steel cable development mainly was focused on funicular railways such as the Cable Car San Francisco, later interest turned to aerial ropeways. At the beginning aerial ropeways were mainly used for transportation of material and goods especially in the mining industry. As in Figure 12 is depicted, aerial ropeways are of two types of construction:

- monocable aerial ropeways
- bicable aerial ropeways

In this connection two important inventors can be named:

- 1856 Englishman Robinson, patent for monocable aerial ropeway (English system)
- 1861 German Freiherr von Dücker, patent for bicable aerial ropeway (German system)

The patent of Robinson was used by Hodgson in 1867 to built successfully ropeways for the mining industry in England between 1869 and 1880. On the other hand Freiherr von Dücker installed in 1872 a continuously circulating bicable ropeway for the construction of a fort at Metz, France. The carriages were equipped with screw type coupling grips, patented by an Austrian engineer Theobald Obach [8]. At the same time the German engineers Bleichert and Otto established a firm to construct ropeways. They were very successful, since this company put more than 5000 ropeways into operation throughout the world. The first one was set into operation in 1873 in Teutschenthal, Germany (see Fig. 16).

Fig. 16. Bicable Ropeway for transport of coal in Teutschenthal, Germany constructed by Bleichert in 1873 [8]

Aerial ropeways for passenger transport came into operation shortly before 1900. The very first aerial ropeway exclusively designed for passenger transport was constructed by Ritter in 1866 in Schaffhausen, Suisse. It had four track ropes and a 2-person cabin used only for the inspection of turbines in a small water power plant on the Rhine. The hauling rope was driven by a hand winch.

Some years later in Blackpool/New Zealand a 6-persons aerial ropeway driven by a steam engine was put into operation across a river (operating only until 1890).

A fine example of a historic passenger ropeway is the reversible aerial ropeway in San Sebastian, Spain opened 1907 (see Fig. 17). An open carriage is running with 12 pulleys over 6 track ropes. A similar ropeway provided with a cabin for 46 persons was put into service in 1916 near the Niagara Falls. After several upgrades in design and function this ropeway is still in use [15].

Fig. 17. Reversible aerial ropeway in San Sebastian, Spain, 1907 [1]

The first ropeway of the alpine mountains was built in Gindelwald, Switzerland in the year 1908, the "Wetterhorn Bahn". The track consisting of two track ropes had a maximum gradient of 200%. A single cabin carried 16 people. The ropeway was only in service until 1914, the beginning of the First World War.

The world's oldest twin-cable ropeway which is still functioning is the Predigtstuhlbahn in Bad Reichenhall, Germany, which was built in 1928 by the well know company Bleichert [16].

Fig. 18. Predigtstuhlbahn in Bad Reichenhall, Germany, the world oldest ropeway still in service

Due to the rapid development of the touristic industry also the first applications of ski lifts and chairlifts came in use. This evolution was suddenly interrupted by the beginning of the Second World War.

5 Conclusion

The earliest application of ropes for transportation of goods or passengers dates back to ancient times. The history of ropeways can be divided into two periods, i.e. the time before the invention of steel ropes 1834 and the time after it. From the period before 1834 only a few applications are documented. The invention of the steel cable initiated a rapid development of ropeway systems over the decades, which is further accelerated and diversified by the change from steam engines to electric motors. At present the main demands for further development of ropeways are higher capacity, energy saving, noise reduction and sustainability.

References

[1] Schmoll, H.D.: Weltseilbahngeschichte, Band 1 und 2. Steidl Verlag, Eugendorf (2000)
[2] Weber, W.: Kurze Geschichte der Seilerei. Lenzburg, Suisse (2001)
[3] Quibell, J.E.: The Ramesseum and the Tomb of Ptah-Hetep. Egyptian Research Account 2, London (1896)
[4] Turner, J.C., Van de Griend, P.: History of Science of Knots. World Scientific Pub. Co., Singapore (1996)
[5] Buchheim, G., Sonnemann, R.: Geschichte der Technikwissen-schaften. Birkhäuser Verlag (1990)
[6] Fontana, D.: Della Trasportatione dell' Obelisco Vaticano. Appresso Domenico Basa, Rome (1590)
[7] Guan, X.-W.: An Interpretation of a Shui Lun Fu (Rhapsody on Waterwheel). In: Yan, H.-S., Ceccarelli, M. (eds.) International Symposium on History of Machines and Mechanisms, pp. 215–224. Springer, Heidelberg (2009)
[8] Schneigert, Z.: Aerial Ropeways and Funicular Railways. Pergamon Press, Oxford (1966)
[9] Stephan, P.: Die Luftseilbahnen. Verlag Julius Springer, Leipzig (1907)
[10] Lorini, B.: Fünf Bücher von Festung Bauen, David Beckern, Frankfurt am Main (1607)
[11] Beck, T.: Beiträge zur Geschichte des Maschinenbaues. Springer, Berlin (1899)
[12] Dieterich, G.: Die Erfindung der Drahtseilbahnen. Verlag Hermann (1908)
[13] Hefti, W.: Schienenseilbahnen in aller Welt. Birkhäuser Verlag, Basel (1975)
[14] Hoffmann, K.: Recent Developments in Cable-drawn Urban Transport Systems. In: Proceedings of the XVIII International Conference on Material Handling, Constructions and Logistics, Belgrade, Serbia (2006)
[15] `http://www.niagaraparks.com/attractions/whirlpool-aero-car.html`
[16] `http://www.predigtstuhlbahn.de`

The First Hydroelectric Power Plant in the Balkans Built on the Basis of Tesla's Principles

Svetislav Marković[1], Tatjana Lazović[2], Ljubica Milović[3], and Bratislav Stojiljković[4]

[1] Technical College Čačak, Serbia
svetom@open.telekom.rs
[2] Belgrade University, Faculty of Mechanical Engineering
tlazovic@mas.bg.ac.rs
[3] Belgrade University, Faculty of Technology and Metallurgy
acibulj@tmf.bg.ac.rs
[4] Nikola Tesla Museum, Belgrade, Serbia
bratislav.stojiljkovic@tesla-museum.org

Abstract. At the end of the 19th century, Serbia was among the first European countries to start electrification. A very bad economic situation, caused by centuries-old slavery and lack of capital, did not prevent the Serbs to start introducing electricity into settlements and a few industrial plants. They had the idea of introducing electricity implemented before many other more developed countries of Europe and the world. A great merit for this belongs to the Serbian scientist and inventor Nikola Tesla, as well as to his good friend Djordje M. Stanojević, the professor and rector of Belgrade University. The first electric power plant in Serbia, which produced direct electric current by coal combustion, was put into operation in Belgrade, the Serbian capital, in 1893, twelve years after Edison's first such plant in the world. Also, the first hydroelectric power plant with three-phase alternating electric current in Serbia began operating in 1900, four years after the first power plant of its kind in the world (Niagara Falls, 1896). This hydropower plant, the first in the Balkans and one of the first in Europe, is described in this paper.

1 Introduction

During the centuries-old slavery under the Ottoman Empire, very difficult economic conditions were created in Serbia. The Turks abandoned Serbian villages in 1833, but stayed in the cities until 1867. In 1834 Serbia had a surface area of 37,511 km^2 with a population of 678,192 inhabitants (18 people/km^2), of which 93.5% lived in the villages. The territory was extended after the Serbian-Turkish wars (1876-1878) and before the Balkan Wars (1912), and it was 48,303 km^2. In 1910, about three million inhabitants lived on this territory, or 60 people/km^2. The share of urban population was increased to 18%. The concentration of population is best explained by comparison with the developed European countries. In 1910, the population density was 240 inhabitants/km^2 in Belgium, 167 in The Netherlands, 122 in England. The population

T. Koetsier, M. Ceccarelli (Eds.): Explorations in the His. of Machines & Mech., HMMS 15, pp. 395–406.
springerlink.com

in Serbia was mostly engaged in agriculture, trade and handcraft shops [1]. The educational structure of the population was worrisome, because at the end of the 19th century, 77% male and 95% female population was illiterate. There were only 135 physicians throughout the entire country [2].

The conditions of infrastructure were very poor: a network of roads had only three main roads, with a very small number of usable secondary roads. The first railway began operating in 1884, from Belgrade to the Austrian border and to the Southeast of the country. Other cities got rail traffic much later (1887 to 1912) [1]. The conditions of agricultural technology in the early 20th century looked devastating: one iron plow came to two people, a wooden plow and a harrow to three, a roller to 427, a planting machine to 520, a threshing machine to 661, and a reaper to 19,137 inhabitants. Power plows, mowers and other agricultural machines did not exist in Serbia [2]. The crafts were developing very slowly, so the first baker in Serbia appeared in 1823, the tailor in 1827, the bellmaker in 1830, the pressman in 1831, the bookbinder in 1834, the carpenter in 1836. Data that 208,000 looms were in Serbia show that the textile craft was the most developed [1]. At the beginning of the 20^{h} century there were only 153 industrial enterprises in Serbia, mostly mills and sawmills, with 3,200 workers. By the beginning of World War II, industrialization had got a good momentum, and then there were more than 500 industrial enterprises in Serbia [1].

2 Beginnings of Serbian Electrification

The first bulb In Serbia was lighted in the Artillery Technical Factory in Kragujevac (Central Serbia) in 1884. Electricity was used only for the needs of cartridge workshop and its night operation. Electrification of Kragujevac as the settlement was realized more than three decades later. The first Serbian public power station was built in Serbian capital - Belgrade, with three steam engines with 441.3 kW power (602.5 HP). It started operating in 1893. Three horizontal steam engines of the “Corliss” type were used for electricity production. They worked with 60 rpm, driven by steam from three multitubular boilers. Power was transmitted to nine dynamo-electric machines, which produced direct current. Each of the three steam engines drove three dynamo generators of the “Edison” type. One of them was producing electricity of 2000 V voltage, and the other two – electricity of 120 to 150 V voltage [3].

The first Serbian hydroelectric power plant built in the town of Užice (Fig.1) started operating in 1900, only four years after the hydroelectric power plant in Niagara Falls. Also, this is the first hydroelectric power plant in the Balkans based on Tesla’s principles of polyphase alternating current. It was built for the local weaving factory. The initiator of the polyphase alternating current application was a friend of Tesla’s, Djordje M. Stanojević, professor of physics at the Great School (later Belgrade University).

Fig. 1. Užice's hydroelectric power plant "Under the town" (photo 1901) [6]

3 One of the Oldest Hydroelectric Power Plants in Europe Built on Tesla's Principles

The need for electricity in the Serbian town of Užice (200 km southwest of Belgrade) in the 1880's was imposed by the local weaving workshop. A management board planned to build a new mechanical workshop with water powered machines. Professor Stanojević suggested them to build a hydroelectric power plant, because they could make a workshop anywhere in the town, instead on the expensive land by the river [4]. Moreover, electric power would be used for the town's electric lighting. Professor Stanojević recommended building a hydroelectric power plant on Tesla's system of polyphase alternating current, because that system was a real world's future of electricity production and transmission. The corner-stone for the plant was set up by King Aleksandar Obrenović in 1899 [5]. Užice's hydropower plant was built on the river Djetinja under the old town-fortress. Thus, it was named hydroelectric power plant "Under the town".

Since the river drop is very small at the power plant building site, a dam was built upstream (Fig.2,3). The water was led from the dam to the power plant through the hydraulic canal 800 m long and 2.5 m wide (Fig.3,4). In this way, vertical drop of about 12 m was achieved. The dam (Fig.5,6) is arch-curved, with curvature radius 44.2 m. The dam length is 29.5 m, the height is 4.9 m and the width is 3.7 m at the bottom and 2.3 m at the top. Behind the hydroelectric power plant building, water was falling from the canal onto two turbines through steel pipes [7].

Fig. 2. Dam in Užice's gorge on the river Djetinja [6]

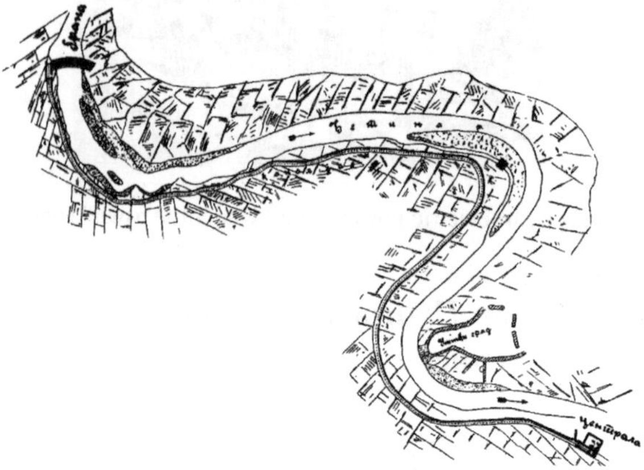

Fig. 3. Layout plan [3]

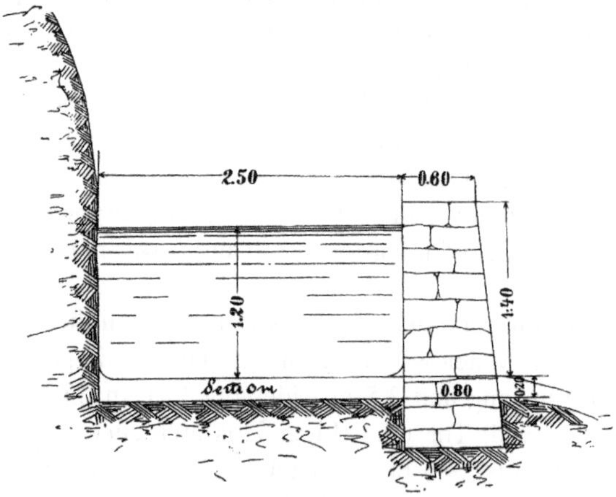

Fig. 4. Cross-section of the hydraulic canal [3]

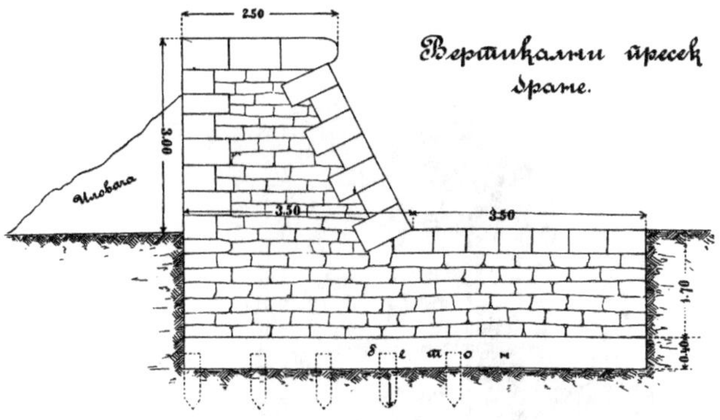

Fig. 5. Vertical cross-section of the dam [3]

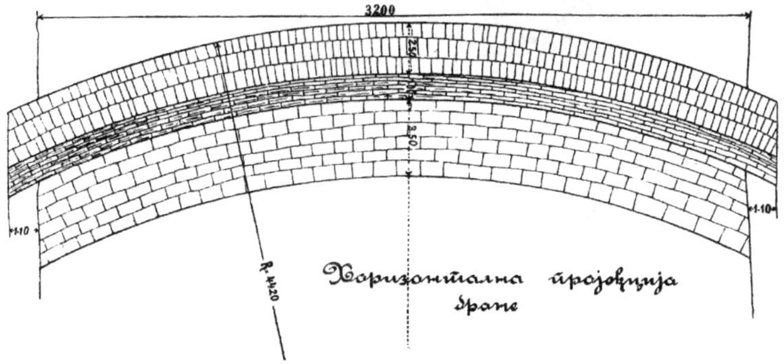

Fig. 6. Horizontal cross-section of the dam [3]

Generators and equipment were purchased from Siemens & Halske Company, Berlin. The control panel with the firm's name written in Serbian Cyrillic is shown in Fig.7. The turbine manufacturer was "Danubius – Maschinen Hartmann" i.e. "Ganz & Co.Danubius Budapest" [5].

A complete equipment for the power plant, town lighting network and 1500 bulb connections arrived by train in Kragujevac, in October 1899 [5]. Because of bad roads, a specially constructed ox car with full wooden wheels, pulled by six pairs of oxen (Fig.8), was used to deliver the equipment to Užice in February 1900.

The ox car strengthening was done by the masters from the Artillery Technical Factory in Kragujevac. Furthermore, special tools were made for equipment loading and unloading. Transport went over the town of Čačak and Jelica Mt. causing huge wondering of the local people [7].

Fig. 7. Control panel (Siemens & Halske) [3]

Fig. 8. Transport of power plant equipment [6]

Fig. 9. Transmission belts [3]

Two Francis turbines with horizontal shafts were mounted in the plant. The turbines were made at "J. M. Voith – Heidenheim" company with built-in "J. M. Voith – St. Pölten" regulator, having 50 HP power at 300 rpm. The turbines were connected with generators with transmission belts (made of goat or camel hair [6]) (Fig. 9). The generators (Fig. 10), made at "Siemens – Werke Wien" company, were of the WID 64/750 type (Fig. 11). They were producing three-phase alternating currents with 50 Hz frequency, 2060 V voltage, 18 A current and 64 kW power at 750 rpm [3].

Fig. 10. Generator [3]

Appropriate switchgears and a control panel with instruments and controllers were located beside the machines. The levers for water flow control in the turbines were located next to the control panel. Electrical works were carried out by the German company "Siemens – Halske" from Berlin. Electric equipment for electricity distribution network was purchased from the same company [7].

Fig. 11. Generator data table

The power plant "Under the town" was put into operation in August 1900. Besides the weaving workshop, the electricity consumers were many Užice's private houses, and street lighting network.

Electricity produced in the power plant was directly transmitted from generators to the transformer houses in the town and from there, under the voltage of 120 V, to the consumers. All town street lamps were divided into two groups: one half of them were lighting to 22 h, and the other half all night. In order to turn on and off the street light lamps from one place, the relay system was used (Fig.12). In the holder base of

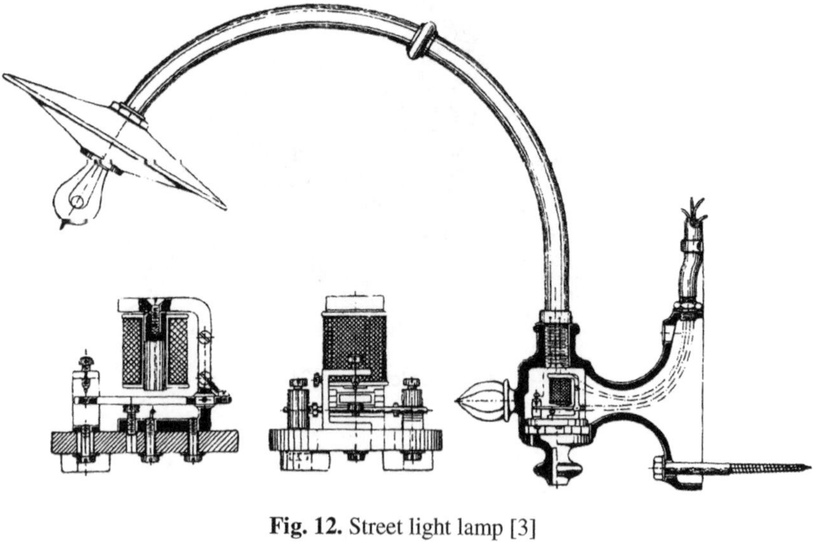

Fig. 12. Street light lamp [3]

each street lamp there was a switch relay powered by direct current from a small dynamo machine. The same principle was used for power plant indoor lighting. Nine Bogen lamps of 10 A and 120 light bulbs with 16 candelas were set for street lighting [3]. For the protection against atmospheric discharges, the network was provided with a sufficient number of horn gap lightning arresters (Fig.13).

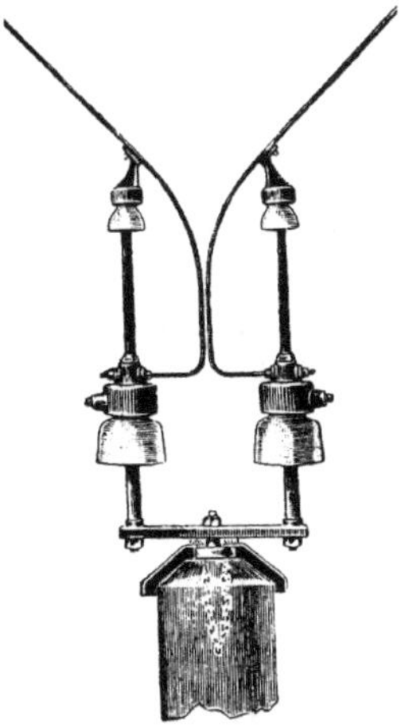

Fig. 13. Lighting arrester of horn gap type [3]

The impact of electricity on the development of Užice's textile industry was immeasurable. At the beginning, the current was used to drive electric motors of 17 weaving looms (Fig.14). Very soon, thereafter, there were 50 new looms in the weaving workshop [7].

Fig. 14. Looms in Užice's weaving workshop [3].

Due to network overload, the third turbine was installed at the power plant in 1904. It was produced by "J. M. Voith – Heidenheim" company and had 120 HP power at 300 rpm and 0.9 m^3/s flow rate. The turbine was connected by transmission belt to "Siemens – Halske" generator with 2100 V voltage, 28.3 A current and 90 kW power [7].

In order to increase the stability of electric power supply to the weaving workshop and to other consumers in the town, especially during water shortages, a steam loco mobile with 50 HP power at 200 rpm was purchased from English manufacturer "Richard Garrett – Lenstal". The water-pipe two-chamber boiler was made in the First Czech-Moravian Machine Works in Prague (I českomkoravská továrna na stroje v Praze) and had 60 m^2 of heating surface and 6 atm vapor pressure. By transmission belt the loco mobile was connected with "Siemens – Schuckert" generator with 32 kW power, 2100 V voltage, 8.8 A current and 50 Hz frequency at 750 rpm.

One of the oldest power plants for alternating current production in Europe ceased to operate in 1970. Nowadays, "Under the town" – the first Serbian and Balkan hydroelectric power plant is a tourist attraction of Užice's region. The Museum of Engineering and National Museum of Užice, as well as Electric Power Industry of Serbia take care of it [8]. The building of power plant and its immediate surroundings were declared a cultural heritage site of great importance.

4 Nikola Tesla (1856–1943)

Nikola Tesla (Fig.15) was born in 1856 in Smiljan (Austria, today the Republic of Croatia). He studied at Graz Polytechnic School and Charles University in Prague. He intended to specialize in physics and mathematics, but soon he became fascinated with electricity. He began his career as an engineer at a telephone company in Budapest in 1881. Before going to America, Tesla joined Continental Edison Company in Paris. After that, he accepted an offer to work for Thomas Edison in New York. Young Nikola Tesla came to the United States in 1884. He spent the next 59 years of his productive life living in New York. Tesla died in 1943 at the Hotel New Yorker, where he had lived for the last ten years of his life.

The main results of Tesla's work were condensed in over one hundred most important patents of his inventions registered with USA, Great Britain and Canada Patent Administrations, as well as in more than 20 other countries. Those patents could be selected into the following eight fields: rotating magnetic field; motors and generators; transformation and electric power transmission; lighting; high frequencies devices and regulators; radio; telemechanics; turbines and similar equipment [9]. Some of his inventions in the field of electricity changed the world: rotating magnetic field, induction motor, polyphase motor, generator and transformer, as well as triangle-star connection and the other elements of the polyphase system for electricity production, transmission, distribution and use. Nowadays, Tesla's electrical energy production, transmission and distribution system is used without essential modifications [9].

Fig. 15. Nikola Tesla

Fig. 16. Djordje M. Stanojević

5 Djordje Stanojević (1858 - 1921)

Djordje Stanojević (Fig.16) was born in 1858 in Negotin, Serbia, where he completed primary school and high school. After that, he completed higher grades of Gymnasium and the Great School (Natural-Mathematics Department) in Belgrade. In 1881 he started to work as an assistant at the Department of Physics at the Great School (later Belgrade University). In 1883 the Ministry of Education appointed him the professor of physics at First Belgrade Gymnasium. As a fellow of the Ministry of Defense he spent the period 1883-1887 studying and improving his knowledge at European science centres [4,10]: University of Berlin, Astrophysical Observatory of Potsdam, Hamburg Meteorology Centre, Paris observatory for physical astronomy in Meudon – Sorbonne, Greenwich – London etc. After returning to the country, in 1887, he was appointed full professor of physics and mechanics at the Military Academy, and in 1893 full professor of experimental physics at the Great School in Belgrade. In 1913, he became the rector of Belgrade University and remained in office until his death, in 1921 [11].

Professor Stanojević was one of the organizers of the first and only one Tesla's stay in Belgrade in 1892 [11]. He published the book "Nikola Tesla and His Inventions" in 1894. It was the first book on Tesla in Serbia and the second one in the world. His friendship with Nikola Tesla was of great importance for the electrification of Serbia: in electric lighting in Belgrade, instead of chemists' idea to light Belgrade with gas, as well as in building hydroelectric power plants using Serbian rich natural water resources. Thus, he placed Serbia among the first European countries which used electric energy. This produced an important influence on industry modernisation and general progress of Serbia at the end of 19th century.

6 Conclusion

The application of electricity in Serbia started in 1884 for the purpose of indoor lighting in the first Serbian industrial plant – Artillery Technical Factory in Kragujevac. However, the first public power plant was built in Serbian capital - Belgrade, almost a decade later. It was a thermal power plant. Later, in 1900, a hydroelectric power plant was built in Užice, the town of western Serbia. This power plant, named "Under the town", was the first hydroelectric power plant built on Tesla's principle of polyphase alternating current in Serbia and in the Balkans and also one of the first in Europe. It started operating only four years after the first hydroelectric power plant in the world at Niagara Falls, where Tesla applied his principles for the first time. Electrification of Serbia was caused by the vision that "in addition to bread and water, electricity will become a day-to-day need" as Professor Djordje M. Stanojević, an electrification pioneer in Serbia, used to say [8]. There are not so many European countries that have had generated electricity for the past hundred years, while the first Serbian and Balkan hydroelectric power plant was built 111 years ago.

This paper was written with the intention to be an idea and "one piece in a mosaic" of potential extensive studies on the history of electrification in Europe and in the Balkans region.

Acknowledgment. This work has been performed within the project TR 35011. This project is funded by the Republic of Serbia, Ministry of Education and Science, whose help is gratefully acknowledged.

References

[1] Electric Power Industry of Serbia, From Djetinja to Djerdap, electrification of Serbia 1893-1941, Electric Power Industry of Serbia – Belgrade (1979) (in Serbian)
[2] PR Dept. of Electric Power Industry of Serbia, Century of Electricity , Electric Power Industry of Serbia, Belgrade (1993) (in Serbian)
[3] Stanojević, D.M.: Electric Industry in Serbia. State press of Kingdom of Serbia, Belgrade (1901) (in Serbian)
[4] Dimitrijević, M.S.: Djordje Stanojević – Astronomer who electrified Serbia. Planeta – Magazine for Science, Investigations and Discoveries VI(33) (2009) (in Serbian)
[5] http://www.nmuzice.com/en_power_plant
[6] Spasojević, A.K., Gluščević, M.: A hundred years of electrification of Užice 1900-2000. Electric Power Industry of Serbia, Užice (2000)
[7] Vučetić, M., Radslavčev, S., Bečejac, L.: Užice power plant "Under the town". Electric Power Industry of Serbia, Belgrade (2005) (in Serbian)
[8] PR Dept. of Electric Power Industry of Serbia, And there was light over the river, Electric Power Industry of Serbia, Belgrade (2011)
[9] Spaić, G., Spiridonović, S.: The Greatest Recognition to the Scientist Nikola Tesla – Award of the Unit for Magnetic Induction (T) in the International System of Units (SI). Serbian Journal of Electrical Engineering 3(2), 217–222 (2006)
[10] Pejović, A.: Digitisation of book "Central Forces in Nature" by Djordje Stanojević. Review of the National Center for Digitization (19), 69–73 (2011)
[11] http://www.negotin.rs/article.php?lg=en&id_article=124

Tesla's Research in the Field of Mechanical Engineering Focused on Fountains Design

Aleksandar Marinković[1], Nenad Zrnić[1], Bratislav Stojiljković[2], and Milan Vulićević[1]

[1] Mechanical Engineering Faculty University of Belgrade, Serbia
{amarinkovic,nzrnic}@mas.bg.ac.rs, v_milance@hotmail.com
[2] Nikola Tesla Museum, Belgrade, Serbia
bratislav.stojiljkovic@tesla-museum.org

Abstract. Nikola Tesla (1856-1943) made numerous inventions and discoveries in the field of electricity that form the basis of new scientific and technological breakthroughs, started in the late 19th century, without which our present life could not be imagined. Tesla's abundant and interesting creative activity in the field of mechanical engineering has only sporadically been quoted in the greatest number of biographical depictions of his life and opus. Accordingly, the terms such as e.g. turbines, pumps, fountains, speedometers or aeroplanes, often act confusing even on a connoisseur, deemed to be at least averagely informed about Tesla's inventions and discoveries. One of those less known achievements is his patent for a fountain. This invention of Tesla's is different from usual solutions applied to fountains and aquariums of that time, where the fluid was sprayed or jetted with appropriate devices for decorative purposes. The abundant documents in the archives preserved at the Nikola Tesla Museum in Belgrade demonstrate Tesla's great engagement in this, at first glance less significant, segment of his creative activity. This paper presents Tesla's work and achievements in the field of fountains. Using the original archival documents, design solutions, technical drawings, calculations, correspondence with companies and individuals, magazine articles and various brochures, the authors examine the scope of Tesla's research in this less known field. The authors also provide CAD models based on some of Tesla's original drawings of the fountains archived as a part of his legacy at the Nikola Tesla Museum in Belgrade.

1 Introduction

Nikola Tesla, a scientist, an engineer and inventor, lived and worked in the period that covers both the 19th and 20th century. He belongs to those rare inventors whose inventions have not ceased to attract the attention of many scientists and researchers for more than a hundred years. It was a series of new inventions and discoveries, from those in the field of electrical engineering, lighting technology, radio technology, to numerous applications of high-frequency currents in industry and medicine and unusual, original inventions in mechanical engineering and aviation.

T. Koetsier, M. Ceccarelli (Eds.): Explorations in the His. of Machines & Mech., HMMS 15, pp. 407–420.
springerlink.com

One of his inventions, which differs considerably from everything else this great scientist dedicated the largest part of his creativity to, was a fountain. For several years during the second decade of the 20th century, with some interruptions, Tesla worked in this field and developed several models of fountains.

The abundance of documents in the archives preserved at the Nikola Tesla Museum demonstrate Tesla's great engagement in this, at first glance less significant, segment of his creative activity. Numerous sketches and technical drawings, a long list of calculations related to the dimensions of fountain elements, fluid flow parameters and the characteristics of motors, as well as precisely defined modifications and improvements in construction, provide enough evidence that Tesla dedicated all his inventive energy and passion to this segment of his opus.

2 Fountains at the Beginning of the 20th Century

At the beginning of the 20th century, fountains in the USA are no longer just objects designed for the city squares and parks, or a privilege of the rich. They are noble, luxurious details of interior decoration in public buildings and private apartments. Wonderful electric fountains decorated the stage of many well-known theatres and cinemas: the Riverside on Broadway, the Strand, the Regent, the Washington Heights, the Family Theatre, etc., Figure 1. Magnificent spectacles of dancing water sprays, spurts and cascades, lighted with automatically controlled multicoloured lights, entertained the spectators during the intermissions in plays. "The latest adaptation of the electric turbine fountain - one that must charm the fancy of the housewife and the consulting professional who makes a specialty of interior decoration and design – is to

Fig. 1. The Stage of the Strand Theatre with its two self-contained electrical fountains.

the interior of homes, where it appears on hall pedestals, in grass-carpeted dens, and on the library table. It is arranged most attractively in combination with a reading lamp, the Tiffany and art-glass shades now so popular lending a rainbow of shaded colouring to the water spray", wrote a journalist for the Illustrated World magazine in his article about fountains at the beginning of 1915 [1].

Even today, one century later, new luxurious, extravagant, and often spectacularly designed fountains are not only the result of the creativity and skill of the author to apply contemporary materials and scientific achievements, but also the confirmation of the old and wonderful truth of their persisting and enduring beauty.

3 Tesla's Fountain – Creation of an Object of Beauty

Nikola Tesla devoted himself to machine engineering at the beginning of the 20th century, at the time when his grandiose Long Island project failed due to the lack of funds. He desperately needed a new invention at the time, similar to the induction motor, required by industry, of a simple construction and cheap to produce, in order to fund his further research. The next twenty years of his life Tesla dedicated exclusively to researches in the field of mechanical engineering, making also use of his inventions in the field of alternating currents (motors and generators) in order to secure the necessary funds. Thanks to his specialist knowledge and exceptional creativity, he was able to solve some technical problems in mechanical engineering and to realize a wide range of new inventions and design solutions based exclusively on physical laws [2].

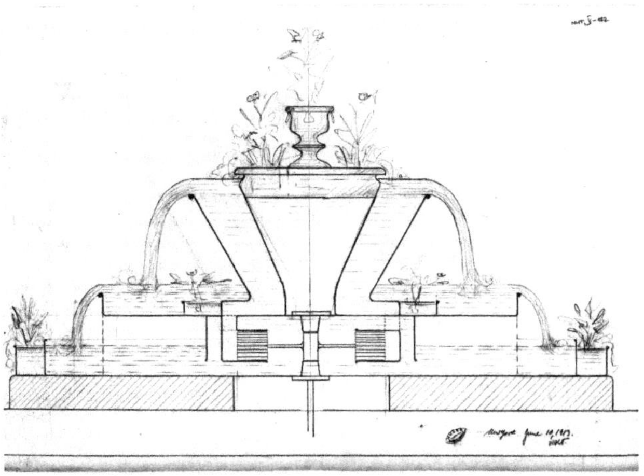

Fig. 2. The drawing of the vertical section of a fountain, dated June 10, 1913, is the oldest preserved document of Tesla's work in the field of fountains [3].

One of these less known solutions is his patent for a fountain. This invention is different from usual solutions applied to fountains and aquariums of that time, where the fluid was sprayed or jetted with appropriate devices for decorative purposes. His idea was to create a fascinating spectacle of a large mass of water in motion, in order to give the impression of great power, Figure 2. With additional lighting and the distribution of overflowing cascades, the achieved effect had to be even more beautiful and luxurious, with relatively low energy consumption in the creation and maintenance of waterfalls.

4 From the Idea to the Patent

Tesla's extraordinary intuition for new technological ideas and his unbelievable capability of visualization distinguishes him among great inventors. His faculty to see his ideas in his imagination as real devices helped him to design them, follow their functioning, detect and correct deficiencies, create different versions based on the same principle, determine the dimensions and types of material and composite parts, and the like. He spared himself long hours of hard work on the construction of apparatuses, because their outlines, drawings and complete composition, were created in the workshop of his creative imagination [4].

The results achieved by Tesla – thanks to his unique method in his virtual research laboratory - were extraordinary and impressive. That is why it is difficult to reconstruct the development of many of his inventions from the idea to the patent. Among the documents preserved, Tesla's legacy contains a number of sketches of that invention. Usually done in pencil, on the paper of different quality and size, they are most often rough sketches, while some of them offer considerably more details, Figure 3. The drafts preserved enlighten clearly and unambiguously the process by which Tesla arrived at this invention and developed it further into a patent and materialization.

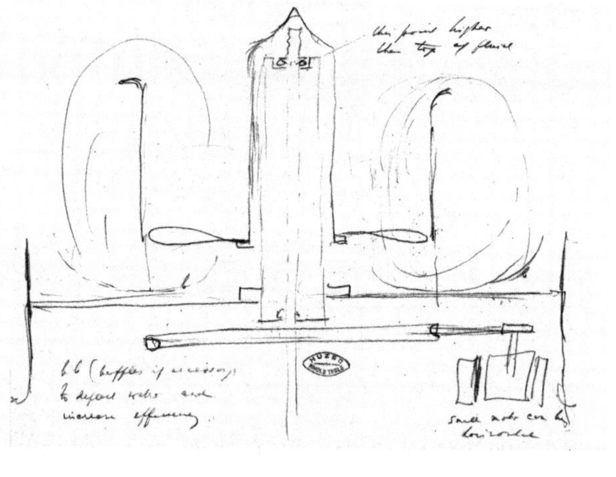

Fig. 3. Preliminary design of a fountain drawn by Tesla himself [5].

The structure of a complete device is provided in its vertical projection, where the main elements are marked with numbers, Figure 4. The reservoir (1) is filled with water or another fluid up to a certain level. The motor, i.e., the fans of the propeller

(9), start rotating and pushing the fluid through the nozzles (4), raising it to horizontally expanded top of the overflow (3). The fluid spills over from the top like a water cascade impressing the viewers' eye with a new and magnificent sight. The device was meant to be simple to construct and operate.

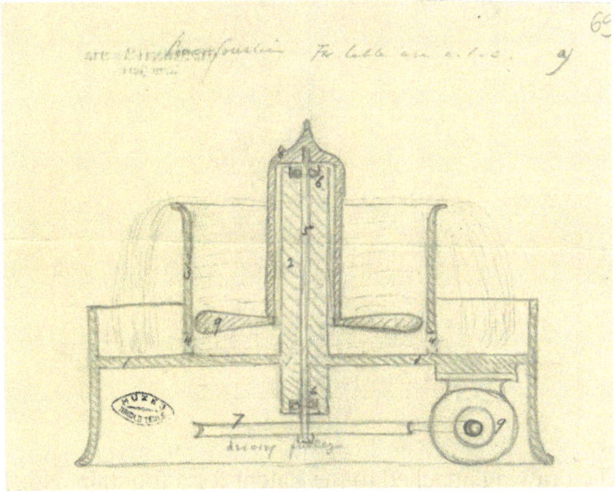

Fig. 4. An outline of the fountain mechanics after which Tesla shaped his invention [6].

The device was presented in the top view and in its vertical cross-section, Figure 5. All the elements of the device are elaborated and positioned giving a complete layout of the new fountain. A shaft (5) is inserted in the central nozzle (2) with a friction wheel or gear at the lower end (7) and a fixed housing (8) on the upper, with fans of the propeller. (9) The motor (10) is positioned so as to transmit the necessary power by friction or otherwise via the wheel (11) to the wheel gear (7). The openings (4) can be protected by replaceable filters. Water or other fluid would be lifted from the fountain receptacle to the horizontal flared out top of a conduit, to overflow it in the form of a unique circular cascade, making a splendid sight for an observer's eye. By applying only 1/25 of a horsepower (~30W) to the shaft and assuming a lift of eighteen inches (~45.7 cm) more than one hundred gallons (~378.5 l) per minute could be propelled to flow over the flared top of a conduit, one foot (~30.5 cm) in diameter with the depth of the overflowing fluid of approximately one-half inch (1.27 cm). As the circulation would be extremely rapid, the total quantity of liquid required would be comparatively small.

Tesla's application for his model of fountain was filed at the United States Patent Office on October 28, 1913, and the letters patent No.1.113.716 was issued one year later, on October 13, 1914. In the text of his patent application for a fountain, Tesla wrote that the invention "has an unlimited field of use in private dwellings, hotels, theatres, concert halls, hospitals, aquaria, and particularly, in squares, gardens and parks in which it may be carried out on a large scale so as to afford a magnificent spectacle far more captivating and stimulating to the public than the insignificant displays now in use" [9].

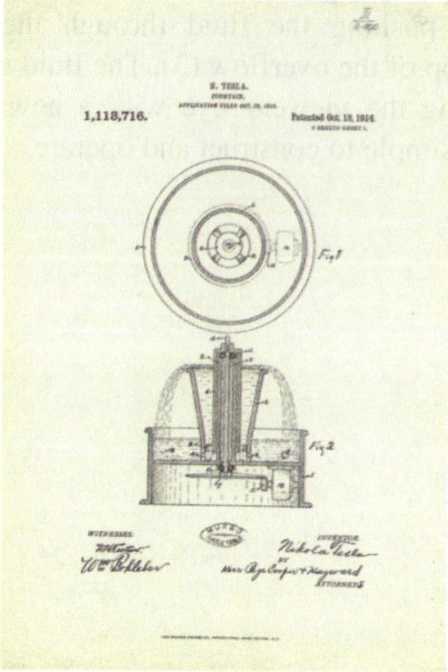

Fig. 5. The drawing attached to the Patent for a fountain, No. 1.113.716 [8].

Over more than eight years, with shorter or longer breaks, Tesla developed numerous models of fountains, but without significant commercial success. The preserved archive documentation, e.g. general designs, technical drawings, calculations of dimensions, characteristics and parameters of fountains, correspondence with companies and individuals, magazine articles, various brochures and the like, provide an insight into the scope of his research in this field of mechanical engineering.

5 Nikola Tesla's Cooperation with Tiffany Studios

After submitting his patent application for "certain new and useful improvements in fountains" at the end of October 1913, Tesla started searching for the ways and options to realize his invention. He offered the model of his cascade fountain, based on the new principle, to Louis Tiffany, an inventor of a special stained glass technique, a famous designer and the owner of Tiffany Studios. The review of the preserved archive documents (the correspondence between these two important men and between Tesla and Tiffany's associates, as well as calculations, drawings of a fountain and other documents), shows that this cooperation lasted from December 4, 1913, to January 9, 1916 [10].

At the end of 1913 and the beginning of 1914 Tesla intensively cooperated with Joseph Briggs, Manager of Tiffany Studios, on the development of particular structural parts of the cascade fountain and the manner of its illumination. Joint venture was formalized by signing a contract on business cooperation between Nikola Tesla and Study Tiffany on February 17, 1914. For the Tesla – Tiffany cascade fountain, Tesla elaborated a complete design of the plant and its components. He also specified characteristics of the propelling motor and proposed a very interesting solution for complete electric illumination. For circulation of water and creation of "the fascinating spectacle of a large mass of fluid in motion", the classic propeller was planned, with blades directly fixed to the driving shaft of electromotor. The propelling assembly was placed vertically along the central axis of a fountain in the interior of the elevated conical overflow.

In the design of this fountain Tesla did not make use of his new principle of energy transmission from the fluid and to the fluid, i.e., of his invention of a turbine with bladeless discs, patented on May 6, 1913. For the illumination of a fountain and realization of a more beautiful and more fascinating spectacle, Tesla used different automatically controlled coloured lights. Electric lighting of the Tesla – Tiffany cascade fountain consisted of two lighting systems independent of one another, differing in the number and power of electric bulbs used, Figure 6. One system, installed on the outer side of the fountain, was composed of four sets, two bulbs of different colours in each, arranged at an angle of 90° and placed vertically under the top of the elevated conical overflow. Two sets had blue bulbs, while the remaining two were equipped with violet and yellow-green bulbs, respectively. The other lighting system was placed in the interior of the fountain, under the glass bell of a particular design. This system consisted of three bulbs of greater power, arranged at an angle of 120° and installed horizontally above the driving assembly of a fountain in the middle of the elevated conical overflow.

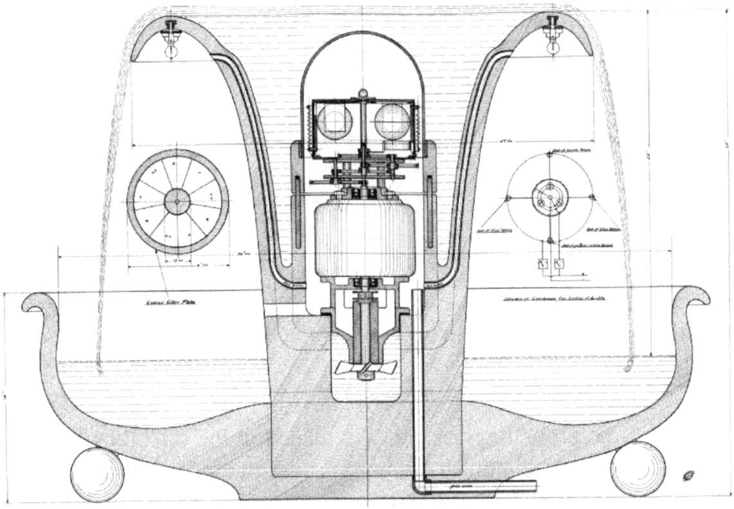

Fig. 6. Vertical cross section of the Tesla-Tiffany Cascade Fountain, New York, December 22, 1913 [11].

A circular glass plate, composed of nine equal segments, was placed above the horizontally built-in bulbs. Each one of these segments was of stained glass: yellow-green, blue or violet. The glass segments of the same colour were fixed at an angle of 120°, thus forming a uniformly coloured set [12]. Alternation of colours of electrical lighting during the work of a fountain, as requested by Louis Tiffany, had to be realized according to the following sequence: starting from yellow-green, the colour had to be changed into blue, from blue to violet, from violet into blue and, finally, from blue into yellow-green. The time required for these alternations had to be about three minutes [13]. The original CAD model and simulation of this fountain is performed by the authors of this paper at the Mechanical Engineering Faculty, University of Belgrade, within the project TR-35011 supported by Serbian Ministry of Education and Science, Figure 7.

Six drawings, made in the period from January 31 to February 14, 1914, represent five variants of a new model, named the Tesla Water Fountain. However, it would be interesting to note that two variants of assembly drawings of a new fountain were named the T.T. (i.e. Tesla-Tiffany) Water Fountain. This way of titling the drawings probably resulted from his intention to realize this model as a joint enterprise with Tiffany Studio. All the five variants of this model have similar shape and design of the receptacle for fluid and elevated conical overflow. In all the variants the electric motor with customary propeller as a unit, was placed in the lower part of the receptacle, under the elevated overflow.

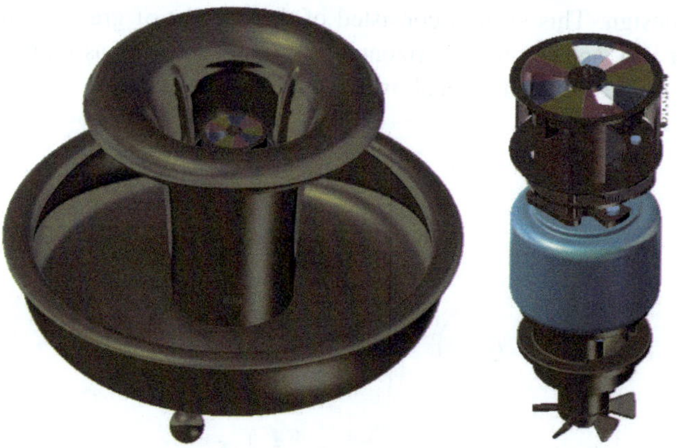

Fig. 7. CAD Model of the Tesla-Tiffany Cascade Fountain and the driving system with electric motor.

The shaft of the propelling assembly is placed in the direction of the vertical axis of the fountain. To accomplish a more beautiful and fascinating spectacle of a fountain in operation, electrical lighting was proposed in all versions of this model. Electric lighting was placed in the interior of the lower part of the receptacle. Through its glass bottom it had to illuminate agitated water with multicoloured light. In four variants, only two bulbs were placed either horizontally or vertically under the glass bottom of the receptacle, Figure 8. In the fifth variant, electric lighting was modified. It consisted of two bulbs horizontally placed under the glass bottom of the receptacle, and ten bulbs, of lower power, mounted on the vertical girder in a separately protected part of the interior of the elevated conical overflow.

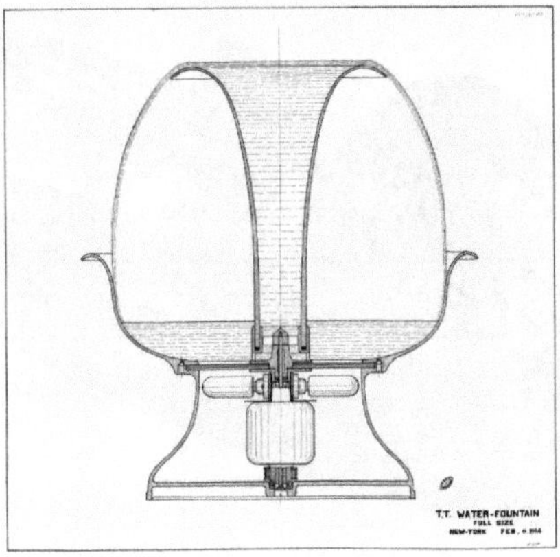

Fig. 8. Vertical cross-section of the T.T. Water-Fountain, New York, February 6, 1914 [14].

At the end of February and the beginning of March 1914, Tesla continued his work on designing and elaborating new variants of light regulator for the Tesla-Tiffany cascade fountain. Two preserved technical drawings form this period indicate clearly a new design of the light regulator together with elements and details for its manufacturing, Figure 9. It may be assumed that with this solution, i.e., with a worm gear, the inventor aimed to make the lighting more attractive and more captivating for the observer's eye and, on the other hand, to simplify the manufacturing of the subassembly of the system.

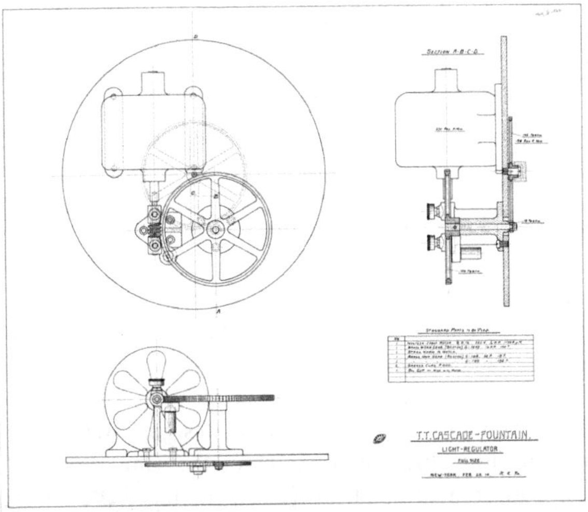

Fig. 9. Subassembly of the light-regulator of T.T. Cascade Fountain presented in three views, New York, February 28, 1914 [15].

Fig. 10. CAD Model of the subassembly of the light-regulator of T.T. Cascade Fountain with his electric motor.

This solution of the transmission with worm gear, shown in Figure 10, is also performed by the authors in the form of CAD model and operating simulation, within the project TR-35011 supported by Serbian Ministry of Education and Science.

Delay of the US Patent Office in issuing the letters patent for Tesla's fountain, as well as his extremely bad financial situation and impossibility to finance independently the construction of a prototype cascade fountain, on one hand, as well as the transfer of patent rights to the Tiffany Studio, together with the possibility of their estimation of the feasibility of joint enterprise, on the other hand, led to subsequent divergences of opinions and termination of cooperation. Although the cooperation between NikolaTesla and Louis Tiffany ended without accomplishing the planned projects and expected commercial success, it represents a part of Tesla's opus in the field of fountains and makes evident his broad interests, extreme creativity and ease in solving complex engineering problems in his own, original way.

6 A Model of Tesla's Fountain from 1917

During 1916, after the termination of cooperation with Louis Tiffany because of unsuccessful realization of his cascade fountain, Tesla dedicated himself to new researches in other fields of mechanical engineering, e.g. speedometers, flow meters, valvular conduits etc. At the beginning of the next year, Tesla turned again to research and realization of fountains, this time of new designs. Between January 18 and 25, 1917, he worked out four different versions of the same model of his fountain. In a particularly designed compartment, at the top of the elevated conical overflow, he planned to arrange a rich and interesting decorative arrangement of plants. In the design of this fountain model he applied, for the first time, his patented pump with discs. The pump would raise the fluid from the receptacle to the horizontal, broadened top of the overflow, thus forming a unique circular water cascade. The variants of this model differed from one another in shape and height of the fluid receptacle, position of a propelling motor and position and way of embedding of the additional electrical lighting.

In the period from May 11, 1917 to June 1, 1917, Tesla designed four different models of his table fountains. The preserved drawings present new concepts of fountains, each one shown in a view and vertical section. At all of these four models, Tesla's pump with discs was placed in the lower part of the receptacle and had to be used for water circulation. The propelling assembly was placed vertically, along the central axis. All these models of fountains were of unique design, with peculiar, exceptionally beautiful decorations on reservoirs, elevated overflows and other constructional elements. Electric lighting was planned to be in all the variants of table fountains, in order to make the spectacle of their operation as magnificent, fascinating and impressive as possible, Figure 11.

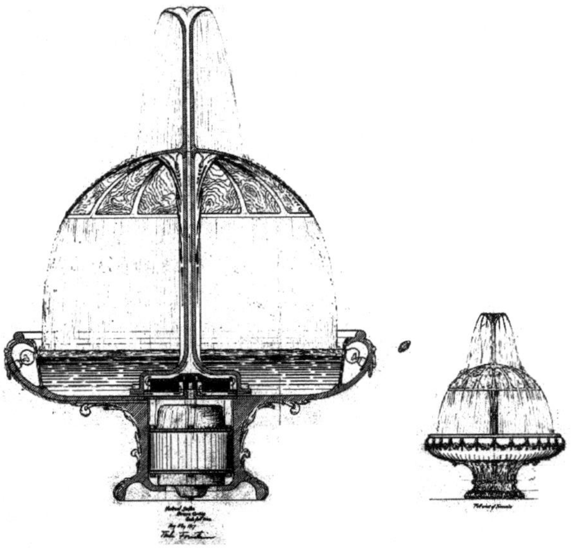

Fig. 11. Tesla's table fountain, shown in a view and in vertical cross-section, New York, May 11, 1917 [16].

7 A Model of Tesla's Fountain from 1920

After several unsuccessful attempts to realize the Fountain patent, Nikola Tesla attempted it for the last time at the end of the second decade of the 20th century. He worked on the project and realization of two new, very similar models of the fountain, using the principle of operation, improvements and construction of his earlier models. Both models of the fountain used Tesla's pump with discs to move water.

The review of the preserved archive materials – design solutions, sketches, technical drawings, calculations, brochures, order forms, collected and systematized data about the manufacturers of fountains and their parts, as well as other documents – leads to the conclusion that Tesla's research lasted throughout the period from February 20, 1919, to September 20, 1920, but again without any greater commercial success. The assembly drawing and several dimensioned workshop drawings of constructional elements have been preserved for one of these Tesla's table fountain models from this period. The propelling electric motor is placed in the direction of the vertical central axis of the fountain, under the elevated conical overflow in the lower part of the receptacle. For this model, electric lighting had not been planned [17].

For another model of Tesla's table fountain, only some sketches and unfinished drawings, showing its basic design in several modifications, have been preserved, Figure 12. Water circulation and elevation to the level required is accomplished by the use of Tesla's pump with discs, which is placed in the direction of the vertical axis of the elevated conical overflow. The propelling electric motor is placed either horizontally or vertically under the bottom of the receptacle. Electric lighting consists of three electric bulbs arranged at an angle of 120º and placed horizontally under the glass surface of the bottom of the receptacle. Glass surfaces are of circular form of a specified diameter (first variant); two bulbs, placed vertically in a particularly protected compartment in the receptacle. The protected compartment is manufactured of cast glass in the form of closed cylinder, preventing the break-through of water to electrical system (second variant).

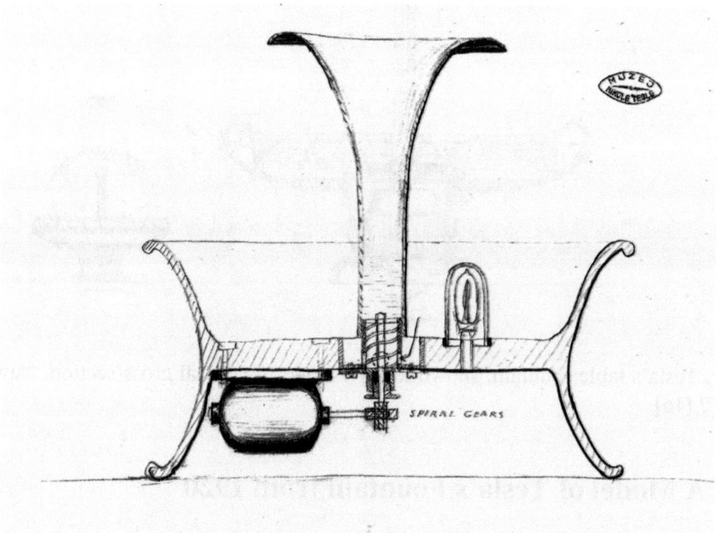

Fig. 12. The sketch of the vertical cross-section, drawn in pencil on paper, was a preliminary design for the development of other types of Tesla's table fountain [18].

It might be of interest to note that in this period Tesla managed to manufacture one working model of his table fountain, aiming to perform adequate tests and obtain sufficient data required for its realization and production on a larger scale, Figure 13. Several preserved photographs of his table fountain in operation confirm this assumption. The date written on the back of some photographs is: "June 7, 1920".

A working model of Tesla's fountain was realized by combining the constructive elements of both of his models from that period. The receptacle for the fluid and the subassembly of the electric motor and Tesla's pump were taken from one model of the fountain, and the elevated conical overflow from another one. However, there are no reliable data that could confirm its subsequent manufacture and any financial success for Tesla.

Fig. 13. Testing of Tesla's table fountain in the laboratory.

8 Conclusion

This paper focuses on highlighting Tesla's research in the field of mechanical engineering that a broader scientific community is not so well acquainted with, compared to the knowledge of Tesla's researches in the field of electrical engineering. The paper also reveals how wide the scope of interests of this ingenious inventor was.

The abundance of documents in the archives preserved at the Nikola Tesla Museum in Belgrade facilitates a comprehensive survey of the extent of Tesla's researches in the field of fountains. They also afford a more distinct perception of the time and circumstances in which this great scientist and inventor lived and created. During more than eight years – with shorter or longer interruptions – Tesla developed numerous models of his fountains, but without any commercial success. Nevertheless, viewed as a whole, Tesla's researches in the field of fountains demonstrate his broad interests and inexhaustible inspiration which permeates his researches, and his endeavours to attain new knowledge, accomplish discoveries and inventions.

Acknowledgement. This work has been performed within the project TR-35011. This project is funded by the Republic of Serbia, Ministry of Education and Science, whose help is gratefully acknowledged.

References

[1] Illustrated World. Illustrated World Press, Chicago (June 1915)
[2] Jovanovic, B.: Tesla – Spirit, Work, Vision, Freemental, Belgrade (2001)
[3] Technical drawing, MNT, V, I-4, 78, Nikola Tesla Museum, Belgrade (June 10, 1913)
[4] Electrical Experimenter - science and invention, vol. VI, Whole No. 70, pp. 744–745. Experimenter Publishing Company Inc., New York (February 1919)
[5] Technical drawing, MNT, CCCLXIV, 75 A, Nikola Tesla Museum, Belgrade
[6] Technical drawing, MNT, CCCLXIV, 69A, Nikola Tesla Museum, Belgrade
[7] Technical drawing, MNT, CCCLXIV, 58A, Nikola Tesla Museum, Belgrade
[8] The drawing attached to the Patent for the fountain No. 1.113.716, MNT, CDXCVII, 46 A, Nikola Tesla Museum, Belgrade (October 13, 1914)
[9] Nikola Tesla, Patent IV, Institution for Textbooks and Teaching instruments, Belgrade (1996)
[10] Stojiljkovic, B., Vujovic, M.: Tesla's Fountain. Nikola Tesla Museum, Belgrade (2007)
[11] Technical drawing, MNT, V, I-4, 77, Nikola Tesla Museum, Belgrade (December 22, 1913)
[12] Marinković, A., Ćoćić, A., Stojiljković, B., Vulićević, M.: Design of Tesla-Tiffany cascade fountain as a simple of Tesla's research creativity in field of mechanical engineering. In: 10th Anniversary International Conference on Accomplishments in Electrical and Mechanical Engineering and Information Technology, DEMI 2011, pp. 117–122. University of Banja Luka, Mechanical Engineering Faculty, Banja Luka, Bosnia and Hercegovina (2011)
[13] Letter of J. Briggs, Tiffany Studious to Nikola Tesla, New York, MNT, CCCLXV, 156 A, Nikola Tesla Museum, Belgrade (Decembar 24, 1913)
[14] Technical drawing, MNT, V, I-4, 82, Nikola Tesla Museum, Belgrade (February 6, 1914)
[15] Technical drawing, MNT, V, I-4, 86, Nikola Tesla Museum, Belgrade (February 28, 1914)
[16] Technical drawing, MNT, V, I-4, 93, Nikola Tesla Museum, Belgrade (May 11, 1917)
[17] Stojiljković, B., Marinković, A., Ćoćić, A., Vulićević, M.: Tesla's research in the field of fountains. In: The 24th International Conference of Efficiency, Cost, Optimization, Simulation and Environmental Impact of Energy System, ECOS 2011, pp. 161–168. Novi Sad, Serbia (2011)
[18] Technical drawing, MNT, CCCLXV, 19 A, Nikola Tesla Museum, Belgrade
[19] Photograph, MNT, VI/VIII, 15, Nikola Tesla Museum, Belgrade

Mechanical Models for Anti-Rhomb Linkage

Simona-Mariana Cretu, Gigi-Dragos Ciocioi-Troaca, Emil Soarece, and Eugen Marian Paun

University of Craiova, Faculty of Mechanics, Romania

Abstract. This paper presents a brief evolution of the linkages and focuses on the antiparallelogram mechanism and its main features. In one of Bricard's books it is presented a particular case of the contraparallelogram – the anti-rhomb linkage –, which is applied to open a door both ways. We realized some mechanical models that use this linkage, providing different functional characteristics of the door: the opening inside and outside, on the left and/or right side of it.

1 Short Evolution of Linkages

People have used linkages in their work from early ages. Scheiner's remarkable pantograph (Fig. 1a), invented in 1603, is highlighted among the achievements of the hinge linkages.

Christoph Scheiner (1573-1650) was born in Wald, Swabia (Germany) (Fig. 1b). He was known as a theologian, physician, mathematician and astronomer, but the pantograph invention (1603) – a linkage which can copy a figure at an adjustable scale – brings him celebrity; he described this mechanism 27 years later, in [1].

James Watt (1736-1819), a great designer of mechanisms, was born in Greenock, Scotland (Fig. 2). Although he had no formal studies in the field of mechanisms, he is recognized as a great inventor whose work contributed to the progress of the world.

At the age of 28, James Watt was working as a mechanic at the University of Glasgow, and had the duty to repair a Newcomen steam machine. Due to the lower efficiency of the machine (0.63%), he started studying this problem, and obtained some remarkable inventions for that period. This marked the beginning of the industrial revolution. One of these inventions was the antiparallelogram mechanism, patented in 1784, whose midpoint of the rod – joined with rod of the piston – traces the curve named lemniscate. The rectilinear part of this curve was used to move approximately straight the rod of the piston machine.

In 1784 Watt wrote in a letter to Boulton about one of his inventions (Fig. 3a), and explained it in another letter to his son, 24 years later (1808): 'The idea originated in this manner. On finding double chains, or racks and sectors, very inconvenient for communicating the motion of the piston-rod to the angular motion of the working beam, I set to work to try if I could not contrive some means of performing the same from motions turning upon centres, and after some time it occurred to me that AB and CD being two equal radii revolving on the centres B and C, and connected together by

T. Koetsier, M. Ceccarelli (Eds.): Explorations in the His. of Machines & Mech., HMMS 15, pp. 421–430.
springerlink.com

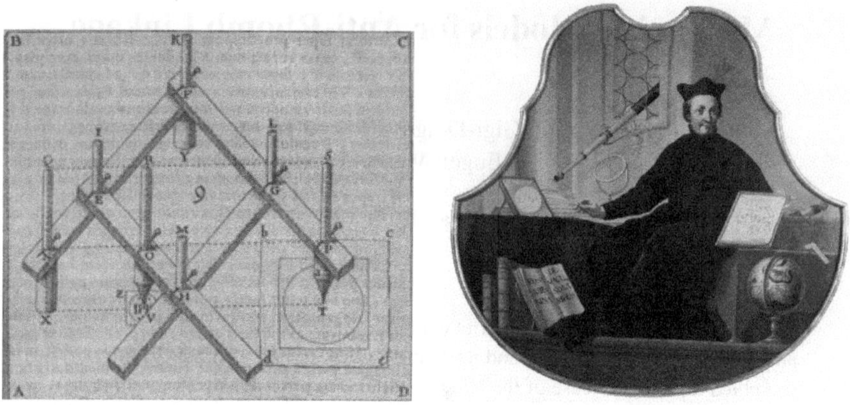

(a) (b)

Fig. 1. a) Scheiner's pantograph b) Christoph Scheiner

Fig. 2. James Watt

a rod AD, in moving through arches of certain lengths, the variation from the straight line would be nearly equal and opposite, and that the point E would describe a line nearly straight, and that if for convenience the radius CD was only half of AB, by moving the point E nearer to D the same would take place, and from this the construction, afterwards called the parallel motion, was derived. [Fig. 3b]... Though I am not over anxious after fame, yet I am more proud of the parallel motion than of any other invention I have ever made.'[2]

Reuleaux wrote that Watt's invention process is not entirely explained: 'We quite appreciate the motives as well as some of the final results of Watt's exertions, but we obtain no indication of a methodical train of ideas leading up to them.' [2]

We consider that the sparkle for this invention resides in Watt's sentence: 'the variation from the straight line would be nearly equal and opposite'; this idea is very well captured by Willis in [3]: 'A parallel motion is a term somewhat awkwardly applied to a combination of jointed rods, the purpose of which is to cause a point to

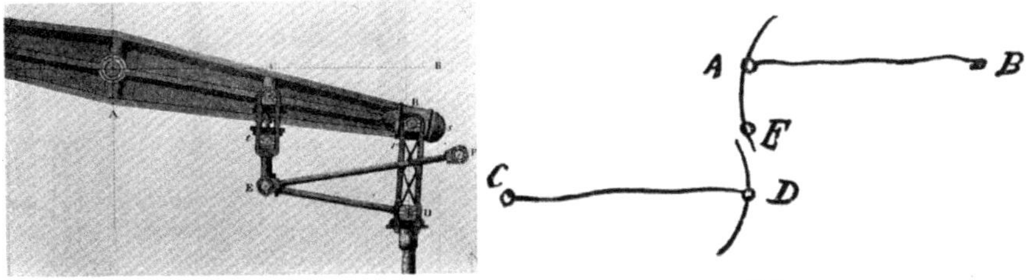

(a) (b)

Fig. 3. a) Watt's pantograph b) Watt's sketch for pantograph

describe a straight line by communicating to it simultaneously two or more motions in circular arcs, the deviations of these motions from rectilinearity being made as nearly as possible to counteract each other. The rectilinear motion so produced is not strictly accurate, but by properly proportioning the parts of the contrivance, the errors are rendered so slight that they may be neglected.'

Pafnuty Lvovich Chebyshev (1821-1894), a well-known Russian mathematician, was born near Moscow (Fig. 4a). He worked 39 years in analysis and synthesis of approximately straight-line mechanisms. For two of them he obtained the Nobel Prize: the walking wooden horse (Fig. 4b – realized by the students of the scientific association of history of machines and mechanisms, Faculty of Mechanics, Craiova, Romania, 'Istoria masinilor si a mecanismelor' – AIMM) and the boat-linkage.

Chebyshev tried to reduce the errors of the straight line by combining many mechanisms – such as Watt's and Evans' linkages (Fig. 5) –, but they are too complicated.

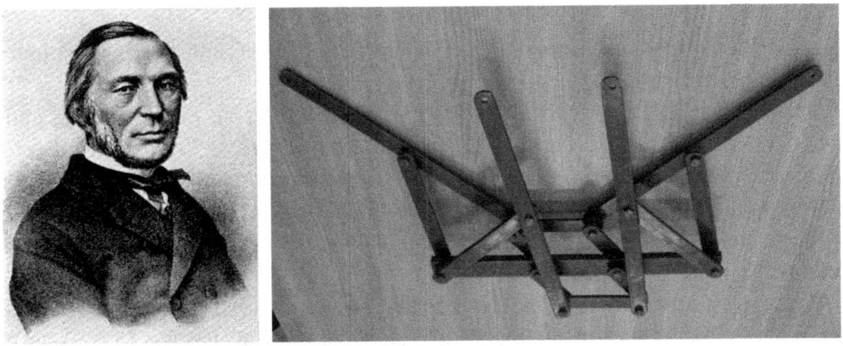

(a) (b)

Fig. 4. a) Pafnuty Lvovich Chebyshev b) Mechanism used for Chebyshev's 'horse'– AIMM

Another example of Watt-type linkage is Roberts' linkage (Richard Roberts, 1789-1864), which also approximates a straight line (Fig. 6).

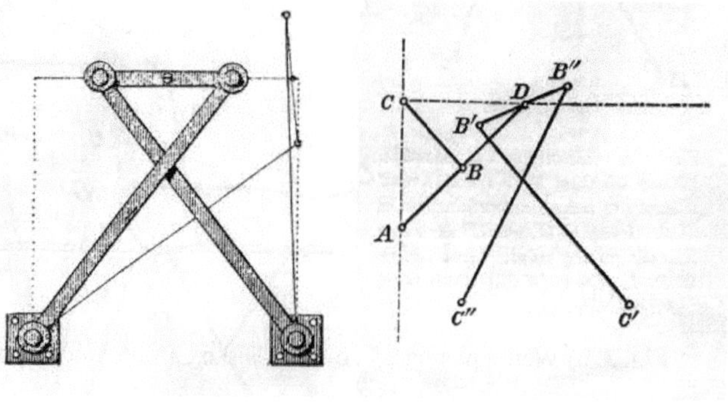

Fig. 5. Chebyshev's straight-line mechanism

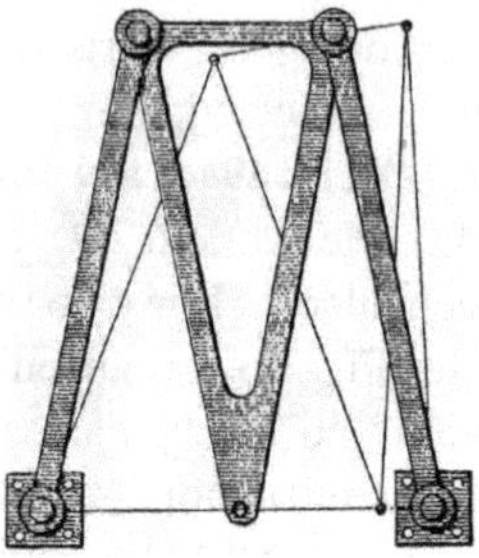

Fig. 6. Roberts' straight-line linkage

Peaucellier (1864) and Lipkin (1871) were those who found the exact solution to obtain the perfect straight line [4].

Charles Nicolas Peaucellier (1832-1913) graduated École Polytechnique and became captain in the French army. He wrote in 1864 in *Nouvelles annales de mathématiques* about a class of linkages named 'compound compasses', including Watt's parallel motion, pantograph and the polar planimeter. He obtained the solutions to design linkages capable to draw a straight line, a circle of any radius and conic sections [5].

Although mathematicians neglected his article at that time, the year 1864 represents the beginning of the theoretical study of the linkages. In 1873 Peaucellier sent the solution of the problem to the same journal [5]. As a linkage generates a curve with an algebraic expression, a given curve can be generated by a suitable linkage. Also he gave a geometric proof for the exact straight-line 'compound compass' (Fig. 7).

Yom Tov Lipman Lipkin, a Lithuanian Jew, was Chebyshev's student. He studied the straight line motion and discovered, independently of Peaucellier, the pantograph. Mathematician James Joseph Sylvester (1814-1897), who was working at the Royal Institution in London, covered in one of his lectures the subject: 'Recent discoveries

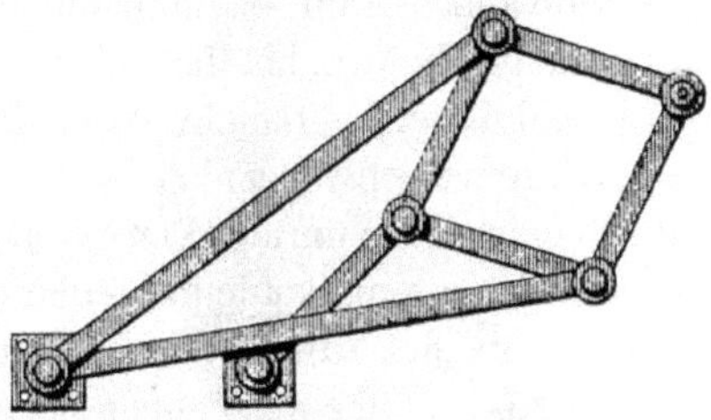

Fig. 7. Peaucellier's exact straight-line linkage

in mechanical conversion of motion' (1874). Afterwards, the interest of other mathematicians for this important discovery increases; some of them were: Hart, Clifford, Roberts, Cayley, Kempe (from England) and Saint Loup, M.Laisant, M.Lemoine, M.Leaute (from France) [4].

Also, Leaute treated the four-bar linkage and obtained given curves with an accurate approximation. Darboux published some studies about the deformation of the four-bar linkage in his journal, and M. Neuberg published the abstracts of Darboux's conferences.

After Watt discovered the antiparallelogram, other scientists studied its properties.

If one fixes the shorter side of the antiparallelogram, the crossing point of the non-parallel sides of it describes an ellipse, which has the distance between the foci equal to the smaller side. The same point, in regard to the movable smaller side, describes also an ellipse which rotates without sliding on the other fixed ellipse, during the process of motion of the mechanism. The common tangent at both ellipses is the angle bisector with the vertex in the instantaneous point of rotation. If the larger side of antiparallelogram is fixed, the crossing point describes a lemniscate during the motion [6].

2 Anti-rhomb Linkage

Raoul Bricard (1870-1944) was a great French engineer and mathematician. All his activity was rewarded with the Poncelet Prize in Paris by National Conservatory of Arts and Crafts (1932). He realized important discoveries, especially in three fields: descriptive geometry, kinematics and mechanical linkages. Referring to mechanisms, he wrote studies with mechanical applications, as well as some books. One of them is *Cinématique et Mécanisme* (1926) [7].

This book has two parts, one about theoretical kinematics and the other about mechanisms. In the first part he pointed clear differences between finite displacements (which utilize only geometry) and continue movements (which use analytical geometry and infinitesimal analysis). The author considers that the second part of his book is far from being a complete trait of applied kinematics, studying only the mechanisms of a real practical interest.

In the chapter entitled Linkages, Bricard presents the four-bar linkage, ABCD, formed of two equal non parallel sides and of diagonals of an isosceles trapezoid (named contraparallelogram, or crossedparallelogram, or antiparallelogram). The antiparallelogram has the opposite sides equal two by two, but they are not parallel (Fig. 8a).

During the deformation of the linkage, the triangles ABC and CDA remain congruent – they have three equal sides –, BD remains parallel to AC – the two triangles have the same height –, and the angle BAC is congruent to the angle DCA, all the time.

Bricard analyses the anti-rhomb, a remarkable particular case of antiparallelogram, with all sides congruent (Figs. 8b, 8c, 8d). In this case, two opposite vertices are always coincident. In a special position, the mechanism becomes completely flattened (Fig. 8b). During the motion of mechanism, which has AD as driving side, both A and C vertices rest coincident, as well as B and D rest coincident if the driving element is BC. Bricard proposes as application the use of this linkage to realize the hinges for a door which is opened in both senses.

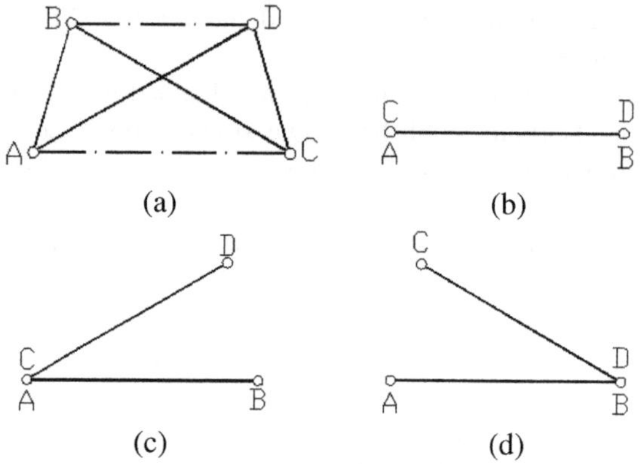

Fig. 8. a) Crossed parallelogram; b), c), d) Anti-rhomb linkage

We propose some constructive variants of this mechanism which provide different functional characteristics (Figs. 9, 10, 11, 12, 13 and 14).

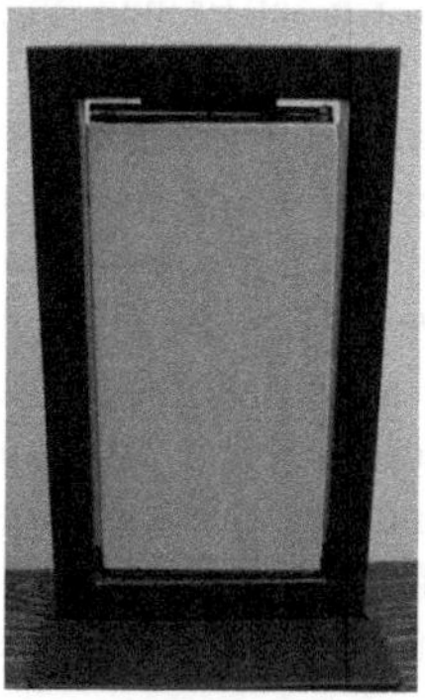

Fig. 9. Four-bar linkage door – version 1 –, AIMM

Fig. 10. Four-bar linkage door – version 1 –, detailed images

The mechanism from Figs. 9 and 10 has the door linked with both cranks by hinges, which are positioned on one of its diagonal, so it becomes rod in mechanism.

This linkage has the possibility to open the door inside and outside, from left and right sides (Fig. 11).

Another model (Fig. 12) has the linkage placed on the lower or upper side of the door, or both. In the second case, the load of the door stresses more the bolts from the frame. As in the previous model, the door is the rod of the mechanism and it is linked with both cranks by hinges.

Fig. 11. Four-bar linkage door – version 1

Two bolts are splited as semi pipes (Fig. 12a), so that the door is completely closed (Fig. 12b). The door could be opened on both sides from interior (Figs. 12c, 12d) and one side from exterior (Fig. 12e), only if we switch on/off a mechanical system.

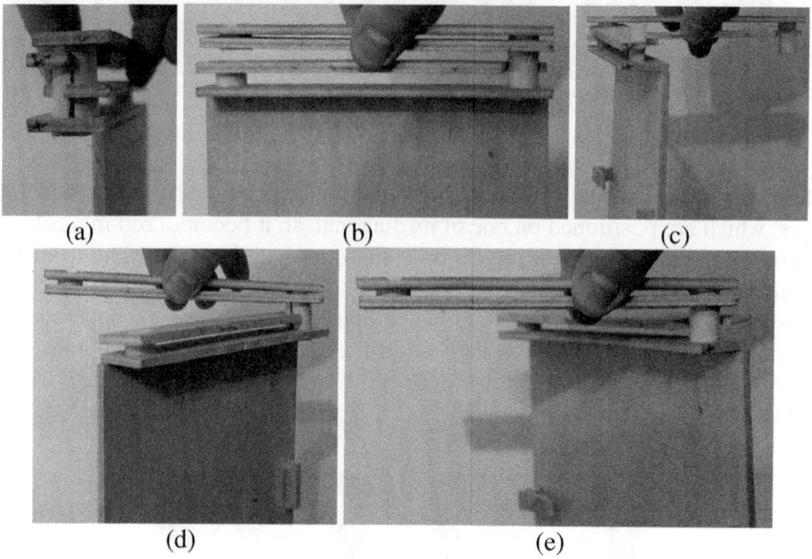

Fig. 12. Four-bar linkage door – version 2 –, AIMM

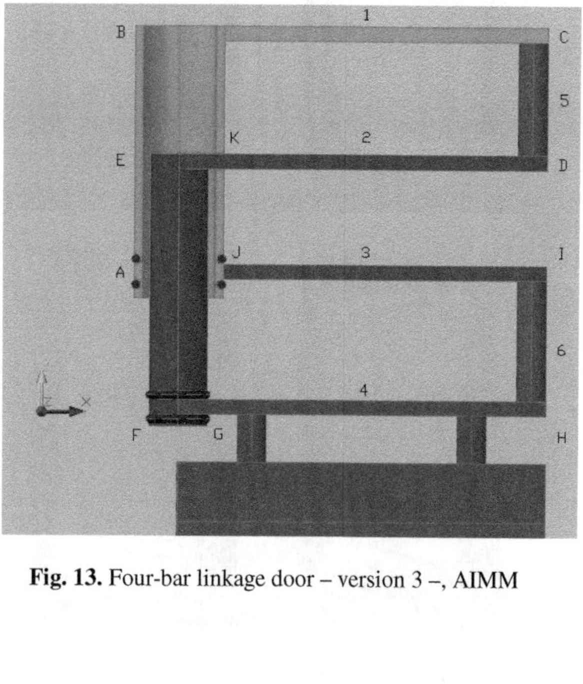

Fig. 13. Four-bar linkage door – version 3 –, AIMM

Figs. 13 and 14 present another model which has also the door as rod (element 4) of the anti-rhomb linkage. A cut (in the area K) is made in a pipe (element 1), so that a crank (element 2) can move inside it. This kind of door could be opened on both sides from exterior and one side from interior.

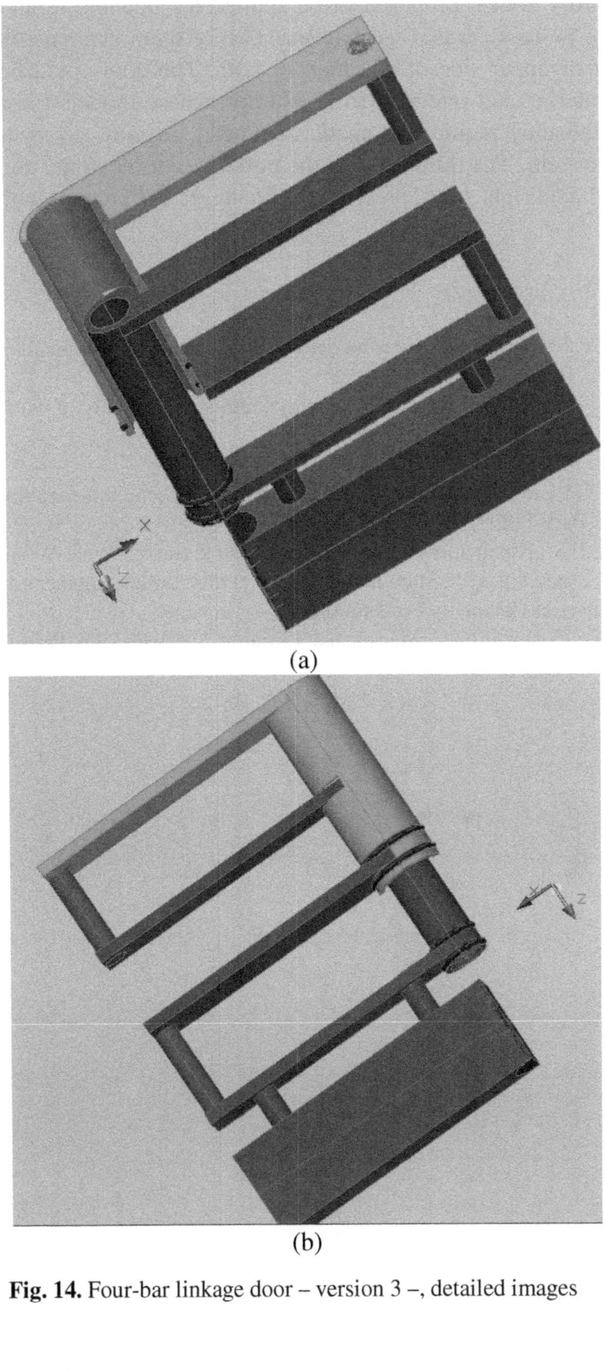

(a)

(b)

Fig. 14. Four-bar linkage door – version 3 –, detailed images

3 Conclusions

First, we shortly presented the evolution of linkages, considering that some of them contributed to the progress of humanity. We analyzed the anti-rhomb linkage from one of Bricard's books, proposed to be applied for opening a door both ways. We realized three models for this mechanism. Two of them can have the linkage placed on the lower or upper side of the door, or both. The door could be opened on both sides from interior and one side from exterior. These models present technological difficulties and they require automatic command, in order to switch on/off certain parts of the system. The third one has the possibility to open the door inside and outside, from left and right sides, presenting advantages with regard to the other two.

References

[1] Scheiner, C.: Pantographice seu Ars delineandi res quaslibet per parallelogrammum lineare seu cavum, Rome (1631)

[2] Reuleaux, F.: The kinematics of machinery: outlines of a theory of machines. In: Kennedy, A. (ed.) MacMillan, London (1876)

[3] Willis, R.: Principles of Mechanism, Cambridge (1841)

[4] Koenigs, G.: Leçons des cinématiques. Cinématique théorétique, Librairie scientifique A. Hermann, Paris (1897)

[5] Ferguson, E.S.: Kinematics of mechanisms from the time of Watt, Washington (1962)

[6] Bryant, J., Sangwin, C.J.: How round is your circle? Where Engineering and Mathematics Meet. Princeton University Press (2008)

[7] Bricard, R.: Cinématique et Mécanisme. Librairie Armand Colin, Paris (1926)

History of Human Powered Threshing Machines: A Literature Review

K.S. Zakiuddin[1], H.V. Sondawale[2], J.P. Modak[1], and Marco Ceccarelli[3]

[1] Department of Mechanical Engineering, Priyadarshini College of Engineering, Nagpur-440001, India
qszaki1@rediffmail.com, jpmodak@gmail.com
[2] Department of Mechanical Engineering, Smt. Rajshree Mulak College of Engineering for Women, Nagpur-440009, India
hvsondawale@rediffmail.com
[3] LARM, Laboratory of Robotics and Mechatronics, DiMSAT, University of Cassino, Italy
ceccarelli@unicas.it

Abstract. The objective of this paper is to present the recent developments in the field of agricultural technology with respect to human powered threshing machines. This paper covers the form and existence of human powered threshing machines as well as manual methods from last decades. The different types of designs and progressive developments with operating parameters and mechanisms used can be overviewed from the literature review to find out the scope for further developments in human powered threshing machines.

1 Introduction

Area of agriculture is leading towards the commercialization in economic point of view. Emphasis is on the mechanization of pre- harvesting and post- harvesting equipment's. Post harvesting equipment's including threshing, have promising future towards the industrialization and local employment if the equipment is available at low cost and with the ease of operation. Then the recommendation will be given to human powered threshing machine which is eco friendly as well as operated with the human power which is recognized as a renewable source of energy.

The early methods of threshing includes hand threshing, bullock threshing, human powered thresher and the modern one of using a power operated thresher or tractor, and intermediate one where part of the output is threshed by bullocks and part by a thresher or a tractor. The paper focus on human powered threshing machine so taking a brief history with the traditional methods of threshing.

2 Materials and Methods

This study was designed on the basis of Investigative Survey Research Approach (ISRA) as developed by Anazodo (1975, 1983). The investigative survey research approach for operating data entails the history and forms of the human powered

T. Koetsier, M. Ceccarelli (Eds.): Explorations in the His. of Machines & Mech., HMMS 15, pp. 431–445.
springerlink.com

threshing machines worldwide. The kind of data sought for year wise available. The data collected and sought for is in the same form available in the literature as it is meaningless to change the composition of the valuable history and research.

3 Discussion

Definition of Threshing:
Threshing or thrashing is the process of separation of grain from the stalk on which it grows and from the chaff or pod that covers it (answers.encyclopedia.com/question/threshing-410357.html).

Principle of Threshing
Grains can be separated from the panicle by a combination of stripping, rubbing and impact action on plant panicle head. This action resulted in application of tensile, compressive, bending and twisting forces and their combination on plant panicle.

Simple Threshing Methods (1969) (1)
Primitive threshing was done by spreading the crop on a threshing floor where it is beaten either with sticks or flails (Fig.1) or trampled by the hoofs of animals. The sheaves were made the grain in often threshed out by beating the sheaves against shielded grates.

Fig. 1. Simple threshing devices Chinese threshing flail

Threshing Sleds (1969) (1)
Threshing sled is an animal-drawn implement consisting of two wooden boards, slightly raised in front which are fitted with short pegs, serrated knives or hard stones, inserted into holes on the underside of the boards. The operator stands on the implement, to add weight, and it is dragged over the crop spread on the floor, the knives or pegs rubbing out the grain and bursting the straw.

Threshing Rollers (1969) (1)
Another way of threshing was to pass ribbed rollers (Fig.2) or two or three parallel sets of discs mounted on a frame over the laid out crop (Fig.3). The disc thresher, known as the *norag* in the near East, and Olpad thresher in India, had serrated iron discs about 45 cm in diameter and about 15 cm apart , with a seat, and it is drawn by a

pair of animals. The crop is spread in layers 30 to 50 cm deep and turned over with wooden forks while being threshed.

The grain obtained is not clean and had to be separated not only from chaff but also from dust and dirt. Grain is also likely to be broken and some not threshed, or eaten by the draught animals. The whole process was rather slow, but it had the advantage that the straw was bruised. Threshing rollers were used through-out the Near East and in parts of Spain and Portugal.

Fig. 2. Simple threshing devices Stone Threshing Cylinder

Fig. 3. Olpad Thresher, made in India on the pattern of thresher; diameter of discs about 15 cm, number of discs 20.

<u>Hand-operated drum threshers (1969) (1)</u>

The hand-operated thresher shown in sectional view in Fig.4 is a type in which the whole sheaf is passed through the machine. The output is 200 to 300 kg of grain per hour according to the proportion of grain to straw. This low output, with two men needed at the drive wheels, makes the machine uneconomical; therefore its manufacture has been discontinued in Europe.

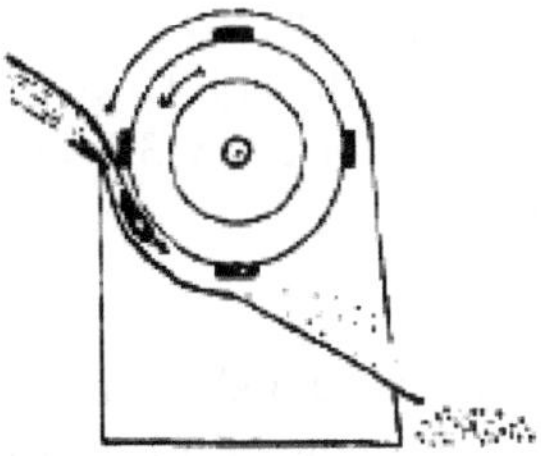

Fig. 4. Sectional View of drum thresher

Pedal-operated drum thresher (1969) (1)
The pedal-operated drum thresher is common in rice growing countries. A revolving threshing drum fitted with wire teeth and driven by a pedal, connecting rod and a crank is mounted in a framework. The sheaf is held at the butt end with the head against the revolving drum until all the grain is beaten out (Fig. 5). The drum has a speed of approximately 300 to 400 revolutions per minute, and the output is about 150 kg of grain per hour. Two man treadle machines with a considerably higher output are also available. These machines are inexpensive and portable, averaging 35 to 70 kg in weight, but are satisfactory only when the grain is easily detachable from the stalks, such as rice, but not wheat or barley.

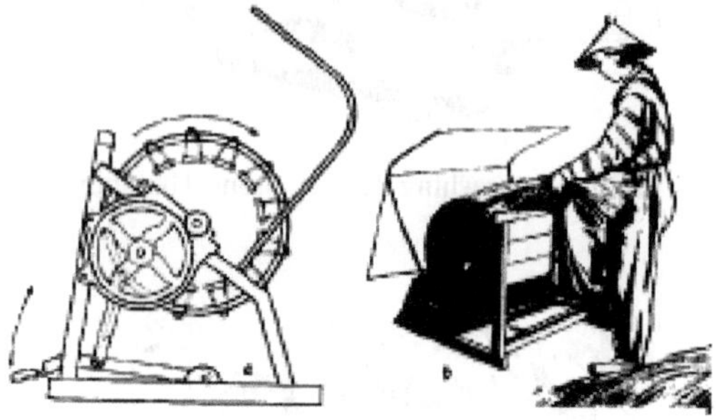

Fig. 5. Japanese Rice Thresher (a) Sectional View of Rice Thresher (b) Rice Threshing with a Pedal Operated Machine

Manual Thresher (1972) (19)
C.C. Lee stated that the manual thresher worked by two persons in which one keeps the drum turning and the other feed the grain is the most usual type of thresher in Korea.

Pedal Driven Thresher by Dynapod (1976) (2)
It was found suitable to operate the pedal driven thresher by Dynapod. Dynapod comprises a frame comprising pedal cranks, saddle, handlebars and power take-off to be clamped to machine. It is a universal device to operate the pedal driven machinery.

Pedal Threshers(1981) (3)
In 1950 Japan a agriculture was just entering the early stages of mechanization with many small pedal threshers, some 13,000 power tillers and an equal number of power sprayers, but essentially no other machinery. Early data on threshers is not available, but in 1955 there was one thresher for every 3 hectares and in 1960s there was one thresher for every 2.5 hectares.

Bicycle Thresher (1985) (22)
A quite efficient manual threshing method is employed by machines based on the bicycle. One or two operators are required to feed the hopper while providing the

Fig. 6. Bicycle Thresher

power via v-pulley to the threshing mechanism. This thresher is suitable for rice and most other cereals. The unit is operated by two men. A sloping sieve at the discharge point helps to separate the straw from the grain which can be collected in a shallow skip or a sheet. The thresher is equipped with twin wheels and has an output of up to 200 kg/h. (See Fig. 6)

Pedal Powered Thresher(1999) (4)

The paper confirmed the existence of pedal powered wheat threshers in Bangladesh. It had some problems with the gear slipping but it was seemed to work well with smallish sheaves. It was by the visited to Bangladesh and desired to get a copy of it for Nepal.

Pedal paddy thresher (2001) (5)

In Tarai(Nepal), there is a trend towards the use of improved manual tools, e.g. corn sheller, pedal thresher etc. Pedal paddy threshers are becoming popular in the Kathmandu valley. It is estimated that about 3000 pedal threshers are currently in use. The mechanization related ongoing R&D activities of AED are to study the relationship between cylinders speed and moisture content of wheat for better performance of the pedal rice thresher. Paddy Pedal Thresher (2004) (6)

In manual threshing by beating, one can separate about 100 kg of grain/h, whereas by using a pedal thresher (oscillating or rotary mode) one can separate about 150 kg of grain/h from medium sized paddy/wheat plants. However, the pedal threshing is a strenuous activity with high muscular strain due to speedy pedalling and holding of paddy plants on the rolling drum. The ergonomics improvement in the pedal thresher might allow a rhythmic legwork in sit stand position; in addition, the weight of the rolling drum at about 8 kg might be comfortable to the user. (See Fig. 7.)

Fig. 7. Paddy Pedal Thresher in Operation

Pedal Thresher or Treadle Thresher (2005) (20)
The pedal thresher or treadle thresher consists of threshing drum, base, transmission unit and a foot crank. When pedaled, the threshing drum rotates and rice can be threshed when panicles are applied against the threshing drum. Because small straws, chaff, and foreign matter drop along with the threshed grain, whole grains must be separated using a flail, sieve or by winnowing. (See Fig. 8.)

Fig. 8. Pedal or Treadle Thresher

Hand and Pedal Operated Sunflower Thresher (2007) (7)
Manual threshing of sunflower found tedious, time consuming and labour intensive work. It was decided to mechanize the threshing of safflower trial plots by developing a mini plot thresher. Earlier, a hand-operated single plant thresher was designed and built to see its suitability for threshing safflower trial plots and later on a pedal-operated system was incorporated in the thresher for reducing the human effort required to run it.

However, since the machine was expected to be operated continuously for 7-8 hours daily, hand or pedal-operated threshers could prove to be very strenuous for a labour. Also there were practical limitations to a frequent replacement of the labours, so it was decided to design a mini plot thresher running on a battery-powered motor.

Manual Pedal Thresher (2007) (8)
Beside the existence of power tillers and power driven pumps more than about 350,000 about 200,000 manually operated pedal threshers were also being practiced in different districts of Bangladesh (Anon, 2004). Farmers in Dhaka and Jessore are not interested in this expensive power operated closed drum thresher because they are habituated with less expensive pedal thresher. Due to low cost pedal threshers were excluded from the economical analysis in Bangladesh.

Pedal Threshers used in the Region North-Western Himalayan region (NWHR) of India (2008) (9)
The traditional pedal paddy thresher, shown in Fig. 8, is used very rarely in hills. A mild steel sheet has been used as the covering material in this thresher, but this increases the total weight of the machine to around 50 kg.

Fig. 9. Traditional Paddy thresher

Developed VL paddy thresher (2008) (9)
The seat and hand rest of the VL paddy thresher was designed in such a way that the spinal column (15751 from the vertical plane) and arms (angle between the upper and lower arm 1351) of the operator remain in a comfortable position (Fig. 10) during operation. The height of the hand rest and seat can be adjusted according to the need of the operator. The power was transmitted from the pedal to the threshing cylinder through a chain and sprocket system having a speed ratio of 1:7. The threshing drum diameter was 350mmat the tip of wire loops. A safety cover was also provided in the thresher to protect the operator as per norms (IS: 9020, 1979) and to save the grain losses during operation. The MS sheet in the body of the pedal thresher was replaced by a polycarbonate sheet of 1mm thickness to reduce the total weight. The machine was designed adhering to the National Standards and guidelines on working posture (Hunang and Suggs, 1967). The material used for the threshing drum was as per the Indian Standard Institute (IS: 3327, 1962). The frame of the machine was made using the combination of a standard MS angle (40_40_5mm) and an MS pipe (40/38mm diameter). The machine was manually operated by a single person. A chain and sprocket system was used for transmitting power from the

pedals to the threshing drum. The drum of the thresher was made of MS flat (40_5mm), MS sheet (1mm) and MS angle (30_30_5mm). For beating action, V-shaped wire loops made of 5mm MS round bars were welded in a staggered manner on the MS angle. These angles with wire loops were fitted on the periphery of the threshing drum with the help of a nut and a bolt.

Fig. 10. Developed VL Paddy Thresher

Pedal Operated Sunflower Thresher (2009) (10)

It is a hold on type thresher operated by one person. It consists of a threshing wheel of 660 mm dia., which is placed horizontally over a main shaft. The power transmission system includes a bevel gear unit, which gets power through a chain and sprocket arrangement attached to the pedal assembly. It has a blower to clean the threshed seeds. Two persons excluding the operator hold the cubs over the rotating wheel through its feeding chutes. The seeds detached from the cubs due to impact force of the threshing wheel and are cleaned by a blower unit and collected at the outlet (Fig. 11).

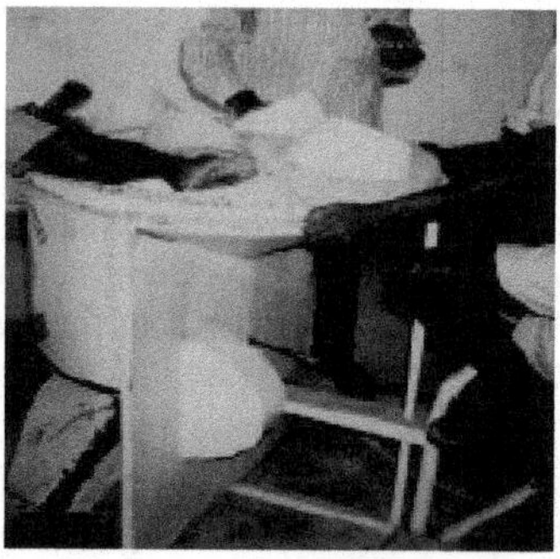

Fig. 11. Hold on Type Thresher

Wire mesh type Sunflower thresher (2009) (10)
It consists of a square frame having each side of 600 mm and a height of 200 mm and is made up of 30 x 30 x 3 mm MS angle. Wire mesh with mesh size of 10 x 10 mm is welded with the main frame and serves as the threshing surface (Fig.12). The workers rub the cubs over the wire mesh and the seeds are dropped at the ground, which is collected later and cleaned manually.

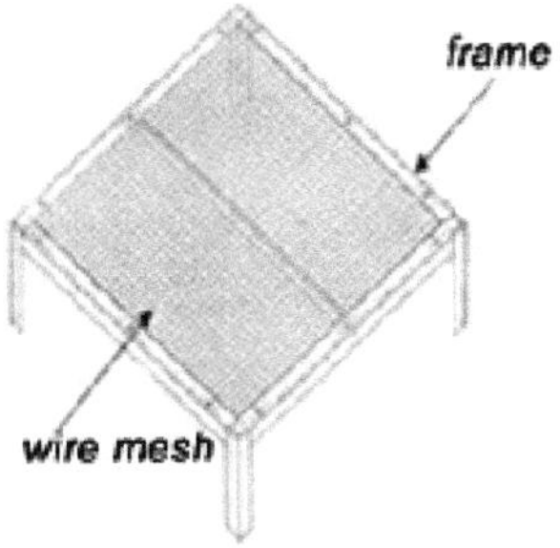

Fig. 12. Wire mesh type thresher

Perforated GI sheet type Sunflower thresher (2009) (10)
It is similar to the wire mesh type thresher but here except the wire mesh threshing surface, perforated GI sheet with perforation of 10 mm dia. was used as threshing surface. Four treatments namely pedal operated sunflower thresher (T1), wire mesh type thresher (T2), perforated GI sheet type thresher (T3) and conventional manual threshing (T4) were included in the experiment and each treatment was replicated five times. The data were analyzed in accordance with single factor ANOVA.

A Pedal Driven Threshing Machine (2010) (11)
The model an United State Patent Application describes a pedal driven machine and methods for processing grain using the physical exertions of a single individual feeds a panicle of grain in to a thresher, drives the thresher to thresh grain, winnows the threshed grain with a winnower, and collect the threshed grain, all concurrently. The panicle may be withdrawn from the thresher after being threshed without the panicle having passed entirely through the thresher. The machine that enables the foregoing processing by a single individual has a pedaling mechanism for driving both the thresher and winnower via a single drive member. (See Fig. 13)

Manual Cowpea Thresher (2010) (12)
Design of a manual cowpea thresher has been carried out based on outcome of threshing cowpea with a conventional thresher the alvan blanch (Maunde, 2008). (Maunde et al., 2005b, 2006, 2007). The following parameters of the thresher were determined. Principal drum parameters which include drum length, drum diameter (for primary and secondary threshing), crop feed rate, drum speed and power required to rotate the drum (Bossoi et al.,1990) and (Caruthers and Rodriquez, 1992). Power transmissions for manual thresher was designed based on (Rasnikov, 1991) and (Aiyeleni, 1993). System parameters which include shaft diameter, key type, bearing

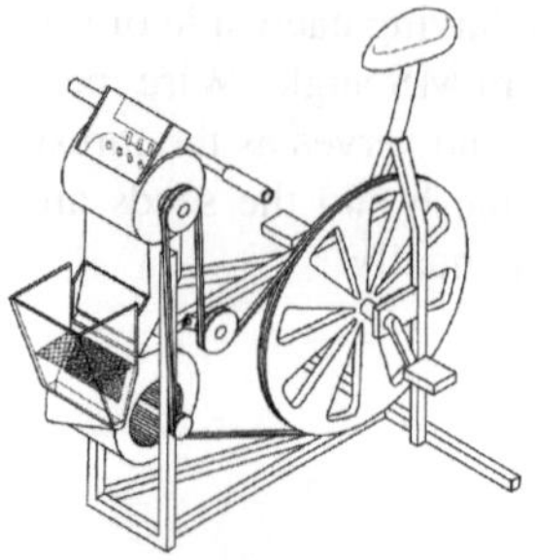

Fig. 13. Pedal Operated Thresher

type, width, inner and outer diameter were determined based on calculations (Maunde et al., 2008). Structural frame included the hopper design, shape, length, bottom and top width, volume and angle of inclination were designed according to (Maunde et al., 2006) and (Maunde et al., 2007). The stress on the components parts and assembly of the thresher was determined using auto desk inventor version 10 software at Cranfield University at Silsoe, UK. The software works with the principles of input data of some of the calculated parameters while some of the unknown data was automatically determined by the software. (See Fig. 14)

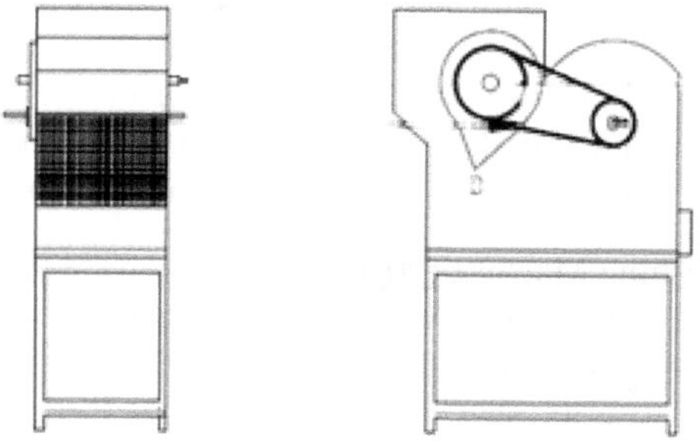

Fig. 14. Manual Cowpea Thresher

Pedal - Operated Paddy Thresher (As Per BIS Standard) (2010) (13)

Sealed offers were invited by 'The Madhya Pradesh State Agro Industries Development Corporation Limited Bhopal, India' with the Specification of Pedal Operated Paddy thresher according to BIS Standard IS 3327:1982 with latest amendments with a brief description such as.

1. Material to be used for various parts with applicable standard.
2. Constructional requirements included Body frame, Base, Side frames, front grain shield, rear grain shield, cylinder, slat, cylinder end discs, threshing teeth, drive, gear housing, crank, pedal frame fulcrum, pedal frame, Pedal Board, Axles.

3. Other Requirements included clearance between the frame and threshing teeth, lubrication facility and safety arrangement.
4. Dimensions included with the overall height of thresher, height between the ground level and centre of cylinder, height of pedal board from ground level, and the distance between the centre of pedal board and the pedal fulcrum.
5. Workmanship and finish requirements included with the painting of metallic parts with anticorrosive rust preventive paint.

Pedal Operated paddy Thresher (2011) (14)
Author stated under the study of mechanization of post-harvest operations that the pedal operated paddy threshers reduces the drudgery over traditional threshing and have became popular in India.

Foot Powered Thresher (15)
It consisted of the plates to hold the drum firmly to the shaft (Fig.14). The shaft is in two sleeves, with a connector for ease of assembly. Inside the thresher at upper right can be seen the adjustable plate for the large sprocket shaft assembly. On top of the thresher is shown the tool for inserting wire hooks in to the drum. Undersize holes are drilled, and preformed loops are driven into holes with the tool and hammer.

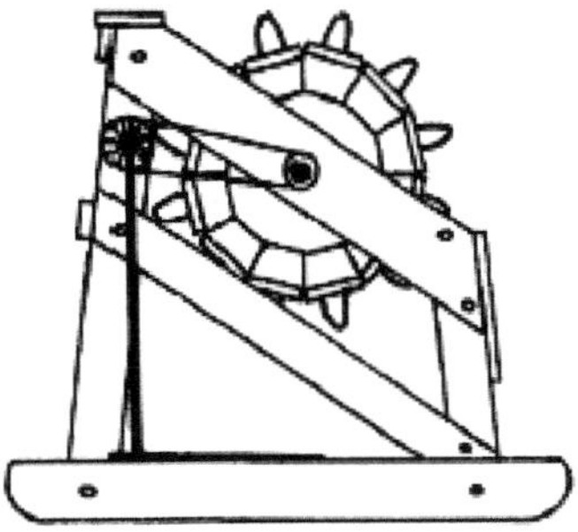

Fig. 15. Foot Operated Thresher

Pedal Operated Thresher (16) (M M Pandey)
It consists of wire-loop type threshing cylinder operated by foot pedal. It is suitable for threshing rice. This machine is based on IRRI design and was adopted by IIT Kharagpur during 1984-90. It costs Rs 2100/- and its cost of operation is Rs 28/q. It was extensively evaluated and adopted by the farmers of West Bengal State for threshing rice. The output capacity, threshing efficiency and labour requirement were 44 kg/h, 98.8% and 5.0 man-h/q, respectively.

Phule Sunflower Thresher (16) (M M Pandey)
It is a pedal operated hold on type sunflower thresher suitable for separating seeds from sunflower heads. It consists of a threshing wheel having spokes and a blower fan operated by foot pedal through chain and sprocket. Four holes are provided on the cover for holding Sunflower heads to the threshing wheel by upto four persons. This implement was developed at MPKV, Pune, during 1987-90. It costs Rs 5700/- and its

cost of operation is Rs 0.40/kg compared to Rs 0.60/kg with conventional method. The output capacity, threshing efficiency and labour requirement were 40 kg/h, 100% and 10man-h/q, respectively.

Treadle Peg-Drum Threshers (16) (M M Pandey)
Treadle operated drum threshers are widely available and are used primarily for rice but also for other crops. The threshing drum is covered in pegs, spikes or wire loops and is supported, in bearings and on a horizontal axis, by a free-standing framework. Sheaves or bundles of the crop are held against the pegs as the drum is rotated by a foot-operated treadle mechanism. Output capacity varies 70-100 kg/h.

Manual Thresher (17) (NPS Sirohi)
Under the 'Long Term Strategies and Program for Mechanization of Agriculture in Agro Climatic Zones in India' the paper favors introduction of manual threshers with motorized thresher on large scale to enhance the quality and output of work.

Treadle Powered Machine (23)
The most basic thresher is a treadle powered machine FT 370(Fig. 15). Threshing drum turns when operator uses one foot to pedal the machine just as early sewing machines were powered. Bundle of the crop is held with grain heads over threshing drum while teeth on drum strip grain off and drop onto ground behind machine. Technical Specification is given as

Dimensions 31.5" x 21.6" x 34"
Net Weight 60#
Drum 14"diameter x 14.5" widths
Capacity 4 bushels / hour
Shipping 25" x 22" x 26"
Gr. Wt. 72#
Cube 8.28 cu ft.

Fig. 16. Treadle Powered Machine

4 Conclusion

In this paper we have concentrated on what we see as the core developments in the early history of machine and mechanism. The emphasis has been very much on human powered threshing machines development. In this paper we dealt primarily with threshing as an area of knowledge and we have not devoted much attention to the institutional level, which concerns textbook. From the literature review it is found that the human powered threshing machines are available with following mechanisms.

1.Hand Operated Threshers are found with crank and chain mechanism.
2.Foot Powered Threshers are found with crank and chain with bevel gear and some with treadle mechanism.
3.Pedal Operated Threshers are found with bicycle mechanism.

What all investigator have investigated is onload condition but we have come up with offload condition. The evolved machine system comprises of three subsystems namely (i) Energy Unit , Comprising of a suitable peddling mechanism, speed rise gear pair and Flywheel conceptualized as Human Powered Flywheel Motor (HPFM) (ii) Suitable torsionally flexible clutch and torque amplification gear pair and (iii) a process unit. The suggested machine system uses human energy achieved by peddling and stores this energy in a flywheel at an energy-input rate convenient to the peddler. After storing the maximum possible energy in the flywheel (peddling time could be 1-2 minutes) the same can be made available for the actuation of any process unit by making available the energy stored in the flywheel through a suitable clutch and torque-amplification if needed. Thus the flywheel will decelerate depending on the actual resisting torque offered by the process. It implies that the peddler does not pedal while the flywheel is supplying energy to the process-unit.

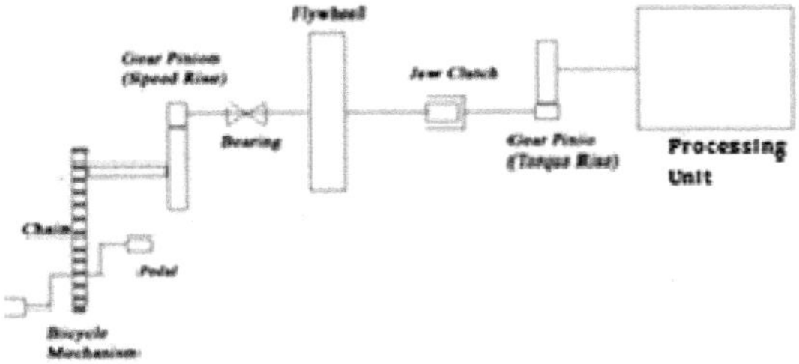

Fig. 17. Human Powered Flywheel Motor Concept

References

[1] Hopfen, H.J.: Farm Implements for Arid and Tropical Regions, Food and Agriculture Organization of the United Nations, Via delle Terme di Caracalla 00100 Rome. Published in AT Microfiche Reference Library, Italy (1969)

[2] Weightman, D.: Design for a Pedal Driven Unit for Transport and Machine Uses in Developing Countries. Lanchester Polytechnic, Industrial Design (1976)

[3] Herdt, R.V.: Mechanization of Rice Production in Developing Asian Countries: Perspective, Evidence and Issues, Consequences of small Rice Form Mechanization Project, Working Project No. 15, pp. 1–35. International Rice Research Institute, Los Banos (1981)

[4] Meisner, C.A., Justice, S.: Chinese Hand Tractor (CHT) Seeders in Bangladesh: An Update on progress on promotion, manufacture and utilization. In: CIMMYT Bangladesh and CIMMYT Nepal, May 28-June 4, p. 8 (1999)

[5] Pariyar, M.P., Shrestha, K.B., Dhakal, N.H.: Baseline Study on Agricultural Mechanization Needs in Nepal, Rice-Wheat Consotorium, Paper Series 13, Rice-Wheat Consotorium for the Indo-Gangetic Plains, New Delhi, India (2001)

[6] Nag, P.K., Nag, A.: Drudgery, Accidents and Injuries in Indian Agriculture. Industrial Health 42, 149–162 (2004)

[7] Nimbkar, N.: Annual Research Report (2006-2007), Unpublished, Nimbkar Agricultural. Research Institute (NARI), Phaltan, India. pp 1–29 (2007)

[8] SAARC Agriculture Centre. SAARC Journal of Agriculture 5(1) (June 2007)

[9] Singh, K.P., Pardeshi, I.L., Kumar, M., Srinivas, K., Srivastava, A.K.: Optimisation of machine paramemters of pedal-operated paddy thresher using RSM. Science Direct, Biosystem Engineering 100, 591–600 (2008)

[10] Goel, A.K., Behera, D., Swain, S., Behera, B.K.: Performance Evaluation of a Low Cost Manual Sunflower Thresher. Indian J. Agric. Res. 43(1), 37–41 (2009)

[11] Moser, et al.: Pedal Operated Threshing Machine, United States, Patent Application Publication, Pub. No. US 2010/0298037 AL, Sheet 1-8, pp 1-4 (November 2010)

[12] Maunde, F.A., Yalmas, S.M., Richard, T.E.: Design of Manual Cowpea Thresher. African Journal of Agricultural Research 5(12), 1419–1423 (2010)

[13] The Madhya Pradesh State Agro Industries Development Corporation Ltd., Bhopal, India: Open Ended Rate Contract Offer (RCO) for supply of ISI Marked Power Threshers (IS:9020:2002), Pedal Operated Paddy Threshers & Rotavator (2010)

[14] Manchikanti, P., Sengupta, M.: Agricultural Machinery in India: IPR Perspective. Journal of Intelectual Property Rights 16, 163–169 (2011)

[15] ITDG Agricultural Equipment and Tools No. 37: Foot Powered Thresher, Intermediate Technology Publications, Ltd., 9 King Street, London WC2E, 8 HN, UK, AT Microfiche Reference Library

[16] Pandey, M.M.: Present status and Future Requirement of Farm Equipment for Crop Production, pp. 69–113. Made Avail by Central Institute of Agricultural Engineering, Bhopal (Unpublished)

[17] Sirohi, N.P.S.: Long Term Strategies and Programme for Mechanization of Agriculture in Agro Climatic Zone-III: Plains Region. Indian Agricultural Research Institute, New Delhi, (Unpublished)

[18] Pathnopas, R.: The Economics of the Rice Threshing Machines in Thailand: A Case Study of Chachoengsao and Supanburi Provinces, Thesis, Thammasat University, Bangkok, Thailand (May 1980) (Unpublished)

[19] Lee, C.C.: Economic and Engineering Aspects of Mechanization of Rice Harvesting in Korea. In: Southworth (ed.) Farm Mechanization in East Asia, Agricultural Development Council, New York (1972)

[20] Resource Manual on Integrated Production and Pest Management (IPPM) in Rice: World Education (INGO) Philippines, Inc., 2999 National Road, Anos, Los Baños, Laguna, Philippines (2005), `http://www.worlded.com.ph`

[21] Simonyan, K.J., Yilijep, Y.D.: Investigating Grain Separation and Cleaning Efficiency Distribution of Conventional Stationary Rasp-Bar Sorghum Thresher. Agricultural Engineering International; the CIGR Ejournal Manuscript PM 07 028 X (August 2008)
[22] Caruthers, I.: Tools for Agriculture: A buyers guide to appropriate equipment, 3rd edn. IT publication in association with GTZ/GATE, London (1985)
[23] A treadle powered machine FT 370, Gridley, CA 95948, `http://www.ferrari-tractors.com`

Section VI

Concepts and Theories

At the Origins of the Concepts of Máthema and Mekhané: Aristotle's *Mekhanikà* and Archimedes' *Tropos Mekhanikòs*

Giuseppe Boscarino

96010 Sortino(SR, Italy)
gpp.bos@libero.it

Abstract. For Newton, modern science began its trend when it is connected to that tradition of thought of the Greek world, which had had its heart in the close rapport of **mechanics, mathematics and philosophy**, leaving that other tradition of thought of the Greek world, which had had his heart in the philosophical and epistemological concepts of **substantial form** and **occult quality**.

We think it is true that both the mathematic thing, the **máthema**, and **its model**, the **mekhané**, have been blocked from using the epistemological concept of **substantial form and occult quality of Platonic-Aristotelian tradition.**

In our study we want to investigate the **internal aspects**, the onto-epistemological ones, in our opinion little investigated by traditional historiography, attentive to external aspects of the concept. For this we compare two patterns of the two traditions: Aristotle's *Mekhanikà*, in our opinion, instead, an eclectic written of Platonic-Aristotelian tradition, and Archimedes' *tropos mekhanikòs* , in which the concepts of **máthema** and of **mekhané** with a clot of onto-epistemological concepts, developed from a tradition of thought with a strong philosophical, cognitive, not merely instrumental and utilitarian mark, come to sublime synthesis.

Only inside a different conception of "**theory**", of "**máthema**" and of "**mekhané**" and only inside a different conception of **philosophy** that Pappus, in the passage cited of Newton, called "**physiology of material elements**", could develop what has been then the modern and fruitful conception of máthema and mekhanè.

1 The Problem of the Stagnation of Concepts of Máthema and of Mekhané in the So-Called Greek Science

Reading philosophers, scientists and historians of culture of different periods, a common opinion strikes about a key reason about the so-called stagnation that would hit the Greek science from the fourth century B.C., it is written, due to **the continuing domination of the concept of substantial form and quality of Platonic-Aristotelian tradition.**

T. Koetsier, M. Ceccarelli (Eds.): Explorations in the His. of Machines & Mech., HMMS 15, pp. 449–461.
springerlink.com

At the same time in which this tradition of thought is accused as responsible for the stagnation, another one instead is indicated in the ancient Greek world, that it is considered to have to take, but which historically has been a loser and suffocated.

This reflection is manifested especially in the early philosophers and scientists of the modern world. Bacon wrote that **Plato and Aristotle have had a lot of damage in the field of natural philosophy with their search for abstract forms and final causes, of metaphysical and theological nature, not understanding that the efficient cause is only the vehicle's form**. And even in opposition to this tradition of thought of abstract forms and final causes to natural philosophy of Pythagoras, Philolaus, Parmenides and Democritus was given more attention. (See Francis Bacon, *Of the Dignity and Advancement of Learning*, Book III)

For Newton, modern science began its trend when it connected to a tradition of thought of the Greek world, which had had his heart in the close rapport of **mechanics**, **mathematics** and **philosophy**, leaving that other tradition of thought of the Greek world, which had had his heart in the philosophical and epistemological concepts of substantial form and occult quality. So he can write early in his first preface to his 1686 masterpiece, *Philosophiae Naturalis Principia Mathematics : Since the ancients (as we are told by Pappas), made great account of* ***the science of mechanics*** *in the investigation of natural things; and the moderns, lying aside* ***substantial forms and occult qualities****, have endeavoured to subject the phaenomena of nature to the laws of mathematics, I have in this treatise cultivated* ***mathematics*** *so far as it regards* ***philosophy****.*

The criticism of the domain of the form on the side of mathematical research as a reason for stagnation in the Greek world is then emphasized by the expert Greek historian of mathematics Boyer, adding, however, at the first judgement, in our opinion, a second, however, prejudice, common in the historiography of mathematics of the twentieth century.

He writes that the calculus has not been able to develop at the beginning of the Greek world, for an **epistemological block**, as dominated by the concept of **form** (" *Greek geometry was concerned with form rather than witch variation* "), being neglected the *concepts of variation, of real number, of limit of an infinite series, of infinity.* (Boyer,1939).

Prisoner of the prejudice , as other historians, he writes again that the calculus has not been able to develop as it has been degraded by its presumed rational purity sensed by Platonism, to thing impregnated with intuitive, sensitive and mechanical elements, and turned away from its rigor for the spurious use of concepts of **indivisible and infinitesimal commingled with mechanical something.** (Boyer, 1939).

Now we think it is true that both the mathematic thing, **the máthema**, and its model ,**the mekhané,** have been blocked by using the epistemological concept **of substantial form and occult quality of Platonic-Aristotelian tradition**, when mechanics and its product, the machine, have been degraded by mathematical thing, to purely empirical knowledge, to simply imitative construction, that it deserves, this means, the contempt of Archimedes.

In our study we investigate the **internal aspects**, that onto-epistemological one , in our opinion, a little investigated by the traditional historiography, attentive to those outside, of the concept of mekhanè and mingled with it the concept of máthema, the result of such a tradition of thought , which had its beginnings in Pythagoras and Parmenides, Eudoxus, and its development in Architas, and therefore in Democritus, its cultural climate in Sophistic, its culmination Archimedes' **tropos mekhanikòs**, and continued by Hero and Pappus, recovered in modern age, as has been mentioned by Newton, and at first by Galileo, the fathers of modern science.

2 Traditions of Thought in Greek Science and End of a Presumed Unity of the So-Called Greek Science

Meanwhile, for us, it is wrong and misleading to speak in **a general sense** of blockage or stagnation of the concepts of calculus, machine and of its use in the ancient Greek world, without instead first making a due distinction between traditions of thought(See G.Boscarino,1999) with onto-epistemologies, in deep contrast and competition between them, but not without eclectic attempts of mediation and confusion-misrepresentation in the course of history of the Greek world from VI-V century BC until the fourth century AC., to exhausting oneself of the so-called Hellenistic period.

Therefore Koyré and Vernant do not convince us, to quote a few historians of science and of Greek culture, when in a vague way they talk of Greek science and Greek mechanism, putting them under the mantle of Plato and Aristotle, when in fact these, too their own say, have distorted and blocked them with their onto-epistemologies.

Therefore the following expressions of Koyré and Vernant, which we quote, do not seem to us to be correct and shareable by the historical point of view.

"*The Greek science has not constituted a real technology, because it has not developed a physics. This is because it did not believe that it was possible.* "(A. Koyré,1948). "*The Greek science focuses on immutable essences, does not quantify the becoming, does not establish connections between mathematics and physics. Greek science is pure logic and deductive theory.*" (ibid.)

Vernant then states categorically: *The machine does not yet appear as the universal model of the physical structures (!?).* (J. P. Vernant, 1957).

While some of the following expressions, which we quote, seem to us really shareable inside our research and our thoughts.

"*For Greek science, the daily reality is not mathematicizable. It is the domain of motivation, of inaccurate, of <more or less>,of <about>.* "(A. Koyré,1948). "*Plato, doing of mathematics, or geometry, a pure science, perfect, highest than sensible world, and Aristotle a <abtract> science, too it, far from sensible world , have erected a gap between mathematics and physical reality. The result was been that if we want to apply mathematics to the study of nature make a mistake and a nonsense* "(Koyré, ibid.)

For Koyré then: The Greek science did not recognize the value of the measurement and calculation, or the world of precision in the physical field, because thought impossible; it has not changed the empirics into episteme, the tekne into technology, the utensil into machine. Believing rightly insufficient the sociological or external theory he finally demands: *'Why the inventors of episteme (Plato and Aristotle!)not applied it to the Greek praksis, why science has not developed a technology of which it had developed the idea* ?> (Koyré, ibid.)

Quite fitting then, insightful and indicative of a track of research, which is ours, Vernant's reflections seem to us about the role played by the concept of Platonic form or idea (1) in the block not only of the machine concept and of the development of possible Greek scientific technology in the ancient world, but also, in our opinion, the same concept of máthema, if we want to remain on the recall of Boyer.

We ask ourselves at this point: But are Plato and Aristotle truly the inventors of the episteme or scientific practice with its ontology and its conception of máthema and of mekhané, as will also be practiced by modern scientists, such as Galileo and Newton or instead we must look elsewhere for a different scientific practice, in a different tradition of thought, that ancient Greek thought develops, but, according to the previously mentioned Bacon, will not be pursued and even stifled?

With our analytical work then we want to dwell on the internal aspects, or onto-epistemological one, which blocked the growth process of the concepts of máthema and mekhané, in the ancient Greek world, not underestimating the external ones, highlighted by historians.

Surely the existence of an abundant and cheap slave labor, the lack of a mercantile economy and exchange value, the exaltation of the values of contemplation, of liberal and idle life, the contempt for manual work, etc.. etc. determine the stagnation of a possible evolution from tekne to technology, or to a systematic use of applied science, but not less important, in my opinion, the onto-epistemological reasons, that we want to highlight in our study, because with regard to it there is a lot of philosophical and epistemological confusion, spread strong misunderstandings, if not actual historiographical misrepresentations

3 The Meanings of Máthema and Mekhané in the Traditions of Thought and Archimedes' *Tropos Mekhanikòs*

Let's see first how the concept of máthema evolved and how was structured within the tradition of Platonic-Aristotelian thought, broadly.

If Plato and Aristotle, as rightly pointed out by the previously mentioned Koyré, find impossible to apply mathematics to physics, that we can build a mathematical theory of natural things, it follows that they claim about the epistemic battle, or a **mathesis**, which keeps separate physical quantities and numbers, which deals only with abstractions, that cannot be applied to the natural world, not being able to innovate it, but only to imitate it (Aristotle), or a **mathesis**, which makes it the same philosophy or almost, that true knowledge tout court, but not vehicle to nature knowledge.

This second choice occurs, as can be seen by Proclus, with Plato and Neoplatonism. Proclus writes: ***Mathematics** deals with intelligible objects. ... It seems that this name cannot have been made by chance, as it is the case for most of the names, but, just as they say, comes from the Pythagoreans, who well knew that the* ***whole mathesis, so called, is a reminiscence inherent in souls**. Plato says ... If, in fact, we give attention to geometric figures, just proves, in the most manifest way , that* ***mathesis is reminiscence***. (Proclus, *Comment on the first book of Euclid's Elements, prologue,*)

With this episteme, the concept of knowledge disappears as excavation within the natural things, to assert that one of knowledge, as interior excavation, as reminiscence, or **mathesis = anamnesis**, and, again, the concept of number as vehicle to knowledge, as a means of calculation, ratio of physical quantities, which the Pythagoreans called **mathematical number**, disappears, but to affirm the Platonic one, called **ideal number**, considered the true number, which exists in a world apart, the supercelestial, as Aristotle was witnessed in this regard. (See Aristotele. *Metaphysics,* M). It is clear that within this rigid separation of **mathesis and physis**, **magnitude and number**, could not be possible to meet another concept, that of mekhané, if it is true what Koyré says, that, without the concept of calculation, precise measurement, could not develop the concept of the machine.

But there is another tradition of thought, which begins with Pythagoras, and which makes the Pythagorean Philolaus of Croton, V sec. a. C., the first clear expression of **another way of understanding the episteme and with it the mathesis** : "*All things that are known have number (**aritmòn**): without a number would not be possible to think or know anything*" (*Presocratics*, 44, B, 4).

Here in this episteme: there are things, they have a number, or a ratio that identifies as such; if we does not know this ratio, neither we think nor know.

Within this onto- episteme for Pythagorean Archytas it is natural to apply the principles of mathematics to mechanical things if he wants to think and to know them properly.

It is witnessed in fact : "*For first he discussed with method mechanical things (ta mekhanikà) using principles of mathematics*" (D. Laertius, *Lives of Eminent Philosophers* ,Archytas of Tarentum).

But the true knowledge cannot be taken from sensible things, which are and are not, but from the **theory**, another term coined by the Pythagoreans, that is, by what observed by reason through its entities, of which only we can say and think, if we respect the prohibition that the absurd, the contradictory cannot be said and thought, because it cannot exist. This is Parmenides' principle, a new pillar to build the theory. **The world of episteme has to be a rationally constructed world without contradictions.**

We pass from Pythagoras-Philolaus, through Archytas, to Parmenides and then to Democritus, on the plane of ontology of episteme theory, however, after the philosophical sophistic storm. This upsets the cultural limits within which **the philosophical reaction represented by Plato's philosophy** wants to keep closed the tekne, and with it the concept of machine, a symbol of power and wit. Tekne is not just imitation of nature, as argued by Plato and Aristotle, then repeats, but also **innovative power, first of all in the philosophical and political field.**

Against some wicked philosophers, who say that **the gods are the result of men's tekne**, that nature and chance make natural things, Plato critics what these say, or that "*politics has little contact with nature, that tekne prevails in it; so even all legislation is not by nature, but for tekne, and the laws it lays down are not true*>. (Plato, *Laws*, X, IV)

Within this culture of progress and technology as an innovation, especially political, Democritus can make the advancement of Pythagorean-Parmenidean episteme and of the episteme of mechanics and mathematics relation. He deals, according to testimony, with the mathematical problems and technical problems. In fact he was the first to apply the name <*problems*> to technical research, according to testimony. (D. Laertius, ib., *Democritus*)

Principles of things are for Democritus the ideas.

According to a precise testimony, Democritus believed that "*The principles of entities are indivisible bodies, but as* ***theoretical elements*** *(logoi theorethà).* (See S. Luria, *Democritus*, 197)

Here for Democritus the indivisible body, or element, it is not such for a presumed metaphysical, absolute nature, but for the report that it maintains within a theory, within which is defined and finds its only meaning.

We are well within a conception of element, not of metaphysical nature as it will in Aristotle (Aristotle, *Metaphysics,* V, 3), **but epistemic.**

The elements or atoms can be collected by only two ideas or principles, **the full** and **the vacuum**.

The elements are of infinite number, may be successively, may be put in biunique correspondence, may then be compared, measured and added together, giving them a finite number.

On the other side with Eudoxus, taken by Euclid, is built the **mathesis** of entities bodies, which are magnitudes and numbers. By this we compare commensurable magnitudes and incommensurable magnitudes, and assign them numbers. Peano finds already in the definition V of Book V of Euclid, translating from Greek into Interlingua the definition by abstraction of a real number, even if he tends to point out that Euclid's always talks about number as ratio between magnitudes and never of the existence of mere numbers (Peano, *Opere,III, p.385*). **In short it is constructed a precision mathesis able to link mathematics and physics, magnitudes and numbers, the máthema at the mekhané**. With this onto-epistemology the machine sublimates her role as human capacity powering, as has been handed down through the centuries from the imaginary literary, not only in the material field, but above all in that one of **mathesis, as knowledge of the truth.**

In short we are at **Archimedes' tropos mekhanikòs** , in which we find Eudoxus, initiator with Archytas of mechanical episteme, and Democritus, the only philosopher-mathematicians mentioned in all the work of Archimedes, through his famous *Letter to Eratosthenes* (T.L. Heath, 1912) in which Archimedes is addressed to Eratosthenes, as a zealous teacher of philosophy, which then can appreciate in mathematical things the **theory** by which we discover and know **the truth and**

principles that are used in it. These principles are the full and the vacuum. **There are no substantial metaphysical forms or abstract qualities, they are also geometric one,** but geometric figures, imagined vacuum, of spheres, spheroids, conoids, which, filled with infinite solid elements, or by **circles-sections-weight** could be placed in biunique correspondence, that of one with those of another, can be accumulated and compared through mechanical laws, until we find the ratio, the exact number. Thanks to this and the theory, which expresses it, manifests himself the true knowledge, **the máthema through the mekhané.** Omen of the Pythagorean Philolaus seems to come true.

If we interpret this as a Pythagorean- Democritean-Archimedean onto-epistemology, then we can deny the false statements and correct the partial and forced truths of Koyré, Vernant and Boyer, to name a few prominent historians, quoted by us.

It is reduced the revolutionary nature of Archimedes' tropos mekhanikòs, compared to the traditional Greek concept of machine, as a way to overcome material difficulties not easily overcome, through the use of **metis** (cunning, sagacity) when it is claimed, for example, according to expert Micheli, that it merely extends the meaning, even to trick of calculation, measurement, simple means of mathematical discovery.

The **tropos mekhanikòs,** as theoretical complex structure of magnitudes and numbers, as onto-epistemological principles, not only uses machines, such as lever, but is itself **a machine**, as it writes Archimedes "**gives the power to observe in mathematical entities through mechanical things** "(*eis tò dunasthai theorein en tois mathemasi dià ton mekhanikòn*).(Heath, 1912, p.3).

Mekhanè has taken on many meanings for the Greeks, but in its most profound and constant meaning over time was considered synonymous with power (**dynamis**) as well as **wonder**.

Mekhanè is no longer just for Archimedean then a material mean of power, by its laws, but **a theoretical framework, which , as it gives to us the power observing the truth, shows us the way, the verse, the *tropos*, on which we have to walk to search for true**. On the dialectic of true and false in Archimedes see Boscarino. (*The onto-epistemological background of Archimedes' máthema* International Conference – Bergamo, 2010).

Therefore, we can certainly say that with the tropos mekhanikos Archimedes subverts Patonic and Neo-Platonic philosophy, like that of Aristotle. The **tropos mekhanikòs,** as mekhanè is the true màthema, **the model, the paradigm, the exemplar** of knowledge; is this in fact that brings into being, to manifest the true, the mathesis, purely geometric, reproduces, **imitates**, in his showing, what **to its has been indicated by mekhané**. Plato mimics this philosophy, which matures later with Archimedes, in his *Timaeus*, when he imagines his Demiurge **mekhanasthai**, or that he imposes the forms, the ideas, which he contemplates, to formless matter. (Bourgne R., *Mechane-Mekhanasthai chez Platon* , 1986).

Perhaps that explains why Pappus at the beginning of Book VIII of his *Collection* tells us that the mechanical theory has always enjoyed the great confidence of philosophers and has been sought after by all the mathematicians, because it deals with **the physiology of the material elements in the cosmos** (Pappus d'Alexandrie, *La collection mathematique*, 1982).

To Platonic and Aristotelian theory of substantial forms the mechanical theory opposed the physiology of mechanical elements, i.e. an **epistemic theory**, organized on physical magnitudes and their ratios, the numbers, but supported by **physical elements**, which we called with the Archimedes' tropos mekhanikòs.

In the next section, in particular, we will compare two onto-epistemologies with two different concepts of máthema and of mekhané, about the mechanical thing, called lever, in the two traditions of thought, represented by Aristotle and Archimedes, or by Aristotle's *Mekhanikà* and by Archimedes' *Stoikheia ton mekhanikòn.*(Now called *On the Equilibrium of Planes*, I, II)

4 The Lever: A Physiology of Elements (Archimedes) or a Representation of Forms or Qualities (Plato and Aristotle)?

It is in the following passage of Hero that we want to bring from his work *Les mécaniques,* were it is highlighted in a palpable way the contrast between the two onto-epistemologies with forcing and mystifying interpretations attached to the eclectic way in which they overlap, that we here write in formulas, but in a note (2) reported in its French version.

Let be
BA, AE = distances from the center A and B, E extremes
T, Z = weights and
T> Z, then if
1) BA : AE = T : Z, for the law of Archimedes, we get the equilibrium.
If we want to raise T, it is enough to put
BA> AE, and
BA : AE > T : Z

or the ratio between the distances from the center A greater than the ratio between the "weight" and the "power", applied at the extremes B and E.

This means that we get the lifting of T by multiplying weights for distance, according to the quoted relation, if we separate the magnitudes in their unitary magnitudes and their numbers, which multiply them,

$$(BA.\ X.\ Z) > (AE.\ X.\ T), \text{ or } n.\ x.\ m(BA_u\ .X.\ Z_u = M_u) > l.\ x.\ t\ (AE_u.\ X.\ T_u = M_u'),$$

were n, m, l, t, are numbers, x is an operation between numbers, X is an operation between magnitudes, BA_u, AE_u, Z_u, T_u are the unitary magnitudes, M_u, M_u' are the unitary moments.

Thanks to more number the moment M can determine the lifting of the weight T, or **it can change the form of the machine**, that is its state of equilibrium.

How one interprets this passage. Since Aristotle in *Mekhanikà* to explain the wondrous power of the machines makes use of a geometrical entity, the circle, as their principle or cause, is quick to point out that "*Aristotle is, in natural philosophy, the master of Mécaniques' author*", while it is denounced the ingratitude of the author of the passage quoted, Hero, in the fact that Aristotle is not invoked as a duty. (Héron d'Alexandrie,1988, p 42.)

In reality, in the quoted passage, as in other parts of his work, the only one mentioned by Hero is Archimedes, so he should be acknowledged master of *Les mécaniques'* author!

Not only so it is made a forcing, but also a mystifying interpretation. It was pointed out by critics, as indeed you can see by reading the passage, that the circle in Hero' exposition is alien to the context, superficial, redundant, whereas, in our opinion, it is instead visible another onto-epistemology in it, which is headed right to Archimedes, which is what we called the **physiology of the elements**, following the language of Pappus about **mechanical theory**.

From testimonies of the same Archimedes it results that he called his work of mechanical theory: *Elements of mechanical things*. That does not seem random or insignificant but very likely to all our talk.

In fact if we compare the two theories of the lever of the two authors, Aristotle and Archimedes, we see as in the case of Aristotle become established as an explanatory cause of the power of the lever, an abstract geometric form, such as a circle, with its amazing qualities, for us really contradictory and absurd, expressed in an obscure and ambiguous language of natural philosophy, about what is **by nature** and what is **against nature**, where it is not clear if the machine imitates or deceives nature, if one speaks about geometric entities, which refer to a specific theory of magnitudes, or if one is simply parroting the language of the magnitudes and their ratios, when in the same text then one does reference mainly to empirical objects.

Not for nothing Maurolico with regard to this rightly opposed the precise geometric language of Archimedes to Aristotle's empiricism with its ingenuous **sensibilibus experimentis.** (See Gatto R, Galileo Galilei, *Le mecaniche*, 2002, pp. XXXIX-XL.)

Let us see briefly how the language is developed by Aristotle, then putting it in comparison with that Archimedean one of *Elements of mechanical things*, and then to demonstrate how in the concrete of the quoted passage there can be no presence of Aristotelian philosophy right starting from its philosophy of number and its concept of element, but rather just a different philosophy of science, which will expressed by Galilean science in the wake of the Archimedean one.

For Aristotle mekhané is a subclass of tekne, or that particular tekne which assists in the difficulties that nature stands in what is useful to men. With the lever, men overcome a particular type of difficulty, that one to lift big weights with small forces. It is due to the nature or quality of the circle this power to dominate the forces of nature, inspiring **wonder**. Only a more wonderful thing can be cause of a wonderful thing. And the first of all the wonders is the circle, as a synthesis of opposites.

If it is arbitrary arguing that a thing only because it would be more wonderful than another is cause of another, it is certainly ambiguous and absurd, then arguing that the opposites are the cause of anything, unless you specify what is meant by synthesis or interpenetration of opposites. For Aristotle, if balance issues can be traced to the qualities of the circle, and that ones of the lever to proprieties of the balance, all mechanic issues can be traced to the mechanical properties of the lever.

Now what is the wonderful quality of the circle that explains the balance and the lever?

This resides in the fact that the point of the diameter of the circle that is farther from the center of the circle moves faster than the point that is less distance, although both moved by the same central force. Aristotle finds the cause of this apparent anomaly in the fact that the points, lying at different distances, are actually subject to two different dynamics, being subject to forces of different natures, against nature and by nature, that determine different geometric configurations, or straight line, when their ratio is constant, or a curved line, when their ratio varies. The farther is the point from the center the lesser suffers the force against nature toward the center, becoming faster in its movement, instead remaining constant the motion by nature of the two points in a straight line, at different distances from the center.

This explains why two equal weights placed at equal distances, are in equilibrium in the balance, having the same speed, but placed at unequal distances are not in equilibrium, and a smaller weight placed at a larger distance balances instead a greater weight placed at a lesser distance , in the lever.

Aristotle then concludes: *The ratio between the weight moved and what moves it is inverse to the ratio of their distances from the center* (Aristotle, *Mekhanikà)*

The reading of this proposition has led some historians of science to believe that here Aristotle advances, in dynamic terms, what then Archimedes proves in static terms. .(G.Vailati, *Il principio dei lavori virtuali da Aristotele a Erone d'Alessandria,* in Scritti,1911, p.91).

In reality in all discussion of Aristotle there is a strong ambiguity: the forces at work in determining the circular motion, if on the one hand they seem real, so this is a compound, not simple, then a not natural motion, and therefore what of wonderful it produces cannot be its property, on the other hand they seem forces purely apparent, fictitious, because the natural, true motion is still the circular one, to which they are subject. In short, **it is the geometric substantial form circle, with its qualities**, which determines the law, not any **simpler elements** that determine it.

Lacking a precise theory of magnitudes and numbers to which Aristotle refers, it seems forced, therefore, the interpretation of some authors (See S.T. Heath,1948) about an alleged anticipation of the inverse proportionality between weights and distances in Aristotle of Archimedes matrix. There seems plausible reading of Micheli in this respect (G. Micheli, 1995, p. 83).

The work seems, therefore, not so much a work of Aristotle, what an eclectic one, in short, a work of Platonic-Aristotelian tradition. For Aristotle, a point cannot describe a line, neither a line a surface. So the work should not be an Aristotelian one . The principle of opposites is then of Pythagorean origin. The myth of the perfect circle is also of Plato, nor strong influences of Democritus are excluding, as the main interlocutor of Aristotle on natural philosophy, but also on questions of philosophy of tekne, was Democritus(Aristotle, *Physics*, II, 2).

However, that this statement of Aristotle or of Platonic-Aristotelian tradition does not possess any shadow of **máthema** and **mekhané,** that is of geometric and physical magnitudes , of their ratios, and of theoretical elements, but it expresses a generic reference **to simple verbal opposition** of weights and relative distances, this becomes clear if we compare the passage quoted by Hero with the theory of numbers and of element concept of Aristotle.

For Aristotle
The numbers are neither substance (ousia) or causes of the form (aìtioi tes morfes). ... The number is always certain number of things, or of parts of fire or earth or arithmetic units. ... The number what is general and what is composed of units, is neither efficient cause nor matter nor the form concepts of thing. ... The ratios that are found in mixtures are expressed by the addition of numbers and not by mere numbers, for example by 3 +2 and not by 3 × 2. (Aristotle, *Metaphysics, N, 1092b 0.15 to 35).*

This philosophy of the number and of the mixture is absolutely incompatible with the quoted Hero expressed by us in mathematics formulas with reference to the law of Archimedes, where it is clear that

1) **Weights** and **distances as elements** in the law of the lever do not give rise to a mixture, where magnitudes and numbers can only combine and mingle themselves, however remaining different, but give rise to a **new theoretical element, the moment**, separable in a **unitary moment** and in a **new number** resulted from the multiplication of numbers. In this case

2) The new number, along with new theoretical element, is the **cause** that determines **the new form**, the **efficient cause**, which lets the lighter weight lift the heavier one. Therefore the element is not absolute, metaphysical something, but is given by the theory, by episteme, where elements combine giving rise to new elements, **the moment.**

3) Only the **exact ratio** between the physical magnitudes, **the máthema**, can generate the mekhané, that represents it, manifests it, lets it **be observed rationally** (this is the meaning of **theorein** in **Archimedes' tropos mekhanikòs**) in **its essence** or s**ubstance**, if we want to use the language of Aristotle, however, reversed again, where it is the quantity that generates the form, the quality, and not vice versa.

5 Conclusions

Therefore in the Archimedean law worth are **the elements, weight and distance**, and **the quantitative ratios** between them, not the geometric forms, with their amazing qualities, even of divine nature. If this episteme is expressed by Archimedes in his sixth postulate of *The Elements of mechanical things*, the philosophy, in terms of elements and quantities, underlying it is clearly expressed by Galileo in one of his passages of *Le mecaniche* , where he writes : *Stante la medesima gravità dei pesi ed i medesimi termini delle sospensioni,quando bene si variassero le loro figure,riducendole in forme sferiche,conforme alle due X, Z, o in altre, non si dubiterà che il medesimo equilibrio sia per seguire;essendo la figura accidente di qualità ed impotente ad alterare la gravezza,che piuttosto dalla quantità deriva"* (G. Galilei , *Le mecaniche* , p. 52,Olski Editore, 2002). ("*Given the same gravity of weights and the same terms of suspension, when their figures were well varied, reducing them in spherical forms, according to the two X, Z, or in other ones , there will no doubt that* *the same equilibrium is to follow; being the figure accident of quality and powerless to alter the gravity, which instead derives from quantity"*, our translation)

In the work the **moment** enters explicitly as a fundamental element, while ambiguity and confusion of Aristotelian or pseudo-Aristotelian issues are deleted or clarified about the relationship between **technique and nature** (the technique does not deceptive nature, nor merely imitates it, but also innovates it with its wonderful inventions), about the relationship **technique and science** (science teaches the value and the right knowledge and expertise in the use of the machine to *"poco intendenti ingegneri (o volgari mecanici) ,mentre si vogliono applicare a imprese impossibili".* (ib p.48 ,p.52),. (*"little understanding engineers (or vulgar mechanics), when they want to apply to impossible exploits."*, our translation)

Therefore we understand why Archimedes did not despised the mekhanè, taking it away from philosophy and mathesis, but only he despised its vulgar use and its short-sighted interpretation. We demystify in the same time Plutarch's partisan testimony about a presumed contempt of Archimedes of mechanical thing and a mathematic physics.

Acknowledgements. The author is very grateful to Armando Anzaldo for his assistance and advice about the English translation.

References

[1] Aristotele: Mekhanikà
[2] Aristotele: Physics
[3] Aristotele: Metaphysics
[4] Bacon, F.: Of the Dignity and Advancement of Learning, Book III
[5] Boyer, C.B.: The concepts of calculus, New York (1939)
[6] Bourgne, R.: Mechane- Mekhanasthai chez Platon (1986)
[7] Boscarino, G.: The onto-epistemological background of Archimedes' mathema. In: International Conference, Bergamo (2010)
[8] Boscarino, G.: Tradizioni di pensiero. La tradizione filosofica italica della scienza e della realtà. Mondotre-La scuola italica, Sortino (1999)
[9] Gatto, R., Galilei, G.: Le mecaniche. Olski Editore (2002)
[10] Gianni, M.: Alle origini del concetto di macchina. Olski editore (1995)
[11] Heath, T.L.: The works of Archimedes, with a Supplement The method of Archimedes, Cambridge (1912)
[12] Heath, T.L.: Mathematics in Aristotle, Oxford (1948)
[13] Héron d'Alexandrie: Les Mècaniques ou L'élévateur des corps lourdes. Les Belles Lettres, 126–127 (1988)
[14] Koyré, A.: Du monde de l' à-peu près à l'univers de la precition (1948)
[15] Laertius, D., trans. Hicks, R.D.: Lives of Eminent Philosophers, I. Harvard University Press (1925); Trans. Hicks, R.D.
[16] Newton, I.: Philosophiae naturalis Principia matematica (1686)
[17] Pappus d'Alexandrie: La collection mathèmatique (1982)
[18] Peano: Opere, III, p. 385

[19] Proclus: Comment on the first book of Euclid's elements, Prologue
[20] Plutarch: Marcellus' Life 14, 8–12
[21] Vailati, G.: Il principio dei lavori virtuali da Aristotele a Erone d'Alessandria. In: Scritti, p. 91 (1911)
[22] Vernant, J.P.: Remarques sur les forms e les limites de la pensée technique chez les Grecs (1957)

Notes

(1) *C'est un* ***eidos****, une forme, donnée d'avance à la facon d'une réalité naturelle. L'artisan ne l'a pas inventée; il ne peut pas la modifier; il n'a pas mème,e tant qu'artisan, qualitè puor la connaitre: la science de la forme de l'oject fabriqué appartient, non au producteur, ma à l'usager....*

Supériere à l'ouvrier et à sa technè, la Forme oriente e dirige le travail qui la réalise; elle lui assegne son terme,fixe ses limites,définit son cadre et ses moyens. Dans l'ouvrage de l'art, tout comme dans la production naturalle,c'est la ause finale qui determine et qui commande l'ensemble du processus troducteurLa cause efficace-l'artisan, ses outils, sa technè – n'est que l'instrumen grace auquel une Forme préexistentefaconne la matière.

L'artisien est,pour les Anciens, un homme qui ordenne une matière, opaque a l'esprit, en y icarnant une Forme supérieure a son esprit.. L'ouvrage possède plus perfection que l'ouvrier...

*Il ne lui faut,dans son travail, ni l'esprict d'initiative, ni riflexion. Sa fontion et sa virtu,dira Aistotle c'est d' obéir. (Politique, 1277 b 29.) (*Vernant J.P. *pp. 223- 224)*

(2) II. 7. *Nous avons achevé d'exposer la constrution des cinq machines simples dont la description precede,et d'expliquer leur mode d'emploi. Quant à la cause qui fait que chacun de ses instruments meut des poids considérables avec une trés faible puissance, nous allons mintenant en parler comme il suit. Supposons deux cercles ayant un meme centre A; soient leurs diameters les deux lignes BC, DE; ces deux cecles sont mobiles autour du point A, qui est leur centre commun,et perpendiculaires au plan de l'horizon. Suspendons aux deux points B, C deux poids égaux, désignés par Z et H. Il est evident que les cercles ne penchent ni d'un coté ni de l'autre,puisque les deux poids Z et H sont égaux et les distances BA,AC égales. Faisons di BC un fléau de balance mobile autour d'un point de suspension qui est le point A. Si nous transportons en E le poids qui est appliqué en C, le poids Z incline avers le bas ,et il fera tourner les cercles.*

Mais si nous augmentons le poids T, il fera de nouveau équilibre au poids Z; et le rapport du T au poids Z sera égal au rapport de la distance BA à la distance AE.. Ainsi la ligne BE joue le role d'un fléau de balance mobile autour d'un point de suspension qui est le poin A. ***Archimède*** *a déjà donné cette poposition dans son livre* ***Sur l'équilibre entre les poids****.*

Il est évident,par là, que nous pouvons mouvoir un corps très lourd avec une faible puissance, porsque, étant donnés deux cercles concentriques et un grand poids applique à un arc quelconque du grand cercle, le rapport de la ligne issue du centre du grand cercle à la ligne issue du centre du petit est plus grand que le rapport du grand poids à la faible puissance qui le meut. La faible puissance l'emport alors sur le grand poids.

Historical Reflections on the Scale Ratio in the Galilean *Trattato Di Fortificazione*

Raffaele Pisano[1] and Danilo Capecchi[2]

[1] 'Cirphles', Ecole Normale Supérieure Paris, France
[2] Università di Roma 'La Sapienza', Italy

Abstract. Based on a research program in history of mechanics and some previous papers on history and epistemology of mechanics we present here a concise case–study on the use of the scale ratio in the Galilean *Trattato di fortificazione* (end of 16th century). We focus in particular on scientific reasonings around the scale proposed, on the figures and on the peculiar lack of effective use of the ratio in the manuscript.

1 A Short Introduction

Galileo Galilei (1564-1642) in addition to addressing technical subjects as the strength of materials in the *Discorsi e dimostrazioni matematiche* [1], was busied also with architecture and fortresses [3]. He did it not only by teaching to military expert architects, but also as lecturer for students. The *Trattato di fortificazione* stems from teaching speeches collected by his pupils and/or on indications of Galileo himself [3–5].[1] In fact, it was usual that scholars and famous researchers were in demand as private lecturers for some particular matters.[2] Recent works investigated on Galileo's teaching activities in Padua, remarking two main activities at home: the workshops and the private lessons. They were absorbing, not to speak of the general administration of the household. So it is reasonable to claim that Galileo wrote these 'Iuvenilia'[3] manuscripts without the care he will dedicate to publications of his later works.

Galileo produced[4] two essays on the military fortifications in different periods, reproduced in the *Edizione Nazionale* of Galileo works. They are assembled from

[1] Favaro wrote: "[…] In addition to keeping around this and other matters, particular lessons, he released his manuscript copies to his auditors and indeed for their transcription he had sometimes a scribe at his wages […]." [26, vol. II, p. 9, line 6; 11]. Translation is ours. See also: *Mss. Galileiani*, Par. III, Tome IV, Car. 81r. In [6]).

[2] Sagredo in "Letter of 1-st sept. 1599" [7].

[3] 'Iuvenilia' is the general title cited in *Edizione Nazionale* (vol. I, p. 7, p. 9). It includes speeches wrote by Galileo during his time in Pisa (fl. 1581–1585). The Aristotelian arguments and the influence of these topics on young Galileo is enough witnessed in 'Iuvenilia'.

[4] Hereafter we use to write: "Galileo wrote…etc..", but as we noted in our previous remark, in fact he only lectured and someone else assembled his speeches. This is very important from a historical point of view.

T. Koetsier, M. Ceccarelli (Eds.): Explorations in the His. of Machines & Mech., HMMS 15, pp. 463–473.
springerlink.com

several manuscripts [10, p. 9]. The first work edited by Favaro is *Breve instruzione all'architettura militare* [10]. The correspondent original two manuscripts archived at *Biblioteca Ambrosiana* are *Trattato di fortificazioni e modo d'espugnare la città con disegni e piante di fortificazioni* [9] and *Trattato delle fortificazioni, con disegni diversi* [12]. The second work edited by Favaro is *Trattato di fortificazione* [11]. The correspondent original manuscript[5] is also archived at *Biblioteca Ambrosiana* and classified as *Trattato di fortificazioni, con disegni e figure (acephalous)* [13].[6]

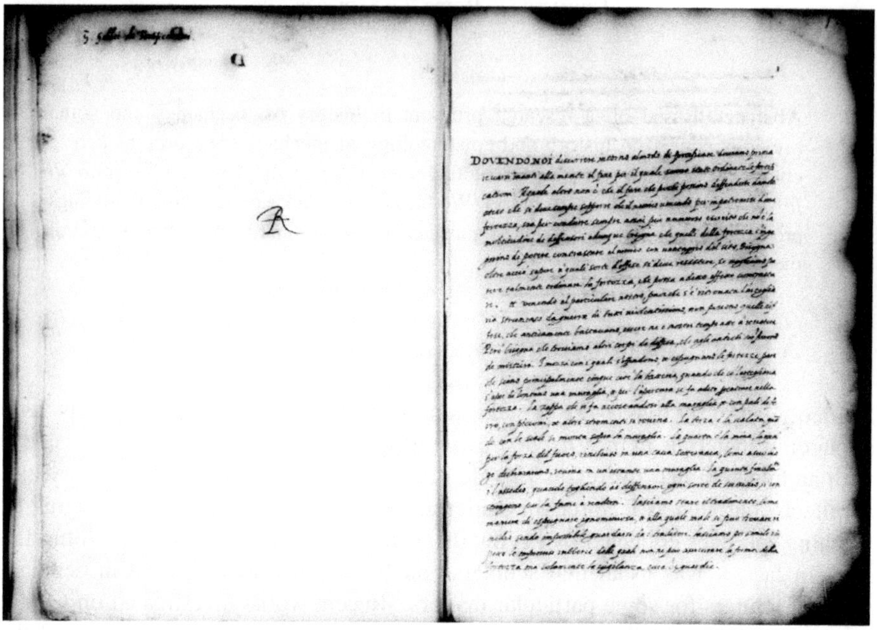

Fig. 1. A Galilean manuscript page [12]

Trattato di fortificazioni, con disegni e figure (acephalous)[7] [13] contains already the characteristic signature of Galileo's way of moving for problem in the process of scientific knowledge. Let us remark that none of the two mentioned Galilean

[5] We note that another corresponding manuscript named by Favaro "*n*" [10, pp. 13–14] is classified as Ms. B ("D 296 Inf.") since it is included in the same folder as the up cited Ms B. Nevertheless they are slightly different. Moreover, *m* is better conserved than manuscript *n* [10, pp. 12-13], so we decided to use Mss. A [9], B [12] and *m* [13].

[6] The secondary literature on these two manuscripts is not large. General considerations of them are present in the philological and historical works by Antonio Banfi (1886–1957) and Ludovico Geymonat (1908-1991); but especially in those by Antonio Favaro (1874-1922) and Stilmann Drake (1910-1993). Banfi and Geymonat [22, 23] centred mostly around the operations of *Operazioni del Compasso geometrico e militare* (1606) [24, 25].

[7] We studied these three manuscripts and correlated them with the two printed versions by Favaro. We can suppose, at first glance, that they were written by two different pupils and/or in different moments and maybe place.

manuscripts on fortifications is autograph, and since *Trattato di fortificazione, con disegni e figure (acephalous)*, as a matter of fact, more complete than *Breve instruzione all'architettura militare*, it is possible to claim that this latter is antecedent to the former. Let we note that 'Breve instruzione all'architettura militare' is exactly the last sentence that appears at the end of the first manuscript [9] (as well as at the end of he *National Edition* edited by Favaro): "E questo basti per questa breve instruzione dell'architettura militare" [10, p. 75].

In fact *Breve instruzione all'architettura militare* is effetely a collection of Galilean lessons (around the second half the XVI sec.) at the *University of Arts. Trattato di fortificazione* is instead a collection of Galilean private lessons during Galileo's stay in Padua (1592–93)[8]. On the other hand "Living of the chair only was almost impossible and private lessons charge was necessary to be paid [...]"[9] [15]. It is possible then to support a Padua lecturing activity by Galileo highlighting the educational character of the opera [8, 16] and underlining that these writings were due by a Galileo not yet scientifically mature [5].

The manuscripts start with elements of Euclidean geometry without neither demonstrations nor procedures with a rigid axiomatic development [11, p. 83; 4, 17]. To note the absence of the use of an advanced calculus. Neither decimal fractions appear, despite they were introduced in Europe by Simon Stevin (1548-1620) in 1585 [18][10].

2 A Relationship Science Technique

Reflections on scale ratio juts before *Trattato di fortificazione*

In the Galilean manuscript in a paragraph, it is presented a discourse on the use of the scale, that is what usually is done today and that Tartaglia did in the *Quesiti* [21, Book VI, Q. I, pp. 70-71]. In particular, in the *Gionta al Libro sesto*, Tartaglia deepens the *third* and *fourth* 'quality' on the geometry and composition of the perimetric walls:

> N [Niccolò Tartaglia]. But since to show in the drawing all those forms into a sudden way should generate confusion at your Eccellentia, so I will be showing them one by one, and I want to start from the more complicated, as the traders do, who want to sell their merchandise. [...][11]. [21, *Gionta,* Q. I, p. 70, line 34].

There it follows a long discourse on the bastions and curtains in order "[...] to follow the modern use of strengthening [...]" [21, *Gionta*, Q. I, p. 71, line 27], and on the

[8] For the accuracy of the data seems to not grant much historians, then we carry the opinion of a committed historian as Stlillman Drake (1910–1993) that confirmed that on 1593-1594 Galileo wrote speeches on mechanics and military architecture for his private teaching students; and no of the was university lectures [14].

[9] The translation is ours.

[10] Although there is no doubt that Galileo-teacher had such knowledge having been reader of mathematics in Pisa, he was always tied to the use of ratios and proportionalities, both for physical and mathematical quantities. Perhaps, this way of dealing with Archimedean theories came from the teachings of Ostillo Ricci (1540-1603).

[11] Translation is ours.

importance of the "parianette"[12] to be built with a certain thickness ("grossezza"), that are often built to absorb the energy of the balls ("balle") of cannon. These arguments are just prior to Tartaglia presentation of walls with obtuse angles for the drawings of which the ratio of scale mentioned above is associated with the idea of building walls having oblique rather than medieval vertical parameters.

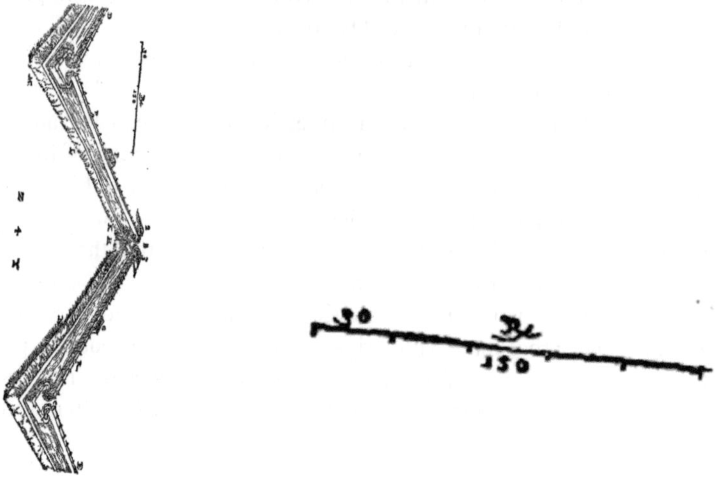

N [Niccolò Tartaglia]. Because I want too that, in the top of each curtain, many *parianette* are made, of joists planted and good planks, quite high over the height of a man, which *parianette* traverse the whole top of the curtain, but this crossing should not be orthogonal, but I want they proceed with the outer part somewhat toward the city, and the inner part toward the country side as you see drawn in this figure. It is true that the *parianette* want to be somewhat more oblique than the figure for the same reasons that I say below. Being this made. I want from the side that looks towards the country of each of these *parianette*, a small earth embankment of such a size, which cannot be damaged by enemies with their artillery, under each such small embankment, I want there, a falconetto[13] with 6 or 3 lbs balls [...]. [21, *Gionta,* Q. I, p. 71, line 10].[14]

We note that in the previous quotation the figure is actually in scale. This is a point historically important for this writing.

The scale ratio in the *Trattato di fortificazione*

In *Trattato di fortificazione*, as previously mentioned, it appears, in a short paragraph on the relationship about scale, "Della scala" [11, p. 102; 12; 13]. These are arguments about units of measure and its proportionality ratio, highlighting a nontrivial problem. Consider,

[12] The *parianette*, also called *traverse*, are structural elements placed along the walls of the curtains. They are usually arranged vertically. The aim was to limit the effects of enfilade fire. As is clear from the text, Tartaglia has personal news for the construction of the *traverse* by assuming an inclination of and a height greater than that of a man.

[13] A small gun.

[14] The image in the quoted text is suitably enlarged and rotated.

e.g., the case of a designer who, far from their own country, was going to draw a certain plant in lands that could not adopt his own unit of measure. Below we present images of the Galilean paragraph on scale both from the two Ambrosian manuscripts and in the Favaro's printed text which clarify the matter:

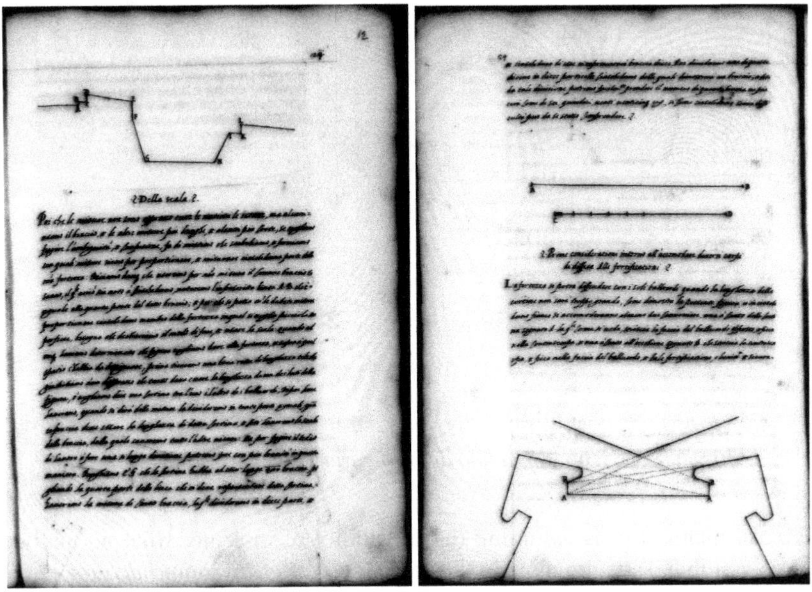

Fig. 2. Galilean wording on ratio scale [12]

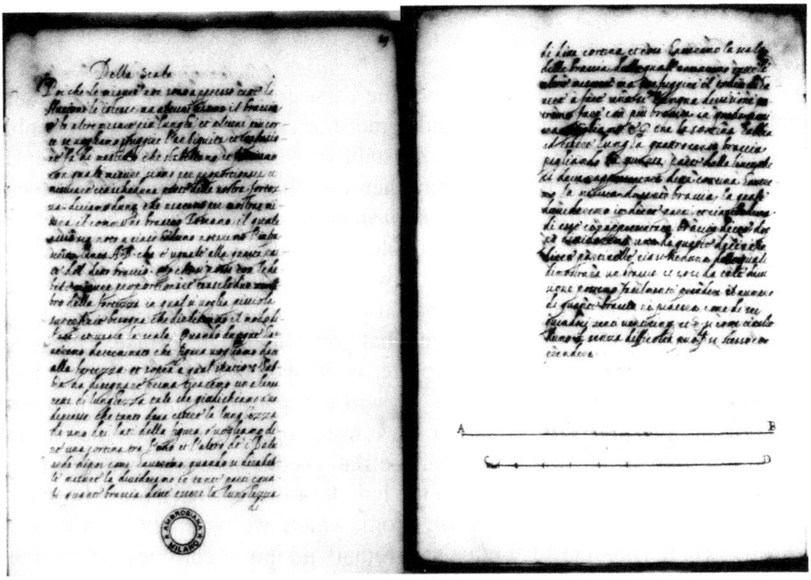

Fig. 3. Galilean wording on scale ratio [13]

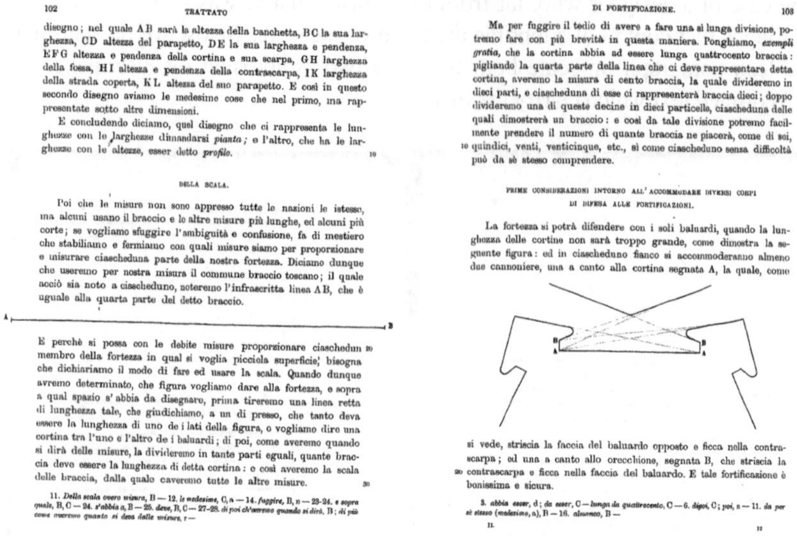

102 TRATTATO

disegno; nel quale AB sarà la altezza della banchetta, BC la sua larghezza, CD altezza del parapetto, DE la sua larghezza e pendenza, EFG altezza e pendenza della cortina e sua scarpa, GH larghezza della fossa, HI altezza e pendenza della contrascarpa, IK larghezza della strada coperta, KL altezza del suo parapetto. E così in questo secondo disegno aviamo le medesime cose che nel primo, ma rappresentate sotto altre dimensioni.

E concludendo diciamo, quel disegno che ci rappresenta le lunghezze con le larghezze dimandarsi *pianta;* e l'altro, che ha le larghezze con le altezze, esser detto *profilo.*

DELLA SCALA.

Poi che le misure non sono appresso tutte le nazioni le istesse, ma alcuni usano il braccio e le altre misure più lunghe, ed alcuni più corte; se vogliamo sfuggire l'ambiguità e confusione, fa di mestiero che stabiliamo e fermiamo con quali misure siamo per proporzionare e misurare ciascheduna parte della nostra fortezza. Diciamo dunque che useremo per nostra misura il commune braccio toscano; il quale acciò sia noto a ciascheduno, noteremo l'infrascritta linea AB, che è uguale alla quarta parte del detto braccio.

A ——— B

E perchè si possa con le debite misure proporzionare ciaschedun membro della fortezza in qual si voglia picciola superficie, bisogna che dichiariamo il modo di fare ed usare la scala. Quando dunque avremo determinato, che figura vogliamo dare alla fortezza, e sopra a qual spazio s'abbia da disegnare, prima tireremo una linea retta di lunghezza tale, che giudichiamo, a un di presso, che tanto deva essere la lunghezza di uno de i lati della figura, o vogliamo dire una cortina tra l'uno e l'altro de i baluardi; di poi, come averemo quando si dirà delle misure, la divideremo in tante parti eguali, quante braccia deve essere la lunghezza di detta cortina: e così averemo la scala delle braccia, dalla quale caveremo tutte le altre misure.

11. *Della scala overo misura,* B — 12. *le medesime,* C, A — 14. *fuggire,* B, n — 23-24. *e sopra quale,* B, C — 24. *s'abbia a,* B — 25. *deve,* B, C — 27-28. *di poi ch'averemo quando si dirà,* B; *di più come averemo quanto si deva dalle misure,* r —

DI FORTIFICAZIONE. 103

Ma per fuggire il tedio di avere a fare una sì lunga divisione, potremo fare con più brevità in questa maniera. Ponghiamo, *exempli gratia,* che la cortina abbia ad essere lunga quattrocento braccia: pigliando la quarta parte della linea che ci deve rappresentare detta cortina, averemo la misura di cento braccia, la quale divideremo in dieci parti, e ciascheduna di esse ci rappresenterà braccia dieci; doppo divideremo una di queste decine in dieci particelle, ciascheduna delle quali dimostrerà un braccio: e così da tale divisione potremo facilmente prendere il numero di quante braccia ne piacerà, come di sei, quindici, venti, venticinque, etc., sì come ciascheduno senza difficoltà può da sè stesso comprendere.

PRIME CONSIDERAZIONI INTORNO ALL' ACCOMMODARE DIVERSI CORPI DI DIFESA ALLE FORTIFICAZIONI.

La fortezza si potrà difendere con i soli baluardi, quando la lunghezza delle cortine non sarà troppo grande, come dimostra la seguente figura: ed in ciascheduno fianco si accommoderanno almeno due cannoniere, una a canto alla cortina segnata A, la quale, come

si vede, striscia la faccia del baluardo opposto e ficca nella contrascarpa; ed una a canto allo orecchione, segnata B, che striscia la contrascarpa e ficca nella faccia del baluardo. E tale fortificazione è bonissima e sicura.

3. *abbia esser,* d; *da esser,* C — *lunga da quattrocento,* C — 6. *dipoi,* C; *poi,* n — 11. *da per sè stesso (medesimo,* a), B — 16. *almanco,* B —

II. 12

Fig. 4. Scale ratio in *Trattato di fortificazione* in Edizione Nazionale[15]

Galilean manuscript is assuming the possibility of inserting what normally today one does and that Tartaglia already did in his *Quesiti et Inventioni diverse* [21, Book VI, Q. I, p. 71][16]. The plan of a fortress was draw on paper, showing to the reader, near it, a unit of measure: it is a graphical scale. The graphical scale is not

[15] "On the scale. As the measures are not the same for all nations, but some use the *braccio* and other longer measures and some shorter. If we want to avoid ambiguity and confusion, it is necessary that we establish with what measures we take and measure each part of our fortress. I say therefore that we will use for our measures the common Tuscany *braccio* [*Tuscany braccio* is ca. 58,36 cm. The length of the segment AB printed in Edizione Nazionale is about 14,6 cm [11, vol. II, p. 102]]; to make it be known to everybody, you will notice the line AB, which is equal to the fourth part of the *braccio*. And because you can with the necessary measures provide each member of the fortress in what you want small surface, I must declare how to make and use the scale. Therefore, when we have determined the form we want to give to the fortress, [...] first we will get a straight line of length equal to the length of one side of the figure, or if we want, a curtain between the one and the other of the bulwarks; then, as we will tell about the measures, we will divide in many equal parts, how many *braccio* the length of this curtain should be and so we will have the scale of the *braccio*, from which we will obtain all the other measures" ([11, 102, line 11. Let us also remark a kind of currency arbitrariness. In fact the figure seems arbitrary due to proportion does not correspond).

[16] Please see Tartaglia's figure in the text.

autographically reported regularly in the *Galilean manuscripts* [12, 13]. To be precise, on admission of same Antonio Favaro, editor of *Opere di Galileo Galilei*, one can see that:

Among the figures, which in the two copies are perhaps due to the same pen, the best there seem to be those of B. Since the drawings have simply demonstrative purpose, we omitted the scale, which sometimes is given in the manuscripts, because it almost always did not correspond to the proportions of drawings. [As a note Favaro writes:] By way of exception, we applied the scale to the figure of page 123, because, being incomplete and incorrect in all the versions of manuscripts, we had to redraw it, by means of data supplied by the text and elements of the various codes. [...] Because the figures [of the *Trattato di fortificazione*], are often wrong in the various codes, we preferred in general to take them by the Ambrosiana, however with correction any time we considered it necessary.[17]

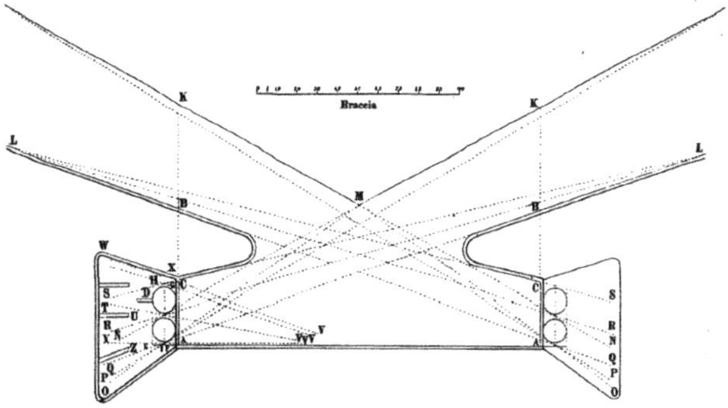

Fig. 5. The occurrence of the scale[18]

Then in both the manuscripts on fortifications [12, 13], there are not consistent scale ratios and the only one present in the national edition [26, p. 123] is not Galileo's autograph. The figures of the national edition [12, 13] are, for Favaro's admission, drawn with a scale factor that does not correspond to the original figures. Then it lacks a control parameter to ensure the reader the possibility of comparison.

Reflections on scale ratio just after *Trattato di fortificazione*

Some years later also Buonaiuto Lorini (1550–1611) after presenting the essential geometry useful for the treatment of the fortresses in his work, *Le fortificationi di Buonaiuto Lorini nobile fiorentino nuovamente ristampate, corrette & ampliate con*

[17] [26, vol. II, pp. 12–13, line 16; line 14, line 3]. (*Italic style* and brackets from the author). In the *Breve instruzione all'architettura militare,* the same argument – "Della scala" – is reported at pages 31–32 [26].

[18] Let us remark that figure drawn by Favaro [26, p. 123] is here rotated by 90 degrees clockwise.

l'aggiunta del sesto libro (hereafter *fortificationi*)[19], establishes a unit of measure for the plants and figures [3, pp. 261–336; 27; 28]:

> But we must first decide the measure, with which we will form and measure the entire body of the fortress; in order to take it more familiar, we serve of Florentine *braccio*, then of the Venetian *passo*. The *passo* contains three of these *braccio*, and up to five feet, as in the following two figures you see for AB, the measure of a Venetian foot, divided into twelve ounces, and for CD, middle Florentine *braccio*; which measures will be trained all drawings of this work [27, p. 7, line 29; Image, p. 8]. [...] Since with the proposed scale we should measure and proportion the members with the whole body of the fortress [...]. [27, p. 9, line 3; Image, *Ivi,* 8].

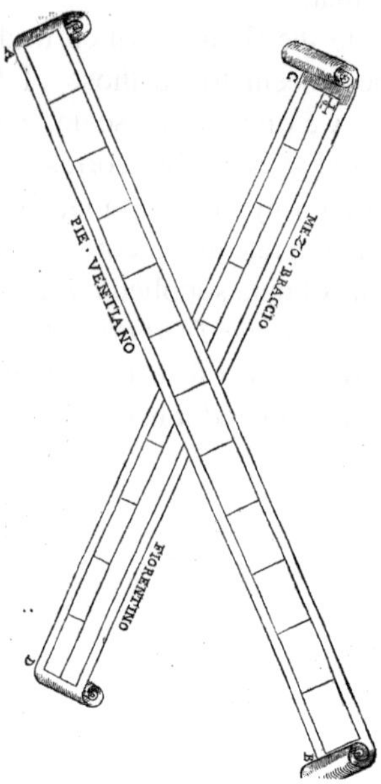

Lorini uses both the Tuscany *braccio* (like Galileo) and the Venetian *passo* (pass),[20] both as reference of the scale ratio.

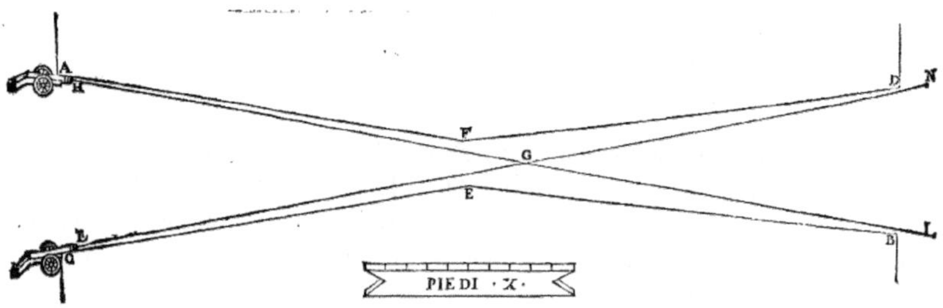

Fig. 6. Lorini's scale ratio in *Le fortificationi* [27, p. 48, line 20]

[19] Lorini's *fortificationi* was published in two editions: 1) 1596. *Delle fortificationi di Buonaiuto Lorini nobile fiorentino Libri Cinque. Ne' quali si mostra con le più facili regole la Scienza con la ratica, di fortificare le città, & altri luoghi sopra diversi siti..., Nuovamente dati in luce,* 2) 1609. *Le fortificationi di Buonaiuto Lorini nobile fiorentino nuovamente ristampate, corrette & ampliate con l'aggiunta del sesto libro* [27]. We studies the second one since more complete than first. For a full historical investigation see [3, pp. 261–336].

[20] The Venetian *passo* was composed by 5 feet, for a length of about 1.74 m.

3 Final Remarks

The paper aims to show a particular aspects belonging to 'Iuvenilia' corpus of Galileo. His works on fortress even if (fortunately) edited in Edizione Nazionale are very few known. Moreover, the limited amount of secondary literature is a matter of fact, testifying the modest interest in these topics with respect to (the more relevant) Galileo's researches on mechanics, instead largely commented by many scholars.

Galilean speeches on fortresses are presented both with the strict language of geometry (typical of scientists), and with the synthesis expressed by the drawing (typical instead of architects). Galileo suggests the possibility of including a graphical scale (but he does not refers that Tartaglia already did it before him in the *Quesiti et inventioni diverse* [21]): that is, when you have found the object to be reported on paper, it is drawn in proportion indicating to the reader(-technician), on the same page and next to the figure, a unit of measure to which to refer for the scale ratio. But this relationship in scale made by Galileo is not present in the *Trattato* di *fortificazione*.[21]

We think that these 'Iuvenilia' works are historically crucial to better understand both philogical aspects and early foundational Galilean convictions-doubts-methods before his definitive study on parabolic trajectories (1608-1609) and his mechanics [2]. In fact a cautious attitude sometimes shines, sometimes scientifically immature, is present also in other parts of the *Trattato* di *fortificazione,* e.g. where they are presented the trajectories of projectiles with drawings that show rectilinear directions for the trajectories [3, pp. 241-260].

Finally, although the two treaties are of unusual conception – at least according to [11, 12, 13] – they are nonetheless finite texts, although necessary of clarification.[22] One of the objectives of this research program is to understand the scientific role played by Galilean cautious attitude: is it a scientific cautions solely or is it something related with others, e.g., social facts in that period? On the other hand everybody know the end of history.

References

[1] Galilei, G.: Discorsi e dimostrazioni matematiche. In: Favaro, A. (ed.) Opere di Galileo Galilei, Edizione Nazionale, vol. VIII, pp. 4–458. Barbera Editore, Firenze (1890)

[2] Galilei, G.: Le mecaniche. In: Favaro, A. (ed.) Opere di Galileo Galilei, Edizione Nazionale, vol. II, pp. 155–191. Barbera Editore, Firenze (1909)

[3] Pisano, R.: The role played by mechanical science in the architects and engineers in the Renaissance, vol. 2, pp. 241–260. University of Roma La Sapienza, Roma, Ph.D. dissertation (2008) (in Italian)

---

[21] Even if, generally speaking, it is possible to think to design a scale ratio object without knowing its owner ratio scale.

[22] In fact next will deeply focus on the strength lines of the trajectories designed in the *Trattato di fortificazione* concerning the bullets.

[4] Pisano, R., Capecchi, D.: Galileo Galilei: Notes on Trattato di fortificazione. In: Altamore, A., Antonini, G. (eds.) Galileo and the Renaissance Scientific Discourse, pp. 28–41. Edizioni nuova cultura, Roma (2010)
[5] Pisano, R.: On method in Galileo Galilei' mechanics. In: Hunger, H. (ed.) Proceedings of ESHS 3rd Conférence, pp. 147–186. Austrian Academy of Science, Vienna (2009)
[6] Favaro, A.: Galileo e lo studio di Padova, vol. I, II, p. 209, 175. Le Monnier, Firenze (1883)
[7] Galilei, G.: In: Favaro, A. (ed.) Opere di Galileo Galilei, Edizione Nazionale, vol. X, p. 77. Barbera Editore, Firenze (1890-1909)
[8] Renn, J., Buttner, J., Damerow, P., Schemmel, M., Valleriani, M.: Galileo and shared knowledge of his time. The Max Planck Institute for the history of science pre–print, Berlin, p. 49, n. 228 (2002)
[9] Galilei, G.: Ms. A – Trattato di fortificazioni e modo d'espugnare la città con disegni e piante di fortificazioni. Biblioteca Ambrosiana D: 328 Inf.– 38640
[10] Galilei, G.: Breve instruzione all'architettura militare. In: Favaro, A. (ed.) Opere di Galileo Galilei, Edizione Nazionale, vol. II, pp. 15–75. Barbera Editore, Firenze (1890)
[11] Galilei, G.: Trattato di fortificazione. In: Favaro, A. (ed.) Opere di Galileo Galilei, Edizione Nazionale, vol. II, pp. 77–146. Barbera Editore, Firenze (1890)
[12] Galilei, G.: Ms. B – Trattato delle fortificazioni, con disegni diversi. Biblioteca Ambrosiana D: 296 Inf.– 38499
[13] Galilei, G.: Ms. m – Trattato di fortificazioni, con disegni e figure (acephalous). Biblioteca Ambrosiana N: 281 Sup.– 2– 80277
[14] Drake, S.: Galileo: pioneer Scientist. The University of Toronto Press, Toronto (1990)
[15] Galilei, G.: Lettera del 1° sett. 1599. In: Favaro, A. (ed.) Opere di Galileo Galilei, Edizione Nazionale, vol. X, p. 77. Barbera Editore, Firenze (1890-1909)
[16] Renn, J., Valleriani, M.: Galileo and Challenge of the arsenal. Nuncius XVI(2), 481–503 (2001)
[17] Pisano, R.: Il ruolo della scienza archimedea nei lavori di meccanica di Galilei e di Torricelli. In: Giannetto, E., Giannini, G., Capecchi, D., Pisano, R. (eds.) Proceedings of XXVI SISFA Congress, Da Archimede a Majorana: La Fisica nel suo Divenire, pp. 65–74. Guaraldi Editore, Rimini (2009)
[18] Stevin, S.: The principal works od Simin Stevin. In: Dijksterhuis, E.J. (ed.), vol. I. N.V. Swets & Zeitlinger, Amsterdam (1955)
[19] Pisano, R., Capecchi, D.: La meccanica in Italia nei primi anni del Cinquecento. Il contributo di Niccolò Tartaglia. In: Tucci, P. (ed.) Proceedings of XXV SISFA Congress, Milano, pp. C17.1– C17.6 (2008), `http://www.brera.unimi.it/sisfa/atti/index.html`
[20] Capecchi, D., Pisano, R.: Tartaglia's science of weights. Mechanics in XVI Century. Springer, Heidelberg (forthcoming, 2012)
[21] Tartaglia, N.: La nuova edizione dell'opera Quesiti et inventioni diverse de Nicolo Tartaglia brisciano Riproduzione in facsimile dell'edizione del 1554. In: Masotti, A. (ed.) Commentari dell'Ateneo di Brescia, Tipografia La Nuova cartografica, Brescia (1959)
[22] Banfi, A.: Galileo Galilei. La Nuova Italia, Firenze (1966)
[23] Geymonat, L.: Galileo Galilei. Einaudi, Torino (1980)
[24] Galilei, G.: Le operazioni del compasso geometrico e militare. In: Favaro, A. (ed.) Opere di Galileo Galilei, Edizione Nazionale, vol. II, pp. 343–424. Barbera Editore, Firenze (1890)
[25] Galilei, G.: In: Brunetti, F. (ed.) Opere, vol. I, pp. 189–261. U.T.E.T., Padova (2005)

[26] Galilei, G.: In: Nazionale, E., Favaro, A. (eds.) Opere di Galileo Galilei, Edizione Nazionale, vol. II, pp. 12–13. Barbera Editore, Firenze (1890)
[27] Lorini, B.: Le fortificationi di Buonaiuto Lorini nobile fiorentino nuovamente ristampate, corrette & ampliate con l'aggiunta del sesto libro. In Venetia, presso Francesco Rampazetto (Lorini B.: Delle fortificationi di Buonaiuto Lorini nobile fiorentino Libri Cinque. In: Ne' quali si mostra con le più facili regole la Scienza con la ratica, di fortificare le città, & altri luoghi sopra diversi siti, Nuovamente dati in luce. In: Vinegia, appresso Gio. Antonio Rampanzetto. BNCF - Palatino 11. 3.7. 37) (1596) (1609)
[28] Pisano, R., Capecchi, D.: Il ruolo della meccanica ne Le fortificationi di Buonaiuto Lorini. In: 3° Convegno di Storia dell'ingegneria, vol. II, pp. 797–808. Cuzzolin editore, Napoli (2009)

Poncelet and the Development of Applied Mechanics

Agamenon R.E. Oliveira

Polytechnic School
Department of Applied Mechanics and Structures
Federal University of Rio de Janeiro

Abstract. A serious problem faced by Rational Mechanics until the end of the eighteen century was the difficulty of creating a general theory consistent with machines [1, 2], despite the enormous amount of empirical knowledge about machines accumulated since antiquity. In addition, during this period the theory of mechanics made a considerable advance following the 1788 Lagrangean formulation [3]. This paper highlights the participation of Jean-Victor Poncelet (1788-1867) in the solution of this dilemma in the context of the development of applied mechanics [4]. In this context the concept of work and the use of living forces principle applied to machines by the polytechnicians engineers established the fundamental step to that accomplishment.

1 Introduction

It is a well known fact, and this paper will try to highlight some important aspects of this question, that until the end of the eighteen century there was an extraordinary rich store of practical knowledge about machines, including their design and construction [5]. These were historically accumulated forms of knowledge with an empirical nature, and which from time to time had undergone some type of systematization, although without ever having reached the stage of a more broad ranging theory covering all machines.

Alongside this practical knowledge of machines, following the publication of Newton's *Principia* in 1687 the theory of motion became Rational Mechanics in the form of pure mathematical knowledge [6]. In addition, mechanics theory reached its highpoint with the formalization proposed by Lagrange (1736-1813) in his *Méchanique Analytique* [7]. Despite these favorable conditions, a general theory for machines still had to be proposed. A solution to this problem started to be found in the works of several engineers, mainly the French *polytechniciens*, at the beginning of nineteenth century [8].

The development of applied mechanics is a complex social and collective construction in which some important achievements need to be emphasized. Navier (1785-1836) practically rewrote Belidor's famous book *Architecture hydraulique* by adding a number of remarks to the main work. In these remarks and additional observations he demonstrated the *vis viva* conservation principle for a discrete mass. He also made the

T. Koetsier, M. Ceccarelli (Eds.): Explorations in the His. of Machines & Mech., HMMS 15, pp. 475–490.
springerlink.com

generalization to a system of material points by using d'Alembert's principle. The *vis viva* principle was of fundamental importance for Coriolis' theory of machines which can be considered to be a development of Carnot's general theory [9].

Carnot's basic approach to machines used the available mathematical tools of Rational Mechanics, starting with d'Alembert's principle, while also accepting the empirical ideas of Newton's mechanics and being strongly influenced by Locke's philosophy. Other important developments are Coulomb's theory of friction [10] and the work of Poncelet directly related to applied mechanics, as will be shown in this paper.

In 1825 Poncelet became a professor in Metz and worked on several new aspects of machine science in a form on applied mechanics courses. He gave the first course in Metz between 1825-1834, later transformed into a book and published in 1826. The second course was given in Sorbonne from 1838 to 1848. The subject of both courses will be discussed in detail in this paper [11].

2 The Knowledge of Machine Constructors

During the final decades of the eighteenth century an exceptional quantity of knowledge was accumulated in practical mechanics. This had been built up since the ancient Greeks. According to the historian of mechanics René Dugas, the founder of mechanical theory was Aristotle (384 BC - 322 BC) who, amongst numerous other works, wrote one entitled *Physis* [12]. In relation to practical mechanics another work attributed to Aristotle is known as *Problemata Mechanica* or *Quaestiones Mechanicae* [13]. This can be considered to be the first study which tries to explain how machines function. It is of a very special interest, because unlike the approaches of Euclides (330 BC - 260 BC) and Archimedes (287 BC – 212 BC), who studied situations of equilibrium in machines; the work attributed to Aristotle makes a dynamic analysis of machines [14].

In the so-called Alexandrian School studies of mechanics applied to machines achieved great progress. The main findings and developments are associated with Archytas (428-347 BC), Ctesibius (285-222 BC), Philon of Byzantium (20 BC-50 AD) and Hero of Alexandria (10-70 AD), with the latter being the great constructor of automata [15]. In 1693 four of Hero's works were published: *Treatise on Machines of War*, *Pneumatics*, *Automata* and *Chirobalistics*.

During the Arab period, which followed immediately after the Alexandrian School, other interesting advances worth highlighting occurred. Many of these are the work of Al-Jazari (1136-1206), an engineer and constructor of machines with much in common with Hero. Al-Jazari left a work published in 1206 entitled *Treatise on the Theory and Practice of the Mechanical Arts*. It is a type of theoretical manual which describes machines which really can be constructed and gives, in addition to an explicative text, drawings and plans to guide their construction.

Due to the repercussions they would have on industrial processes, the most important machines from the Arab period are water wheels, watches and later the steam engine.

The water wheel, ancestor of the hydraulic turbine, made rapid progress from the eleventh century onwards due to its importance for the economic system of Western Europe. From the technical point of view these water wheels did not differ much from those used by the Greek and Romans several centuries before. Their importance was due more to the fact that they functioned as an essential part of a productive unit, located in the field or in the city. The classic work on the study of the origins of the water wheel is Vitruvius' treatise called *On Architecture*, published in the first century BC.

The origin of the mechanical watch can be dated to around the thirteenth century. The most probable date for its invention is between 1280 and 1300, by an unknown maker. But the invention of the watch as a precision instrument would only happen in the second half of the seventeenth century [16]. It is not necessary to outline the enormous cultural and scientific impact of this instrument.

In relation to the steam engine, it emerged from the experiments of Otto von Guerick (1602-1686) carried out in 1654. Nonetheless, its origins are related to the works of scientists and machine operators in the seventeenth century, such as Huygens (1629-1695), Denis Papin (1647-1712), Leibniz (1646-1716) [17], and others who systematically worked with the problem of atmospheric pressure and the use of heat as a form of energy. Papin tried to construct an atmospheric machine in 1707, with the aim of pumping water into a reservoir. Thomas Savery (1650-1715) built a machine with the same purpose in 1698.

After several attempts, the third steam engine constructed was successful. It is known as Newcomen's machine. Its motive power came from atmospheric pressure while a piston was being pushed down. The first of these machines was put into operation in Staffordshire in 1712. It is known that it produced 12 strokes per minute, each one of which lifted 10 gallons of water from a depth of 51 yards, in other words, this machine had an approximate power of 5.5 hp.

In 1763 James Watt (1736-1819) received a Newcomen (1664-1729) steam engine to repair. He worked on the idea for several months and then modified its structure so that the steam used passed through a condenser separated from the main cylinder. Watt's improvements to the steam engine in 1769 and 1784 converted a machine of limited use to an efficient one with multiple applications. He thereby contributed to the transformation of an agricultural society into an industrial one.

3 The Foundation of the Polytechnic School

In the aftermath of the French revolution, in order to change how engineering was taught, a group of scientists and engineers under the leadership of Gaspar Monge (1746-1818) proposed that a new type of engineering school should be organized to replace those which had existed under the old regime. The proposition was officially approved in 1794 and at the end of that year the revolutionary government approved the commencement of teaching in the new school, which was allocated 12,000 pounds and opened with 400 students [18]. In 1795 the school was given its present name, *École Polytechnique*.

Among the staff of the new engineering school were the mathematicians Lagrange (1736-1813) and Laplace (1749-1827) and the chemist Bertholet (1748-1822). Other famous scientists took part in the teaching of mathematics and mechanics, such as Monge and Prony (1755-1839). Soon afterwards Fourier (1768-1830) and Poisson (1781-1840) joined them.

Within a few years Polytechnic School became a great center for the creation and dissemination of knowledge in Europe [19]. Paris also became the undisputed capital of scientific thought in Europe and at the beginning of nineteenth century the scientific spirit was firmly established in France.

A reform of engineering teaching was necessary because the war with other European countries stimulated the construction of fortifications, roads and bridges and the development of artillery. The new context propelled France to apply scientific principles to industry, with the result that the new engineering school had to provide universal scientific knowledge as well as tools and methods applicable in a diversified range of practical situations.

Fig. 1. Old Polytechnic School Building

Other schools appeared afterwards, such as the *École d´Application de l´Artillerie et du Génie* in Metz, the largest of the pre-World War I applied schools, as well as other much smaller schools. The *École Polytechnique* thus became a central component of a more complex training structure which many historians refer as the *polytechnique system* [20].

4 Lagrangean Mechanics

At the end of the eighteenth century, Newtonian mechanics underwent a profound change through the formalization carried out by Lagrange. We can say that through the work of Lagrange, rational mechanics was significantly renovated, enriched and transformed [21].

The reading by Lagrange of Euler´s treatise "Methodus inveniendi lineas curvas maximi minimine proprietate gaudentes", published in 1744, has led Lagrange in 1755 to discover his "method of variations" which Euler (1707-1783) had sought in vain. Lagrange communicated his method to Euler who recognized its enormous importance. At the same time, Lagrange drew his attention to the principle of least action, enunciated for the first time in a vague way by Maupertuis (1698-1759) in 1744, and in 1759 he informed Euler that he had almost finished a treatise concerning his method of variations and deduction of mechanics starting with this principle. Euler apparently showed no interest in the treatise.

Before the publication of "Analytical Mechanics", Lagrange published in 1760 two memoirs in Turin: "Essai d´une nouvelle method pour determiner les *maxima* et les *minima* des formules integrals indefinies" and "Application de la methode exposée dans le mémoire precedent a la solution de differents problems de Dynamique".

Fig. 2a Pierre-Louis Lagrange

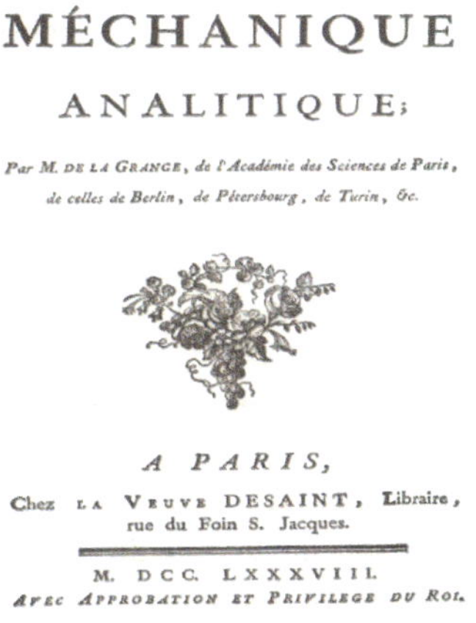

MÉCHANIQUE

ANALITIQUE;

Par M. DE LA GRANGE, de l'Académie des Sciences de Paris, de celles de Berlin, de Pétersbourg, de Turin, &c.

A PARIS,

Chez LA VEUVE DESAINT, Libraire, rue du Foin S. Jacques.

M. DCC. LXXXVIII.

AVEC APPROBATION ET PRIVILEGE DU ROI.

Fig. 2b

Lagrange´s "Analytical Mechanics" was published in 1788, crowning a series of Lagrange works and other important contributions previously developed by d´Alembert (1717-1783) and Euler [22]. In relation to the new mechanics, Lagrange stated: "We already have various treatises on mechanics, but the plan of this one is completely new. I propose to reduce the theory of this science, and the art of solving problems related through to formulae, by means of general formulae, and in this way simple development gives all the equations necessary for the solution of each problem. I hope that the method which I have just developed will achieve this objective, not leaving anything wanting". And he continued: "Figures cannot be found in any part of this book. The methods outlined here do not need constructions, nor geometrical or mechanical reasoning, but rather only algebraic operations, subject to a regular and uniform pace. Those who love analysis will, with pleasure, see mechanics as a new branch, and will be grateful to me for thus having extended its domain".

5 The Laplacian Project

In the first decades of the nineteenth century Laplace (1749-1827) and some of his followers began to defend the point of view that optics refraction, the cohesion of solids, the capillarity effect and chemical reactions were the result of attractive forces exercised by particles of matter in a similar manner to Newton's law of universal gravity. This attempt to extend and universalize the ideas of Newtonian mechanics to other domains of the physical world became known as the Laplacian project.

This kind of programme was presented in a more developed form by Denis Poisson (1781-1840) as a theory of 'physical mechanics' with the objective of substituting Lagrange's 'analytical mechanics.' The development and progress of the natural sciences in the nineteenth century, especially in the areas of heat, electricity, optics and magnetism, resulted in the defeat of this project [23]. In other words, it proved impossible to unify all those natural phenomena around Newtonian theory. In spite of this, the Laplacian project stimulated many developments in physical sciences.

With the decline of the Laplacian project, the mechanics hegemonically represented by Lagrangean mechanics became the paradigm of the physical world. In this approach the originally Cartesian ideas and the project of a very formalized mechanics, which dispensed with the results of experimentation and drew solely and exclusively on reason, became dominant. As a result a general theory of machines coming from Rational Mechanics would have to wait from several more years.

While, on the one hand, the hegemony of analytical mechanics satisfied one school of thought, on the other it created and worsened the problem and the dilemma of an absence of a theory for machines in general in the context of the expansion and large-scale use of machinery in the ongoing Industrial Revolution. The solution of the dilemma faced by Rational Mechanics at the end of the eighteenth century would arise from, in addition to other developments, the 'physical mechanics' transformed in Poncelet's courses.

6 The Main Developments in Applied Mechanics

Applied mechanics and further developments of industrial mechanics are a complex result of the work of many places and institutions in several different countries. In this paper, as we are focusing on Poncelet, a famous French *polytechnicien*, we will explore the contribution of French engineering to applied mechanics, as follows:

Charles-Augustin Coulomb (1736-1806) – Coulomb's major work on friction forces appeared in his *Théorie des Machines Simples* [24], which won him the Grand Prix from the *Académie des Sciences* in 1781. He investigated both static and dynamic friction of sliding surfaces, as well as friction in the bending of ropes and in rolling. Because of this contribution Coulomb can be considered the founder of the science of friction. The discoverer of laws of friction provided solutions for many of the important problems of applied mechanics, such as his elegant solution to the problem of torsion in cylinders and the use of torsion balance in physical applications.

Fig. 3a Charles-Augustin Coulomb

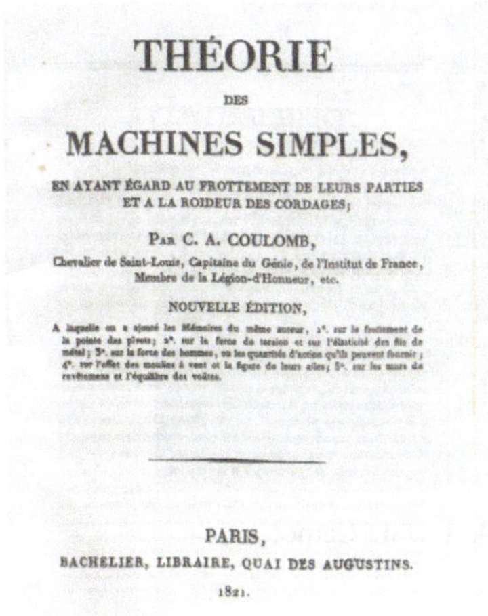

THÉORIE

DES

MACHINES SIMPLES,

EN AYANT ÉGARD AU FROTTEMENT DE LEURS PARTIES ET A LA ROIDEUR DES CORDAGES;

PAR C. A. COULOMB,

Chevalier de Saint-Louis, Capitaine du Génie, de l'Institut de France, Membre de la Légion-d'Honneur, etc.

NOUVELLE ÉDITION,

A laquelle on a ajouté les Mémoires du même auteur, 1°. sur le frottement de la pointe des pivots; 2°. sur la force de torsion et sur l'élasticité des fils de métal; 3°. sur la force des hommes, ou les quantités d'action qu'ils peuvent fournir; 4°. sur l'effet des moulins à vent et la figure de leurs ailes; 5°. sur les murs de revêtemens et l'équilibre des voûtes.

PARIS,

BACHELIER, LIBRAIRE, QUAI DES AUGUSTINS.

1821.

Fig. 3b

Coulomb is also responsible for the introduction of economic studies in relation to machines which can be considered as a precursor of industrial mechanics. This task mechanically measured the human body as a kind of machine in order to economically measure machine performance. The memoir *Sur la Force des Hommes,* published in 1778, is an attempt to understand human work mechanically and is the first publication on physiology and ergonomics. He proposed to measure the quantity of action (work) that a man can expend in a day of work by different ways of applying forces. Coulomb studied the old problem of human mechanical capacity, also dealt with by Desagulliers and Daniel Bernoulli (1700-1782).

By introducing the question of human work as a machine operation, it was possible to study the work carried out by a machine economically. This double analogy created the conditions to construct the concepts of net work and global work applied to a given machine in order to measure its efficiency, costs, energy consumption, and provided the necessary theoretical tools to compare two different machines.

Lazare-Nicolas-Marguerite Carnot (1753-1823) – Carnot started to develop a general theory of machines at the end of the eighteenth century. His basic approach consisted of applying d'Alembert's principle and the mathematical tools provided by Rational Mechanics. In this way he found the origin of the difference between the concept of *vis viva* (kinetic energy) and the moment of activity (work) which operates with these two concepts for a machine and uses a kind of equivalence between them. This is similar to the application of the principle of energy conservation 60 years before its discovery in 1847. Obviously, this application is restricted to mechanics.

Fig. 4a Lazare Carnot

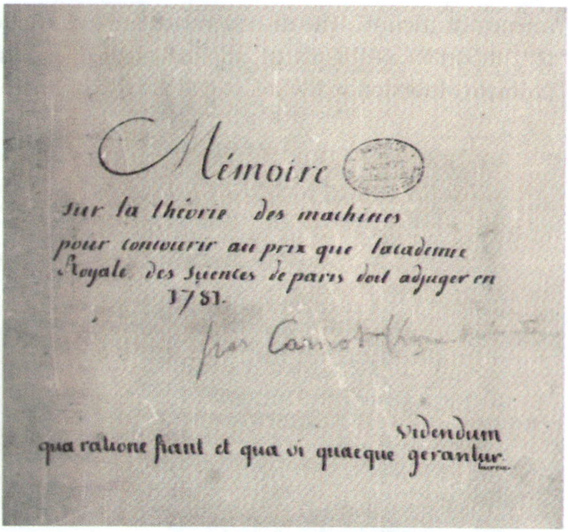

Fig. 4b

Carnot's theory of machines appeared in his *Principes Fondamentaux de l'Equilibre et du Mouvement* [25], published in 1803, a continuation of two previous *Essays*, published in 1779 and 1781, respectively. The general approach found in the theory of machines from the *Essays* onwards is derived from his mechanics based on impacts and fast changes in the velocities of bodies. Consequently, the method based on virtual velocities as used by Lagrange cannot be applied in his case, even though Carnot made a generalization of this method changing infinitesimal velocities for finite velocities with the concept of geometrical motion. These motions can appear in any part of the system and do not depend upon the action and reaction of bodies but only on the constraints. This concept of geometrical motion has two important meanings. It is an attempt to generalize the principle of virtual works for dynamical systems without the application of d'Alembert's principle in its modern version. It is generally accepted that Sadi Carnot owed to Lazare Carnot the idea of reversibility in the comparison of the two states of a system – initial and final [26].

Gaspard-Gustave de Coriolis (1792-1843) – After Lazare Carnot, the most important step towards the construction of an applied mechanics for machines was made by Coriolis in his 1829 work *Du Calcul de l'Effet des Machines* [27], published in 1829. In the twenty-five years which separate it from Carnot's *Principes* no important intermediate work was published, apart from the notes and additions made by Navier (1785-1836) and *Architecture Hydraulique* by Bélidor (1693-1761). Coriolis' book is a complete manual on mechanics where the problems raised by the construction and use of machines are treated with rigor and theoretical refinement. The results of mechanics and differential and integral calculus are used with great skill.

In his above mentioned book, Coriolis corrected the expression for kinetic energy introducing the constant ½ in the old expression mv^2 and adopted the term 'work' to represent the product of a force by its displacement along the force trajectory, replacing the previous nomenclature.

Fig. 5a Gaspard-Gustave de Coriolis

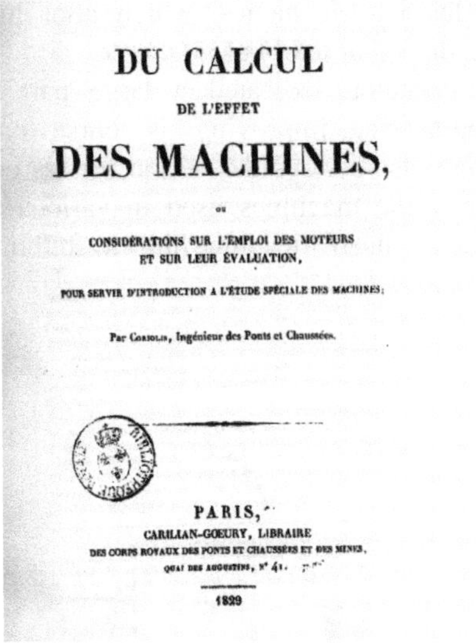

DU CALCUL
DE L'EFFET
DES MACHINES,
OU
CONSIDÉRATIONS SUR L'EMPLOI DES MOTEURS
ET SUR LEUR ÉVALUATION,
POUR SERVIR D'INTRODUCTION A L'ÉTUDE SPÉCIALE DES MACHINES;
Par Coriolis, Ingénieur des Ponts et Chaussées.
PARIS,
CARILIAN-GOEURY, LIBRAIRE
DES CORPS ROYAUX DES PONTS ET CHAUSSÉES ET DES MINES,
QUAI DES AUGUSTINS, N° 41.
1829

Fig. 5b

What is new in Coriolis' approach is the extensive use of the concept of work with its mathematical formalism which is quite different from Newtonian concepts. The approach he adopted can be better understood if we look at the two general approaches used to solve a given mechanical problem. The first considers variations of motion as the result of forces, in a Newtonian manner. The alternative is to discover the amount of work generated by the system and thus try to find the 'living forces' involved. In other words, we can use Newton's second law or the work-kinetic energy principle.

Coriolis discovered the term, the so-called 'fictitious force' which would later receive his name, in 1831 when he was studying the motion of a water wheel applying the principle of the conservation of living forces to the relative motion of these machines [28]. The work was published the following year with the title: *Memorial on the Principle of Living Forces in the Relative Motions of Machines*. A year after his death, his 1829 book was republished with the title *Treatise on the Mechanics of Solid Bodies* and had an important impact in scientific circles.

7 Poncelet´s Contribution to Applied Mechanics

Jean-Victor Poncelet was born in Metz on 1 July 1788, the son of a rich land owner and lawyer [29]. In 1807 he entered the Polytechnic School, where his teachers included Monge, Lacroix (1765-1843), Ampère (1775-1836), Poinsot (1777-1859) and Hachette (1769-1834). He graduated at twenty-two and was admitted as a professor in Metz where, using the work of Navier and Coriolis as well as his own discoveries, he gave the first course on *Applied Mechanics to Machines*, which would be transformed into a book in 1826. In 1829 Poncelet published the Course of Industrial Mechanics, which after his death became an important reference and fundamental textbook.

Poncelet dedicated a large part of his professional life to the teaching of engineering, improving his courses on theoretical and applied mechanics. Machines were also always the reason for his concerns. In 1825 he made a tour of Europe to study the machines used in various countries. He visited Germany, Belgium and France itself, leading him to rethink the application of machines to industrial processes.

Fig. 6a Jean-Victor Poncelet

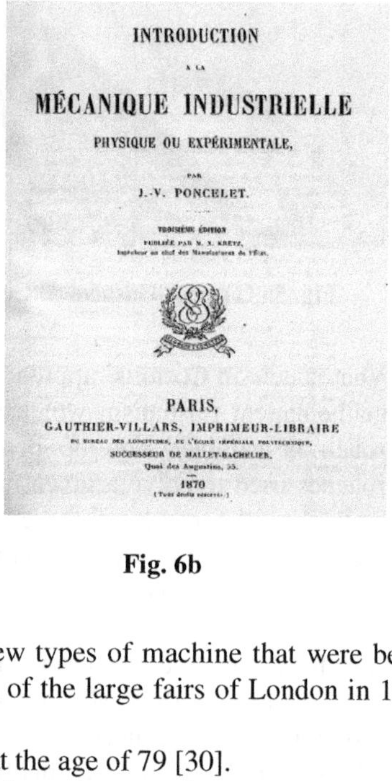

INTRODUCTION

À LA

MÉCANIQUE INDUSTRIELLE

PHYSIQUE OU EXPÉRIMENTALE,

PAR

J.-V. PONCELET.

TROISIÈME ÉDITION

PUBLIÉE PAR M. X. KRETZ,

PARIS,

GAUTHIER-VILLARS, IMPRIMEUR-LIBRAIRE

SUCCESSEUR DE MALLET-BACHELIER,

Quai des Augustins, 55.

1870

Fig. 6b

In order to keep himself up to date on the new types of machine that were being developed, he took part as one of the organizers of the large fairs of London in 1851 and Paris in 1855.

Poncelet died in Paris on 22 December 1867 at the age of 79 [30].

The most important characteristics of Poncelet's work on applied mechanics can be summarized as follows:

Poncelet-*polythecnicien*: The project of the experimental and physical mechanics, which is a restricted version of the Laplacian project, mobilized a group of French *polytechniciens* during the first-half of the nineteenth century, including Poisson, Navier, Coriolis, Cauchy (1789-1857), Saint-Venant (1797-1886) and others. The most representative name in the field of the new mechanics is obviously Poncelet. It is possible to separate many aspects of Poncelet as a *polytechnicien*:

Inventor: a model of a bridge presented to the Fortification Committee in 1820, constructed and much used; the hydraulic wheels with his name, whose design was published in1827 with the title *Mémoire sur les roues hydrauliques a aubes courbes* [31]; a dynamometric device to measure the work of motors and machines; an apparatus to experimentally discover the laws of non-uniform motion of machines.

Theoretician of the new physical mechanics: he proposed a geometric representation of the molecular constitution of matter which can explain the macroscopic phenomena such as elasticity and deformation of bodies.

Experimentalist: he idealized and performed several research programs with an experimental nature about water flow through orifices;

Lecturer: working on theory and experiments, Poncelet developed a modern theory of machines. Applying this knowledge in his courses he made this knowledge known to engineers, students, industrialists, craftsmen and workers.

Poncelet- theoretician: Facing the new problems presented by 'physical mechanics,' Poncelet had to formulate a theory that could explain the constitution of matter and allow him to make conclusions about the behavior of natural bodies. Poncelet was one of the few theoreticians who adopted the concept of bodies as systems of molecules maintaining small distances between each other. These systems are subjected to attractive and repulsive forces, similar to the general proposition of the Laplacian project. In addition to these assumptions he adopted a geometric representation in which he discusses the problems which appear in macroscopic scale.

On the other hand, Poncelet postulated a philosophical point of view based on a scientific realism, which analyzed the natural world using models. It means that he postulated the existence of theoretical entities that were not directly observable but which were nonetheless as real as the physical world.

In relation to the general model of the molecular constitution of matter, Poncelet proposed some specifications. Similar to Poisson, Poncelet also proposed that action between two molecules, being attractive or repulsive, depends upon the respective distances and that a kind of law could be derived. With the expansion of distance the forces decrease as also prescribed by the Laplacian project.

Poncelet was the first to propose a graphic representation of the variation of the intensity of forces as a function of the distance that separates two molecules. Based on this model he explained macroscopic phenomena as elasticity. The difference between different materials can be explained by different geometric curves representing their molecular action.

Poncelet-experimentalist: Working with the military engineer Joseph-Aymé Lesbros (1790-1860), Poncelet carried out a series of experiments trying to understand the laws of water flow through orifices. Based on numerical findings the final report of this investigation is a good manual for experimentalists. With its help it is possible to reproduce and to perform the same experiments. In addition they proposed a kind of experimental method within this field of investigation.

Also common in Poncelet is an experimental procedure that is a kind of 'mental experience,' in other words a mental construct from possible experimental situations. In an example of a similar situation, in 1798 Lagrange used an imagined pulley system to prove the virtual velocities principle which belonged to basic rational mechanics.

In relation to the experiments in the applied mechanics field it is important to explain that this operation cannot be reduced to a production activity. In order to produce a set of valuable results by means of experiments it is necessary to have experimentation protocols which establish the conditions of findings. On the other hand, it is worth emphasizing that the experimental works carried out by eighteenth century engineers were neither sufficient nor rigorous enough to create valuable scientific knowledge.

8 Poncelet's Applied Mechanics Courses

Poncelet participated significantly in the spreading of mechanical knowledge in different forms. Probably the most important was teaching applied mechanics courses. From 1825 to 1834 Poncelet gave a course on machines in the Metz School of Artillery and Engineering to engineers who had finished the Polytechnic School and intended to follow a military carrier. He also gave another course with the title of industrial mechanics to workers in Metz from 1827 to 1830.

In 1874, François Xavier Kretz published a book that is part of Poncelet´s lessons in École d´Application de l´Artillerie et du Génie de Metz. The course, "Cours de mécanique appliqué aux machines", which grew out of lectures that Poncelet had given from the above mentioned period. The course follows the general topics presented below:

9 Applied Mechanics to Machines

SECTION I: GENERAL CONSIDERATIONS ABOUT MACHINES IN MOTION

SECTION II: MAIN METHODS TO CONTROL THE ACTION OF FORCES APPLIED TO MACHINES AND HOW TO TRANSMIT VELOCITIES WITH GIVEN RATIOS

SECTION III: HOW TO CALCULATE THE RESISTANCES OF PIECES IN UNIFORM MOTION

SECTION IV: ON THE INFLUENCE OF VELOCITIES VARIATIONS OVER THE RESISTANCES

Some years later this course was complemented with more two sections including practical hydraulics, gas motion, water wheels and steam machines. In Section 1, Poncelet had the objective to explain the general principles applied to the theory of machines in motion and how the mechanical properties can prescribe the conditions to be satisfied by machines. He also established the general equations governing machines in motion by applying the living forces principle. In Section 2, Poncelet develops the study concerning the systems of regulation of the action of forces in order to assure uniformity to motion. This section was dropped out after 1826.

In Section 3, different kinds of friction were studied based on Coulomb findings which were taken into account by Navier and Belidor. In addition, Poncelet also used the approaches indicated by Lazare Carnot in his "Principes" regarding motion with high changes of velocities.

The Sorbonne course given from 1838 to 1848 represents the final period of his life as professor. The title of this course is applied and experimental mechanics. Poncelet uses elementary mathematics such as geometry, arithmetic, elementary algebra, trigonometry and logarithms. This course is heavily illustrated and rich in descriptions of measurement instruments and similar apparatus with detailed instructions for their use. The program reproduced below refers to a course in Sorbonne in 1840-41 [32].

10 Physical and Experimental Mechanics

PART I: SUMMARY OF FUNDAMENTAL NOTIONS RELATED TO THE CONSTITUTION AND PHYSICAL PROPERTIES OF BODIES

SECTION I: ON MOTION FROM GEOMETRIC POINT OF VIEW

CHAPTER I: Fundamental preliminary notions

CHAPTER II: The laws of motion

CHAPTER III: Geometric transformations of motion

CHAPTER IV: Composition and decomposition of absolute and relative motion

SECTION II: ON FORCES AND THE ABSTRACTION OF MOTION THAT THEY PRODUCE OR DESTROY

SECTION III: ON THE RESISTANCE OF SOLIDS

SECTION IV: ON THE COMMUNICATION OF MOTION BY FORCES

The above topics are only the main ones, because the sections are further divided into chapters and the chapters into lessons. Figures showed below are illustrations presented in that course.

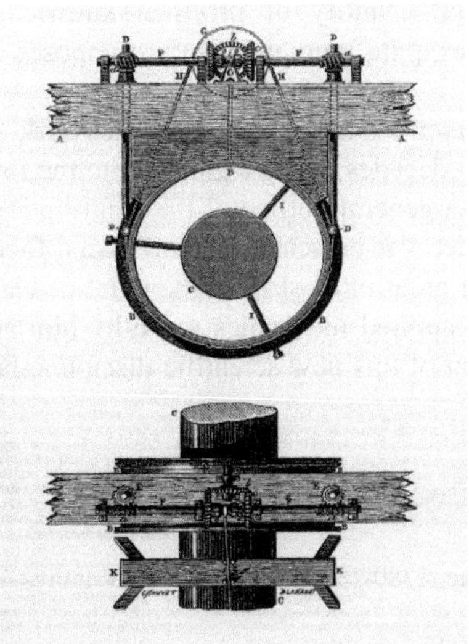

Fig. 7. Dynamometric brake

Fig. 8. An alternative motion mechanism

Some remarks about Poncelet's course in Sorbonne need to be made. Part I, dedicated to summarizing 'Fundamental notions and physical properties of bodies,' is similar to the corresponding part of his 1841 'Introduction to Industrial Mechanics.' In relation to Section 1 on 'Motion considered from the geometric point of view,' this was the first course on theoretical kinematics either in France or elsewhere in the world. In 1834, Ampère suggested the creation of a science of motion abstracting the conditions of its production, in other words the study of motion without any consideration of forces. After Poncelet this new science was definitively incorporated in mechanics.

Returning to the course program, after Section 1 Poncelet analyses statics which can be visualized from the point of view of theoretical mechanics as well as the effects of forces implied in motor studies that are part of applied mechanics. These two approaches are unified by a systematic utilization of the concept of work. In addition to the study of the theoretical aspects of work, Poncelet develops methods and instruments to measure this.

In relation the Section 3, on the 'Resistance of Bodies,' this is actually the same as strength of materials [33]. We can find tensile, compression and shear forces, bending and torsion applied to prismatic sections. The course is of an advanced level and proposed several new concepts and methods at that time.

Poncelet also referred to a course given in 1837-38 by Saint-Venant in *École des Ponts et Chaussées* entitled Mechanical Lessons. It was the first course on the strength of materials that used the new concepts of the theory of elasticity.

11 Final Remarks and Conclusion

Throughout this paper we have tried to show that the development of applied mechanics and afterwards of industrial mechanics was part of a complex social, economic and cultural process which combined the dynamics of the industrial and scientific revolution in the context of a large quantity of practical knowledge accumulated over the centuries. This process was quite long and also presupposes an important advance in the theory of mechanics.

An important part of the development of applied mechanics was the work of polytechnic engineers in the period from the last decades of the eighteenth to the first decades of the nineteenth century. In this paper a general outline of the contributions of Coulomb, Lazare Carnot, Coriolis and Poncelet was presented with a special focus on the latter. The study of Poncelet's work is of great interest from the point of view of the history of science. The developments in applied mechanics made by him are not restricted to the formulation and construction of this new scientific discipline but also to its diffusion.

References

[1] Vantin, F.: Le travail: Économie et physique 1780-1830. Presses Universitaires de France, Paris (1993)

[2] Oliveira, A.R.E.: The evolution of the physical concept f work in machines context. Prestígio Editorial, Rio de Janeiro (2006) (in Portuguese)

[3] Landriani, G.S., Curzio, A.: Sfogliando la méchanique analitique. Edizioni Universitarie di Lettere Economia Diritto, Milan (2006)
[4] Chatzis, K.: Jean-Victor Poncelet et la "science des machines" à l'École de Metz: 1825-1870. Actes de La Journée d'étude du, 32–43 (Novembre 2, 1995)
[5] Lilley, S.: Hombres, máquinas y história. Editorial Ciência Nueva S.L., Madri (1965)
[6] Blay, M.: La naissance de La mécanique analytique. Presses Universitaires de France, Paris (1992)
[7] Lagrange, J.L.: Mécanique analytique. Éditions Jacques Gabay, Paris (1989)
[8] Gillispie, C.C.: Lazare Carnot savant. Princeton University Press, Princeton (1971)
[9] Oliveira, A.R.E.: Coriolis' theory of machines and mechanisms. In: IFToMM World Congress, Guanajuato, Mexico (2011)
[10] Oliveira, A.R.E.: The contribution of Coulomb to applied mechanics. In: HMM Symposium, pp. 217–226. Kluwer Academic Publishers, Cassino (2004)
[11] Chatzis, K.: Jean-Victor Poncelet ou le Newton de la mécanique appliqué: quelques réflexions à l'occasion de son cours inédit à la Sorbonne. Bulletin de la Societé des Amis de la Bibliothèque de l'École Polytechnique (19), 69–97 (1998)
[12] Aristote: La physique, Librairie Philosophique J. Vrin, Paris (1999)
[13] Oliveira, A.R.E.: Some origins of TMM arisen from pseudo- Aristote and Hero of Alexandria. In: HMM Symposium, Taiwan, China (2008)
[14] Dugas, R.: A history of mechanics. Dover Publications, Inc., New York (1988)
[15] Dálexandrie, H.: Les mécaniques ou l'élévateurs des corps lourds. Texte árabe de Qustã Ibn Luqã établi et traduit par Carra de Vaux. Imprimerie F. Paillart, Paris (1988)
[16] Leopold, J.H.: L'invention par Christiaan Huygens du resort spiral réglant pour les montres. Librairie Philosophique J. Vrin, Paris (1982)
[17] Elster, J.: Leibniz et la formation de l'ésprit capitaliste. Éditions Aubier Montaigne, Paris (1975)
[18] Oliveira, A.R.E.: Challenges to technological education in history. In: Proceedings of the COBEM, Ouro Preto, Brazil (2005)
[19] Belhoste, B.: Les origins de l'École Polytechnique, Histoire de l'éducation, vol. 42, pp. 13–53 (1989)
[20] Chatzis, K.: Theory and practice in the education of French engineers from the middle of the 18th century to the present. American Engineering Education, Reforms and Changes, 1920-1980, Icon, (5), 40–63 (1999)
[21] Rashed, R.: Sciences a l'époque de la Révolution Française. Recherches historiques, Librairie Scientifique et Technique Albert Blanchard, Paris (1988)
[22] Oliveira, A.R.E.: Euler's contribution to classical mechanics. In: Proceedings of the XII IFToMM World Congress, Bésançon, France (2007)
[23] Harman, P.M.: Energy, force and matter: the conceptual development of nineteenth-century physics. Cambridge University Press (1982)
[24] Coulomb, C.A.: Théorie des machines simples. Librairie Scientifique et Technique Albert Blanchard, Paris (2002)
[25] Carnot, L.N.M.: Principes fondamentaux de l'équilibre et du movement (1803)
[26] Charnay, J.P.: Lazare Carnot ou le savant-citoyen. Presses de l'Université de Paris-Sorbonne, Paris (1990)
[27] Coriolis, G.G.: Du calcul de l'effet des machines. Librairie Scientifique et Technique Albert Blanchard, Paris (1990)
[28] Koetsier, K.: A note on Gaspard-Gustave de Coriolis (1792-1843). In: Proceedings of the 11th IFToMM World Congress, Taiwan, China (2003)
[29] James, I.: Remarkable mathematicians. Cambridge University Press, England (2002)

[30] Bell, E.T.: Los grandes matemáticos. Editorial Losada, Spain (2009)
[31] Poncelet, J.V.: Mémoire sur les roues hydrauliques a aubes coubes. Librairie de Me. Ve. Thiel, Metz, France (1827)
[32] Chatzis, K.: Les cours de mécanique appliquée de Jean-Victor Poncelet à l'École de l'artillerie et du génie et à La Sorbonne, 1825-1848. Histoire de l'éducation, 120, 113–138 (2008)
[33] Timoshenko, S.P.: History of strength of materials. McGraw Hill Book Company, Inc., New York (1953)

The Case of Kinematics, the Genesis of a Discipline

Teun Koetsier

Department of Mathematics, Faculty of Science,
VU University Amsterdam
t.koetsier@vu.nl

Abstract. This paper contains a sketch of the history of kinematics from Antiquity until the twentieth century. The word kinematics was proposed by Ampère in the first half of the nineteenth century. First the circumstances surrounding that proposal are discussed. Then a sketch is given of the areas where the word was used after Ampère. Finally a survey is given of the different areas where kinematical results existed before Ampère.

1 Introduction

Imagine three countries called A, B and C, inhabited respectively by Geomonians, Mechanians and Machinians. See figure 1. At a certain moment a prophet appears who teaches that in the area where the three countries meet the inhabitants are not really Geomonians, Mechanians or Machinians, but something else: they are actually Kinamanians. The leader urges that all Kinamanians unite and form a new country: country D. In such a situation all kinds of interesting questions can be asked. What is going to happen in the future? How are the populations of A, B and C going to react? All kinds of scenarios are imaginable: the Kinamanians get restricted autonomy in the three countries, they are exterminated, they take over the whole world, etc. One also wonders what the position of these Kinamanians was in the past. Did they already view themselves as different? The prophet cannot have come out of the blue. Obviously in course of time the borders between A, B and C need not always have been the same and the names of the peoples may have been subject to change. This product of my fantasy gives us a nice way to look at the history of kinematics. Ampère was the prophet who realized that parts of rational mechanics, the theory of machines and geometry deserved a different name: kinematics. At least that is a way to understand his message.

The structure of the present paper is the following. First I will discuss the background of Ampère's proposal to introduce the word 'kinematics'. Subsequently I will briefly look at the development of kinematics after Ampère in three areas: geometry, rational mechanics and mechanical engineering. Finally we will discuss the role of kinematics before Ampère in geometry, astronomy and mechanics. One notices that the three areas before Ampère differ from the three areas after Ampère. In the 17th century and before 'mechanics' was another word for 'theory of machines'

T. Koetsier, M. Ceccarelli (Eds.): Explorations in the His. of Machines & Mech., HMMS 15, pp. 491–504.
springerlink.com

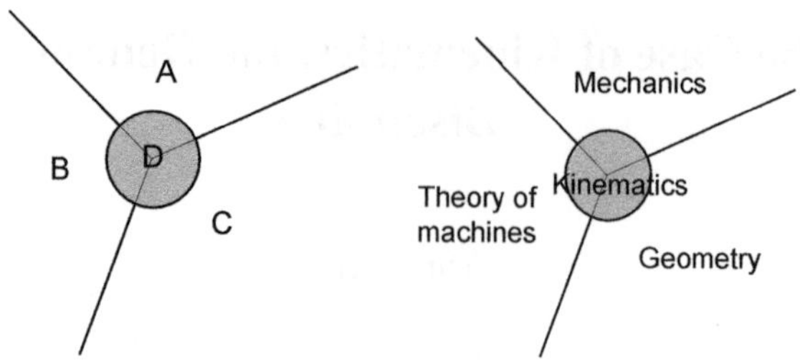

Fig. 1.

and the problem of the motion of the planets, which in the 18th century became part of rational mechanics, was part of astronomy. Borders changed and so did the meaning of names.

2 The Immediate Background of Ampère's Proposal

At the end of the 1820s Ampère coined the word 'kinematics' [1]. He wrote: "This science must encompass everything that can be said of the different kinds of motion, independently of the forces that can produce them". Ampère was at the time developing a new classification of the sciences. In his final classification he distinguished thirty two 1st order sciences. Four of these 1st order sciences were mathematical: arithmology (=arithmetic), geometry, *mechanics* and uranology (=astronomy). In Ampere's scheme, mechanics consisted of two 2nd order sciences: *elementary mechanics* and *transcendental mechanics. Elementary mechanics* consisted of two 3rd order sciences: *kinematics* and *statics,* while *transcendental mechanics* consists of *dynamics* and *molecular dynamics*. Clearly, in Ampere's view kinematics had to be considered as a 3rd order science, as a sub-science of mechanics.

It is very probable that Ampère was inspired by a suggestion of Lazare Carnot, because Ampère referred to Carnot's work. In 1783 *Lazare Carnot (1753-1823*) wrote his *Essai sur les machines en général* [3]. In 1803 he wrote *Principes fondamentaux de l'équilibre et du movement* [4], in which he elaborated on the same ideas. He wrote in the *Principes*: <<*The theory of geometrical movements is very important; it is [...] a kind of science intermediate between ordinary geometry and mechanics. It is the theory of the movements that an arbitrary system of bodies can make in such a way that they do not hinder each other or exert some action or reaction on each other.* This science has never been dealt with in particular. *This science must be created completely and it deserves for its beauty and its utility the full attention of the scholars*; because the great analytical difficulties one meets in mechanics and especially in hydraulics are only caused by the fact that a theory of geometrical movements has not been created at all.>> [4, p. 116; italics are mine – T. K.].

Carnot's geometrical movements are virtual displacements. Carnot considers a machine as consisting of non-elastic hard bodies. Because a distribution of virtual displacements is equivalent to a velocity distribution, Carnot can apply the theory of collisions and derive general results on kinetic energy. Carnot represents a new approach in the theory of machines that I call the *energetic approach*: one looks at machines with the intention to minimize the loss of energy during its functioning. Until the end of the 18th century, the prevailing theory of machines treated the simple machines: the lever, the wheel, the pulleys, the wedge and the screw. More complex machines consisted of simple machines. An analysis of such a machine took place on the basis of the determination of the equilibrium of forces.

It is remarkable that in this period also elsewhere the geometrical aspects of machines were gaining importance as part of a third approach to the investigation of machines. Watt's famous straight line linkages nicely illustrate this trend.

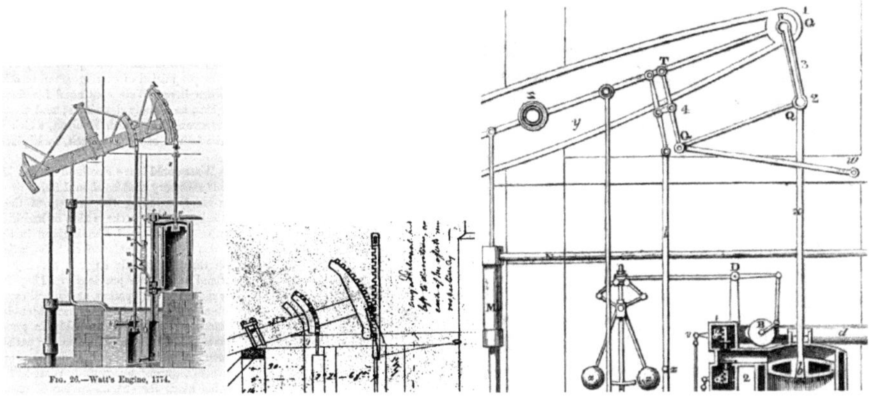

Fig. 2. Watt's 1747 engine; sketch of the simple parallel mechanism and (right) Watt's parallelogram. Source: [27], [28] and [27] respectively.

"I am more proud of the parallel motion than of any other mechanical invention I have ever made" Watt wrote in 1808 [5]. The parallel motion was soon modified by Watt. He added a parallelogram and made the mechanism more compact. In 1834 Arago wrote about "Watt's parallelogram": "At each double oscillation it unfolds itself and closes itself with the softness, I would almost say with the grace that fascinates us in the gestures of an accomplished actor" [5]. Watt's parallelogram transformed a to and fro rectilinear motion in a to and fro rotary motion. Clearly, the way in which in a machine motions were transformed into other motions represented an interesting and important geometrical aspect of machines. In 1784 the Ecole Polytechnique was founded and Monge introduced on the basis of this idea a course called "*Elémens de machines*" which was part of the course on descriptive geometry. In this course machine elements or 'mechanisms' were treated as motion transformers. It became clear that a machine could also be considered as consisting of mechanisms. The mechanisms were viewed as mechanical means to transform an input motion into an output motion. Four kinds of motion could be distinguished: circular continuous, circular to and fro, linear continuous and linear to and fro.

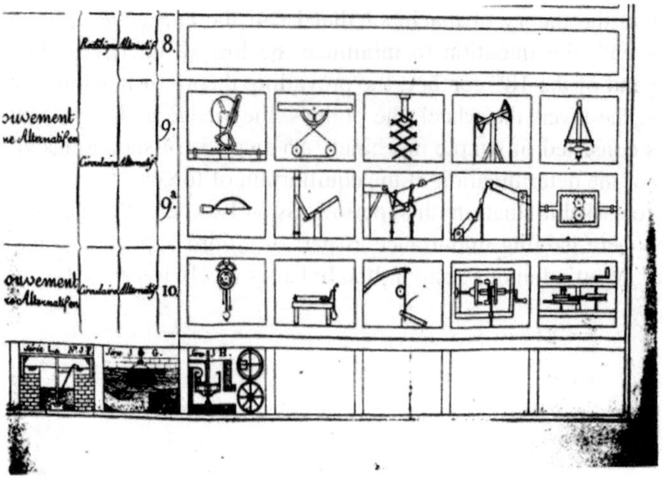

Fig. 3. [6]

Figure 3 shows part of the well-known drawing that was used in the course on machine elements at the Ecole Polytechnique. One notices Watt's mechanisms in the rows 9 and 9ª that show mechanisms that transform a rectilinear to and fro motion into a circular to and fro motion.

After the publication of Ampère's new classification of the sciences slowly the word kinematics gained popularity. Older scattered results in geometry, mechanics and in the theory of machines were recognized as belonging to kinematics and new results were obtained. In principle all results in geometry related to motion, all results on velocities and accelerations in theoretical mechanics and all results in mechanical engineering only dealing with the geometry of mechanisms and the time-dependence of motion could be considered as kinematical.

3 Kinematics and the Geometers after Ampère

The 19th century was a golden age for geometry. Poncelet defined projective geometry and Chasles reworked it. Plücker came up with a theory of the singularities of curves. Non-Euclidean geometries were discovered. Klein introduced the notion of group in geometry. In the 18th century geometry had basically slept, but it woke up in the 19th century with a vengeance. Because kinematics and geometry are very close, many 19th century geometers were fascinated by the kinematics. A complete survey of the work of these geometers in kinematics goes far beyond the scope of this paper. We will restrict ourselves to a few remarks.

A considerable list of French geometers contributed to kinematics. Many of them distinguished between pure and applied kinematics. One of the earliest who did so was Aimé Henry Resal (1828-1896). Pure kinematics encompassed the general kinematical properties of motion: instantaneous centre and polhodes, their spherical and spatial analogues, properties of velocities and accelerations, all properties that are valid for more than specific motions. Applied kinematics dealt with specific motions. Often these specific motions were defined by specific mechanisms. For example, Watt's mechanism directly led to the investigation of the four bar coupler curves. A nice survey of the work by French mathematicians can be found in [7] and [8] by Gabriel Koenigs (1858-1931). They can be considered as a culmination of French 19th century work in kinematics by mathematicians. In Germany among the geometers Ludwig Burmester has a special position. His *Lehrbuch der Kinematik* [9] represents a first synthesis of pure and applied plane kinematics. Figure 4 suggests the complexity that kinematics of mechanisms reached in this period.

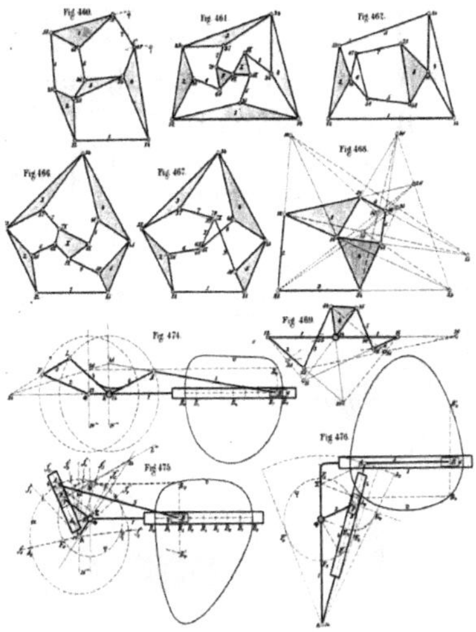

Fig. 4. A page from the atlas that went with Burmester's [9]

4 Kinematics and Rational Mechanics after Ampère

In the middle of the 18th century Newton's law F =m.a or F=d(m.v)/dt was interpreted in two different ways. D' Alembert used it as a definition of force in terms of mass and velocity. Euler saw it as a physical law that relates three different notions: force, mass and acceleration. After Ampère, Resal followed in D'Alembert's footsteps: "As soon as one has postulated the law of inertia and the equality of action and reaction, the physical part of mechanics, that is the investigation of the motion of material systems, solid bodies, is reduced to a simple corollary of kinematics" ([10] Préface, p. x). From this point of view all of rational mechanics was basically kinematics. Following the discussions on the foundations of mechanics goes beyond the scope if this paper. Yet it is remarkable that the special theory of relativity is essentially a kinematical theory. The 19th century kinematical background of Einstein's relativity theory has been studied by Alberto Martinez in 2009 [11].

5 Kinematics and Mechanical Engineering After Ampère

After Ampère in mechanical engineering the word kinematics was used too. In this respect, undoubtedly the German professor of mechanical engineering Franz Reuleaux (1829-1905) is the most conspicuous figure. Reuleaux defined kinematics as the study of those arrangements of a machine by which the mutual motions of its parts, considered as changes of position, are determined. He viewed kinematics as part of the science of machines *where it stands first in many respects*. Basically Reuleaux attempted to turn kinematics into the foundation of the theory of machines. He introduced this approach in his lectures in Zürich in the early 1860s and later in his fundamental book published in 1875 [12]. A revolutionary element in his approach is his proposal to view the pedestal of a mechanism, the parts fixed to the ground, as one of the links in a kinematic chain. In figure 5 on the right side we have a kinematic chain consisting of links. Whenever we fix one of the links we get a mechanism. The different mechanisms that can be generated in this way are kinematically equivalent.

Fig. 5. Front of the lithographed 1865 edition of Reuleaux' classes in Zürich (right). Example of kinematical chain (left).

Reuleaux's definition of kinematics differs from the definition that many others were using. Reuleaux considered Burmester's work not as kinematics but as geometry of motion or 'phoronomy'. This last word had also been used by the philosopher Immanuel Kant. Reuleaux severely criticized Burmester in this respect. He wrote: "[Burmester's ideas] show to which completely unforeseeable dangers a developing theory is exposed, *because* they are *completely fairytales*, *phantasy from A to Z* [...] while one, in Germany, I emphasize, recently says, writes, teaches '*kinematics*' instead of '*geometrical theory of motion*', one also started spreading [...] that for Ampere himself *kinematics* meant *geometrical theory of motion*" [13]. Reuleaux thought that he alone understood what Ampere had meant. In order to understand Reuleaux' reaction one should know that Reuleaux' kinematics was at the time under vehement attack from Alois Riedler (1850–1936), a colleague of Reuleaux, who represented engineering practice. What is the use of Reuleaux' kinematics? he asked. And indeed, the gap between Reuleaux' work and engineering practice was considerable.

6 Kinematics in the 20th Century

In the 20th century in geometry, mechanics and mechanical engineering things changed. The Golden Age in geometry was over and so was the Golden Age of kinematics. Kinematics never became more than a sub-discipline of another discipline. Although some geometers continued to work in kinematics, most mathematicians turned to other more fertile research subjects. In mechanics we had a similar situation. The word was used make clear masses and forces played no role in a particular problem. In mechanical engineering it was discovered soon that a machine is much more than kinematics. Yet in the 20th century, mechanical engineering was the only area where kinematics still flourished.

7 Kinematics and Geometry before Ampère

Motion has always played a role in geometry. One can even without exaggeration say that without motion geometry would not exist. This is remarkable because one often finds the opposite view. For example, Felix Klein wrote (1908): "[the Ancients feared that motion would] bring into geometry an element foreign to it, namely the notion of time." [14]. And Thomas Heath (1921) when discussing Plato's views about geometry: "Constructions, then, or the processes of squaring, adding, and so on, are not of the essence of geometry, but are actually antagonistic to it." [15]. In a similar vein a more recent paper says: "The Eleatics with their paradoxes of motion had shocked mathematics and led mathematicians to try to eliminate all motion from their discipline. Aristotle, for example, had forbidden the use of motion in geometry. Euclid avoided any explicit mention of it."[16]. I disagree with the last statement but it nicely illustrates the ideas about Greek geometry that float around. What is the fact of the matter? Of course Plato was very influential and motion seems to be opposed to the nature of mathematics as may have seen it. On the other hand, every mathematician in his right mind realized that already in the Euclidean foundation of the geometry in the first three postulates motion is essential. That is why Proclus wrote: "How we can introduce motions into immovable geometrical objects and move things that are without parts – operations that are altogether impossible [...]?" [17]. And Proclus answered: the motion involved is not bodily, but in the imagination.

In Antiquity motion was the most important way to generate new geometrical objects. In Book XI of the *Elements* Euclid defines the sphere as the figure comprehended by the revolution of a semi-circle about its diameter (Definition 14). In the same book Euclid defines a cone as the figure comprehended by the revolution of a right-angled triangle about one of its sides containing the right angle (Definition 18). The cylinder is defined by Euclid as the figure comprehended by the revolution of a rectangle (Definition 21). Of course, it is in principle possible to give non-kinematic definitions of these geometrical objects. For example, before Euclid Aristotle defined the sphere without reference to motion. However, the great advantage of kinematical definitions is that the continuity of the surfaces is immediately guaranteed.

It is well known that several problems of construction defined by the ancients cannot be solved by means of compass and ruler. The most famous are: the trisection of the angle, the doubling of the cube and the quadrature of the circle. Several other solutions to these problems have been given in Antiquity. Many of them are based on the introduction of new curves that are kinematically defined. Examples of these curves are the quadratrix, Archimedes' spiral, conchoids and the cissoids.

Against this background it is interesting to consider the role of motion in geometry in the scientific revolution. The introduction of algebra in geometry by Fermat and Descartes eventually led to a situation in which a completely algebraic treatment of geometry with the full exclusion of motion became possible. Yet here again motion played an essential constitutive role. 16th and 17th century mathematicians combined great admiration for the Greek achievements with the willingness to violate the canons of Greek mathematics in order to get results. For most 17th century mathematicians the possibility of idealized motion going beyond the motions of compass and ruler in geometry was a natural starting point. Looking back it is clear that this was a sine qua non for the development of mathematics in the 17th century. Let us consider some examples.

Napier started the booklet in which he introduced the logarithm by means of an arithmetical sequence of exponents and a geometrical sequence that are both defined in a kinematical way by means of two points that are moving along straight line segments. The two movements are coupled. The advantage of Napier's kinematical starting point is that the motions are continuous. Some 20 years later the combination of the two movements led to a new kinematically defined curve, the *logarithmica*. Out of this development the exponential functions and the number e would be defined.

In the 17th century analytical geometry was independently introduced by Fermat and Descartes. For both mathematicians motion played an essential role. We will restrict ourselves to Descartes. In analytical geometry curves are represented by means of equations. However, for Descartes the equations did not define the curve. Descartes argued that curves are only acceptable "provided they can be conceived of as described by a continuous motion or by several successive motions, each motion being completely determined by those which precede; for in this way an exact knowledge of the magnitude of each is always obtainable". ([18], p. 43). In Descartes' research program the kinematical generation of curves was essential. This kinematical generation does not take place in empirical space. Descartes like Proclus argued that the generation of the curves takes place in the "imagination". Descartes saw the imagination as a screen on which the mind could create figures, manipulate them, and inspect the results of these manipulations.

It is remarkable that also when the calculus was defined by Newton motion played an important role. Newton started his "Treatise of the quadrature of curves" (De quadratura curvarum, written in 1676 and published in Latin in 1704) with the following sentences: "I consider mathematical Quantities in this Place not as composed of very small Parts; but as describ'd by a continued Motion. Lines are describ'd and thereby generated, not by the Apposition of Parts, but by the continued Motion of Points; Superficies's by the Motion of Lines; Solids by the Motion of Superficies's, Angles by the Rotation of the Sides, Portions of Time by a continual Flux: and so in other Quantities."

8 Kinematics and Astronomy before Ampère

With Hipparchus Greek astronomy became a scientific discipline in quite a modern sense of the word. On the one hand, astronomical phenomena were recorded with great precision and care. On the other hand geocentric kinematical models were built in order to explain these phenomena. These models were based on the combination of uniform circular motions. Let us consider an example from Ptolemy's work.

Consider the model for the Moon in figure 6 (left). The Moon is describing a small epicycle of which the center describes a big circle around the center of the earth. The arrow pointing at the Mean Sun doesn't point at the actual position of the Sun; it points at the position that the Sun would have if it were executing a uniform rotation about the earth. However in order to couple the motion of other heavenly bodies to the motion of the sun, the Mean Sun can be used. The position of the Mean Sun is

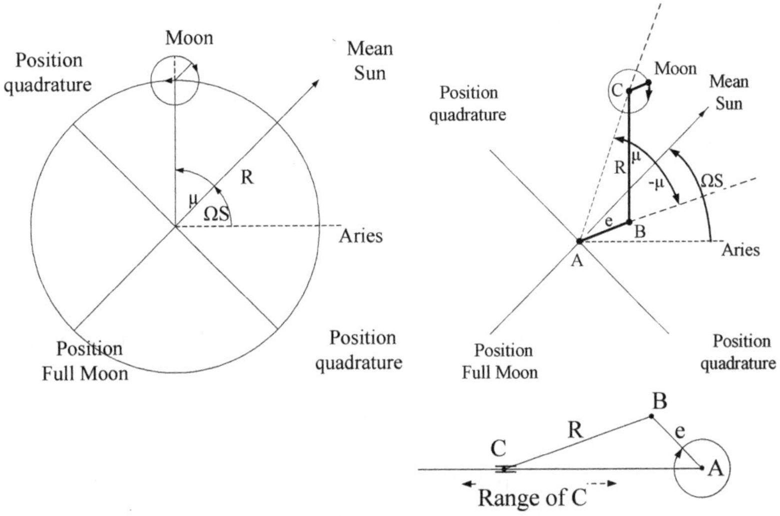

Fig. 6. Epicycle model for Moon (left); Ptolomy's modification (right)

determined by means of the angle ΩS, starting from a fixed point on the ecliptic, for example, Aries. The position of the epicycle of the Moon is determined by means of the angle ΩS+μ, also measured from Aries. Obviously Ω, μ and ΩS+μ are linear functions of time.

The epicycle model for the Moon was fine except for the fact that Ptolemy needed a mechanism to pull the Moon somewhat towards the Earth at so-called quadrature positions from the Sun.

Ptolemy did this in Book IV of the *Almagest* by adding what is from a modern point of view an abstract crank and slider mechanism to the simple epicycle model. See figure 6 right. The connections A, B and C are hinges. Point A is a fixed and so is the line AC along which C can slide during the rotating input of crank AB.

The basic idea is that we start from the crank slider motion. A is the position of a terrestrial observer and C is the center of the epicycle of the Moon. A is fixed but AC rotates about A with the corresponding angular velocity. We also maintain the angular velocity of the Moon about the centre of the epicycle from the first model. The distance of the epicycle from the earth is regulated by means of the crank AB; the crank moves in such a way that its direction, measured from Aries, is given by the angle ΩS+μ.

Astronomical models remained kinematical until the end of the 17th century. But even after the introduction of Newtonian mechanics astronomical research led to interesting results. For example Euler and D'Alembert share the honor for the discovery of the instantaneous axis of rotation in spherical kinematics. D'Alembert showed in his *Recherches sur la Précession* of 1749, on pp. 82-83, that at each instant the locus of points of the earth that are at rest with respect to the center of gravity of the earth is an axis of rotation. D'Alembert was studying the motion of the axis of the earth with respect to its center of gravity. Euler's treatment concerned all rigid bodies and, moreover, Euler added a proof which gave a more general result: Given two different positions of a body with a fixed point, it is always possible to move the body from one of the two positions to the other one by means of a rotation about an axis through the fixed point. Euler discovered the axis of rotation in discrete spherical mathematics.

9 Kinematics and the Theory of Machines in Antiquity

The theory of machines was until the 18th century the theory of the simple machines. Starting with Archimedes the simple machines were usually treated on the basis of equilibrium conditions; motion was excluded from the treatment. Yet sometimes another approach was used. Before Archimedes Aristotle (or one of his pupils) wrote a text that is usually called *Mechanical Problems*. The author studied 35 problems. For example:

1. Why are great weights and bodies of considerable size split by a small wedge, and why does it exert great pressure?
2. Why is it that if one puts two pulleys on two blocks [...] one can draw up great weights, even if the dragging force is small?
3. Why can one easily break nuts without a blow in instruments made to break them?

The basic idea is that all these phenomena can be explained on the bsis of the law of the lever. The text contains its first precise formulation: in equilibrium the ratio of the weights is inversely proportional to the ratio of the distances from the fulcrum. Moreover the author believes that the law of the lever is a consequence of the fact that the velocities of the points on a balance or lever are proportional to the distances to the fulcrum. The basic question that he attempts to answer before turning to the 35 problems is: The natural motion on earth of an object is linear. What brings about on earth circular motion?

First, however, he proves the theorem that the composition of two uniform linear movements is a uniform linear movement along the diagonal. See figure 7 (left). Nota bene: this is the first extant appearance of the parallelogram of velocities. The proof runs as follows. We consider two uniform motions, one from A to B and one from A to G. The ratio of the movements is that of AB to AG. The composition of the two movements yields a movement from A to H. The question is: What kind of movement is this? Imagine that when A in its movement to B has reached D and in its movement to G has reached E. Then, given the definition of uniform motion, we have AD : AE = AB : AG. This implies that the composition of the two movements of A to D and F, respectively, is on the diagonal of parallelogram AGHB.

This result implies that circular movement on earth cannot be the result of the composition of two uniform linear movements. Because in the case of the composition of two uniform movements they possess a fixed ratio in a fixed time, the author concludes that circular movement must be the result of a composition of two linear movements "with no fixed ratio and in no fixed time" [20, pp. 340-341]. This means that at least one of the two movements must be non-uniform. At this point the author refers to figure 7 (right)

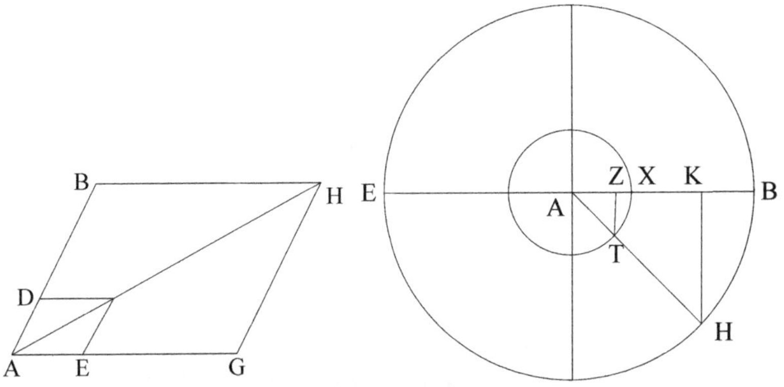

Fig. 7. The parallelogram of (uniform) velocities (left) and circular motion(right)

The diameter EB rotates about the fixed point A. He considers the points X and B when they both move downward from the initial horizontal position of the diameter EB. X moves to T and B moves to H. Their motion is the composition of a linear movement downward, along the tangent, and a translation towards the center. The downward movements of B and X are, respectively, equal to KH and ZT. Their translations are equal to, respectively, BK and XZ. The analysis of the author is essentially restricted to the following remarks.

A point on a radius moves naturally in the direction of the tangent. When X moves to T along an arc of the small circle, B goes further to H. The author says this is because "the natural travel is equal, but the unnatural is less". I guess he means that because B is further away from the centre, the attracting force exerts less influence and the result is that it covers a greater distance in its natural movement – "the natural movement is greater in the greater circle" - and a smaller distance in its unnatural movement [20, pp. 344-345].

What we have here is a remarkable attempt to look at circular motion as the composition of two linear motions. Because the Greeks had no simple quantitative way to describe non–uniform motion this attempt was bound to remain qualitative. They had no other alternative than to do what they actually did and accept uniform circular motion as basic and irreducible to a more primitive motion.

Yet, as we have seen in astronomy they succeeded in treating non-uniform motions by means of combinations of uniform circular motions.

10 Kinematics and the Theory of Machines in the Middle Ages and the Renaissance

In the 14th century at Merton college in Oxford several learned men, Heytesbury, Swineshead and Dumbleton, proved the so-called *Merton-theorem*: In the case of uniform acceleration over a period of time the motion is as for the distance covered equivalent to a uniform motion over the same period of time with the medial velocity, i. e. the velocity in the accelerated motion at the middle instant of the period time. They did so in a purely theoretical and even theological context.

Before 1350 the ideas of the Merton College Group reached Italy and Paris. In 1346 Giovanno Casali gave another proof. It is based upon the idea to represent the degrees of velocity (i. e. the instantaneous velocities) by means of line segments perpendicular to a line segment representing a time interval. Oresme develops this idea much more extensively than does Casali. Moreover, Oresme seems to have understood that the areas of the figures represent the distances traversed ([2], p. 343).

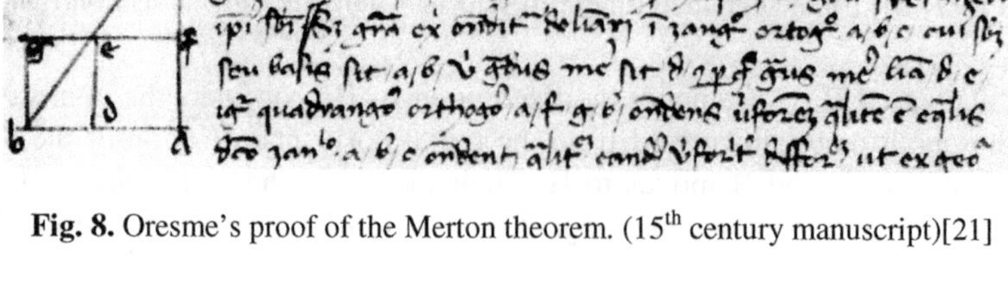

Fig. 8. Oresme's proof of the Merton theorem. (15th century manuscript)[21]

The Merton theorem returns in Galilei's work. In the Dialogue on the *Two Great World Systems* and in the *Discorsi* Galilei applies the Merton theorem to understand the motion of a falling object. Yet, there is a big difference between the 13th and 14th century work of the Merton College group and Casali and Oresme, on the one hand and Galilei's work, on the other hand. The medieval schoolmen were engaged in completely theoretical considerations, while Galilei was executing experiments as well. With Galilei the Middle Ages are over.

It is well known that Galilei discovered that an object thrown away in vacuum under the influence of gravitation describes a parabolic trajectory. He developed this theory as the second new science in his *Discorsi e dimostrazioni matematiche intorno a due nuove scienze*, published in 1638 in The Netherlands (the first new science dealt with the strength of materials). Some have argued that this theory was born from Galileo's head like Pallas Athena was born from the head of Zeus. Recent investigations have made clear that it was to a considerable extent rooted in the practical knowledge of the experts on artillery working in the Venetian Arsenal [22]. Purely kinematical considerations in combination with experiments led Galilei to the parabolic trajectory of an object thrown or shot away in vacuum.

For the Greeks velocity had not been a quantity, length and time yes, but not velocity. With this development starting with the Merton College Group instantaneous velocities had become quantities that could be represented by means of line segments and later by means of numbers. The impact was huge. It led, for example to new kinematical methods to draw tangents. Roberval's work can be mentioned here [23]. Roberval determined the tangent to the cycloid in this way.

The cycloid became a serious object of investigation in the 17th century. This kinematically defined curve in itself had an enormous impact. Also curves described by the points that are not on the circumference but in the plane of a rolling circle which is rolling on another circle were studied: the epi- and hypocycloids. De la Hire studies in his *Traite des Roulettes* the most general situation: arbitrary curves rolling on each other without slipping [24]. In that treatise De la Hire discovered the inflexion circle [26]. Apparently Ole Rømer discovered epicycloidal gearing in 1674. Probably De la Hire discovered it independently. He certainly was the first to systematically study the applications of the epicycloids in mechanics. See [25]. These results are all part of the same development in which Huygens in 1659 discovered that the cycloid is isochronous and introduced the evolute and the involute. Involute gearing was discovered by Euler [19].

References

[1] Ampère, A.-M.: Essai sur la philosophie des sciences, ou, Exposition analytique d'une classification naturelle de toutes les connaissances humaines, Paris, vol. 1/2 / (1834/1843)

[2] Merleau-Ponty, J.: Essai sur la philosophie des sciences d'Ampère. Revue d'histoire des Sciences 30, 113–118 (1977)

[3] Carnot, L.: Essai sur les machines en général, Paris (1783)

[4] Carnot, L.: Principes fondamentaux de l'équilibre et du movement, Paris (1803)

[5] Koetsier, T.: A contribution to the history of kinematics, Part I and Part II. Mechanism and Machine Theory 18, 37–42, 43–48 (1983)

[6] Lanz, Betancourt: Essai sur la composition des machines, Paris. 2nd edn. (1819)
[7] Koenigs, G.: Leçons de cinématique, professées à la Sorbonne; avec des notes... In: Darboux, G., Cosserat, É., Cosserat, F.(eds.) Cinématique théorique, Paris (1897)
[8] Koenigs, G.: Introduction à une nouvelle theorie des mécanismes, Paris (1903)
[9] Burmester, L.: Lehrbuch der Kinematik, Erster Band, Die ebene Bewegung, Leipzig (1888), With a separate atlas that contains the figures
[10] Resal, H.: Cinématique pure, Mallet-Bachelier (1862)
[11] Martinez, A.: Kinematics: The Lost Origins of Einstein's Relativity. Johns Hopkins University Press (2009)
[12] Reuleaux, F.: Theoretische Kinematik, Grundzüge einer Theorie des Maschinenwesens (1875)
[13] Reuleaux, F.: Ueber das Verhältnis zwischen Geometrie, Mechanik und Kinematik, Zeitung des Vereins Deutscher Ingenieure 34, 217–225 (1890)
[14] Klein, F.: Elementary mathematics from an advanced standpoint, Geometry. The Macmillan Company, New York (1939)
[15] Heath, T.: A History of Greek Mathematics, vol. 1, p. 287. Clarendon Press, Oxford (1921) (Reference is to the 1981 Dover edition)
[16] Marchisotto, E.A.: Lines without order. The American Mathematical Monthly 99(8), 738–745, 744 (1992)
[17] Proclus: A Commentary on the First Book of Euclid's Elements (Translated with Introduction and Notes by Glenn R. Morrow). Princeton University Press, Princeton (1970)
[18] Descartes, R.: The geometry of René Descafrtes with a facsimile of the first edition, translated from the French and Latin by D.E. Smith, M.L. Latham. Dover Publications, Inc., New York (1954)
[19] Koetsier, T.: Euler and Kinematics. In: Bradley, R.E., Edward Sandifer, C. (eds.) Leonhard Euler, Life, Work and Legacy, pp. 167–194. Elsevier (2007)
[20] Mechanical Problems. In: Aristotle Minor works, on colours etc., with an English translation by W.S. Hett. London William Heinemann Ltd., Harvard University Press, Cambridge, Massachusetts (1963)
[21] Clagett, M.: Science of Mechanics in the Middle Ages, Plate 7. University of Wisconsin Press, Oxford University Press, Madison, London (1959)
[22] Renn, J., Valleriani, M.: Galileo and the Challenge of the Arsenal, Letture Galileiane, Florence, Max Planck Institute for the History of Science, Berlin (March 21, 2001), `http://www.mpiwg-berlin.mpg.de/Preprints/P179.PDF`
[23] de Roberval, G.P.: Observations sur la composition des mouvemens et sur le moyen de trouver les touchantes des lignes corbe. In: Mémoires de l'Académie Royale des Sciences Contenant les Ouvrages Adoptez par Cette Académie Avant Renouvelement en 1699, Tome III, Ouvrages de M. Roberval, pp. 1–67 (1699)
[24] de la Hire, P., des Roulettes, T.: Histoire de L'Académie des Sciences. In: Année MDC-CVI, Paris, pp. 340–389 (1707)
[25] Woodbury, R.S.: The First Epicycloidal Gear Teeth. Isis 49(4), 375–377 (1958)
[26] Koetsier, T.: The inflexion circle in the history of kinematics. In: Wolfschmidt, G. (ed.) Es gibt für Könige keinen besonderen Weg zur Geometrie, Festschrift für Karin Reich, Algorismus, Heft 60, Augsburg (2007)
[27] Stuart, R.: A descriptive history of the steam engine, 2nd edn., London (1824)
[28] Koetsier, T.: A contribution to the history of kinematics, Part I and Part II. Mechanism and Machine Theory 18, 37–42, 43–48 (1983)

Cognate Linkages the Roberts – Chebyshev Theorem

Egbert Verstraten

Department of Mathematics, Faculty of Science, Vrije Universiteit,
De Boelelaan 1081, NL-1081HV, Amsterdam, The Netherlands
e.verstraten@scalacollege.nl

Abstract. This article deals with investigations by Pafnuty Chebyshev and Samuel Roberts in the late 1800s, which led them independently to the conclusion that for each curve that can be drawn by four bar linkages, there are always three linkages describing the same curve. These different linkages resulting in the same curve can be called cognate linkages.

We will see that Chebyshev and Roberts had very different interests and motivations for studying four bar linkages. Despite of these differences, they came to a similar result concerning cognate linkages.

1 Introduction

In 1784 James Watt patented a major improvement for his steam engines ([1], p. 112). This improvement involved a new link between the cylinder of the engine and the beam that drives the machinery. An example of a steam engine used before 1784 is the engine made by Watt in 1782 and shown in figure 1 ([2], Plate XV, p71). On the left side we see the cylinder which is connected to the big working beam by a rack and sector. On the opposite side of the working beam is another sector, which is connected by a chain to the rod SS which is doing the real work. In this case the engine was used for pumping water.

The general idea of the 1784 patent can be seen in figure 2 ([2], Plate XXI, p 94). The piston rod is in the middle of the figure (DL). Contrary to the earlier machines, the piston rod is not connected directly to the working beam A, but to a piece of iron, which connects the working beam to beam CF. The two beams are describing circular motions, just like the big beam in figure 1. The major difference is that the intermediate point D is moving almost along a straight-line.

Watt was very enthusiastic about this improvement, according to the letter he sent to his partner Matthew Boulton ([3], p. 191): *"I have got a glimpse of a method of causing a piston-rod to move up and down perpendicularly, by only fixing it to a piece of iron upon the beam, without chains, or perpendicular guides, or untowardly frictions, arch-heads, or other pieces of clumsiness ...... I have only tried it in a slight model yet, so cannot build upon it, though I think it a very probable thing to succeed, and one of the most ingenious simple pieces of mechanism I have contrived"*

T. Koetsier, M. Ceccarelli (Eds.): Explorations in the His. of Machines & Mech., HMMS 15, pp. 505–519.
springerlink.com

Fig. 1. Engine by Watt of 1782. Rack and sector are used to transmit the force from piston rod to the working beam

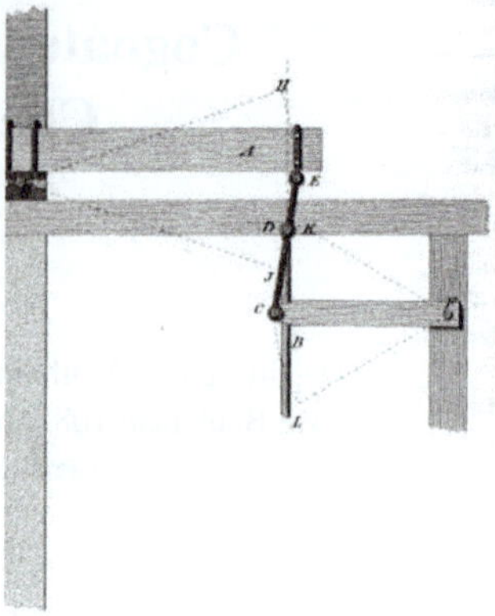

Fig. 2. Watt's mechanism from the 1784 patent. The piston rod is not connected directly to the working beam (A), but to a piece of iron connecting the working beam and beam CF.

Because the beams in figure 2 are not of equal length, the piston rod is not connected to the middle of the connecting iron, but closer to beam A, in a proportion comparable to the lengths of the two beams.

Watt uses figure 3, which appears in a letter to his son ([4], p.294) and in the specifications of the patent ([3], p.89), to explain the situation. Points B and C are fixed with A and D attached to them, so A is rotating around B and D is rotating around C. AB and CD can be considered as the two rotating beams. The intermediate point E is lying on a connecting link between A and D. Because A and D are describing arcs of opposite convexity, point E will describe an approximately straight line.

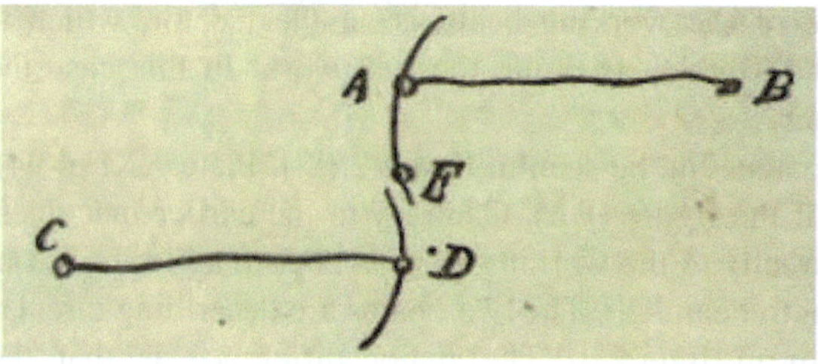

Fig. 3. Figure used by Watt in his explanation how he discovered nearly straight line motion

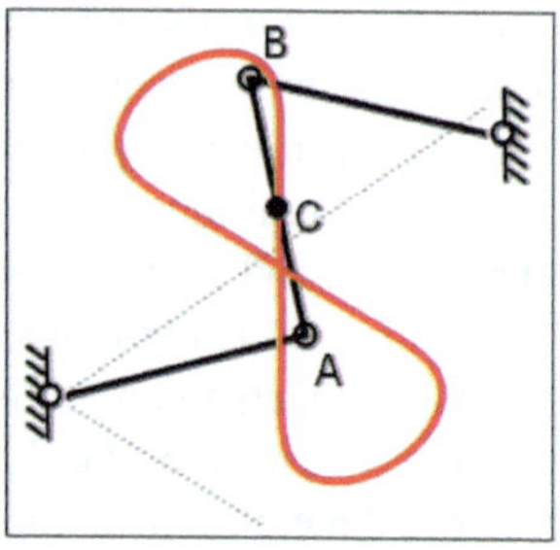

Fig. 4. Eight-like figure described by the coupler point in Watt's mechanism.

What Watt is using here is in fact a four bar linkage. Two beams, both with one fixed point and a connecting coupler bar. The frame to which the beams are attached, is considered the fourth bar. The coupler point in this specific four bar linkage describes an eight-like figure which contains a part that is almost a straight line as can be seen in figure 4.

In 1788 the Spaniard Augustin de Bétancourt Y Molina (1758 – 1824) visited Britain to study steam engines and other machinery ([4], p. 267). During this visit he also went to Birmingham for an investigation of the new engines made by "Wast and Bolton" (as Bétancourt named them in his later work!). Watt and his partner Boulton were not very anxious to share their work with others as is clear from letters they sent each other. As an example Watt writes to Boulton in 1790 *"We must be more and more careful in respect to foreigners"* ([3], p.239). Despite this lack of cooperation, Bétancourt succeeded in describing Watt's renewals and introduced them in France. The Perrier brothers were then the first to build a steam engine according to Bétancourt's articles ([5], p. 572).

The articles of Bétancourt not only led to the introduction of these new steam engines in Europe, but also were a start of mathematical interest in these linkages. The mathematician Gaspard de Prony (1755 – 1839) was one of the first to study this linkwork. The linkwork he is studying is similar to the one shown in figure 2. De Prony is making very precise measurements of the curve described by point D, the point that is describing the straight line movement. De Prony finds the deviations of D from the straight line in several positions of the working beam. If, for example the angle between working beam and horizontal is 30°, the deviation from the straight line is $0{,}0158485 \cdot b$, where b is the length of the working beam. De Prony calculates these deviations for angles between -25° and 30° ([6], p.130).

Watt's interest, as an engineer, was to approach a straight line. This was also the interest of De Prony; calculating the deviation from the straight line. We will see that this was also the main focus of Chebyshev.

Watt was not interested in general four bar linkages. The first to study general four bar linkages and come up with a general equation for the curve that the coupler point of such a linkage describes, was Alexandre Joseph Hidulphe Vincent (1779 – 1868). In his article "d'une théorie du parallélogramme de Watt" (1836) ([7], p. 5 – 27) Vincent gives a rather long and, according his own introduction of the article, boring derivation of the equation. We will see that the work of Roberts is more in line with Vincent. Roberts is also studying general four-bar linkages, without the intention of getting the best approximation of a straight line.

So resulting from Watt's invention we see two different approaches in studying four bar linkages. The first, which we could call the engineer approach, is the line followed by Watt, de Prony and Chebyshev. It is mainly focused on the mechanical part and its goal is to find the best approximation of a straight line. "The best" meaning as close to a straight line as possible. But it also means the technically spoken most useful one.

The second could be called the mathematical approach by Vincent and Roberts. They are studying general four bar linkages and their curves, rather than the applications of it.

In the next two paragraphs we will see that both Chebyshev and Roberts will come up with the idea of cognate linkages: different linkages describing equivalent curves.

2 Chebyshev

The Russian mathematician Pafnuty Lvovich Chebyshev (1821 – 1894) was born in the village of Okatovo in Russia, south-west of Moscow. Chebyshev was first educated at home and later studied at the University of Moscow. In 1847 he promoted at the St. Petersburg University. Chebyshev travelled several times to Western Europe. One of the reasons for these journeys was studying machinery in France and Britain. He was interested in the work of James Watt and linkwork in general.

His first journey to Western Europe was in 1852 from June 21st to November 7th (16, p. VII – XVIII). In his report about this journey he mentions several visits to different machines and mechanisms: windmills in Lille, watermills in Meaux and several steam engines like the ones in Cavé. On October 2nd he arrives in London where he visits the still existing engines of Watt ([8], p. XVI). Writing about Watt and his engines Chebyshev mentions some possible improvements ([8], p. X). He also observes that, although some analytical work is done, a lot of investigation is yet to be done. As a result of this first trip to Western Europe, Chebyshev was interested in approximation theory and the theory of mechanisms. The two are very much related in Chebyshevs work. In his articles about mechanisms Chebyshev is never interested in linkwork in general, but always in linkwork that is used to the best possible approach of a straight line.

One of the first articles on approximation is "Théorie des mécanismes connus sous le nom de parallélogrammes", published in 1854 ([9], p. 111 – 143). The article is about approaching a function by a polynomial on a certain interval. Chebyshev is not interested in the deviation in certain points, e.g. the end points of the interval, but in a uniform approximation. Given a function $f \in C[a,b]$, Chebyshev is looking at the best approximation of degree n of f, which is the polynomial $\inf_{p_n \in P_n} \left\{ \max_{x \in [a,b]} |f(x) - p_n(x)| \right\}$. Here P_n is the set of all polynomials of degree n. The higher the degree of the polynomial, the better the approximation will be. This gives Chebyshev a quality for the approximation of a straight line. In Chebyshevs articles an approximation is of degree n, meaning that it approaches a straight line as good as a polynomial of degree n.

As mentioned before, Chebyshev was constantly looking for a linkage that gives the best approach of a straight line. In the article "Sur un mécanisme" written in 1868 ([8], p. 49 – 57) he introduces the linkage shown in figure 5 ([8], p. 52), which is named after him the Chebyshev linkage.

Fixed points are C and C_1. The bars CA and C_1A_1 are of equal length. The coupler point M describing the desired curve is on the coupler AA_1. This four bar linkage is similar to the Watt linkage, except that the bars CA and C_1A_1 cross each other which is not the case in the Watt linkage.

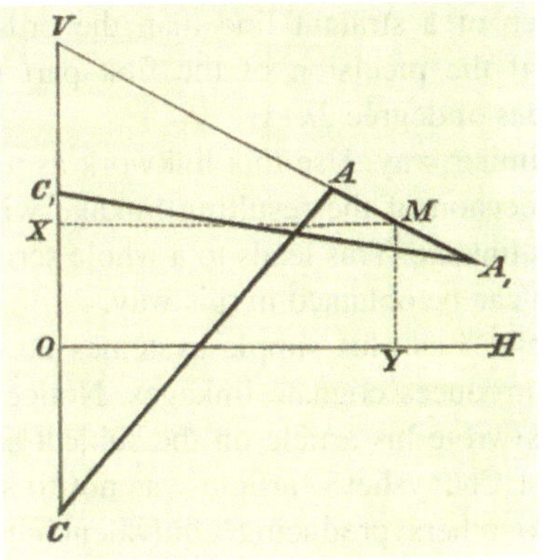

Fig. 5. Chebyshev linkages

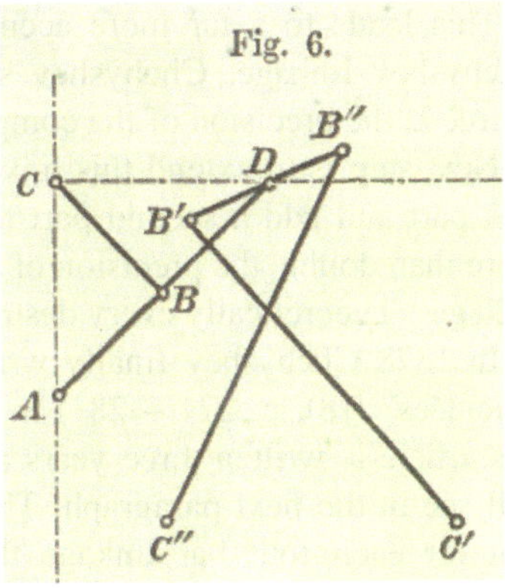

Fig. 6. Extended Chebyshev linkage

Chebyshev had already shown that in order to get the best approximation of a straight line the linkage must satisfy two conditions. First the distance CC_1 must be equal to 1/3 ($AC + A_1C_1 + AA_1$). The second says that the length of the coupler (AA_1) must be at least a quarter of the length of the other beams ($AA_1 \geq \frac{1}{4} AC$).

If the length of the coupler approaches a quarter of the length of the beam, the resulting curve will be better, but shorter.

Chebyshev shows that the distance MX is almost constant. As A circles around C, M will move between two lines parallel to CC_1. To compare his linkage with the Watt linkage, Chebyshev scales his linkage to the one studied by the Prony mentioned before. The latter was moving over a length of 0.64 with a maximum deviation of a straight line of 0.002. The deviation of the Chebyshev linkage is in a comparable situation 0.00029.

Still looking for a way to produce the best straight line approximation, Chebyshev comes up with a very good result in his article "Sur les parallélogrammes" written in 1869 ([8], p. 85 – 106). In this article Chebyshev doesn't restrict himself to four bar linkages, but studies linkwork with an arbitrary number of bars (m), links (n) and fixed points (v). The first and easiest situation he discusses is m = 1, v = 1 and n = 0. The result is an ordinary compass, which leads to a second order equation: an approximation of the straight line of second degree.

The situation in which m = 3 (four bar linkages), v = 2 and n = 2 result in the Watt linkage and the earlier introduced Chebyshev linkage.

Then Chebyshev focuses on the situation where m = 5, v = 3 and n = 4. An example, based on the Chebyshev four bar linkage, is shown in figure 6 ([8], p. 94).

This linkage consists of two connected parts. The first is the Chebyshev linkage C'B'B"C" with fixed points C' and C". The second part consists of a bar connected to the third fixed point and a bar to connect it with the first part. It is connected to the coupler point of the first part. Point A is the point that describes the desired curve. There are some extra conditions on the lengths of the bars.

This leads to a far more accurate approach of a straight line than the ordinary Chebyshev linkage. Chebyshev shows that, if the precision of the first part is of degree λ, the precision of the complete linkage is of degree $2\lambda+1$.

Next step is to extend this linkwork in a similar way. Use this linkwork as a new first part and add a second part to it. The precision of the resulting linkage will be more than double the precision of the previous linkage. This leads to a whole series of linkages. Theoretically every desired precision can be obtained in this way.

In 1878 Chebyshev finally writes the article "Les plus simple systèmes de tiges articulées" ([8], p. 271 – 281) in which he introduces cognate linkages. Notice that this article is written three years after Roberts wrote his article on the subject as we will see in the next paragraph. The purpose of Chebyshevs' article was not to show that for each four bar linkage there are two others producing equivalent curves. Chebyshevs' aim was, as always, to find better or more practical linkwork to produce approximations of a straight line. In this article he finally constructs a linkage with some nice qualities.

In the introduction of the article Chebyshev writes "On les <cognate linkages> obtient par la substitution de certaines tiges à autres du premier mécanisme, faite en vue de ne modifier en rien la trajectoire du point décrivant." What Chebyshev says is to replace certain bars by others, without modifying the curve. This is exactly what he is doing in the article. It is a construction of a cognate linkage based on the original one. Figure 7 shows this construction.

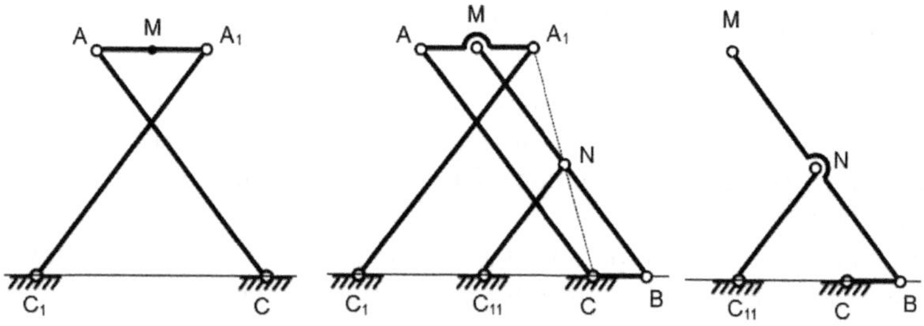

Fig. 7. Chebyshev's construction of cognate linkages

The original linkage is the well-known Chebyshev four bar linkage. Add to the linkage the bars BC and BM, parallel to AA_1 and CA. N is the midpoint of BM. Independent of the position of this new construction, N is also the midpoint of CA_1. Therefor N will describe a similar, but relatively smaller, curve as A_1 and also there direction will be similar at every time. As A_1 describes an arc with center C_1 and radius C_1A_1, N will also describe an arc. The center of this arc must be on a line parallel to C_1A_1 and will be the point C_{11}. Now choose C and C_{11} as fixed points of the new linkage and remove the bars of the original linkage. This leads to a cognate linkage with the same coupler point M describing the same curve. Choosing C_1 and C_{11} as new fixed points would have led to a third cognate linkage, but this is not mentioned by Chebyshev.

Next step in the article is to start with a similar four bar linkage C_1A_1AC, but with a far more arbitrary coupler point M. In this situation M is attached to AA_1, but not necessarily on AA_1. A similar construction as before (shown in figure 8) leads to a linkage that produces an equivalent curve.

Starting with a symmetrical linkage, Chebyshev was able to construct a non-symmetrical linkage which produces the same approach of the straight line. Choosing other bar lengths, this cognate linkage can be reformed in a way that one of the beams can make a full rotation and the total curve the coupler point describes is an almost straight line. An example is given by Chebyshev at the end of his article.

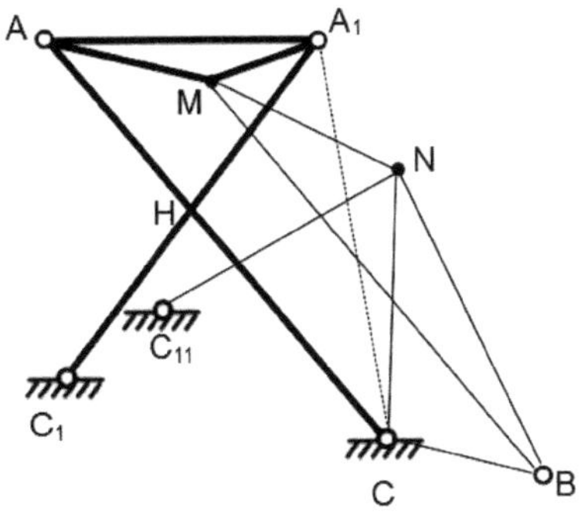

Fig. 8. Chebyshev's construction of a cognate linkage in a more general case

As we see, cognate linkages are a result of Chebyshevs search for linkages to produce straight lines not as a goal in itself. This probably explains some remarkable aspects of the article. First of all, Chebyshev doesn't even have a name or word for these cognate linkages. He just describes them. But far more remarkable is the fact that Chebyshev doesn't mention that there exists a third cognate, which is easy to construct. Starting with the same linkage as in figure 8, but constructing B and N on the left side of the linkage, will give an cognate linkage C_1BNC_{11}.

3 Roberts

Samuel Roberts (1827 – 1913) studied law and mathematics. He became member of the London Mathematical Society shortly after its foundation in 1865. Unlike Chebyshev his interest in linkwork was not inspired by straight lines. His first article on the subject shows that he is interested in instruments to draw third or fourth order equations ([10]). In this article "On the mechanical description of some species of circular curves of the third and fourth degrees" he states that there are only some isolated examples of instruments to do this job. Roberts presents a more general extension of these instruments.

Beside this mechanical interest, Roberts had a mathematical interest in linkwork, as we will see in the next article he wrote on the subject.

It is known to Roberts that four bar linkages (Roberts calls them three bar linkages as he does not count the link between the two fixed points) describe curves of sixth degree. In his article "On three bar motion in plane space" ([11]) written in 1875 he gives a general equation of these curves.

Remarkable is that the coupler point describing the curve does not have to be on the link AB. Roberts considers a plane which is moving with two points on two fixed circles. Each point in the moving plane can be considered as coupler point. We have seen this in the 1878 article of Chebyshev ("Les plus simple systèmes de tiges articulées"), but that was after Roberts wrote this article.

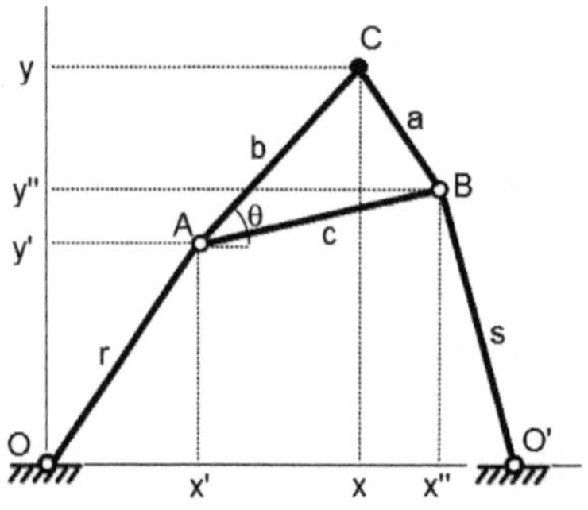

Fig. 9. Construction used by Roberts in his article "On three bar motion in plane space".

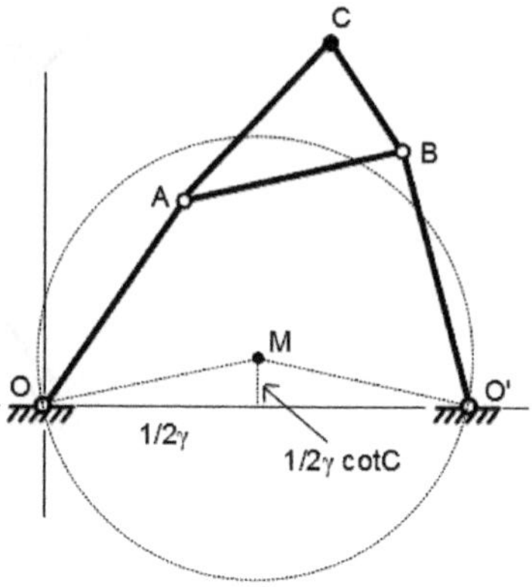

Fig. 10. the circle of foci of the coupler curve

First step is to derive an equation of the locus of point C, which can be found using some basic calculation (see figure 9):

$$\left\{\begin{matrix}\sin A\left[(x-\gamma)\sin C-y\cos C\right]\left[x^2+y^2+b^2-r^2\right]+\\ \sin By\left[(x-\gamma)^2+y^2+a^2-s^2\right]\end{matrix}\right\}^2+$$
$$\left\{\begin{matrix}\sin A\left[(x-\gamma)\cos C+y\sin C\right]\left[x^2+y^2+b^2-r^2\right]-\\ \sin Bx\left[(x-\gamma)^2+y^2+a^2-s^2\right]\end{matrix}\right\}^2= \tag{1}$$
$$4k^2\sin^2 A\sin^2 B\sin^2 C\left[x(x-\gamma)+y^2-\gamma\text{cotan}Cy\right]^2$$

where A, B and C are the angles of the moving triangle, γ is the length of OO' and we use $k=\frac{a}{\sin A}=\frac{b}{\sin B}=\frac{c}{\sin C}$

This equation can be rewritten as

$$(L\varphi+M\psi)^2+(N\varphi+P\psi)^2=K(LP-MN)^2 \tag{2}$$

in which $\varphi = 0$, $\psi = 0$ and $LP - MN = 0$ represent circles.

Circle $LP - MN = 0$ can be rewritten as

$$\left(x - \frac{1}{2}\gamma\right)^2 + \left(y - \frac{1}{2}\gamma\text{cotan}C\right)^2 = \left(\frac{\gamma}{2\sin C}\right)^2 \tag{3}$$

which is the circle with center $M\left(\frac{1}{2}\gamma, \frac{1}{2}\gamma\text{cotan}C\right)$ and radius $\frac{\gamma}{2\sin C}$. Both fixed points O (0,0) and O' (γ,0) are on this circle.

From figure 10 it is clear that

$$\tan\left(\frac{1}{2}\angle OMO'\right) = \frac{\frac{1}{2}\gamma}{\frac{1}{2}\gamma\text{cotan}C} = \tan C \tag{4}$$

so we can conclude that for every point P on this circle the angle OPO' is equal to angle C.

It is also clear that OO' subtends an angle C or 180° – C at every node of the curve. In order to see this, suppose that ABC and A'B'C are two positions that determine a node (see figure 11 – left picture). Than

$\angle ACA' + \angle A'CB = \angle ACB = \angle A'CB' = \angle A'CB + \angle BCB'$, so $\angle ACA' = \angle BCB'$.

Because OACA' and O'BCB' are kites, we also have $\angle ACA' = 2\angle OCA'$ and $\angle BCB = 2\angle BCO'$.

This all leads to

$$\angle OCO' = \angle OCA' + \angle A'CB + \angle BCO' = \angle ACA' + \angle A'CB = \angle ACB = C \tag{5}$$

In the case the node is on the other part of the circle between O and O' as in the right picture of figure 11, a similar calculation leads to $\angle OCO' = 180° - C$

This shows that every node is on the circle $LP - MN = 0$.

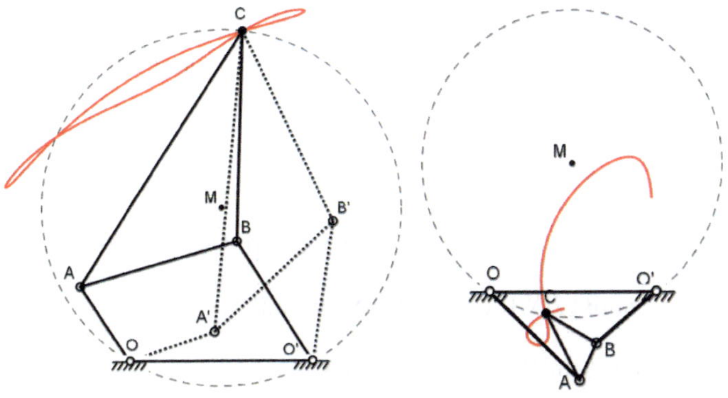

Fig. 11. All nodes of the coupler curve are on the circle of foci.

As a next step we will rewrite the equation we found for the curve in homogeneous coordinates. These are triples (X,Y,W) representing normal points (X/W , Y/W) if W $\neq$ 0 and infinite points in different directions if W = 0.

This leads to the equation

$$\left\{\begin{matrix}\sin A\left[(X-\gamma W)\sin C-Y\cos C\right]\left[X^2+Y^2+b^2W^2-r^2W^2\right]+\\ \sin BY\left[(X-\gamma W)^2+Y^2+a^2W^2-s^2W^2\right]\end{matrix}\right\}^2+$$

$$\left\{\begin{matrix}\sin A\left[(X-\gamma W)\cos C+Y\sin C\right]\left[X^2+Y^2+b^2W^2-r^2W^2\right]-\\ \sin BY\left[(X-\gamma W)^2+Y^2+a^2W^2-s^2W^2\right]\end{matrix}\right\}^2= \tag{6}$$

$$4k^2\sin^2A\sin^2A\sin^2C\left[X(X-\gamma W)W-\gamma YW^2\text{cotan}C\right]^2$$

Putting W = 0 to find the infinite points on the curve we can deduce that

$$\left(X^2+Y^2\right)^3\left[\sin^2A-2\sin A\sin B\cos C+\sin^2B\right]=0 \tag{7}$$

Conclusion is that the infinite points (1,i,0) and (1,-i,0) are on the curve. Curves with this property are called circular. Because the points occur triple on these coupler curves, they are called tricircular. These two circular points at infinity are also called the isotropic points.

Next we observe the intersections of the tangents to the curve in these points. The tangents in the infinite points are the asymptotes of the curve. Real intersections of these tangents or asymptotes are called the singular foci of the curve. To find these it is a good idea to make a coordinate transformation. As new coordinates we choose $(X,Y)=\left(\frac{x+iy}{2},\frac{x-iy}{2}\right)$. The coordinates of the isotropic points after this transformation are $(0,1,0)$ and $(1,0,0)$, so through this transformation, the axes are in the direction of the two isotropic points.

The transformation leads to a new equation for our curve:

$$\left\{\begin{matrix}\sin A\left[(2X-\gamma)\sin C+i(2X-\gamma)\cos C\right]\left[4XY+b^2-r^2\right]+\\ \sin B(-2iX)\left[(2X-\gamma)(2Y-\gamma)+a^2-s^2\right]\end{matrix}\right\}\times$$

$$\left\{\begin{matrix}\sin A\left[(2Y-\gamma)\sin C-i(2Y-\gamma)\cos C\right]\left[4XY+b^2-r^2\right]+\\ \sin B(2iY)\left[(2X-\gamma)(2Y-\gamma)+a^2-s^2\right]\end{matrix}\right\}= \tag{8}$$

$$4k^2\sin^2A\sin^2B\sin^2C\left[4XY-\gamma(X+Y)-i\gamma\cot C(Y-X)\right]^2$$

To derive the tangents at the isotropic points, we only have to examine the coefficients of X^3 and Y^3; the highest degrees in the equation, because the behavior of the curve in infinity is fully determined by these. The lower powers of X and Y can be neglected at infinity. We equate the coefficients of X^3 and Y^3 to zero to obtain the tangents.

For X^3 this leads to:

$$\begin{cases} Y = 0 \\ 2Y - \gamma = 0 \\ 8\sin A Y(\sin C + i\cos C) - 4i\sin B(2Y - \gamma) = 0 \end{cases} \tag{9}$$

Retransforming these to the original coordinates gives

$$\begin{cases} x - iy = 0 \\ x - \gamma - iy = 0 \\ x - \dfrac{\gamma\cos A\sin B}{\sin C} - i\left(y - \dfrac{\gamma\sin A\sin B}{\sin C}\right) = 0 \end{cases} \tag{10}$$

Doing the same for the coefficient of Y^3 gives also three possible tangent lines:

$$\begin{cases} x + iy = 0 \\ x - \gamma + iy = 0 \\ x - \dfrac{\gamma\cos A\sin B}{\sin C} + i\left(y - \dfrac{\gamma\sin A\sin B}{\sin C}\right) = 0 \end{cases} \tag{11}$$

This leads to nine intersections of tangents of which there are only three real. These three real intersections are the singular foci:

The intersection of $x - iy = 0$ and $x + iy = 0$ is $O(0,0)$.

The intersection of $x - \gamma - iy = 0$ and $x - \gamma + iy = 0$ is $O'(\gamma, 0)$.

Finally the third real intersection is $\left(\dfrac{\gamma\cos A\sin B}{\sin C}, \dfrac{\gamma\sin A\sin B}{\sin C}\right)$.

Let us call this one O''.

It appears that O'' is on the circle $LP - MN = 0$. So we can conclude that $\angle OO''O' = C$, which we have for all points on this circle.

We finally come to Roberts's conclusion: "*We are thus led to remark that, since the singular foci are similarly related to the locus, we might have taken as fixed centres the focus O, and the third focus, say O'', and by means of links of suitable lengths we should obtain the same locus. And in like manner we might have taken as centres O' and O''. We conclude, then, that the locus in question can be described in three different ways, by similar traversing triangles moving with an angle on each of two circles about two singular foci as centres, and the angle moving on a circle is equal to the angle subtended at its centre by the two remaining singular foci.*"

4 The Proof by Schor

As in many cases there are far more proofs of the Roberts – Chebyshev theorem. A very nice one was given by J.B. Schor in 1941 [12]. It uses vectors and complex numbers.

It starts with a general four bar linkage O_AABO_B with coupler point C [figure 12].

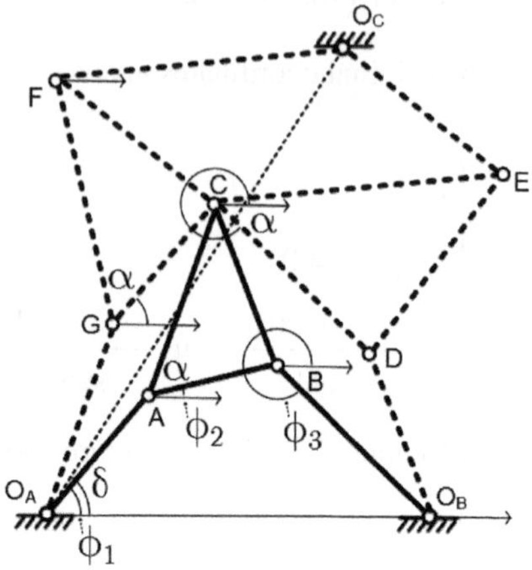

Fig. 12. Proof by Schor

On this four bar linkage two parallelograms are constructed and on these parallelograms two triangles, similar to triangle ABC. In triangle GCF we have $\angle G = \angle A$ and in triangle CDE it's $\angle C = \angle A$. Last step is constructing parallelogram $FCEO_C$. In this way we have constructed three four bar linkages O_AABO_B, O_AGFO_C and O_BDEO_C. They all have the same coupler point C, describing the same coupler curve. It remains to proof the constructed point O_C is a fixed point.

First let us define some angles:

- δ: angle between O_AO_B and O_AO_C
- $\varphi 1$: angle between O_AO_B and O_AA
- $\varphi 2$: angle between O_AO_B and AB
- $\varphi 3$: angle between O_AO_B and BO_B
- α: (fixed) angle between AB and AC

Point O_C can be written as $|O_AO_C|e^{i\delta}$. We will show that this is constant.

As the vector O_AO_C is the sum of three vectors we have

$$|O_AO_C|e^{i\delta} = |O_AG|e^{i(\varphi_2+\alpha)} + |GF|e^{i(\varphi_1+\alpha)} + |FO_C|e^{i(\varphi_3+\alpha)}. \tag{12}$$

These lengths can be rewritten using

$$|O_AG| = |AC| = |AB|\frac{|AC|}{|AB|} \tag{13}$$

$$\frac{|GF|}{|GC|} = \frac{|AC|}{|AB|} \text{ and } |GC| = |O_A A|, \text{ giving } |GF| = \frac{|AC|}{|AB|}|O_A A| \tag{14}$$

$$\frac{|CE|}{|CD|} = \frac{|AC|}{|AB|} \text{ and } |CE| = |FO_C|, \text{ giving } |FO_C| = \frac{|AC|}{|AB|}|BO_B|. \tag{15}$$

Rewriting our previous equation leads to

$$|O_A O_C| e^{i\delta} = \frac{|AC|}{|AB|} e^{i\alpha} \left(|O_A A| e^{i\varphi_1} + |AB| e^{i\varphi_2} + |BO_B| e^{i\varphi_3} \right) \tag{16}$$

Between the brackets is a sum of three vectors, equal to the vector $|O_A O_B|$.

$$|O_A O_C| e^{i\delta} = \frac{|AC|}{|AB|} e^{i\alpha} |O_A O_B| \tag{17}$$

Conclusion is that $|O_A O_C| = \frac{|AC|}{|AB|} |O_A O_B|$ and $\delta = \alpha$. Both are constant which proofs that point O_C is fixed compared to O_A and O_B.

5 Roberts vs. Chebyshev

In 1958 Richard de Jonge presented a paper at the New York Academy of Sciences in which he said the following:

The theorem is known everywhere as the Roberts theorem except in the Union of Soviet Socialist Republics, where in recent years it has been renamed the Roberts-Chebyshev theorem. It is claimed there that Chebyshev found it independently. However, Chebyshev applied it only in 1879, that is, three and a half years after Roberts. Consequently this claim is difficulty to accept, because Chebyshev had not only become acquainted, but very friendly, with Roberts and Cayley during his previous journey to England in 1852. They had become so friendly on that occasion that Chebyshev was made a foreign member of the Royal Society through Cayley's influence, Cayley being a Fellow of that Society. Since Cayley discusses Roberts' paper in his own paper which he read before the London Mathematical Society on March 10, 1876 and which was published in the same volume of that society's Proceedings as Roberts' paper, it stand to reason that Chebyshev knew of Roberts' discovery, more than sufficient time having elapsed for copies of Roberts' and Cayley's papers to have come into his hands. Moreover, nowhere in his writings has Chebyshev ever claimed to have discovered this as a new theorem, a fact that speaks well for his honesty. All he did was to use it for a derivation of other forms from the mechanism invented by him. Thus the claim of present day Soviet scientists that Chebyshev found the theorem independently seems completely unsubstantiated. Consequently the theorem is here, as usual, called Roberts theorem only. ([13], p. 95, footnote).

We have seen how Roberts and Chebyshev both discovered the possibility of multiple generation of coupler curves. Although we cannot exclude with absolute certainty that Richard de Jonge's description of what happened is correct, it seems much more probable that Chebyshev indeed independently arrived at a somewhat less general version of the theorem several years after Roberts' discovery. Roberts' paper gives such a full treatment of the theorem relating the triple generation to the properties of the coupler curve that those who call the theorem "Roberts theorem" can hardly be blamed. On the other hand, those who call it the "Roberts-Chebyshev theorem" also have a point. They honor Chebyshev's independent discovery and his important work on mechanisms. Cayley made Chebyshev foreign member of the Royal Society certainly not only because they had become friends but primarily because he recognized genius.

Remark: The Roberts-Chebyshev theorem is part of the research area of coupler curve synthesis. In 1974 Nolle gave in [14] a nice historical survey. Worth reading is also [16]. A recent application of Chebyshev's approach to straight line generation can be found in [15].

References

[1] Thurston, T.: A History of the Growth of the Steam-engine. Appleton and Company, New York (1878)
[2] Muirhead, J.P.: The Origin and Progress of the Mechanical Inventions of James Watt, vol. III. John Murray, London (1854)
[3] Muirhead, J.P.: The Origin and Progress of the Mechanical Inventions of James Watt, vol. II. John Murray, London (1854)
[4] Muirhead, J.P.: The Life of James Watt. John Murray, London (1858)
[5] de Prony, G.: Nouvelles Architecture Hydraulique, vol. I. Firmin Didot, Paris (1790)
[6] de Prony, G.: Nouvelles Architecture Hydraulique, vol. II. Firmin Didot, Paris (1796)
[7] Vincent, A.J.H.: Mémoires de la Société royal des Sciences, de l'Agriculture et des Arts, de Lille (1836, 1837, 1838)
[8] Markoff, A., Sonin, N.: Oeuvres de P.L. Tchebychef, Tome II, l'académie impériale des sciences. St. Petersburg (1907)
[9] Markoff, A., Sonin, N.: Oeuvres de P.L. Tchebychef, Tome I, l'académie impériale des sciences. St. Petersburg (1899)
[10] Roberts, S.: On the Mechanical Description of some Species of Circular Curves of the third and fourth Degrees. Proceedings London Mathematical Society, 125–136 (1869)
[11] Roberts, S.: On three bar motion in plane space. Proceedings London Mathematical Society, 14–23 (1875)
[12] Schor, J.B.: On the theorem of Roberts – Tchebyshev. Journal of Applied Mathematics and Mechanics 5, 323–324 (1941)

[13] Richard de Jonge, A.E.: The Correlation of Hinged Straight-Line Four Bar Devices by means of Roberts Theorem and a New Proof of the Latter. Annals of the New York Academy of Sciences 84, 77–145
[14] Nolle, H.: Linkage Coupler Curve Synthesis: A Historical Review -I and II. Mechanism and Machine Theory 9, 147–168 (1974)
[15] Ceccarelli, M., Vinciguerra, A.: Approximate Four-Bar Circle-Tracing Mechanisms: Classical and New Synthesis. Mechanism and Machine Theory 35, 1579–1599 (2000)
[16] Koetsier, T.: A contribution to the history of kinematics, Part I and Part II. Mechanism and Machine Theory 18, 37–42, 43–48 (1983)

The Work of Otto Fischer and the Historical Development of His Method of Principal Vectors for Mechanism and Machine Science

Volkert van der Wijk and Just L. Herder

University of Twente, Department of Mechanical Automation,
P.O. Box 217, 7500 AE, Enschede, The Netherlands
{v.vanderwijk,j.l.herder}@utwente.nl

Abstract. This article gives an overview of the distinctive work of Otto Fischer (1861-1916) on the motion of the human musculoskeletal system. In order to be able to derive the individual muscle forces for human in motion, he invented *the method of principal vectors* to describe the motion of the centers-of-mass and the inertias of body segments. This method has proven to be successful, not only for the studies on biomechanics but in particular also for mechanism and machine science. A historical development of the application of the method is presented for today's potential.

Keywords: Otto Fischer, center-of-mass, biomechanics, static balancing, dynamic balancing.

1 Introduction

With time the amount of knowledge increases at a rapid pace. For research it is a challenge to both remain up-to-date and to not forget the past. Particularly this is a challenge when old research results and ideas lose interest for a certain time, for decades or for ages. To retrieve the results, and more importantly to retrieve the philosophy with which they were obtained, often investigation of the original sources is needed. Citations of old literature in current articles often tend to become rather 'automatic' or indirect, e.g. because of language differences or unavailability, and therefore do not reveal the essence and the potential of the original sources. This however is necessary to discover the value old methods may have for contemporary research goals.

With this perspective, the work of the physiologist, physicist, and medical doctor (Physiologe, Physiker, Arzt) Otto Fischer[1] is summarized and the philosophy and application of his *method of principal vectors* is investigated. In addition to his still relevant results on the mechanical properties of the human musculoskeletal system and of his investigation of the human gait, this method has been essential for studying the individual muscle forces for human motion in the pre-computer era. Although invented for biomechanics, Fischer emphasized the potential of his method for mechanism and machine science, which was taken over by various researchers. This article includes a historical development of his method in mechanism and machine science to illustrate its current potential.

[1] *26-04-1861 Altenburg (Thüringen), Germany †16-12-1916 Leipzig, Germany.

T. Koetsier, M. Ceccarelli (Eds.): Explorations in the His. of Machines & Mech., HMMS 15, pp. 521–534.
springerlink.com

2 Otto Fischer and his Works

[1][2][3] Otto Fischer studied mathematics and physics at the German universities of Jena, München, and Leipzig and obtained his PhD-degree (Dr. phil.) in mathematics from the University of Leipzig in 1885, supervised by Prof. Felix Klein. In 1887 he became teacher of mathematics and physics at the business school (Handelslehranstalt) of Leipzig. He continued as a teacher at the Petri-Realgymnasium in Leipzig in 1895, where he would teach until the end of his life, from 1912 onward at the position of rector.

At the same time Fischer worked on medical and biological research together with the anatomist Prof. Christian Wilhelm Braune at the Anatomic institute of the University of Leipzig. In 1893 he obtained his habilitation on physiological physics (Physiologische Physik) at the University of Leipzig and became teacher at this topic at the Faculty of Philosophy. In 1896 he became Professor extraordinarius on medical physics (Medizinische Physik) at the Medical Faculty of the University of Leipzig where he remained until his early death in 1916.

Together with Braune, Fischer was pioneering the research on biomechanics of the human musculoskeletal system in motion. While anatomists at that time investigated the mechanics of the human body empirically, he understood the necessity of theoretical investigation. He stated that only with an accurate kinematic model of the human skeletal system the functioning of human motion and of the muscles could be investigated to achieve full understanding. Being a mathematician, he stressed in probably his first publication in 1885 that a mathematical method was needed to gain exact and reliable results in human kinematics [4].

The investigation in [4] consists of three-dimensional measurements of the motion of the forearm. At three non-collinear points at the bone wooden needles were attached. The motions of the tips of the needles were recorded by drawing their projections on millimeter-paper and by recording the distance from the needle to the plane of the paper. With these measurements he became the first to make three-dimensional motion analysis of human. A mathematical description of the measured motions was derived by screw motion, for which he determined the screw-axes of each body segment. Based on the motion of the screw-axes the real motion could be interpreted accurately.

His results led to important corrections on the results from purely two-dimensional empirical analysis by others. For example in 1887 he reexamined the motion of the elbow and the hand with a 'rigorous mathematical analysis' gaining new insight at a topic that had been considered known [5].

In [6] Fischer considered the importance of measurements on the living hand as compared to a cadaver hand, because of the significant influence muscles have on the real motion capabilities in addition to the joint geometry. To measure the motions, he placed metal tubes around the finger to which the wooden needles and steel wires for measuring were attached. In 1888 he started analyzing the combined motion of multiple body segments by considering the shoulder with the humerus [7] and in 1889 the functioning of the flexor muscles of the elbow was investigated [8].

For the first time in 1889, Fischer considered the motion of the center of mass (CoM) of the complete human body [9]. As compared to statics and mechanics of the human body at a certain pose, which can be analyzed from individual body segments, he stated that for deriving the muscle forces for certain human motion knowledge is needed about the motion of the body CoM. One argument was that while each body segment has various possible motions, the body CoM always moves in a very specific manner. Fischer and Braune determined the positions of the CoM of human bodies and of parts of human bodies by using rigidly frozen cadavers. They verified their results with the results of anatomists for individual body segments and presented them as ratios or coefficients in order to account for differences of individuals. Their results remained important well into the computer era [10,11].

In 1891 advanced kinematic measurements and analysis of the knee joint were published [12] followed in 1892 by measurements of the inertia values of body segments [13].

One year after the death of Braune, Fischer published in 1893 his habilitation work on determining the muscle forces for human motion [14]. He noted that the human body had become a mechanically known object. At anatomic level the properties of each body segment were well known and at kinematic level the relative motion of the body segments had become well known too. With the rich knowledge on the functioning of individual muscles at an anatomic level, the next new step had to be the investigation of the combined muscle forces that cause human motion. He distinguished human motion by the absolute motion of the body CoM and by the relative motion of the body segments with respect to the body CoM. For the analysis of the relative motions and to derive muscle forces he invented 'The method to derive kinetic energy' ('Die Methode der Ableitung der lebendigen Kraft'), of which a part would be named later 'The method of principal vectors' [15]. With this method, Fischer investigated the muscle forces of the lower arm in 1895 [16] and he made other contributions to the statics and dynamics of muscles in 1896 [17] and 1897 [18].

From 1895 to 1904 Fischer published a series of six works on human gait (Der Gang des Menschen). The first part consists of the measurements of the motion of unloaded and loaded humans within a spatial co-ordinate system with high accuracy [19]. The results of these measurements were used for analysis in the following five parts for which Otto Fischer is said to be the first to conduct three-dimensional gait analysis [20]. In part two in 1899 the precise motion of the body CoM based on the CoM of individual body segments is investigated in relation with the external forces that apply [21]. In part three in 1900 the lower extremities are investigated [22], while part four in 1901 treats the forces and moments in the foot [23]. In both part five in 1903 [24] and part six in 1904 [25] the motion of the upper-leg is treated with which he became the first to give the exact proof that for walking the upper-leg does not solely swing forward as a passive pendulum.

In 1902 Fischer studied the influence of muscles on one another to stress, again, the importance of considering the motion of the complete human musculoskeletal system in order to investigate the functioning of the individual muscles [26]. In 1905 he extended his method of principal vectors to derive the spatial equations of motion of spatial chains [27].

In 1906 Fischer published his book 'Theoretical fundamentals for mechanics of moving bodies' ('Theoretische Grundlagen für eine Mechanik der lebenden Körper') [28]. In the first part of what would become a classical work [1] all of his achievements on the method of principal vectors are presented and developed from the very beginning. The second part summarizes his investigations on human motion to which the method had been applied. The book ends by illustrating the potentials of his method for mechanism and machine science.

Fischer published also a book on descriptions of human joints from a rather kinematical point of view instead of from the common anatomical point of view in 1907 [29]. In 1909 he published a work on spherical kinematics of Listing's law (das Listinsche Gesetz) [30] and of the humeroradial joint [31].

Most of the cited publications of Otto Fischer were published in the proceedings of the Mathematisch-Physikalischen Classe der Königlich-Sächsischen Gesellschaft der Wissenschaften in Leipzig of which he was extraordinary member since 1893 and ordinary member as of 1905. All cited literature is freely accessible on the internet in the DMG-library[2] or in the SLUB-library of Dresden [3].

3 The Method of Principal Vectors

To derive individual muscle forces for a human in motion, the motion of all body segments needs to be considered. For instance the motion of the elbow affects the forces at the shoulder or motion of the arm may affect to forces in the leg. However, calculations with the inverse dynamic model of the human at each instant are cumbersome to do by hand. Therefore Fischer developed a method that is capable of reducing the mechanics of the human mechanism to solely the element of interest, e.g. the elbow, knee, or foot. From the resulting reduced mass and inertia model the equations for the kinetic energy and the equations of motion could be derived.

The essential choice Fischer made was to investigate the motion of the body CoM independently from the relative motions of the body segments with respect to the body CoM. Since the motion of the body CoM is determined by external forces while the relative motion of body segments is determined by internal (muscle) forces, then also these can be investigated independently. The relative motion of the body segments about the body CoM is determined from the kinematics for which the dynamic model can be reduced. It is exactly this reduction step which is named *the method of principal vectors*, which is part of Fischer's complete 'method to derive kinetic energy' [14].

Fischer explained his method for the linkage of three elements shown in Fig. 1. The elements are linked with revolute pairs at $G_{1,2}$ and $G_{2,3}$ and each element has a mass m_i centered at position S_i. It is possible to define *principal points* H_i at each element. At element 1 principal point H_1 is defined as being the CoM of mass m_1 at S_1 and mass $m_2 + m_3$ at joint $G_{1,2}$. This is a projection of the mass of elements 2 and 3 at element 1. Equivalently, at element 3 principal point H_3 is defined as being the CoM of mass m_3 at S_3 and mass $m_1 + m_2$ at joint $G_{2,3}$. At element 2 principal point H_2 if defined as being the CoM of m_2 at S_2, m_1 at $G_{1,2}$, and m_3 at $G_{2,3}$.

With the principal points, lengths $H_1G_{1,2} = d_1$, $H_2G_{2,3} = d_2$, $G_{1,2}H_2 = c_2$, and $H_3G_{2,3} = c_3$, are determined which are the principal dimensions of the linkage. With

[2] http://dmglib.org/dmglib/handler?biogr=34004 (Dec. 2011).

[3] www.slub-dresden.de/sammlungen/digitale-sammlungen (Dec. 2011).

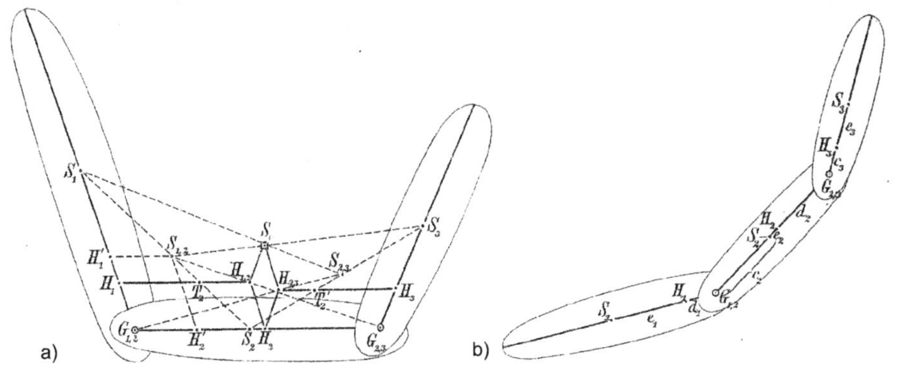

Fig. 1. Derivation of the principal points H_i that define the principal dimensions of a linkage of three elements with their CoM at S_i. The linkage CoM S_0 then is found geometrically [28].

these lengths the CoM of all three links S_0 can be geometrically found by parallelograms as indicated in Fig. 1a. Since the principal dimensions are independent of the motion of the linkage, these parallelograms trace the linkage CoM for all motion of the links. With respect to a fixed reference frame, the vectors describing the principal dimensions are named the principal vectors.

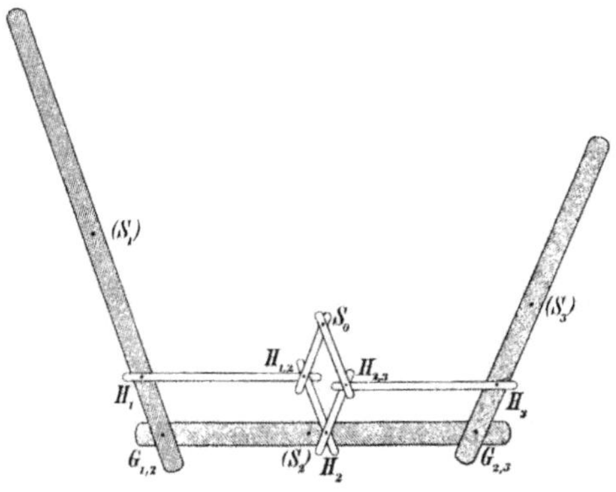

Fig. 2. Mechanism by Fischer to trace the CoM of three links at S_0 by additional links [28].

The geometric result of Fig. 1 can be seen as a linkage as shown in Fig. 2. Assuming the linkage CoM S_0 to be stationary, the position of the linkage is completely defined with the rotation of each element φ_i and its motion by their derivatives. Then for $\dot{\varphi}_2 = \dot{\varphi}_3 = 0$ element 1 rotates about H_1 while elements 2 and 3 solely translate. The inertia of this motion is the reduced inertia of the mechanism with respect to element 1. For each of the three elements this can be done with which the motion of the linkage consists of three reduced subsystems. For $\dot{\varphi}_1 = \dot{\varphi}_2 = 0$ element 3 rotates about H_3 while elements 1 and 2 solely translate and for $\dot{\varphi}_1 = \dot{\varphi}_3 = 0$ element 2 rotates about H_2 while elements 1 and 3 solely translate.

Based on these subsystems Fischer was able to formulate the kinetic energy equation of the motion of the complete linkage to depend solely on the total mass $m_0 = m_1 + m_2 + m_3$, the three reduced inertias k_i, and the principal lengths. The kinetic energy of the linkage relative to the linkage CoM then is written as

$$T_r = \frac{m_0}{2}(k_1^2\dot{\varphi}_1^2 + k_2^2\dot{\varphi}_2^2 + k_3^2\dot{\varphi}_3^2) + m_0 d_1 c_2 \cos(\varphi_1 - \varphi_2)\dot{\varphi}_1\dot{\varphi}_2 + m_0 d_1 c_3 \cos(\varphi_1 - \varphi_3)\dot{\varphi}_1\dot{\varphi}_3 + m_0 d_2 c_3 \cos(\varphi_2 - \varphi_3)\dot{\varphi}_2\dot{\varphi}_3 \quad (1)$$

Including also the kinetic energy of the motion of the linkage CoM, which depends on m_0 and on the velocity of the CoM $\dot{x}_0$ and $\dot{y}_0$, results in the equation of the kinetic energy for both relative and absolute motion of the linkage

$$T = \frac{m_0}{2}(\dot{x}_0^2 + \dot{y}_0^2 + k_1^2\dot{\varphi}_1^2 + k_2^2\dot{\varphi}_2^2 + k_3^2\dot{\varphi}_3^2) + m_0 d_1 c_2 \cos(\varphi_1 - \varphi_2)\dot{\varphi}_1\dot{\varphi}_2 + m_0 d_1 c_3 \cos(\varphi_1 - \varphi_3)\dot{\varphi}_1\dot{\varphi}_3 + m_0 d_2 c_3 \cos(\varphi_2 - \varphi_3)\dot{\varphi}_2\dot{\varphi}_3 \quad (2)$$

Deriving the equations of motion with Lagrange then results in five differential equations, two for the motion of the linkage CoM and three for the motion of the links with respect to the linkage CoM. The force components of the first two equations correspond with the external forces applied to the linkage CoM. The force components of the latter three equations correspond to the resultant moment on each link about the principal point H_i. These components are the resultants of all internal and external forces applied at the linkage that apply to an individual element from which the muscle forces can be derived.

4 Applications by Otto Fischer

Fischer applied his method to investigate the muscle activity of various parts of the human body. Figure 3 shows the method applied to the right arm. In this case the forces in the shoulder and elbow were considered with the trunk being stationary and the hand being rigidly attached to the lower arm. In Fig. 3a the position of the CoM of the arm with hand is $S_{8,10}$. Figure 3b shows the same arm but with the hand holding a mass for which the principal points are located closer to the hand.

Equivalently, Fig. 4a shows the method applied to a leg, assuming the trunk to be stationary and the foot to be rigidly attached to the lower leg. Figure 4b shows the upper leg for the investigation of the forward swing of a leg during walking. Here the motion of the complete human was considered as being reduced to the motion of the leg. The reduced system was regarded as a linkage of three elements with the lower leg with rigidly attached foot being element 1, the upper leg being element 2, and the upper body being element 3 as compared to Fig. 1.

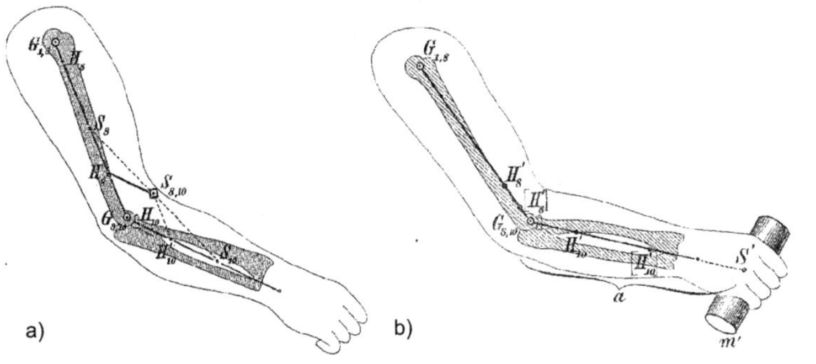

Fig. 3. Fischer's method applied to (a) the right arm, (b) the right arm holding a mass [28].

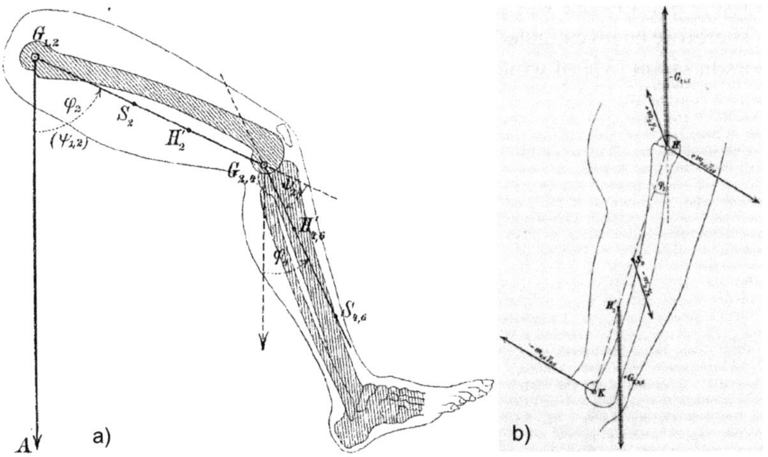

Fig. 4. Investigation of (a) forces in the leg [28] and (b) the swing phase of the leg [24,25].

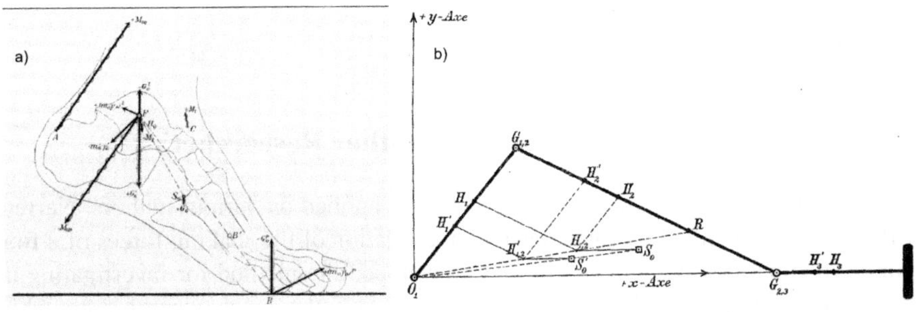

Fig. 5. Investigation of (a) the foot [23] and (b) the shaking force balancing of a crank-slider mechanism [28].

Fischer also studied the forces and moments in the foot during walking. Figure 5a shows the foot being in contact with the ground just before making a step [23]. In [28] Fischer illustrates the potential use of his method for machines and mechanisms by applying it to a crank-slider mechanism as shown in Fig 5b. He investigated the shaking forces, i.e. the resulting dynamic forces at the base, and derived the conditions for which they become zero and the mechanism is shaking force balanced. In [28] he also shows the benefit of his method for deriving the equations of motion with a pendulum used as example. From the reduced system the equations of motions are readily obtained, 'without the need of taking the indirect route by the kinetic energy and the general Lagrange differential equations of motion', with which he concludes his book.

Fischer developed his method in [27] for more complex linkages such as a linkage with six elements in series and a linkage with twenty elements. Also he showed that his method applies to spatial mechanisms for which he investigated a linkage of two elements with spatial motion.

It is likely that Fischer was going to apply his method for the investigation of spatial joints. Therefore he investigated the spherical kinematics of Listing's law [30] and of the humeroradial joint [31] of which Fig. 6 shows a mechanism model and an illustration.

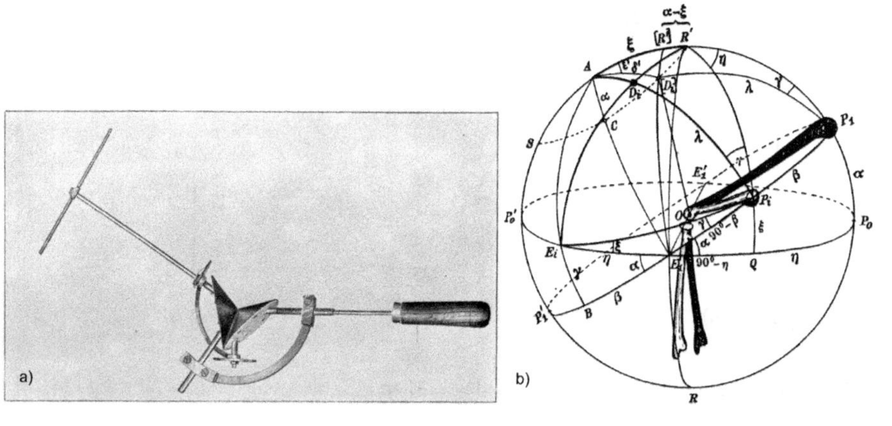

Fig. 6. (a) Spherical mechanism model of the kinematics of Listing's Law [30], (b) Study of the relative motion of the radius at the humeroradial joint [31].

5 Development and Application by Other Researchers

The method of principal vectors is described and applied by various authors. Nerreter in 1912 used the method for an extensive investigation of the shaking forces of a four-cylinder motor [34]. Wittenbauer in 1923 developed the method for investigating the motion of the linkage CoM and applied it to more complex mechanisms among which a parallel linkage [32]. He also applied the method to the linkage with three elements of Fig. 2, but with each element having an arbitrary CoM location as shown in Fig. 7a. Summaries of the method were given by, among others, Beyer in 1931 and 1960 [35,36] and Federhofer in 1932 [37].

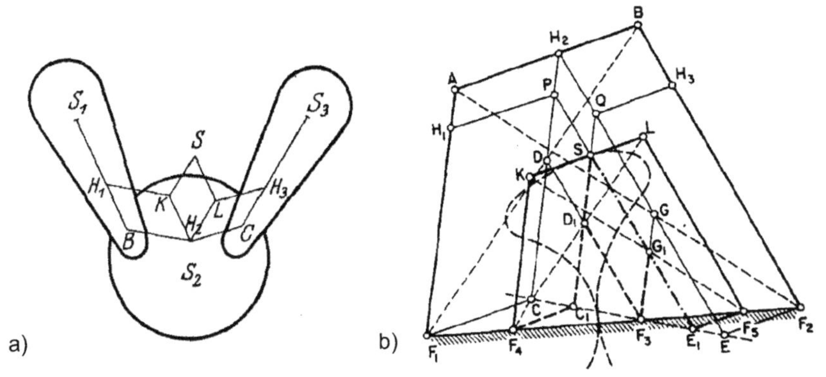

Fig. 7. (a) Application of the method by Wittenbauer for an arbitrary mass distribution of each of the three links [32]. (b) Illustration by Kreutzinger to show that the trajectory of the CoM of a 4R four-bar linkage is similar to a coupler curve of this linkage [33].

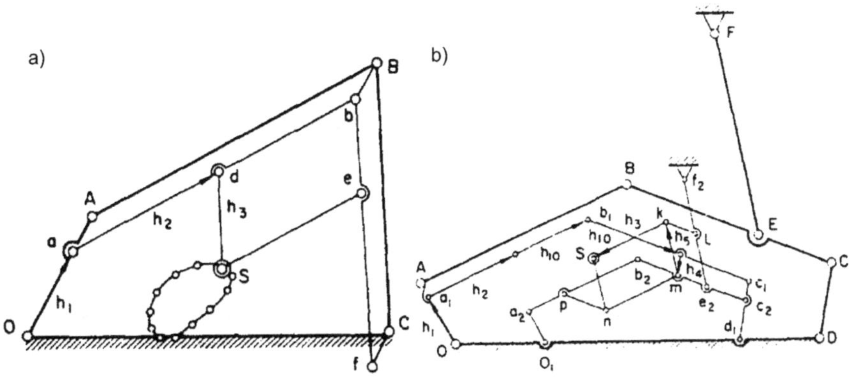

Fig. 8. Alternative geometric solutions by principal vectors to trace the CoM (a) by Artobolevskii and (b) by the *double contour transformation* on a six-bar linkage by Shchepetil'nikov [38].

Kreutzinger in 1942 applied the method of principal vectors to show that the trajectory of the CoM of a 4R four-bar linkage is a curve similar to a coupler curve of this linkage [33]. His solution is illustrated in Fig. 7b.

For the purpose of dynamic balancing, Shchepetil'nikov in 1957 extended the method of principal vectors to the method of *double contour transformation* [38]. He based his research on the findings of Artobolevskii in 1951 [39] who had proposed an alternative geometric solution to trace the CoM with principal dimensions, shown in Fig. 8a. Since for mechanisms with multiple closed chains the geometric solution by parallelograms quickly leads to excessive bulkiness, Shchepetil'nikov showed how the CoM can be traced by similar auxiliary linkages being jointed to the original linkage. One example of the application to a six-bar linkage is shown in Fig. 8b. His approach towards balancing was to have the linkage CoM move along a circular trajectory with a

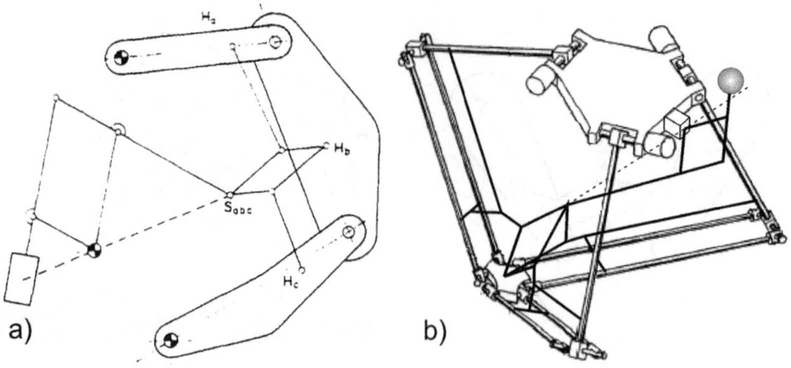

Fig. 9. (a) Hilpert's solution to balance the CoM of a 4R four-bar linkage by a pantograph with countermass [41]. (b) Delta Robot of which the CoM is made stationary by a pantograph with countermass by Van der Wijk and Herder [42].

rotation synchronous to the the driving crank. Then with a single counter-mass moving along with the driving crank the mechanism could be force balanced. By placing the counter-mass at an additional link elsewhere at the base, also the first harmonic of the shaking moment could be balanced. Having applied his method solely to linkages with mass symmetric links, he extended his method in 1975 to linkages with general mass distributions [40].

Hilpert in 1965 considers the balancing of a 4R four-bar linkage for which he uses Fischer's geometrical solution together with a pantograph with counter-mass to bring the linkage CoM at a stationary position with respect to the base as shown in Fig. 9a [41].

When the geometric solution to trace the CoM becomes a real linkage as in Fig. 9a, the masses of the auxiliary links have to be considered too. Agrawal et al. in 2001 showed by experiments that the mass of the auxiliary links of Fischer's original linkage of Fig. 2 can be included for which this linkage traces the CoM of all links together [43]. Then by having the CoM be stationary with the base results in a balanced manipulator as shown in Fig. 10a. In 2004 Agrawal and Fattah showed that also for a spatial manipulator the mass of all links can be included, of which a prototype is shown in Fig. 10b [44]. In this case the linkage CoM was statically balanced by a spring.

Van der Wijk and Herder in 2009 showed how the approach by Hilpert can be applied to balance a Delta Robot [42]. Figure 9b shows how with a spatial linkage the CoM of the moving arms and platform of the robot is derived and made stationary with the base by a pantograph with countermass.

Van der Wijk and Herder in 2010 considered Fischer's linkage of Fig. 2 with an arbitrary mass distribution of all links [45]. Since Fischer's approach on deriving the principal points and the principal dimensions then becomes cumbersome, they proposed a more fundamental approach based on linear momentum equations. Linkages with three and four elements in series were investigated of which the latter is illustrated in Fig. 11. Fig. 12a shows the mechanism model of a balanced 5-DoF chain with five arbitrary elements and mass symmetric auxiliary links which was presented.

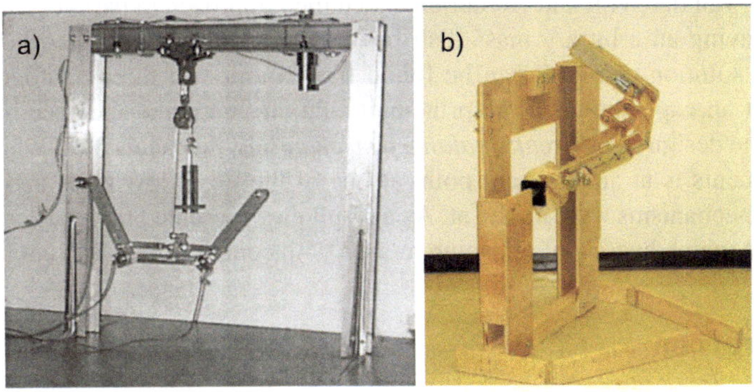

Fig. 10. Applications by (a) Agrawal et al. to include the mass of all links and [43] (b) by Agrawal and Fattah to derive a statically balanced spatial manipulator [44].

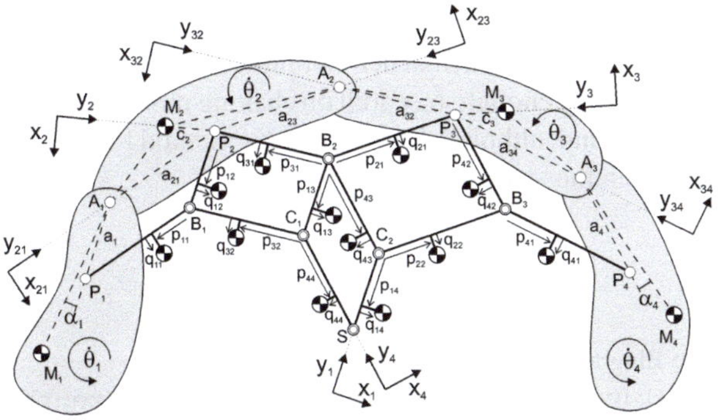

Fig. 11. A 4-DoF chain of 16 links with arbitrary CoMs of which invariant link point S is the CoM of the complete mechanism [45,46].

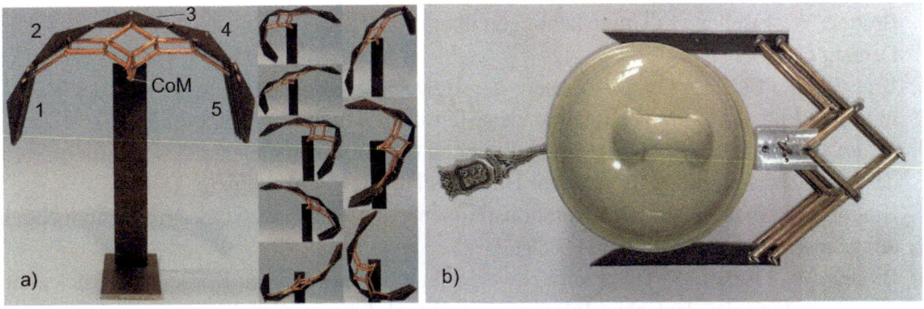

Fig. 12. (a) Mechanism model of a balanced 5-DoF chain [45] and (b) Prototype of a 2-DoF dynamically balanced grasper mechanism [46] by Van der Wijk and Herder.

In 2011 Van der Wijk and Herder extended their approach to parallel chains of links with all having an arbitrary mass distribution [46]. They also showed how Fischer's geometric solution of Fig. 2 can be found from a union of three pantographs which resulted in an explanation of the principal points from another perspective. With the focus on the design of *inherently balanced mechanisms*, mechanisms of which the CoM of all elements is at an invariant point at one of the links, low mass and low inertia balanced mechanisms were aimed at. As a result they presented the grasper mechanism of Fig. 12b which has 2-DoF grasping motion with complete dynamic balance.

6 Conclusion

An overview of the work of Otto Fischer was presented showing his contributions to three-dimensional human gait analysis and in particular the investigation of muscle forces for human motion. His method of principal vectors was summarized and some of Fischer's applications of the method were shown. The historical development of the method for mechanism and machine science was investigated, showing a clear focus on the motion of the center of mass of mechanisms for the purpose of static and dynamic balancing. Mechanisms based on Fischer's solution have led to inherently balanced manipulators. Surprisingly enough, research on shaking moment balancing by Fischer's method has not been found, although he invented his method for calculating the reduced inertia moments within the system.

References

1. Neue Deutsche Biographie, Berlin, vol. 5, p. 202 (1961)
2. (December 2011), `http://histvv.uni-leipzig.de/dozenten/fischer_o.html`
3. Biographisches Lexikon hervorragender Ärzte des neunzehnten Jahrhunderts, Berlin, Wien, p. 513 (1901)
4. Braune, W., Fischer, O.: Die bei der Untersuchung von Gelenkbewegungen anzuwendende Methode. erläutert am Gelenkmechanismus des Vorderarms beim Menschen. Hirzel, Leipzig (1885)
5. Braune, W., Fischer, O.: Untersuchungen über die Gelenke des menschlichen Armes. Hirzel, Leipzig (1887)
6. Braune, W., Fischer, O.: Das Gesetz der Bewegungen in den Gelenken an der Basis der mittleren Finger und im Handgelenk des Menschen. Hirzel, Leipzig (1887)
7. Braune, W., Fischer, O.: Über den Antheil den die einzelnen Gelenke des Schultergürtels an der Beweglichkeit des menschlichen Humerus haben. Hirzel, Leipzig (1888)
8. Braune, W., Fischer, O.: Die Rotationsmomente der Beugemuskeln am Ellenbogengelenk des Menschen. Hirzel, Leipzig (1889)
9. Braune, W., Fischer, O.: Über den Schwerpunkt des menschlichen Körpers mit Rücksicht auf die Ausrüstung des deutschen Infanteristen. Hirzel, Leipzig (1889)
10. Zarrugh, M.Y., Radcliffe, C.W.: Computer generation of human gait kinematics. Journal of Biomechanics 12, 99–111 (1979)
11. Rohlmann, A., Graichen, F., Bergmann, G.: Influence of load carrying on loads in internal spinal fixators. Journal of Biomechanics 33, 1099–1104 (2000)

12. Braune, W., Fischer, O.: Die Bewegungen des Kniegelenks, nach einer neuen Methode am lebenden Menschen gemessen. Hirzel, Leipzig (1891)
13. Braune, W., Fischer, O.: Bestimmung der Trägheitsmomente des menschlichen Körpers und seiner Glieder. Hirzel, Leipzig (1892)
14. Fischer, O.: Die Arbeit der Muskeln und die lebendige Kraft des menschlichen Körpers. Hirzel, Leipzig (1893)
15. Lowen, G.G., Berkof, R.S.: Survey of investigations into the balancing of linkages. Journal of Mechanisms 3, 221–231 (1968)
16. Fischer, O.: Beiträge zu einer Muskeldynamik. In: Erste Abhandlung: Über die Wirkungsweise Eingelenkiger Muskeln. Hirzel, Leipzig (1895)
17. Fischer, O.: Beiträge zur Muskelstatik. In: Erste Abhandlung: Ueber das Gleichgewicht zwischen Schwere und Muskeln am Zweigliedrige System. Hirzel, Leipzig (1896)
18. Fischer, O.: Beiträge zu einer Muskeldynamik. In: Zweite Abhandlung: Ueber die Wirkung der Schwere und Beliebiger Muskeln auf das Zweigliedrige System. Hirzel, Leipzig (1897)
19. Braune, W., Fischer, O.: Der Gang des Menschen. In: I. Theil: Versuche am unbelasteten und belasteten Menschen. Hirzel, Leipzig (1895)
20. Baker, R.: The history of gait analysis before the advent of modern computers. Gait & Posture 26, 331–342 (2007)
21. Fischer, O.: Der Gang des Menschen. In: II. Theil: Die Bewegung des Gesammtschwerpunktes und die äusseren Kräfte. Teubner, Leipzig (1899)
22. Fischer, O.: Der Gang des Menschen. In: III. Theil: Betrachtungen über die weiteren Ziele der Untersuchung und Überblick über die Bewegungen der unteren Extremitäten. Teubner, Leipzig (1900)
23. Fischer, O.: Der Gang des Menschen. In: IV. Theil: Über die Bewegung des Fusses und die auf denselben einwirkenden Kräfte. Teubner, Leipzig (1901)
24. Fischer, O.: Der Gang des Menschen. In: V. Theil: Die Kinematik des Beinschwingens. Teubner, Leipzig (1903)
25. Fischer, O.: Der Gang des Menschen. In: VI. Theil: Über den Einflusz der Schwere und der Muskeln auf die Schwingungsbewegung des Beins. Teubner, Leipzig (1904)
26. Fischer, O.: Das statische und das kinetische Maass für die Wirkung eines Muskels. erläutert an ein- und zweigelenkigen Muskeln des Oberschenkels. Teubner, Leipzig (1902)
27. Fischer, O.: Über die Bewegungsgleichungen räumlicher Gelenksysteme. Teubner, Leipzig (1905)
28. Fischer, O.: Theoretische grundlagen für eine Mechanik der lebenden Körper. Teubner, Leipzig (1906)
29. Fischer, O.: Kinematik organischer Gelenke. Vieweg, Braunschweig (1907)
30. Fischer, O.: Zur Kinematik des Listingschen Gesetzes. Teubner, Leipzig (1909)
31. Fischer, O.: Zur Kinematik der Gelenke vom Typus des Humero-Radialgelenks. Teubner, Leipzig (1909)
32. Wittenbauer, F.: Graphische Dynamik. Julius Springer, Berlin (1923)
33. Kreutzinger, R.: Über die Bewegung des Schwerpunktes beim Kurbelgetriebe. Getriebetechnik 10(9), 397–398 (1942)
34. Nerreter, A.: Graphodynamische Untersuchung einer vierzylindrigen Fahrzeugmaschine mit veränderlichen Hub. Dissertation T.H. München, M. Krayn, Berlin (1912)
35. Beyer, R.: Technische Kinematik. Johann Ambrosius Barth, Leipzig (1931)
36. Beyer, R.: Kinematisch-getriebedynamisches Praktikum. Springer, Heidelberg (1960)
37. Federhofer, K.: Graphische Kinematik und Kinetostatik. Springer, Berlin (1932)
38. Shchepetil'nikov, V.A.: The determination of the mass centers of mechanisms in connection with the problem of mechanism balancing. Journal of Mechanisms 3, 367–389 (1968)
39. Artobolevskii, I.I.: Theory of Mechanisms and Machines. Gos. Izdat. Tekh. -Teoret. Lit., Moscow (1951)

40. Shchepetil'nikov, V.A.: The balancing of bar mechanisms with unsymmetrical links. Mechanism and Machine Theory 10, 461–466 (1975)
41. Hilpert, H.: Weight balancing of precision mechanical instruments. Journal of Mechanisms 3, 289–302 (1968)
42. Van der Wijk, V., Herder, J.L.: Dynamic balancing of Clavel's delta robot. In: Kecskeméthy, Müller (eds.) Computational Kinematics, Proceedings of the 5th International Workshop on Computational Kinematics, pp. 315–322. Springer, Heidelberg (2009)
43. Agrawal, S.K., Gardner, G., Pledgie, S.: Design and fabrication of an active gravity balanced planar mechanism using auxiliary parallelograms. Journal of Mechanical Design 123, 525–528 (2001)
44. Agrawal, S.K., Fattah, A.: Gravity-balancing of spatial robotic manipulators. Mechanism and Machine Theory 39, 1331–1344 (2004)
45. Van der Wijk, V., Herder, J.L.: On the development of low-mass force balanced manipulators. In: Lenarčič, J., Stanisic, M.M. (eds.) Proceedings of the IFToMM 12th Int. Symposium on Advances in Robot Kinematics, pp. 411–420. Springer, Heidelberg (2010)
46. Van der Wijk, V., Herder, J.L.: Synthesis method for linkages with center of mass at invariant link point - pantograph based mechanisms. Mechanism and Machine Theory 48, 15–28 (2012)

Spatial Overconstrained Linkages—The Lost Jade

Yan Chen[1] and Zhong You[2]

[1] School of Mechanical and Aerospace Engineering,
Nanyang Technological University, Singapore, 639798
[2] Department of Engineering Science, University of Oxford, Oxford, OX1 3PJ,
United Kingdom

Abstract. This paper is to review all the discovered three-dimensional overconstrained linkage with only revolute joints. There are 15 types well known and several types recently proposed. Although the kinematics study on this family of linkages is in depth, little industry application has been developed. So it is the lost jade in the treasure box of mechanisms, waiting for being recovered and cherished.

1 Introduction

A mechanism considered to be an assembly of mechanical items designed to achieve a specific purpose within a machine. In the long history of mechanism invented by human being, the development of the mechanisms is function-oriented. The geometric shape and dimension of the mechanism components are generally depended on the space and loading conditions. The mobility of a mechanism can generally be determined by the Grübler-Kutzbach mobility criterion [1]. However, there is an exceptional family of linkage emerging since middle of eighteen century. Such linkages are with full-range mobility even though they do not meet the Grübler-Kutzbach mobility criterion. They are called *overconstrained linkages*. Their mobility is due to the existence of special geometry conditions among the links and joint axes that are often referred as the overconstrained conditions. And the invention of the linkages is driven by the kinematic challenges rather than the practical functions, which makes this linkage family known only academically.

In this paper, the attention is drawn to the 3D single-loop overconstrained linkages with only revolute joints. So in general, to obtain a mobility of one, a single closed loop linkage with revolute joints needs seven links. However, the minimum number of links to construct a mobile closed loop with revolute joints has been found to be four as a loop with three links and the same number of revolute joints is either an immobile structure or an infinitesimal mechanism when all of the revolute axes are coplanar and intersect at a single point [2]. Therefore, all of 3D overconstrained linkages studies in this paper are 4*R*, 5*R* or 6*R* linkages. Comparing to the conventional 3D mechanisms, such linkages with less number of components can generate more complicated output, which is a great advantage in nowadays sustainable industry world. Therefore, our target is to recover this lost jade in the treasure box of mechanisms and cherish them into the real application.

T. Koetsier, M. Ceccarelli (Eds.): Explorations in the His. of Machines & Mech., HMMS 15, pp. 535–550.
springerlink.com

The layout of this paper is as follows. The 3D overconstrained linkages which are well recognised by kinematicians are reviewed in section 2. The recent development of new 3D overconstrained linkages and the reason of limited industry application of this family of overconstrained linkages are discussed in section 3.

2 The Historical Review

The first overconstrained mechanism which appeared in the literature was reported by Sarrus[3]. Since then, other overconstrained mechanisms were discovered by Bennett [4], Delassus [5], Bricard[6, 7], Myard [8], Goldberg [9], Waldron [10-12], Wohlhart [13-15], Dietmaier [16], etc.. Considerable amount of research has been done on the general overconstrained linkages. For instance, Pamidi et al. [17] found the necessary and sufficient conditions for the existence of five-link spatial mechanisms with helical, cylinder, revolute, and prism pairs. These conditions were validated against 36 five-link spatial mechanisms with one and two passive constraints. Phillips [2] dedicated an entire chapter in his book (Chapter 20) on the special geometry of overconstrained linkages with 4*R*, 5*R* or 6*R*. Baker [18] provided a comparative survey of the Bennett-based 6*R* kinematic loops. He later listed all 28 types of 6*R* overconstrained linkage to date [19]. Next, we shall review all the 3D overconstrained linkages with only revolute joints under the general classification quasi-chronologically.

2.1 Double-Hooke's-Joint Linkage (1683)

As a widely-used transmission coupling, double-Hooke's-joint linkage was invented by Hooke [20] when connecting two universal joints by an intermediate shaft, in order to eliminate the non-uniform rotational movement of the single universal joint, also known as the Cardan joint or Hooke's joint. This linkage can in fact be obtained from two spherical 4*R* linkages. Figure 1 shows the arrangement of two spherical 4*R* linkages, which consist of joints 5, 6, 1 and 0, and joints 2, 3, 4 and 0, respectively, and have different centres. Replacing the common joint 0 with link 12, and adding link 45 between joints 5 and 4 result in the 6*R*

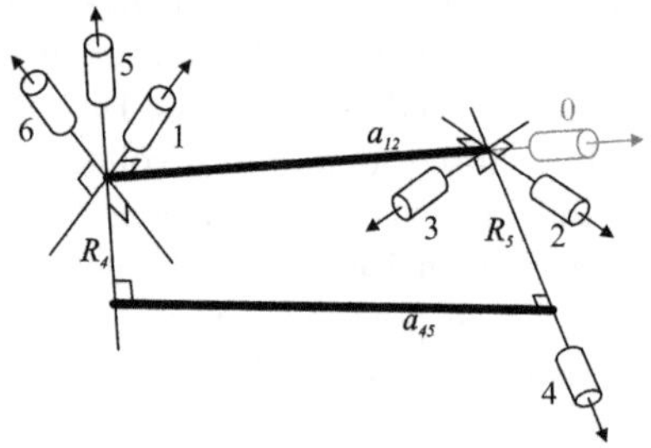

Fig. 1. Double-Hooke's-joint linkage.

linkage with the following parametric properties:

$$a_{23} = a_{34} = a_{56} = a_{61} = 0,\ \alpha_{23} = \alpha_{34} = \alpha_{56} = \alpha_{61} = \frac{\pi}{2},\ R_1 = R_2 = R_3 = R_6 = 0. \quad (1)$$

Considerable amount of further development has been made to the design of the joint components and arrangement. In 1841, Willis deduced that the double Hooke's joint possessed constant velocity properties, even when the input and output shafts were neither parallel nor intersecting. It transmits rotation with a constant angular velocity ratio provided the transmission angles, i.e. the angle between joints 2, 4 and the angle between joints 1, 5, are equal [2].

2.2 The Sarrus Linkage (1853)

Being the first 3D overconstrained linkage reported in literature, Sarrus linkage converts a part circular motion to a linear motion, or *vice versa*, see Fig. 2(a). The four links A, R, S, and B are consecutively hinged by three parallel horizontal hinges, as are the links A, T, U, and B. The directions of the two sets of hinges are different. With such an arrangement, link A can have rectilinear motion, vertically up and down, relative to link B. A model made by Professor Bennett in Cambridge is shown in Fig. 2(b).

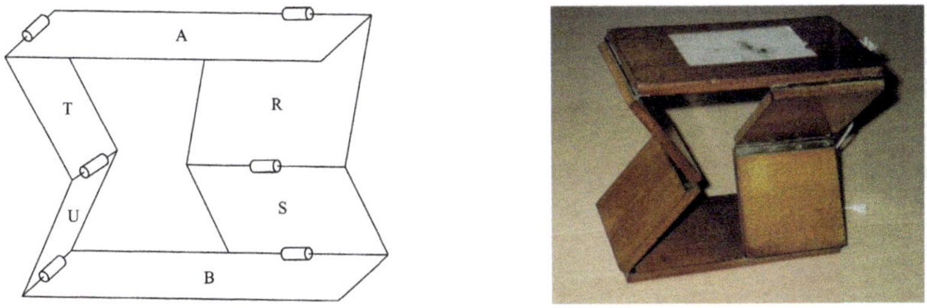

(a) (b)

Fig. 2. Sarrus linkage. (a) Schematic diagram; (b) the model made by Professor Bennett in Cambridge.

Due to the simple linear motion, the Sarrus linkage has been used in many engineering applications, such as the actuation mechanism for the active reflector of FAST (Five-hundred-meter Aperture Spherical Telescope) radio telescope [21], an independent automotive suspension, [22] the positioning mechanism of the saw disc in the new Bosch GCM 12SD mitre saw [23]. This linkage also appears in a number of deployable structures, such as the Hoberman sphere [24], the backbone of deployable antennas or shelters in space proposed by You [25], etc..

2.3 The Bricard's Mobile Octohedra (1897)

Among 6*R* overconstrained mechanisms, the most remarkable are 'Bricard linkages'. Bricard first described three different types of mobile octahedra in 1897 [6], namely the line-symmetric, plane-symmetric and double collapsible cases. Their geometric conditions are as follows.

(a) The line-symmetric octahedral case

$$\begin{gathered} a_{12}=a_{23}=a_{34}=a_{45}=a_{56}=a_{61}=0 \\ R_1+R_4=R_2+R_5=R_3+R_6=0 \end{gathered} \tag{2}$$

(b) The plane-symmetric octahedral case

$$\begin{gathered} a_{12}=a_{23}=a_{34}=a_{45}=a_{56}=a_{61}=0 \\ R_4=-R_1,\ R_2=-R_1\frac{\sin\alpha_{34}}{\sin(\alpha_{12}+\alpha_{34})},\ R_5=R_1\frac{\sin\alpha_{61}}{\sin(\alpha_{45}+\alpha_{61})}, \\ R_3=R_1\frac{\sin\alpha_{12}}{\sin(\alpha_{12}+\alpha_{34})},\ R_6=-R_1\frac{\sin\alpha_{45}}{\sin(\alpha_{45}+\alpha_{61})} \end{gathered} \tag{3}$$

(c) The doubly collapsible octahedral case

$$a_{12}=a_{23}=a_{34}=a_{45}=a_{56}=a_{61}=0,\ R_1R_3R_5+R_2R_4R_6=0 \tag{4}$$

Bennett [26] studied the geometry of the three types of deformable octahedra, and presented the kinematic properties of these mechanisms. One of model made by him is shown in Fig. 3. Interestingly Baker [27] found that a special line symmetric Bricard's octahedral case is equivalent of flexible form of the carbon skeleton of the cyclohexane molecule, known as boat form, and the linkage's stationary configurations are precisely equivalent to the minimum energy conformations of the flexing cyclohexane molecule, known as chair form. Later, Chai and Chen [28] proved that all such special linkages has a rigid closure.

Fig. 3. A line symmetric Bricard Octahedra made by Professor Bennett in Cambridge.

2.4 The Bennett Linkage (1903)

The Bennett linkage is a skewed linkage of four pieces having the axes of revolute joints neither parallel nor concurrent. This linkage was also found independently by Borel [29]. Its behaviour is better illustrated by the schematic diagram shown in Fig. 4. Bennett identified the conditions for the linkage to have a single degree of mobility as follows.

$$a_{12} = a_{34} = a\,,\ a_{23} = a_{41} = b\,,$$
$$\alpha_{12} = \alpha_{34} = \alpha\,,\ \alpha_{23} = \alpha_{41} = \beta\,. \tag{5}$$

$$\frac{\sin\alpha}{a} = \frac{\sin\beta}{b}\,.$$

$$R_i = 0 \quad (i = 1, 2, 3, 4)$$

(a)

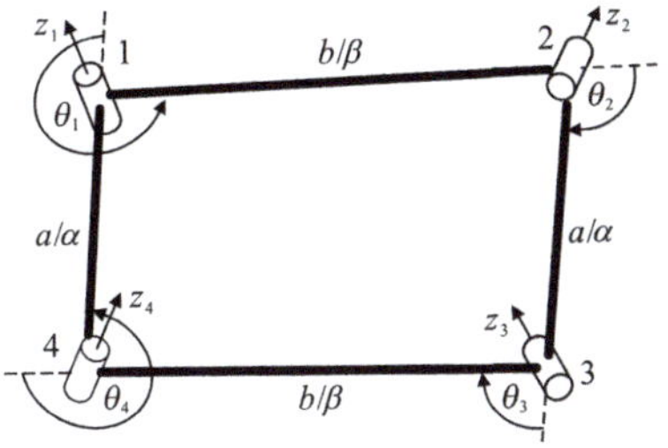

(b)

Fig. 4 The Bennett linkage. (a) Original model made by Professor Bennett in Cambridge; (b) a schematic diagram.

Since its invention, Bennett linkage has been a popular construct element for other overconstrained linkages which will be introduced later. Professor J E Baker and his student have done tremendous work on the kinematics of this linkage [30, 31, 32, etc.]. A robotic arm based on Bennett linkage was introduced by McCarthy [33]. And Chen and You has used Bennett linkage as the basic unit to assembly large scale deployable structures [34-36].

2.5 The Bennett 6*R* Hybrid Linkages (1905)

The 6*R* linkage shown in Fig. 5(a) was discovered by Bennett [37]. Hinges 1, 2 and 3 consecutively connect links A, R, S, and B, and intersect at point X, while hinges 4, 5 and 6 connect links A, T, U, and B, and intersect at point Y. X and Y do not coincide. The geometric conditions are

$$a_{12} = a_{23} = a_{45} = a_{56} = 0 \text{ and } R_2 = R_5 = 0\,. \tag{6}$$

Two further special forms can be derived by taking one or both of the points X and Y to infinity. The former forms a Bennett plano-spherical hybrid linkage as shown in Fig. 5(b), whereas the latter produces a Sarrus linkage.

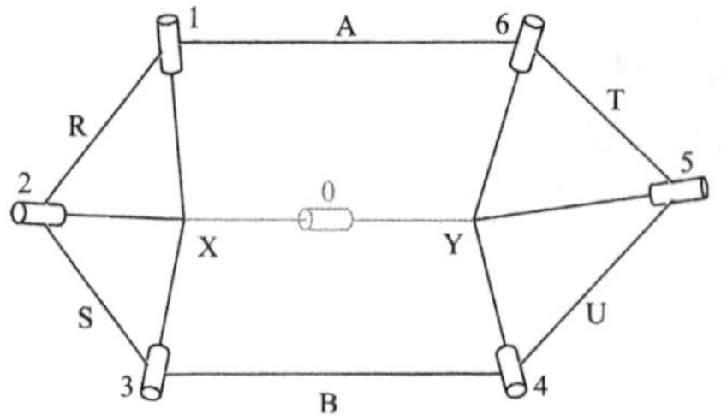

(a)

(b)

Fig. 5. Bennett 6*R* hybrid linkage.(a) The schematic diagram; (b) the plano-spherical hybrid linkage made by Bennett in Cambridge.

2.6 The Bricard 6*R* Linkages (1927)

Three decades later after the discovery of the mobile octahedra, Bricard proposed three types of mobile linkages, namely the line-symmetric, the plane-symmetric and the trihedral linkages in 1927 [7]. They can be summarised as follows.

(a) The general line-symmetric case, see Fig. 6(a),

$$\begin{aligned} &a_{12}=a_{45},\ a_{23}=a_{56},\ a_{34}=a_{61} \\ &\alpha_{12}=\alpha_{45},\ \alpha_{23}=\alpha_{56},\ \alpha_{34}=\alpha_{61} \\ &R_1=R_4,\ R_2=R_5,\ R_3=R_6 \end{aligned} \tag{7}$$

(b) The general plane-symmetric case, see Fig. 6(b),

$$\begin{aligned} &a_{12}=a_{61},\ a_{23}=a_{56},\ a_{34}=a_{45} \\ &\alpha_{12}+\alpha_{61}=\pi,\ \alpha_{23}+\alpha_{56}=\pi,\ \alpha_{34}+\alpha_{45}=\pi \\ &R_1=R_4=0,\ R_2=R_6,\ R_3=R_5 \end{aligned} \tag{8}$$

(c) The trihedral case, see Fig. 6(c) for the model made by Bennett in Cambridge,

$$\begin{gathered} a_{12}^2+a_{34}^2+a_{56}^2=a_{23}^2+a_{45}^2+a_{61}^2 \\ \alpha_{12}=\alpha_{34}=\alpha_{56}=\frac{\pi}{2},\ \alpha_{23}=\alpha_{45}=\alpha_{61}=\frac{3\pi}{2} \\ R_i=0\ \ (i=1,2,\cdots,6) \end{gathered} \tag{9}$$

Again, the kinematics of three linkages was studied by Baker [38] thoroughly.

2.7 The Myard Linkages (1931)

In 1931, Myard proposed four overconstrained linkages, including two 5*R* linkages and two 6*R* linkages. Among them, the best known is the No. 1 5*R* linkage, which is widely referred to as the Myard linkage [8].

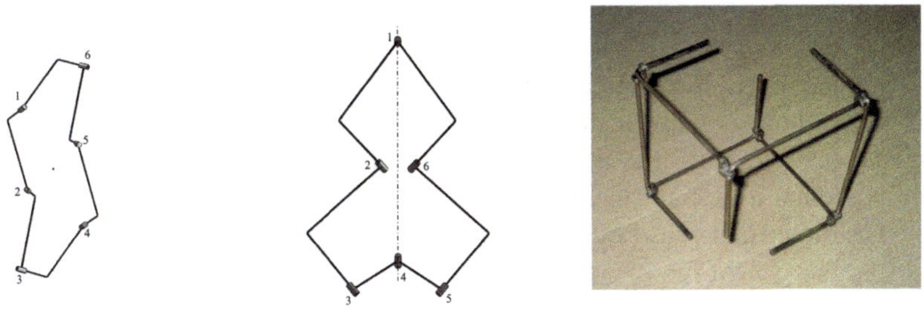

(a) (b) (c)

Fig. 6. The Bricard linkages.

The original Myard No. 1 5*R* linkage is plane-symmetric for which two Bennett linkages, with one pair of twists being $\pi/2$, are symmetrically disposed before combining them. Figure 7 shows the construction of the Myard linkage. The conditions on its geometric parameters are

$$\begin{gathered} a_{34}=0\,,\ a_{12}=a_{51}\,,\ a_{23}=a_{45}\,,\\ \alpha_{23}=\alpha_{45}=\frac{\pi}{2}\,,\ \alpha_{51}=\pi-\alpha_{12}\,,\ \alpha_{34}=\pi-2\alpha_{12}\,,\\ R_i=0\,,\ (i=1,2,\ldots,5)\,,\ a_{12}=a_{23}\sin\alpha_{12}\,, \end{gathered} \tag{10}$$

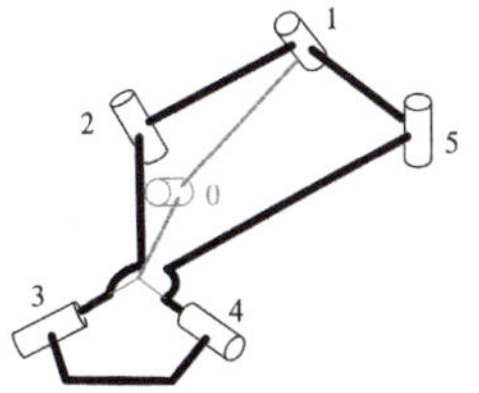

Fig. 7. Construction of the Myard Linkage.

The Myard No. 2 5*R* linkage is the combination of two identical Bennett linkages, which has no symmetric plane. The Myard No. 1 6*R* linkage was obtained in the same way as the Myard No. 1 5*R* linkage, but the requirement that the common link must have a twist of $\pi/2$ was removed, and instead, two Bennett linkages must be complementary rather than identical, i.e., the four twists in two linkages are complementary to each other. The Myard No. 2 6*R* linkage was constructed in a way rather different from his other three linkages, in which one of the revolute joints of a Bennett linkage was replaced by an offset spherical joint. Thus it is generally regarded as an *S-R-R-R* 4-bar linkage [12, 30].

2.8 The Schatz Linkages (1942)

This linkage was discovered and patented by Schatz in 1942 [39], also known by the name Turbula because it constitutes the essential mechanism of a machine by that name, see Fig. 8(a). It is used for mixing fluids and powders. It was considered as a derivation of a special trihedral Bricard linkage[2], see Fig. 8(b).

The dimension constraints of the linkage are as follows.

$$a_{12} = a_{56} = 0\,,\ a_{23} = a_{34} = a_{45} = a\,,\ a_{61} = \sqrt{3}a$$
$$\alpha_{12} = \alpha_{23} = \alpha_{34} = \alpha_{45} = \alpha_{56} = \frac{\pi}{2}\,,\ \alpha_{61} = 0 \quad (11)$$
$$R_1 = -R_6\,,\ R_2 = R_3 = R_4 = R_5 = 0$$

(a)

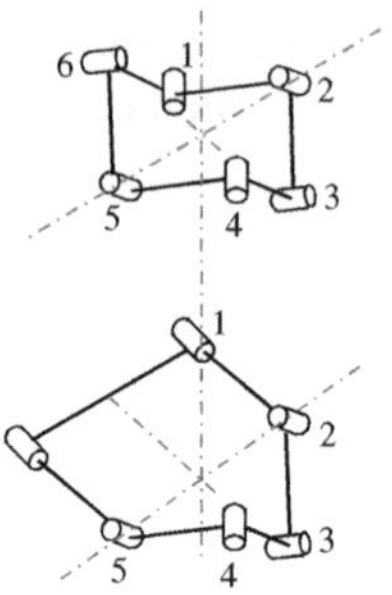

(b)

Fig. 8. Schatz linkage. (a) Turbula machine; (b) related with trihedral Bricard linkage.

2.9 The Goldberg Linkages (1943)

The Goldberg linkages include two types of 5*R* linkages and four types of 6*R* linkages created by summation and subtraction of a number of Bennett linkages [9]. The Goldberg basic 5*R* linkage is obtained by combining a pair of Bennett linkages in such a way that a link common to both was removed and a pair of adjacent links were rigidly attached to each other, see Fig. 9(a). Subtracting the Goldberg basic 5*R* linkage from another Bennett linkage yields the Goldberg syncopated 5*R* linkage, Fig. 9(b). Goldberg conjectured that the syncopated Goldberg 5*R* linkage might be resolved into the combination of two Bennett linkages like the Goldberg basic 5*R* linkage. Later, this conclusion was proved rigorously by Baker [30]. Thus, the Goldberg 5*R* linkage and the syncopated Goldberg 5*R* linkage are essentially the same. The generalised 5R linkage was also outlined briefly by Goldberg [9], the detail of which was later completed by Wohlhart [14].

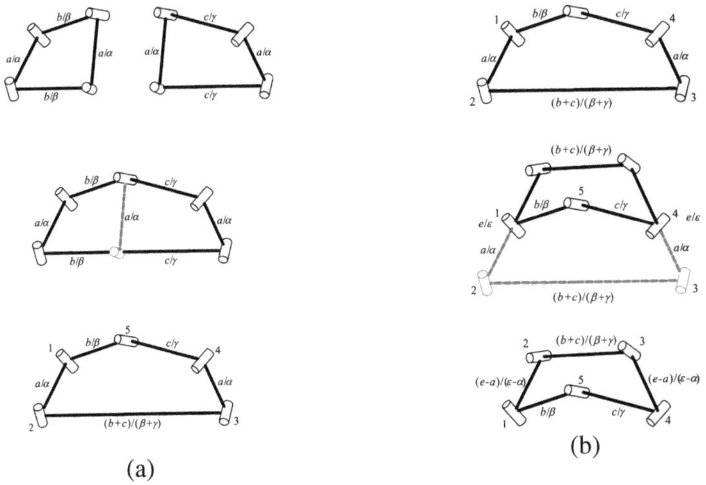

Fig. 9. Goldberg 5*R* Linkages. (a) Basic one; (b) syncopated one.

Similarly, the Goldberg series 6*R* linkage can be produced by combining three Bennett linkages in series, which have a pair of identical links. Three links in series are replaced by a single link while the common links are removed to form a 6*R* linkage. Then, subtracting the Goldberg series 6*R* linkage from another Bennett linkage leads a Goldberg syncopated series 6*R* linkage. The third Goldberg 6*R* linkage is built by an *L*-shaped arrangement of three Bennett linkages, whereas the fourth Goldberg 6*R* linkage is obtained by subtracting the Goldberg *L*-shaped 6*R* linkage from another Bennett linkage, which are known as the Goldberg *L*-shaped 6*R* linkage and Goldberg syncopated *L*-shaped 6*R* linkage, respectively.

2.10 The Altmann Linkage (1954)

Altmann [40] presented a 6*R* linkage as shown in Fig. 9, which actually is a special case of the Bricard line-symmetric linkage. Its dimensional conditions can be expressed as follows:

$$\begin{gathered} a_{12}=a_{45}=a\,,\ a_{23}=a_{56}=0\,,\ a_{34}=a_{61}=b \\ \alpha_{12}=\alpha_{45}=\frac{\pi}{2},\ \alpha_{23}=\alpha_{56}=\frac{\pi}{2},\ \alpha_{34}=\alpha_{61}=\frac{3\pi}{2} \\ R_i=0\ \ (i=1,2,\cdots,6) \end{gathered} \tag{12}$$

2.11 The Waldron's Hybrid Linkage (1968)

Waldron's hybrid linkage is actually a more general linkage than those based on the 4*R* linkages. Waldron [11] drew attention to a class of mobile six-bar linkages with only lower pairs which include helical, cylinder and prism joints in addition to revolute joints. He suggested that any two single-loop linkages with a single degree of

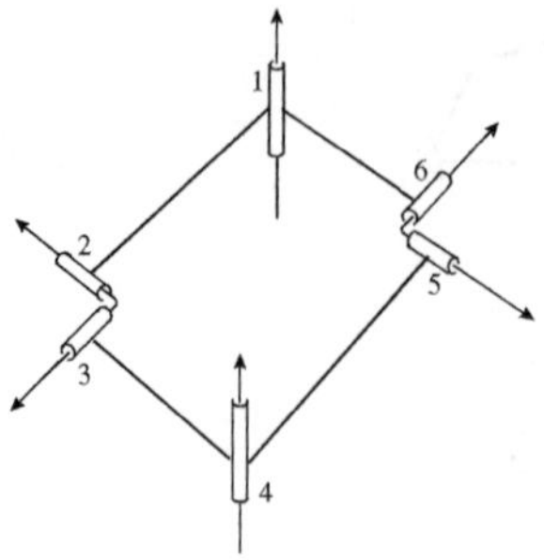

Fig. 10. Altmann linkage.

freedom could be arranged in space to make them share a common axis. After this common joint is removed, the resultant linkage certainly remains mobile. It is obvious that the double-Hooke's-joint linkage and the Bennett 6*R* linkages that we discussed above belong to this linkage family. An example of the Waldron hybrid linkage from two Bennett linkages is illustrated in Fig. 10.

2.12 The Wohlhart 6*R* Linkage (1987)

Wohlhart [13] presented a new 6*R* linkage, which can be regarded as a generalisation of the Bricard trihedral 6*R* linkage. In this linkage, the axes of all joints intersect a straight line, called the transversal line, see Fig. 11. This linkage shows three partial plane symmetries, i.e., three groups of two binary links with a symmetry plane are assembled in such a way that they form a loop. The conditions of its geometric parameters are as follows.

$$\begin{gathered} a_{12}=a_{23},\ a_{34}=a_{45},\ a_{56}=a_{61} \\ \alpha_{12}=2\pi-\alpha_{23},\ \alpha_{34}=2\pi-\alpha_{45},\ \alpha_{56}=2\pi-\alpha_{61} \\ R_6=-R_2-R_4,\ R_1=R_3=R_5=0 \end{gathered} \tag{13}$$

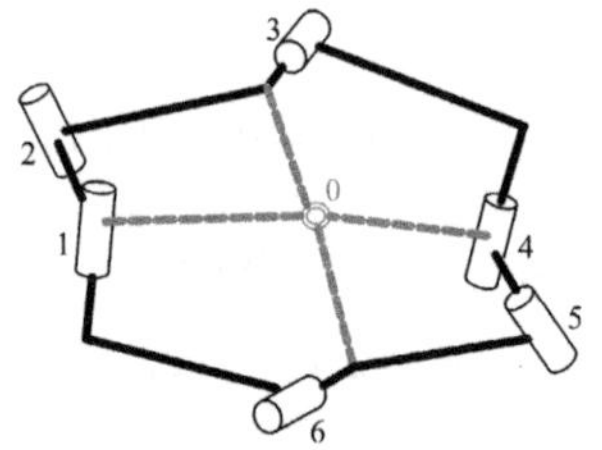

Fig. 11. Waldron hybrid linkage from two Bennett linkages.

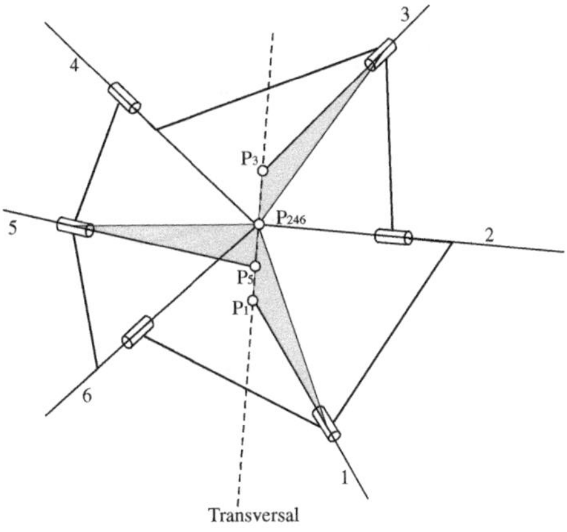

Fig. 12. Wohlhart 6*R* linkage.

2.13 The Wohlhart Double-Goldberg Linkage (1991)

Wohlhart [14] described a 6*R* overconstrained linkage, the synthesis of which was achieved by coalescing two appropriate generalised Goldberg 5*R* linkages and removing the two common links, see Fig. 12(a). The linkage is not line-symmetric generally. Wohlhart further combined his new 6*R* linkage with a Bennett linkage with links and twists being a_{12}, a_{23} and α_{12}, α_{23}, respectively, leading to another line-symmetric 6*R* linkage after removing the common links and joints as shown in Fig. 12(b). This method is called isomerisation.

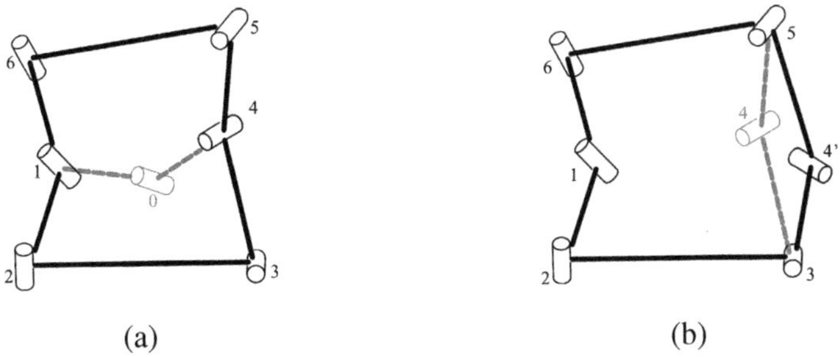

Fig. 13. The Wohlhart double-Goldberg linkage.

2.14 The Bennett-Joint 6*R* Linkage (1994)

This linkage was discovered by Mavroidis and Roth as a by-product of their effort to develop a systematic method to deal with overconstrained linkages [41]. They found that the manipulator inverse-kinematics problem was the same as that of finding the assembly configurations for the corresponding six-link closed-loop linkage. The term,

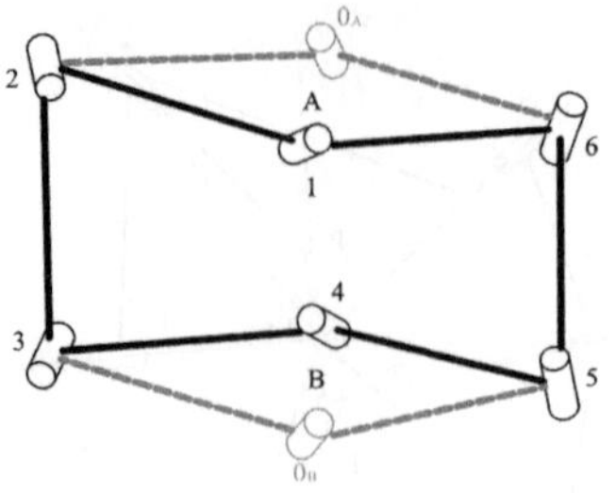

Fig. 14. Bennett-joint 6*R* linkage.

Bennett joint, is proposed as a set of three adjacent revolute joint axes of a Bennett linkage. The constant k_B in the above equation is a parameter of the Bennett joint, and is called the *Bennett ratio.* So for this new 6*R* linkage in Fig. 13, joint axes 6, 1, 2 and 3, 4, 5 form Bennett joints A and B respectively with the same Bennett ratios and with no common axis. Its parametric conditions are

$$\begin{gathered} a_{34} = a_{12},\ a_{56} = a_{23},\ a_{61} = a_{45}, \\ \alpha_{34} = \alpha_{12},\ \alpha_{56} = \alpha_{23},\ \alpha_{61} = \alpha_{45}, \\ R_1 = R_4 = 0,\ R_5 = R_2 \text{ (or } R_3\text{)},\ R_6 = R_3 \text{ (or } R_2\text{)}, \\ \frac{\sin\alpha_{61}}{a_{61}} = \frac{\sin\alpha_{12}}{a_{12}} = \frac{\sin\alpha_{34}}{a_{34}} = \frac{\sin\alpha_{45}}{a_{45}} = k_B\,. \end{gathered} \tag{14}$$

2.15 The Dietmaier 6*R* Linkage (1995)

Dietmaier [16] discovered a new family of overconstrained 6*R* linkages with the aid of a numerical method. The necessary conditions for a 6*R* linkage to be mobile were set up and solved numerically. The relationships of the geometrical parameters can be stated as follows.

$$\begin{gathered} a_{45} = a_{12},\ \alpha_{45} = \alpha_{12}, \\ R_1 = R_3,\ R_6 = R_4,\ R_2 = R_5 = 0, \\ \frac{a_{34}}{\sin\alpha_{34}} = \frac{a_{23}}{\sin\alpha_{23}},\ \frac{a_{61}}{\sin\alpha_{61}} = \frac{a_{56}}{\sin\alpha_{56}}, \\ \frac{a_{23}(\cos\alpha_{23} + \cos\alpha_{34})}{\sin\alpha_{23}} = \frac{a_{56}(\cos\alpha_{56} + \cos\alpha_{61})}{\sin\alpha_{56}}\,. \end{gathered} \tag{15}$$

Note that if the geometric parameters are restricted further by requiring $a_{56} = a_{34}$, $\alpha_{56} = \alpha_{34}$, $a_{61} = a_{23}$, and $\alpha_{61} = \alpha_{23}$, the Bennett-joint 6*R* linkage by Mavroidis and Roth will be obtained.

3 Discussion and Conclusion

All the fifteen types of 3D overconstrained linkages introduced in last section are well recognised in the field of kinematics. However, in the view point of classification, they are not strictly classified, i.e., one linkage could be considered as special case of another linkage from the construction or the overconstrained geometric conditions. Some classification goes into very details such as Baker did in 2002. He listed all 28 types of 6*R* overconstrained linkage in [19]. Meanwhile, among the new linkages appeared recently, some of them are new in the geometric conditions and/or construction method, while others can be considered as special cases of the existing linkages, which leaves large room of arguments among kinematicians. Most of recent contributions are mainly achieved by the construction method, for example, Chen and You's an extended Myard linkage, obtained by combining two Bennett linkages whose the twist angles were not necessary to be $\pi/2$ [42].

A large number of new linkages fall in the big family of Goldberg linkages. A variable called the "kink angle" is introduced. If the kink angle is set to be π, the subtractive Goldberg 5*R* linkage was obtained [43]. By further utilising Goldberg's technique in the choice of links to be locked in Goldberg's 6*R* linkages, two variants of Goldberg's 6*R* linkages were found by Baker [18], which were based on the serial Goldberg 6*R* linkage and the *L*-shape Goldberg 6*R* linkage, respectively. Similarly, but more systematically, Song and Chen [44] constructed all the possible varieties of the Goldberg 6*R* linkage with two subtractive Goldberg 5*R* linkages or one Goldberg 5*R* linkage and one subtractive Goldberg 5*R* linkage. All of them turn out to be the special cases of the 6*R* Goldberg linkages formed from two generalised Goldberg 5*R* linkages. Meanwhile, Recently, Baker [45] used Wohlhart's isomerisation method to construct three variants of Bricard line-symmetric 6R linkage with special geometric condition of the same Bennett ratio on a pair of adjacent links. The same method was applied to find isomeric variants of Dietmaier's 6R linkage [46]. The relationship between two types of 6*R* linkages, reported by Chen and You [47] by combination of two Goldberg 5*R* linkages in a *back-to-back* manner, is also isomerisation.

The research into 3D overconstrained close-loop linkages is centuries old. Besides the industry application of double-Hooke's-joint linkage, Sarrus linkage, and Schatz linkage mentioned earlier, Bennett linkages, Myard linkage, and Bricard linkage has been successfully utilised as building blocks for large deployable assemblies [36]. There is little use elsewhere in engineering applications for the 3D overconstrained linkages. In fact, only double-Hooke's-joint linkage and Schatz linkage were initially designed for the clear purpose of power or motion transmission. The rest were purely studied for the overconstrained kinematic research interest. Evermore, some of which were discovered by mathematicians rather engineering such as Bennett linkage. The design methods, both the construction and numerical methods have been focused on the discovery of overconstrained geometric conditions rather than the trajectory or input-output relationship. That is why there is lack of engineering application except the development of deployable structures, in which the focus is on the geometric shape of mechanism configurations on the kinematic paths[48].

On the other hand, the strict overconstrained geometric condition of such linkages is difficult to achieve in the industrious mass-product environment, which stopped the potential engineering applications. More research work should be devoted to analyse the effect on the tolerance of linkage imperfection. Meanwhile, it has been noticed that the symmetric characteristics play an important role in the mobility and fulfilling the overconstrained geometric conditions [49, 50]. Therefore, it is expected that the symmetric can increase the tolerance of imperfection when the linkage has higher degree of symmetry. More work shall be done urgently to clear the difficulty in the practical application so that to a new life can be revived to these fascinating linkages.

References

[1] Hunt, K.H.: Kinematic geometry of mechanisms. Oxford University Press, Oxford (1978)

[2] Phillips, J.: Freedom in Machinery: Screw theory exemplified, vol. 2. Cambridge University Press, Cambridge (1984)

[3] Sarrus, P.T.: Note sur la transformation des mouvements rectilignes alternatifs, en mouvements circulaires, et reciproquement. Academie des Sciences 36, 1036–1038 (1853)

[4] Bennett, G.T.: A new mechanism. Engineering 76, 777–778 (1903)

[5] Delassus, E.: Les chaînes articulées fermées et déformables à quatre membres. Bulletin des Sciences Mathématiques, 2nd Series 46, 283–304 (1922)

[6] Bricard, R.: Mémoire sur la théorie de l'octaèdre articulé. Journal de Mathématiques Pures et Appliquées 3, 113–148 (1897)

[7] Bricard, R.: Leçons de cinématique. In: Tome II Cinématique Appliquée, pp. 7–12. Gauthier-Villars, Paris (1927)

[8] Myard, F.E.: Contribution à la géométrie des systèmes articulés. Bulletin de la Société Mathématique de France 59, 183–210 (1931)

[9] Goldberg, M.: New five-bar and six-bar linkages in three dimensions. Trans. ASME 65, 649–663 (1943)

[10] Waldron, K.J.: A family of overconstrained linkages. Journal of Mechanisms 2, 201–211 (1967)

[11] Waldron, K.J.: Hybrid overconstrained linkages. Journal of Mechanisms 3, 73–78 (1968)

[12] Waldron, K.J.: Symmetric overconstrained linkages. Transactions of the ASME. Journal of Engineering for Industry 91, 158–164 (1969)

[13] Wohlhart, K.: A new 6R space mechanism. In: Congress on the Theory of Machines and Mechanisms, Seville, Spain, pp. 193–198 (1987)

[14] Wohlhart, K.: Merging two general Goldberg 5R linkages to obtain a new 6R space mechanism. Mechanism and Machine Theory 26, 659–668 (1991)

[15] Wohlhart, K.: The two types of the orthogonal Bricard linkage. Mechanism and Machine Theory 28, 809–817 (1991)

[16] Dietmaier, P.: A new 6R space mechanism. In: Proceeding 9th World Congress IFToMM, Milano, pp. 52–56 (1995)

[17] Pamidi, P.R., Soni, A.H., Dukkipati, R.V.: Necessary and sufficient existence criteria of overconstrained five-link spatial mechanisms with helical, cylinder, revolute and prism pairs. Transactions of the ASME. Journal of Engineering for Industry 95, 737–743 (1973)

[18] Baker, J.E.: A comparative survey of the Bennett-based, 6-revolute kinematic loops. Mechanism and Machine Theory 28, 83–96 (1993)

[19] Baker, J.E.: Displacement-closure equations of the unspecialised double-Hooke's-joint linkage. Mechanism and Machine Theory 37, 1127–1144 (2002)
[20] Schmelz, F., Aucktor, E.: Universal Joints and Driveshafts: Analysis, Design, Applications. Speinger, Heidelberg (2006)
[21] Lu, Z., Yang, G., Wang, Q., Nan, R.: Research on mechanism of actuated active reflectors for FAST. Journal of Beijing University of Aeronautics and Astronautics 32, 233–238 (2006)
[22] Zhao, J.S., Chu, F., Feng, Z.J.: Synthesis of Rectilinear Motion Generating Spatial Mechanism With Application to Automotive Suspension. Transactions of the ASME. Journal of Mechanical Design 130, 065001 (2008)
[23] `http://www.boschgcm12sd.com/12-bosch-miter-saw.shtml`
[24] Hoberman, C.: Radial Expansion/Retraction Truss Structures, U.S. Patent 5024031 (1991)
[25] You, Z.: Deployable structure of curved profile for space antennas. Journal of Aerospace Engineering 13, 139–143 (2000)
[26] Bennett, G.T.: Deformable octahedra. Proceeding of London Mathematics Society, 2nd series 10(1), 309–343 (1911)
[27] Baker, E.J.: Limiting positions of a Bricard linkage and their possible relevance to the cyclohexane molecule. Mechanism and Machine Theory 21(3), 253–260 (1986)
[28] Chai, W.H., Chen, Y.: The line-symmetric octahedral Bricard linkage and its structural closure. Mechanism and Machine Theory 45, 772–779 (2010)
[29] Bennett, G.T.: The skew isogram mechanism. Proceedings of the London Mathematical Society, 2nd series 13, 151–173 (1914)
[30] Baker, J.E.: The Bennett, Goldberg and Myard linkages–in perspective. Mechanism and Machine Theory 14, 239–253 (1979)
[31] Huang, C.: The cylindroid associated with finite motions of the Bennett mechanism. Journal of Mechanical Design 119, 521–524 (1997)
[32] Baker, J.E.: On the motion geometry of the Bennett linkage. In: Proceeding 2 of Eighth International Conference on Engineering Computer Graphic and Descriptive Geometry, Austin, USA, pp. 433–437 (1998)
[33] Perez, A., McCarthy, J.M.: Dimensional synthesis of Bennett linkages. Journal of Mechanical Design 125, 98–104 (2003)
[34] Chen, Y., You, Z.: Mobile assemblies based on the Bennett linkage. Proceedings of the Royal Society A: Mathematical, Physical and Engineering Science 461, 1229–1245 (2005)
[35] Chen, Y., You, Z.: On mobile assemblies of Bennett linkages. Proceedings of the Royal Society A: Mathematical, Physical and Engineering Science 464, 1275–1283 (2008)
[36] You, Z., Chen, Y.: Motion Structures: Deployable Structural Assemblies of Mechanisms. Taylor and Francis (2011) ISBN: 978-0-415-55489-3
[37] Bennett, G.T.: The parallel motion of Sarrus and some allied mechanisms. Philosophy Magazine, 6th series 9, 803–810 (1905)
[38] Baker, E.J.: An analysis of Bricard linkages. Mechanism and Machine Theory 15, 267–286 (1980)
[39] Schatz, P.: U.S. Pat. No. 2,302,804 (1942)
[40] Altmann, P.G.: Communications to Grodzinski, P. and M'Ewen, E, Link mechanisms in modern kinematics. Proceedings of the Institution of Mechanical Engineers 168(37), 889–896 (1954)
[41] Mavroidis, C., Roth, B.: Analysis and synthesis of overconstrained mechanisms. Mechanism Synthesis and Analysis 70, 115–133 (1994)

[42] Chen, Y., You, Z.: An extended Myard linkage and its derived 6R linkage. Journal of Mechanical Design 130, 052301 (2008)
[43] Song, C.Y., Chen, Y.: A spatial 6R linkage derived from subtractive Goldberg 5R linkages. Mechanism and Machine Theory 46, 1097–1106 (2011)
[44] Song, C.Y., Chen, Y.: A family of mixed double-Goldberg 6R linkages, Proceedings of the Royal Society A: Mathematical, Physical and Engineering Sciences 468 871–890 (2012)
[45] Baker, J.E.: Screw replacements in isomeric variants of Bricard's line-symmetric six-bar. Proceedings of the Institution of Mechanical Engineers. Part C: Journal of Mechanical Engineering Science 223, 2391–2398 (2009)
[46] Baker, J.E.: Using the single reciprocal screw to confirm mobility of a six-revolute linkage. Proceedings of the Institution of Mechanical Engineers. Part C: Journal of Mechanical Engineering Science 224, 2247–2255 (2010)
[47] Chen, Y., You, Z.: Spatial 6R linkages based on the Combination of two Goldberg 5R linkages. Mechanism and Machine Theory 42, 1484–1498 (2007)
[48] Pellegrino, S. (ed.): Deployable Structures. Springer, Heidelberg (2001)
[49] Guest, S.D., Fowler, P.W.: A symmetry-extended mobility rule. Mechanism and Machine Theory 40, 1002–1014 (2005)
[50] Guest, S.D., Fowler, P.W.: Symmetry conditions and finite mechanisms. Journal of Mechanics of Materials and Structures 2(2), 293–301 (2007)

Designing, Analysis and Computer Modeling of Straight-Line Mechanisms

Valentin Tarabarin[1], Z. Tarabarina[2], and D. Chirkina[3]

[1] Ass. Prof, docent, BMSTU, Moscow, Russia
[2] Sr. Tutor, BMSTU, Moscow, Russia
[3] Student, BMSTU, Moscow, Russia

Abstract. Some mechanisms connect mechanics and mathematics closely. In the middle of 19-th century scientists' attention was attracted by mechanical linkages providing a transformation from translation into rotation. This led to the search for mechanisms realizing theoretically exact displacement of a target point along a straight line. Dozens of such straight-line linkages have been developed. The majority of them realize the approximate displacement along a straight line and some of them provide exact rectilinear motion. In the creation of these mechanisms J. Watt, France Reuleaux, P. Chebyshev and the others took part. In Reuleaux' famous collection of models about 40 are models of straight-line mechanisms. This article discusses some models of straight-line linkages (further called SLM). Moreover, the results of modeling them graphically and numerically by computer are presented.

1 General Introduction

The task over creation of the linkages providing exact transformation of a rotation into a rectilinear motion have attracted attention of mathematicians and mechanics as in the past [1, 7-9, 11, 12], so nowadays [2-6, 13, 14].

The first theoretically exact SLM by the time of its creation was six-bar spatial mechanism offered in 1853 by Pierre Frederic Sarrus [6]. A photo of this mechanism's model and its structure scheme is presented at Fig.1. Possibly, by a virtue of its "spatiality" it remains a poorly known and mentions seldom among SLM. But despite of this circumstance it applies widely enough in techniques, e.g. in forge furs, jacks and lifts.

Probably the history of SLM shouldn't have been such interesting, if J. Watt hadn't connected device of his beam engine with them. A rod of the piston should be attached in the point of the mechanism moving along a direct axis of the cylinder. Thus the pressure angle of the piston in the point of connection to the mechanism would be close to zero. This path was provided approximately with points of the parallelogram have been offered by Watt. He orders two parallelograms: full and reduced. Including these mechanisms the most significant inventions, Watt very highly estimated creation of these mechanisms.

T. Koetsier, M. Ceccarelli (Eds.): Explorations in the His. of Machines & Mech., HMMS 15, pp. 551–563.
springerlink.com

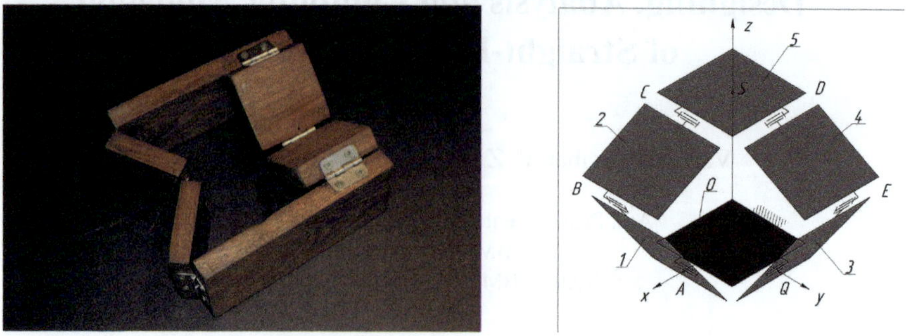

Fig. 1. Model and structure scheme of Sarrus's mechanism.

The most known exact straight-line mechanism is eight-bar planar SLM, offered by Charles Nicolas Peaucellier (1832-1913) in 1864 (Fig.2). The history of creation of this mechanism is interesting: the first mention of it had not been noticed by scientific community. In 1868 this mechanism has been invented at the second time by the student of St-Petersburg University L.I. Lipkin (1854-1875), who was also Chebyshev's follower. In 1871 his paper "About one crank straight-line generating mechanism" was published at the bulletin of Emperors Academy of Science [7], and a year later it carried out at *«Mèlanges mathèmatiques et astronomiques»*. Only in 1873 the detailed decision of this problem was published by Peaucellier on his own [8] and by other French mathematician Lemuan [9].

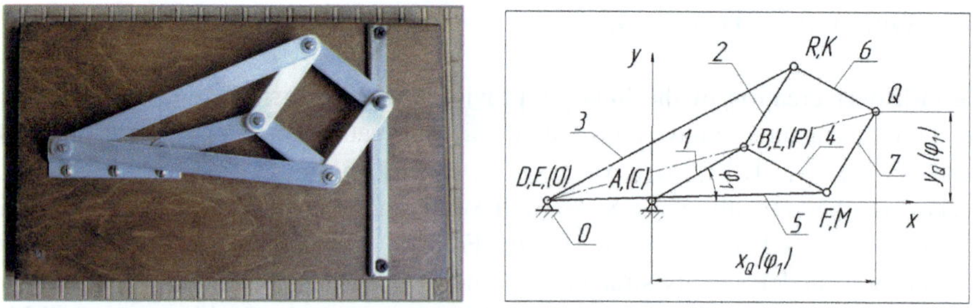

Fig. 2. Photo of model and structure scheme of Peaucellier's mechanism.

The creation of this mechanism had begun the whole section of the applied mechanics, devoted to a reproduction of a straight line by means of crank linkage. In 1874 a mathematician of Woolwich Royal military Academy H. Hart published his papers on SLM [10] and offered exact mechanism (Fig.3), being consisted of six links only [6,11].

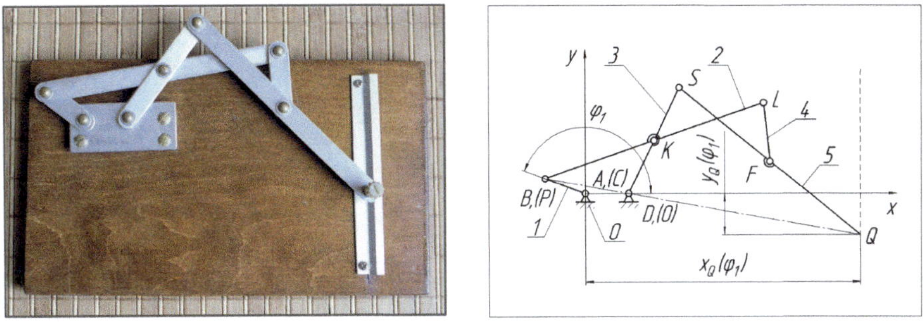

Fig. 3. Photo of model and structure scheme of Hart's mechanism.

2 Inversion like a Method of SLM's Creation

Peaucellier's and Hart's mechanisms are based on mathematical principle of inversion (transformation of curves along radiuses, which are opposite to each other). Let's consider how this principle is used at synthesis of SLM. Let points *Q*, *P* and *O* belong to some straight line (Fig.4), thus point *O* is fixed, and points *P* and *Q* move along the straight line. On fixed plane point *Q* outlines trajectory *q-q*, and point *P* draws path *p-p*. If radius-vectors *OP* and *OQ* of the point's *P* and *Q* relatively point *O* is connected by the Eqn.1:

$$\overline{OP \cdot OQ} = k^2, \tag{1}$$

and *k* is a constant, and then curves *q-q* and *p-p* called opposite to each other or inverse. Thus point *O* is called center of inverse.

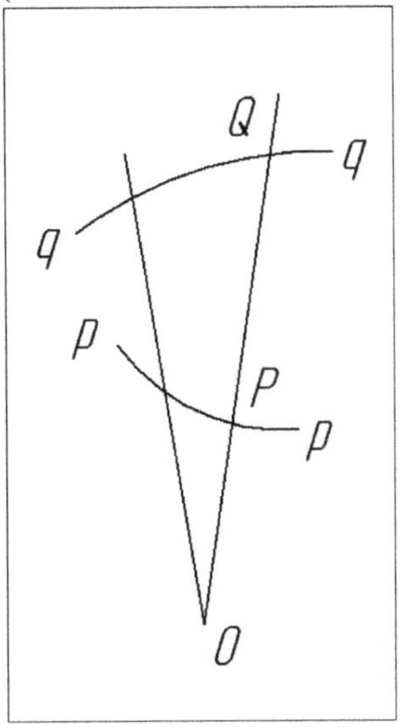

Fig. 4. Schemes of curves' inversions

If curve *q-q* is an arch of a circle, then its inversion *p-p* also is an arch of a circle. The proof resulted in paper [12] shows it. In this case curve *q-q*, turns into a circle with infinite radius or into a straight line, if circle outlined by point *P*, passes through a point *O* (a center of inversion) (Fig.5). Curve *q-q* is an inversion of this circle.

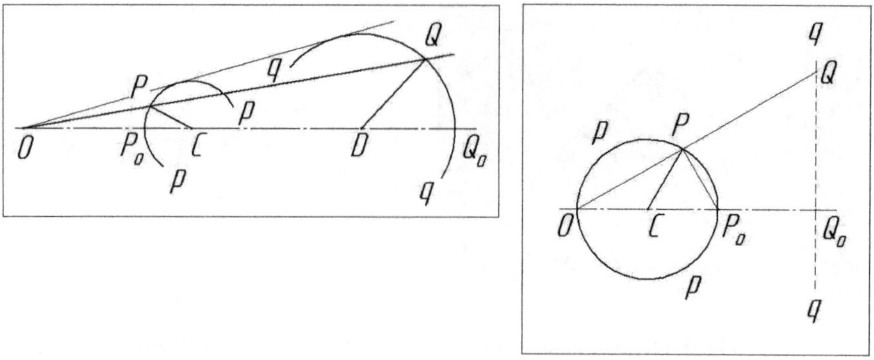

Fig. 5. Schemes of inversions “circle-circle” and “circle-straight line”

Peaucellier’s mechanism (look at scheme on Fig.2) consists of crank rhomb *BRQF* and two levers *DR* and *EF* with equal length. Owing to symmetry of a kinematic chain relative to straight line *OQ* points *O, Q* and *P* settled down on this straight line in any position of the mechanism. Except of it multiplication of radius-vectors *OP* and *OQ* is shown at Eqn.2:

$$\overline{OP} \cdot \overline{OQ} = l^2 - r^2, \tag{2}$$

where lengths of levers are $\overline{DR} = \overline{EF} = l,$ and lengths of rhombus’s edges are $\overline{BR} = \overline{RQ} = \overline{QF} = \overline{FB} = r.$

At the creation of SLM the centers of motionless hinges *D* and *E* is placed in the center of inverse, that is point *O*, and a center of hinge *A* settled down at the middle of line segment *OP*. The length of $\overline{AB}$ of link 1 equals to size interaxal distance $\overline{DA}$, thus $\overline{AB} = \overline{DA} = 0.5 \cdot \overline{OP}.$ Peaucellier’s mechanism consists of a frame, seven movable links and ten kinematic pairs. The mobility of this planar mechanism is equal to one (see Eqn.3):

$$W^{planar} = 3 \cdot n - 2 \cdot p_1 = 3 \cdot 7 - 2 \cdot 10 = 1. \tag{3}$$

It is the second-class mechanism, which contains the primary mechanism (0-1) and three Assur group - dyads RRR (2-3, 4-5, 6-7).

An alternative form of Peaucellier’s inversor would be made, if center of inverse point O will be settled down between points Q and C. In this mechanism lengths of levers *DR* and *EF* will be less, then lengths of rhombus’s *BRQF* links. A photo of this linkage’s model and its structure scheme is presented at Fig.6.

An antiparallelogram *KSLF* with four hinges is put in the basis of Hart’s inversor (see scheme on Fig.3). Lengths of opposite links are equal: $\overline{KS} = \overline{LF}; \overline{KL} = \overline{SF}$, and the largest links are crossed. Antiparallelogram’s links is crossed with straight line, which is parallel to *SL* and *KF*. The proof resulted at paper [12] shows, that points of this intersection define the points of inversor *O, P* and *Q* on this straight line. If centers of hinges is settled down in these points and links' lengths is executed with observance of following equalities: $\overline{KS} = \overline{DK}; \overline{SF} = \overline{FQ}; \overline{BK} = \overline{KL};\ \overline{AB} = \overline{DA} = 0.5 \cdot \overline{OP}$, then six-bar Hart's SLM will be received. Hart's mechanism consists of frame, five movable links and seven rotary kinematic joints. The mobility of this planar mechanism equals to one:

$$W^{planar} = 3 \cdot n - 2 \cdot p_1 = 3 \cdot 5 - 2 \cdot 7 = 1. \tag{4}$$

It is the second-class mechanism, which contains the primary mechanism (0-1) and three dyads RRR (2-3, 4-5 and 6-7) with two final links.

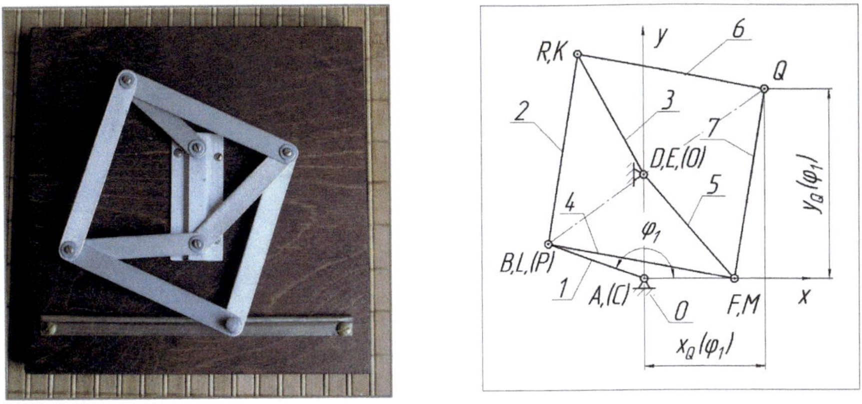

Fig. 6. Model and structure scheme of alternative form of Peaucellier's mechanism

3 The Other Exact Crank SLM

Another exact six-bar crank SLM have been of offered by Hart. At the middle position links of this mechanism form the letter "A" [6], therefore it is named "A-shaped" Hart's mechanism (Fig.7).

Let's consider the mechanism at the average position. Let designate $\overline{AF} = \overline{ED} = \overline{AD} = a,\ \ \overline{FQ} = \overline{EQ} = \overline{BC} = b,$ and $\overline{BF} = \overline{EC} = c.$

Then following equalities (Eqn. 5, Eqn. 6, Eqn. 7) proceed from similarity of triangles $\Delta B'Q'C'$ and $\Delta AQ'D$:

$$\frac{a}{a+b} = \frac{b}{c+b}; \tag{5}$$

$$ac + ab = ba + b^2; \tag{6}$$

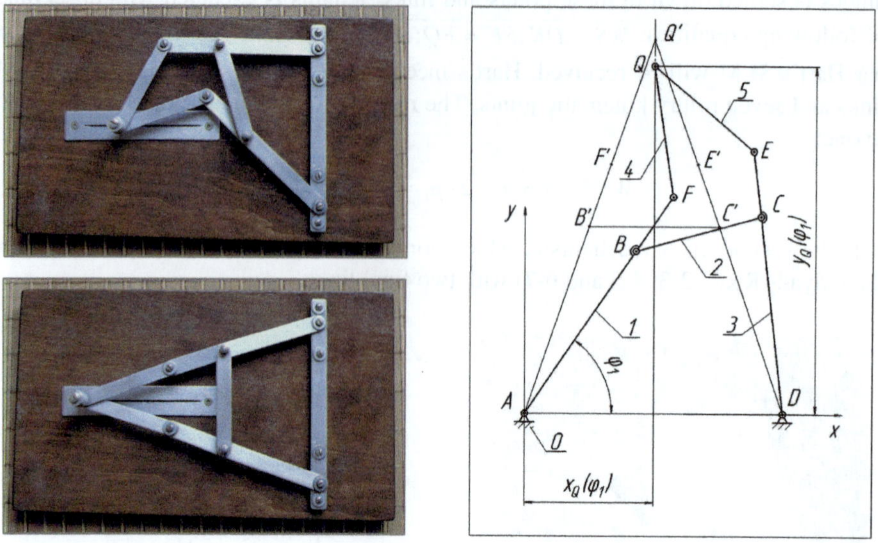

Fig. 7. Photos of model and scheme of "A-shaped" Hart's mechanism

$$b^2 = ac. \tag{7}$$

Defining the dimensions of the Hart's mechanism links it is necessary to follow these equalities.

"A-shaped" mechanism consists of five movable links and seven rotary kinematic pairs, its mobility equals to one (Eqn. 8):

$$W^{planar} = 3 \cdot n - 2 \cdot p_1 = 3 \cdot 5 - 2 \cdot 7 = 1. \tag{8}$$

The mechanism concerns to second class by the Artobolevsky classification and it consists from primary mechanism (0-1) and two dyads RRR (2-3, 4-5).

Cartwright's mechanism also concerns to exact SLM [12]. This mechanism is based on crank kinematic chain *ABQCD* (Fig.8), including four movable links and five rotary kinematic pairs. External pairs *A* and Q of this chain connect it with frame. The mobility of formed mechanism equals two (Eqn. 9):

$$W^{planar} = 3 \cdot n - 2 \cdot p_1 = 3 \cdot 4 - 2 \cdot 5 = 2. \tag{9}$$

The line *q-q* is perpendicular to interaxal line and halves it. The center of hinge *Q* will move through the straight line *q-q*, if conditions of equality of lenghts $l_{AB} = l_{DC}$ for links 1, 4 and $l_{BQ} = l_{CQ}$ for links 2 and 3, also angles $|\varphi_1| = |\varphi_4|$ are satisfied. Thus links 1 and 4 should move under the identical laws but in opposite directions. It is possible to provide it by connection of these links among themselves by external gear pair with

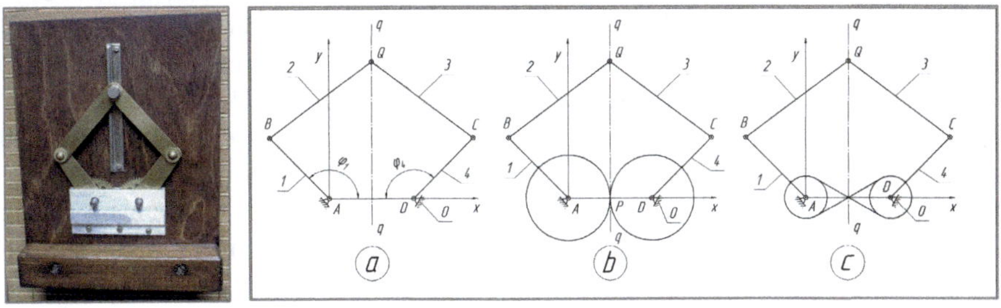

Fig. 8. The formation of Cartwright's exact SLM.

equal gears ($z_1 = z_4$ и $u_{14} = -1$) or gear pair with flexible tape (respectively, Fig.8 *b* and *c*). However, received mechanisms contain gear train with a higher pair or with a flexible connection; therefore they can't be ranked as crank linkage mechanisms:

$$W^{planar} = 3 \cdot n - 2 \cdot p_1 = 3 \cdot 4 - 2 \cdot 5 - 1 \cdot 1 = 1. \tag{10}$$

4 Approximate Crank SLM

Crank linkage replacing gear train doesn't provide permanent transmission ratio, therefore crank Cartwright's mechanism (Fig.9) provides approximate rectilinear translation of point Q only. This mechanism contains five movable links and seven rotary kinematic pairs; its mobility is equal to one:

$$W^{planar} = 3 \cdot n - 2 \cdot p_1 = 3 \cdot 5 - 2 \cdot 7 = 1. \tag{11}$$

The mechanism can be ranked as second-class mechanism; it consists of primary mechanism (0-1) and two dyads RRR (2-3, 4-5).

These mechanisms are widely applied at slip rings of electrical locomotives [13].

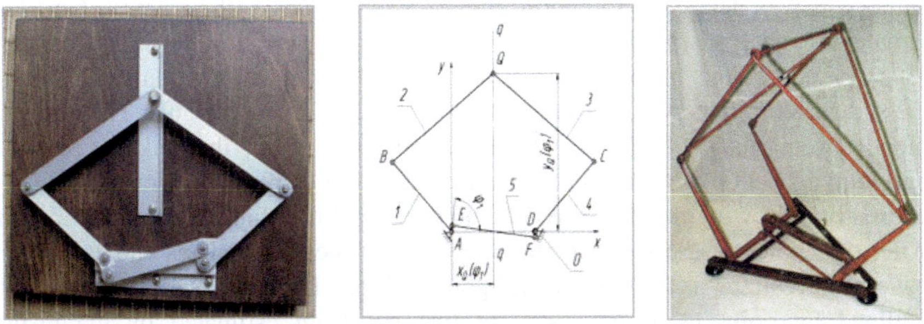

Fig. 9. The model of approximate Cartwright's SLM with its scheme and the model of electrical locomotive's slip ring.

Except of it, one of the variants of approximate Roberts's mechanism was designed and investigated (Fig.10). The Roberts's mechanism, based on Cardan's circles, has been described earlier at [14]. The SLM, considered at this paper, is based on four-bar crank linkage. In this mechanism point *Q* of floating link 2 moves along an approximate rectilinear path between points *A* and *D*, the centers of hinges. The equalities of sizes $l_{AB}=l_{CD}=l_{BQ}=l_{CQ}$ should be realized to provide this motion. As Watt's mechanisms (reduced parallelogram) and Chebyshev's mechanism (straight-line generating mechanism), this mechanism consists of three movable links and four hinges, its mobility equals to one:

$$W^{planar}=3\cdot n-2\cdot p_1=3\cdot 3-2\cdot 4=1. \tag{12}$$

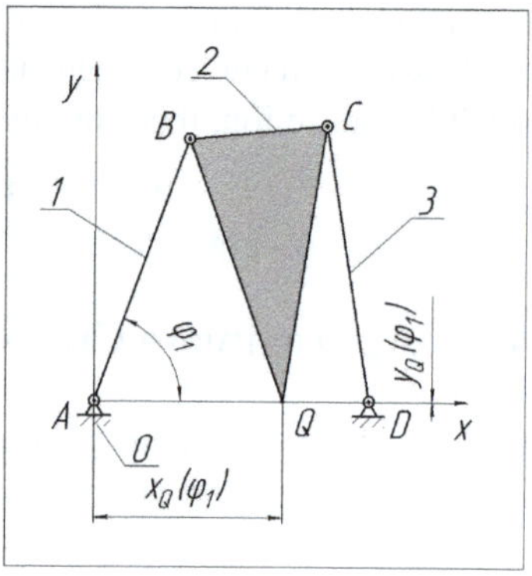

Fig. 10. Approximate Roberts's mechanism and its settlement scheme.

5 Designing and Producing of SLM

All the models of SLM presented on photos have been designed and made by the authors. It was necessary to choose the sizes of the basis (the frame) of model, the lengths of mechanisms' links and operating displacement of links. At creation of rotary pairs the links have been connected by rivets or screws with nuts. The edges of the screws from nuts were split to prevent a self-unscrewing of nuts. The bases of the models were made of plywood by thickness of 10 mm; the links were made of duralumin sheets by thickness of 1.2 mm. Operating displacement of links have been chosen by following reasons: links should overstep the bounds of the basis minimally. After the choice of the sizes graphic modeling of the mechanisms' movement phases have been realized at the Kompas-program (a Fig. 11). The deviation of a trajectory, outlined by point *Q*, from a straight line has been estimated by the results of modeling. The following fact received for all exact mechanisms: the mechanism reproduces a straight line with accuracy providing by the program. Besides, the chosen sizes of model and operating displacement of the links have been checked on phases of movement.

6 Modeling of SLM at MathCAD-Program

Almost all the SLM, considered earlier, are formed by consecutive connection to the primary mechanism with Assur group – dyads of RRR-type. At the first stage of the synthesis we get four-bar crank mechanism of Roberts, at the second stage six-bar crank mechanism (Hart, Cartwright) and at the third stage – eight-bar mechanism of Peaucellier. A position of point *Q* calculates for eight-bar mechanisms by algorithm, which includes the algorithms for four- and six-bar linkages like parts. At first let's apply these algorithms to the simplest four-bar mechanism – to the Roberts's linkage (Fig.10). Let's consider vector loop *ABCD*, written down the following equality for it (Eqn.13):

$$\overline{l_{AB}} + \overline{l_{BC}} = \overline{l_{AD}} + \overline{l_{DC}}. \tag{13}$$

Projections of this contour to axes of coordinates are shown further (Eqn. 14, Eqn. 15):

$$l_{AB}\cos\varphi_1 + l_{BC}\cos\varphi_2 = l_{AD} + l_{DC}\cos\varphi_3, \tag{14}$$

$$l_{AB}\sin\varphi_1 + l_{BC}\sin\varphi_2 = l_{DC}\sin\varphi_3. \tag{15}$$

We'll solve this nonlinear system of the equations concerning angles φ_2 and φ_3 by using a MathCAD - construction "Given-Find". Then we'll define coordinates of points *B* and *C*. Coordinates of point *Q* we'll find as coordinates of point of crossing of two circles, which radiuses are equal $l_{BQ}= l_{CQ}$ and which centers are points *B* and *C*.

Let's synthesize approximate Cartwright's mechanism (Fig. 9) by using a similar algorithm. At first we'll synthesize a four-bar linkage, formed by loop *AEFD*, and define angular coordinates φ_2 and φ_3, and then we'll define coordinates of points *B*, *C* and *Q*.

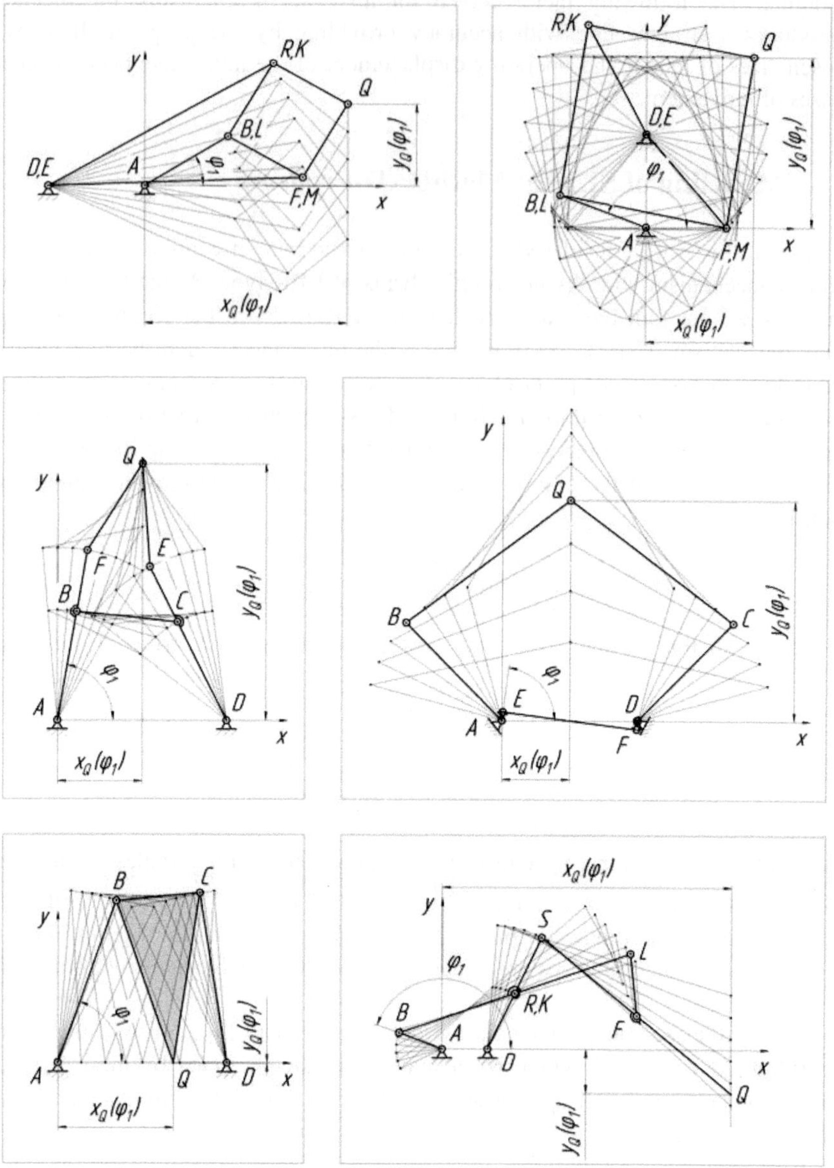

Fig. 11. Modeling of SLM at Kompas-program

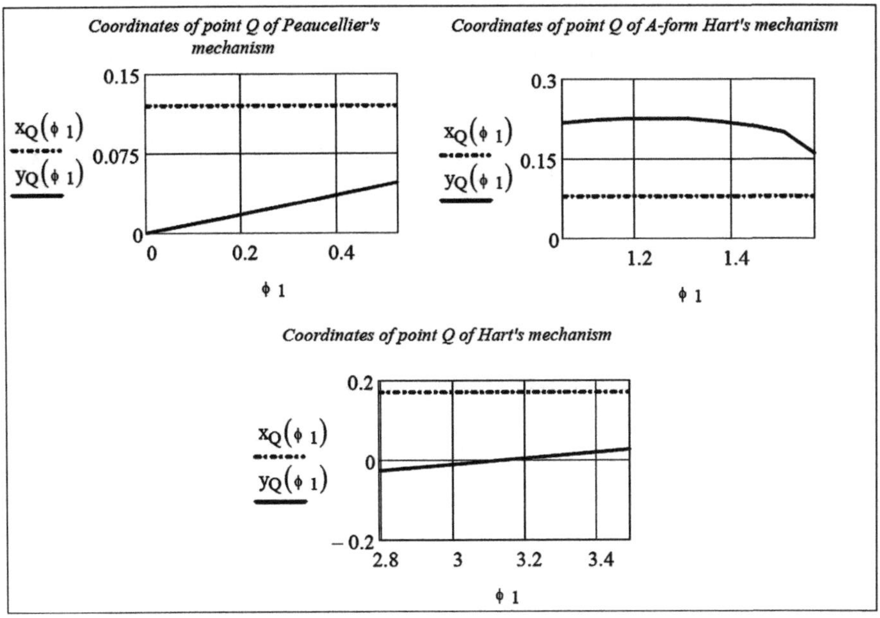

Fig. 12. Modeling of exact SLM at MathCAD-program

In "A-shaped" Hart's mechanism (Fig. 7) coordinates of points *F*, *E* are defined after definition of angles φ_2 and φ_3 from loop *ABCD*, and then coordinates of point *Q*, as a point of crossing of two circles with radiuses l_{FQ} and l_{EQ}.

In other Hart's mechanism (Fig. 3) coordinates of points *F* and *S* are defined after definition of angles φ_2 and φ_3 from loop *ABCD*, and then coordinates of point *L*, as a point of crossing of two circles with radiuses, which radiuses are l_{SL} and l_{FL} and centers are points *S* and *F*. Then angle φ_5 and coordinates of point *Q* are defined.

Then let's define a point's *Q* position at eight-bar Peaucellier's mechanism (Fig. 2). Decision of this problem is shown further: at first the loop *ABKD* was considered. Its angles φ_2, φ_3 and coordinates of its point *K* had been defined. At second the angles φ_4, φ_5 and coordinates of point *F* from the loop *ABFD* had been defined. And so coordinates of point *Q* were found as intersection point of two circles with equal radiuses $l_{KQ}= l_{FQ}$ and with centers, placed in points *K* and *F*. Diagrams of coordinates of point *Q* are shown on Fig.12. The path of point *Q* is a straight line, which is parallel to axes *y*, because it's coordinate, directed on *x*, is constant. Thus the deviation of value of *x* from a constant does not exceed an error of calculations.

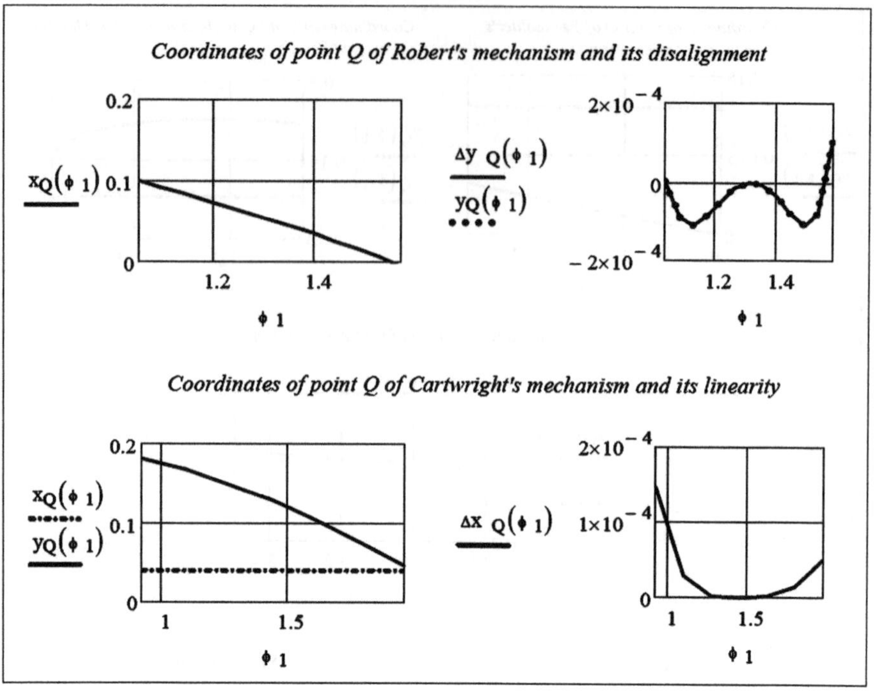

Fig. 13. Modeling of approximate SLM at MathCAD-program

Coordinates of point Q and sizes of its deviation from a rectilinear trajectory have been calculated for SLM. At the Roberts's mechanism (Fig.10) the straight line, from which begins the counting of the deviations, is an interaxal line, which is axis x, therefore the deviation equals to y_Q. At the Cartwright's mechanism (Fig.9) such straight line is parallel to axis y and halves interaxal distance l_{AD}.

7 Conclusion

SLM are an interesting stage, both in development of mathematics, and in development of mechanics. The most part of them has not received the wide practical application, many mechanisms, such as Sarrus's mechanism and Hart's "A-shaped" mechanism, are poorly known today even among mechanics. Computer modeling of SLM in programs Kompas and MathCAD has allowed estimating accuracy of theoretical reproduction of a straight line both exact, and the approximate mechanisms by the means of modern programs.

Physical models of mechanisms have allowed estimating influence of producing errors and of backlashes in hinges. They have shown that six- and eight-bar exact mechanisms have accuracy of reproduction of a straight line less, than approximate four-bar mechanisms. They are more sensitive to mistakes of the lengths of links and to backlashes in kinematic pairs, e.g. in model of the Hart's mechanism (Fig.3) the error appeared because of backlashes in pairs was about 2 mm in the top point of a path of point Q, and in bottom point increased up to 5 mm. The error of producing was more (on one or two stages) than errors of theoretical approximation.

References

[1] Kempe, A.B.: How to Draw a Straight Line. A lecture on linkages, p. 51. Macmillan and Co., London (1877)

[2] Daina, T.: How to Draw a Straight Line. In: Kinematic Models for Design. Degital Library, Cornell University College of Engineering, `http://kmoddl.library.cornell.edu`

[3] Dijksman, E.A.: On the History of Focal Mechanisms and Their Derivatives. In: International Symposium on History of Machines and Mechanisms, vol. 7, pp. 303–314. Kluwer Academic Publishers (2004)

[4] Dijksman, E.A.: A Strong Relationship between New and Old Inversion Mechanisms. Transaction of the ASME. Journal of Engineering for Industry, 334–339 (February 1971)

[5] Kinematic Models for Design. Digital Library. Cornell University College of Engineering, `http://kmoddl.library.cornell.edu`

[6] Bryant, J., Sangwin, C.: How Round Is Your Circle? Where Engineering and Mathematics Meet. Princeton University Press, Princeton (2008)

[7] Veber eine gelenkgeaef ührung von L. Lipkin. Mèlanges mathèmatiques de I'Academie lmpèriale a St-Pètesbourg (1870)

[8] Peaucellier, C.-N.: Note sur un question du gèometrie de compass. Nouvelles Annales de Mathèmatiques, 2 s. 12(2), 71–78 (1873)

[9] Lemoine: Note sur la losange articulè du commandant du Gènie Peaucellier destinè á remplacer le parallelogramme de Watt. Journal de Physique Thèorique et Appliquèe X, 130–134 (1873)

[10] Bogoliubov, A.N.: Theory mechanism and machines in history evolution its ideas, p. 465. Publish house “Science”, Moscow (1976) (in Russian)

[11] On certain conversion of motion. Messenger of Mathematics. Report of the 44 Meeting of British Association for the Advancement of Science. Meeting of Belfast 4, 82–86, 116–120 (1874)

[12] Zernov, D.S.: Applied mechanics. Part1, p. 343. The merged scientific and technical publishing house, Moscow (1937)

[13] Golovin, A., Tarabarin, V.: Russian Models from the Mechanisms Collection of Bauman University. In: Ceccarelli, M. (ed.) History of Mechanisms and Machines Science, vol. 5, p. 238. Springer, Heidelberg (2008)

[14] Klyukin, D., Shchedrin, M., Tarabarin, V.: Stright-line Mechanisms in the Collection of Bauman Moscow Technical University. In: The 13th World Congress in Mechanism and Machine Science, Guanajuato, Mèxico, June 19-25, (ID: A21.451) (2011)

The Transmission of the European Clock-Making Technology into China in the 17th-18th Centuries

Baichun Zhang

CAS Institute for the History of Natural Sciences,
Beijing, P.R. China
zhang-office@ihns.ac.cn

Abstract. Starting in the late 16th century European religious missionaries were sent to China. Such Jesuit missionaries as Matteo Ricci played an important role in the transmission of European knowledge to China. In order to promote their religious mission, they brought European scientific knowledge, instruments and clocks to China. In the 17th century, Jesuits constructed clocks for emperors in Imperial Palace. Chinese craftsmen made copies of European clocks and they established clock-making workshops in such cities as Suzhou and Guangzhou. The main reason why Chinese people accepted European clock-making technology is the fact that European clocks were functionally superior to such traditional timepieces as clepsydra, gnomon, sundial and so on.

1 The Introduction of European Clocks and Watches

After the Reformation in Europe, the Pope actively sent religious missions everywhere in the world. In 1540, Portugal suggested the Church to send religious missions to China. After 1557 Portuguese businessmen were allowed to inhabit Macao. In 1562 Jesuits arrived there and were welcome, but they couldn't promote their religious mission in other places. Some time before 1578, Alexandre Valignani (1538-1606), an inspector from the Society of Jesus who was in charge of missionary business in India and the Far East, arrived in Macao and put forward some new missionary strategies: they would begin their mission by learning Chinese and adjusting to Chinese customs. In 1579, Michel Ruggieri (1543-1607), an Italian missionary, was invited to Macao by Valignani and prepared to go inland by learning Chinese. In order to promote missionary activities, Ruggieri suggested Alexandre Valignani to send also Matteo Ricci (1552-1610) to China. On August 7, 1582, Ricci arrived in Macao from India with a large clock made in Europe. He played an important role in the transmission of European knowledge into China. From the end of Ming China (1368-1644), the Jesuit missionaries in China and their Chinese friends introduced Western machines in China. Much of the exotic mechanical technology was not widely accepted by the Chinese; however, the clocks exerted a similar influence in China as western astronomy did. There is no doubt that the discussion on the transmission of the clock-making technology will be helpful to our understanding of the cross-cultural transmission of technology into the Chinese context.

T. Koetsier, M. Ceccarelli (Eds.): Explorations in the His. of Machines & Mech., HMMS 15, pp. 565–577.
springerlink.com

In December of 1579, Ruggieri visited Guangzhou for the first time on board of a Portuguese commercial ship. In the spring of 1582 Ruggieri went to Guangzhou for the second time. A local official let him stay temporarily at the place where the diplomatic corps from Thailand had lived when they paid tribute to the ruler of China. Ruggieri sent a mechanical clock to Huang Yingjia, a commanding officer in Guangzhou, who promised to introduce him to the court of Ming China ([1], p.190).

On account of the policy of prohibiting sailing, Chen Rui, the newly-appointed satrap in charge of Guangdong and Guangxi, decided to negotiate trade with the Portugal ambassador to have profit. Thus, in the spring of 1582, Ruggieri and Mattia Penella went with precious gifts to Zhaoqing to visit Chen Rui, an old and avaricious official. After getting a courteous reception and making a favorable impression they demanded to stay in Zhaoqing. When they left Zhaoqing, Chen Rui asked them to find precious things for him. Back in Macao Ruggieri fell ill and he asked Mattia Penella to send presents to Chen Rui adding that he Ruggieri would send him a big clock in person after he recovered. Soon after Mattia Penella came back to Macao, the missionaries were "surprised and happy" when they got a letter from Chen Rui in which he eagerly asked for "a big self-sounding clock". In December, 1582, Ruggieri and F. Passio went to Zhaoqing in a hurry, taking triple prisms and the big clock brought by Ricci from India as presents with them. They pleased Chen Rui and were settled in the Tianning Temple. On January 4th, 1583, Ruggieri set the clock and fixed its special cover and various kinds of decoration. In order to follow Chinese habits, he converted the twenty-four-hour system into Chinese twelve-hour system, changed Arabian numbers into Chinese characters, and divided each day into 100 sections and each section into 100 parts ([1], pp.208-209). In March of that year, Chen Rui was dismissed when Ruggieri was planning to find a place to spread the Christian faith in Zhaoqing. Chen Rui ordered all missionaries to leave Zhaoqing, fearing that he would be held responsible for housing foreigners.

Perhaps the new satrap, Guo Yingping, was informed that the foreign missionaries possessed unusual and precious things and there were advantages for him if they lived in Zhaoqing. The satrap ordered the official Wang Pang to bring the missionaries back from Macao [2]. On September 9th, 1583, Ruggieri and Ricci came back to Zhaoqing with a lot of skillful devices made in Europe. Four days later, Guo Yingping agreeably allowed them to choose a place to create a church. When Ruggieri had to go back to Macao to raise money, Wang Pang asked him to make a clock for him. Ruggieri brought a clock and a craftsman from India with him from Macao to the church in Zhaoqing while Wang Pang found two local most skilled craftsmen to help with the construction of a big clock for Wang Pang. Although the foreign craftsman was sent back to Macao ahead of time because of a law suit, the clock was finished in 1584. Several months later, Wang Pang had to send this clock back to the church for there was nobody at his home who could set the clock ([3], p. 182). People regarded the small clocks, triple prisms, maps and the big clock in the church as peculiar devices and were happy to have a look at them. In Zhaoqing the missionaries discovered the great influence that European skillful devices and academic achievements had, so that clocks and triple prisms were sent to very important people as gifts whenever the Jesuits wanted to promoted their mission in a new area. We know that Claudio Aquaviva, the head of the Society of Jesus, in a letter to the mission in China promised to supply two clocks. One of the clocks had a complex structure and could tell hours and quarters.

Alexandre Valignani (1538-1606), the inspector from the Society of Jesus in charge of missionary business in India and the Far East, was not satisfied with the results in Zhaoqing. He appointed Ricci to become the supervisor of the mission in China and proposed that they should go to Beijing to obtain the emperor's authorization of mission. In 1595, Ricci arrived in Nanchang and sent a clock to an official Zhu Duojie. Its time system followed the Chinese style which could indicate the moment at which the sun rises and sets, the length of day and night each month. This present was admired enormously. In early 1598 Ricci and other missionaries went to Nanjing and visited Wang Zhongming, a minister of the Ministry of Rite. They brought two similar big clocks. Ricci sent a clock to Wang Zhongming in order to get his help. In June, 1598, Ricci and the missionaries arrived in Beijing with gifts, but they couldn't find any opportunity to visit the emperor and had to go back to Nanjing. Ricci showed all their clocks in Nanjing. People were eager to have a look at them. Emmanuel Diaz, the headmaster of the divinity school in Macao, sent Ricci a small clock. Thus, they had one more gift to send to the emperor. Ricci decorated this small clock which had a winding mechanism and copper gears. The time was shown with the Chinese characters. The mouth of an eagle was set to the time of day and a delicate dragon was carved on the vaulted top of the clock. In May, 1600, Ricci and other missionaries left Nanjing for Beijing again. When they were detained in Tianjin by Ma Tang, an eunuch, the Wanli emperor remembered a report in which it said that foreigners would send him clocks so that he ordered to let the givers come to Beijing. On January 24th, 1601, Ricci arrived in Beijing. The emperor liked the two clocks very much and took the small one with him. In the next year, he ordered the Ministry of Works to build a wood attic for the big clock with a pendulum. Because of these gifts the emperor began to be interested in clocks and in European civilization as a whole. In order to maintain and use the clocks properly, he appointed four staff members from the Astronomical Bureau to learn from the missionaries. Ricci used the favorable circumstances to ask for the emperor's permission to stay in Beijing repeatedly. With some efforts, they finally found a place to live in the Gate of Xuanwu, and, moreover, they got permission to do the maintenance of the clocks in the palace.

Ricci's success in Beijing consolidated their status in Zhaoqing, Shaozhou, Nanchang and Nanjing. Afterwards, they developed new parishes in Shanghai (1609), Hangzhou (1612), Songjiang (1622), Fuzhou (1625), and Ningbo (1640). Meanwhile, the missionaries introduced clocks and other achievements of European civilization to the Chinese people. Some craftsmen and educated people began to pay attention to the foreign clocks.

After the Qing army took over Beijing, Ming China was replaced by Qing China in 1644 and the Jesuits turned to the court of the Qing China to ask for protection. The Qing emperors and their ministers liked the Western gadgets like clocks very much and treated the missionaries very well and sent them some gifts. The missionaries always selected sophisticated clocks to present to Qing China's emperors. The Kangxi emperor was interested in European science and technology very much and collected a few clocks. The Qianlong emperor was more interested in clocks than the Kangxi emperor; therefore, he collected more clocks. When Yuanming Park was burnt at the tenth year in the reign of the Xianfeng emperor, there were 431 clocks stored in the Forbidden City and 441 clocks in Yuanming Park [4].

In the reign of the Kangxi emperor, the businessmen from Europe began to sell clocks in Guangzhou. The import of clocks gradually increased in the latter part of the reign of the Qianlong emperor. In some places, there was "great enthusiasm for the West", and clocks and watches became necessities in many dignitaries' families. In the first month of the fourth year in the reign of the Jiaqing emperor (1799), He Sheng, a dignitary in the reign of the Qianlong emperor was found guilty and there were 10 big clocks, over 300 small clocks and more than 280 foreign watches at his home [4]. The book, *The Dream of the Red Chamber* written by Cao Xueqin, showed what dignitaries' families looked like in the reign of the Qianlong emperor. The 72th chapter of the book said, one "golden weight-driven clock" was worth 560 *Liangs* silver.

Different kinds of clocks and watches were imported into China during the end of the Ming China. The earliest clocks introduced into China, were driven by weights. Aomen Jilue (*Brief Record of Macao*) wrote about the clocks in the Saint Paul Temple in Macao in early Qing China, and said, there are "self-sounding clocks", wall clocks, table clocks, musical clocks and self-moving watches ([5], p. 757). The six chapter of *The Dream of the Red Chamber* said, "Grandma Liu only hears the sounds of dang, dang, dang… Suddenly, she sees there is a box hung on one of the columns in the sitting-room with a weight-like thing dropping, which keeps on swinging". Obviously, this was a clock with a pendulum. The forty-fifth chapter of *The Dream of the Red Chamber* also said that Baoyu "feels in his front opening and takes out a golden watch the size of a Persian walnut". It indicates that there were pendulum clocks and pocket watches in the houses of medium level dignitaries during the period from the reign of Kangxi to that of Qianlong.

According to the works on clocks and watches written before 1796, wall clocks which were driven with weights were popular in families in China before the end of the 18th century. The pendulum was driven with a winding mechanism, which could be set in any place. An alarm clock, that was, a clock that sounded between hours, was used by the ministers who would go to call on the emperor in the imperial court. A musical clock could make rhythmical music with small hammers to strike several or even tens of bowls in it. Another kind of clock, which was convenient to use by night, could be used to check time at all times, and it worked as soon as it was wound. A watch was something "whose mechanism was as same as that of a clock while it was altered into a small one". One hand of them showed hours and quarters. Two hands showed hours, quarters and minutes. Three hands, hours, quarters, minutes and seconds. With four hands also dates could be included. At that time, the pendulums of the escapement could be divided into six kinds. The components of a clock and a watch included gears, winding mechanism, spring, ratchet wheel, spiral, etc.

The continuous import of clocks and watches eventually led to the trade of clocks and watches and promoted Chinese craftsmen's mastery of practical technology directly. When making clocks and watches through imitation, the Chinese people acquired a rich technological knowledge about them, such as their principles, design, and making techniques. However, this process was usually pretty slow.

2 Construction of Clocks in Imperial Palace

In the reign of the emperor Kangxi, imperial clocks were stored in Duanning Hall. Since 1689, documents called this place the "Self-sounding Clocks Unit". It was included into the Royal Workshop, Yangxin Hall. In the early reign of the emperor Yongzheng, the Self-sounding Clocks Unit had a clocks workshop in Yangxin Hall. In the first month of the tenth year of the reign of Yongzheng (1732), it was named the Clock-constructing Unit [6].

In the imperial palace of the Qing China, the clock-constructing technology mainly depended on the employment of foreigners. The craftsmen from Europe were good at designing, making and maintaining clocks and watches as well as their attached ingenious mechanisms and doohickeys. In the reign of Kangxi emperor, there were some foreigners working in the palace, for example, G. de Magalhaens, a Portuguese, Stad Lin, Jacobus Brocard and Slaviczek, who were Swiss. From 1707 to 1740, Stad Lin constructed various kinds of clocks and improved the making techniques in the palace to a high level. In 1729, Vlen Tinus Chalier, a French expert in clocks and watches, came to the palace and took part in the construction of clocks. Clock-making in the Clock-constructing Unit had its heyday in the reign of Qianlong emperor, who hired more experts in clocks from Europe than any other emperor, including Adeodat, Gilles Thebault and Mathaeus de Ventavon. The latter two were from France. In 1773, the Society of Jesus was disbanded by Pope; therefore, no Jesuits who had knowledge about clocks came to China any more. Thus, the Qianlong emperor decreed to invite other experts in clocks to Beijing. The foreigners who came to the palace to make clocks were Adeodat, Carolus Paus and so on. The emperor was glad to reward who deserved to get presents for their work. Mathaeus de Ventavon said, "Princes and ministers usually invited foreigners to fix their clocks. There are a fair number of clocks in China, but only two people can fix them."([5], p. 761). It was believable that such assessment just described the situation in Beijing when Mathaeus de Ventavon was in the palace and some areas but not the whole country.

The craftsmen in Clock-constructing Unit were more than 100 sometimes, including some Chinese. Among them, the craftsmen sent by the civil and military governor in Guangdong were the most skilled. Some of them were not only good at making clocks but also had knowledge about music and painting. However, the craftsmen from Europe were generally more skilled.

The process of how to construct clocks in the palace was different from those in common workshops. The craftsmen usually made and fixed imperial clocks according to the emperors' requirements. When a clock as a mechanical doohickey was created, firstly, the emperor would describe the style and function which appealed to him. Next, a specific one was appointed to draw it. Only after the emperor examined and approved it, the clock was really made. During the making process or after it was finished, the emperor still could require some changes at any moment till he was satisfied with it. Sometimes, painters, even astronomers would take part in its making.

The clocks made in the palace of the Qing China could be elegant art wares which combined timepieces and mechanical doohickeys together. Besides common mechanism of a timepiece, complex mechanism was adopted to perform many functions of "doohickeys", for example, to show delicate puppets, things and sceneries. They included the puppets who shoot with a bow and arrow, showed time, hold a sword or a shield, and the copper puppets who could write, strike a bell with a gavel, strike a musical instrument with a hammer, included demonstrating celestial motion, showing chirping and flying birds, a self-walking puppet with flowers, a self-walking tiger or a lion, a running horse, a duck on water, a self-moving boat, and included setting off fireworks, playing music and so on. A clock in a rosewood pavilion which was made after 1749 had skillful and complicated construction, including seven winding mechanisms within to control mechanical "doohickeys", to tell time automatically and to play music. The whole process was complex and well-ordered [6]. Some clocks were big, which could symbolize imperial dignity. In 1753, Adeodat and Ir Gilles Thebault constructed a multi-function clock, in which eight winding mechanisms were used [4].

The masters and craftsmen from Europe and China constructed the clocks which appealed to emperors, queens and their favorite ministers with combination of Western techniques and traditional Chinese art. For example, some night-hour clocks were made with Chinese style and set in wooden building to tell time at night in the early Qing China. Their shapes were mainly pavilion, terrace and tower. When to tell time could be set based on the changes of climate. The clock in a rosewood pavilion with carved flowers made in Clock-constructing Unit in the reign of Qianlong, emperor which was driven with a weight, was 3.65 meters tall and over 1000 kilos weight. Thousands of people in the palace worked and rested according to this night-hour clock.

In the early Qing China, the clocks made in the palace were mainly driven with weights which were made of copper outside and lead inside, and were hung with catgut or silk rope. *Ting Xun Ge Yan* (*How to Instruct children at Home*) written in the reign of Kangxi emperor (16621722) said that there was someone to construct clock in the palace, "but they can't set it accurate" because the technique of winding mechanism was very difficult. Afterward, ten clocks and watches were made "after the technique about winding mechanism was learned from foreigners" in the reign of Kangxi [7]. The winding mechanisms used in Clock-constructing Unit were bought from Guangzhou in the reign of Qianlong emperor. For example, in April of the twenty-sixth year in the reign of Qianlong (1761), the Royal Workshop asked the emperor to decree his order that the supervisor at customs in Guangdong will buy steel 2000 *Jin*s in Guangdong. The bissextile May next year, it got steel 2094 *Jins* from Guangdong and constructed 134 winding mechanisms in different sizes [4]. The steel to make winding mechanism should be very elastic, and "the steel from Guangdong" should be imports. The craftsmen in the palace maybe used special, man-driven small lathes and drills, with which foreigners should be very familiar. It was known that machines had been used to make fans and toys in the palace at that time [8]. On account of complicated structure of clocks as well as high requirements of the quality of work, it would take the craftsmen one or two years, even longer than that, to finish one clock. From 1750 to 1759, the prosperous Clock-constructing Unit only constructed 56 clocks [6]. After 1795, Clock-constructing Unit seldom constructed clocks. The craftsmen from Europe and Guangdong left the palace successively.

3 Folk Construction of Clocks

Imitation was the main form for common people to learn foreign technology before the modern age. Before 1626, Wang Zheng had constructed a clock, in which he combined a Western verge-and-foliot weight-drive system for day hours with a Chinese lead-shot scoop-wheel or striking clepsydra for night-watches [9]. It used weight as its impetus and a movable puppet to tell time. From its description, the cross mechanism in the clock should be an escapement with a vertical-axle rod of pendulum. Besides Wang Zheng, there were some people who tried to imitate European mechanical clocks and watches. The chapter 7 of *Xuancheng Xian Zhi* (*Records about Xuancheng County*) said that Rui Yi from Xuancheng of Anhui province was quite creative and he could construct self-sounding clocks" [10]. Ji Tanran from Jiangning (Nanjing) of Jiangsu province constructed a clock through imitation. However, it was a rough-wrought clock which only had the correct shape. It broke when used [11]. There was another person, Zhang Shuochen, who made a weight-driven plate timepiece and a small gun which could shoot twice by himself. Both were as skillful as what foreigners made [11].

In Qing China, with the constant transmission of clocks and watches, early clock-making handicraft industry came into being in some places. As a port, Guangzhou was the main terminal market for imported clocks and watches. The workshops to make clocks in Guangzhou were established not later than during the early reign of Qianlong emperor (1736-1795) and perhaps during the reign of the emperor Kangxi [8]. In 1815 a foreigner, who sold clocks and watches in Guangzhou, admitted that he met with the competition of Chinese products [12]. This proves that the Chinese workshops to make clocks in Guangzhou and environment had a certain production. Clock making in Nanjing was started in the last days of Ming China. However, there were only four workshops before 1853. Till the end Qing China, there were more than 20 workshops [13]. An investigation of cultural relics indicates that Suzhou could manufacture different kinds of clocks [8][14]. In the early 19th century, there were many craftsmen who constructed clocks in Suzhou, and quite a few of them came from Nanjing. In 1851, there were 30 clock-making workshops in Suzhou, 9 in Hangzhou, and 7 in Ningbo [15]. In Yangzhou and Zhenjiang, there were some workshops to make clock. Generally, a clock-making workshop had two or three workers. "Wife operates a bellow", and "a son works as an apprentice". A big one only had four or five people. A craftsman could, perhaps, made one or two clocks each month.

The most famous local clocks were "the clocks made in Guangdong" and "the clocks made in Suzhou". The clocks which were made in Guangdong stored in the Palace Museum were usually about one meter tall with a square base of about 40 centimeters [16]. When it came to the shape, they often reflected the traditional Chinese style, such as a Buddhist tower, a calabash, a deer, a kylin or a white ape presenting a peach. The clocks, made in Guangzhou, stored in the Palace Museum still had attached devices like movable puppets (such as persons, birds and animals), flowers, streams, banners with Chinese characters and musical instruments. This showed the craftsmen from Guangdong had capability to design and make complicated mechanical transmission devices.

The clocks made in Suzhou during the reign of Kangxi adopted Chinese "day hours" to tell time. A clock would strike 96 times every day and night. In the 18th century, some clocks made in Suzhou already adopted the 24-hour system and had two hands, which were driven with winding mechanisms or weights. Some craftsmen still carved Roman letters on the dials which they didn't know through imitation. They, perhaps, imitated foreign clocks in order to sell well. The demonstrate devices and puppets on the clocks made in Suzhou were childlike puppets, copper puppets striking bell, trotters, tweeting birds, puppets shooting with bows and arrows, music performance, the eight immortals crossing the sea and the like. In the latter Qing China, Suzhou gave priority to the production of clocks in big, medium and small sizes. The construction of these products was compact without any attached devices. They adopted chain and step pulley to control the energy produced by winding mechanism so that they improved the accuracy to tell time efficiently. This kind of practical clocks became a famous brand then, whose shape was like ancient Chinese folding screen. China also had capability to make pocket watch. Sun Ruli from Fujian was good at making "a one-*Cun* self-sounding watch" [10]. After examining *The Dream of the Red Chamber*, Hu Shi thought that during the middle 18th century, there must be someone who could fix clocks and watches in Beijing ([5], p. 759). In the reign of emperor Tongzhi (1862-1874), the workshop of clocks in Beijing, Defengzhai, produced pocket watches under the brand of chrysanthemum, which could be divided into three, five, seven and nine chrysanthemums and were named chrysanthemum watch. The clocks and watches made in the Qing China were collected in some museums in the United States, Great Britain and France.

The clocks and watches, especially the skillful ones with additional devices, were composed of many parts. Therefore, it was difficult for small workshops to improve their technology and quality thoroughly and increase productivity. With production increasing, craftsmen chose to entrust some parts to others. Thus, division of labour started in this field. Special workshops were established in Suzhou, Nanjing and Guangzhou to produce such special devices as mechanisms of clocks, bells, chains, rosewood cases, dials, porcelain facing and to engrave flowers. This kind of division of labour made the handcraft industry of clock-making improve technique in competition. Famous products with high quality were produced. The workshop named with Zhang Rong in Suzhou specially made premium clock bells around the early 19th century. It produced over 60 varieties of the bells. The caliber of the biggest bell was 20 centimeters while the smallest one was about 2 centimeters. The Zhang Rong workshop developed a set of techniques to produce clock bells. The characters of these techniques were: cast copper alloy bells with the raw materials in proper proportion. When drilling, fill the clock bell with cold water, and poured the water and refilled it repeatedly several times to decrease the bad effect on the timbre produced by the heat of the drill. Thus, the clock bells made in the workshop were known for their clear and dulcet tones, and were sold in Shanghai, Nanjing, Yangzhou, Hangzhou and Beijing. During the middle and last stages of the Qing China, the mechanisms of clocks made in Suzhou were gradually divided into big, medium and small sizes. They were assembled for clocks by many workshops.

Xu Chaojun from Songjiang studied astronomy, mathematics and clock-making technologies for a long time. In his book, *Zi Ming Zhong Biao Tu Shuo* (*the Illustrations and Descriptions of Self-sounding Clocks and Watches*), he summarized the technology and experience of his own and from other craftsmen, which showed the level of clock-constructing technology in China before the 18th century. This book was divided into ten parts, which describe all the methods to making clock parts. For example, the methods to select gear ratio and number of teeth as well as the techniques and key points to make such parts as axles, gears and winding mechanisms. The methods to make teeth of a gear wheel are to divide evenly, to file accurately and to burnish well. The methods to make spring are: scrape the two sides of a steel thread with a sharp knife and roil it up; burn it with low fire to cyan, the color of a swallow; select one that becomes properly hard and keep it as standby. The methods to make winding mechanisms are: beat steel out to a thin piece as a saw; file and scrape it well; scissor it regularly and roll it well; burn it red and quench it; melt it into lead and change its nature; fill it into bowels and keep it as a standby. Valuable techniques were introduced in every part, including various kinds of faults and methods for the fault remedy, such as the method to fix stopping pendulum, the methods to assemble and dismantle clocks, and the methods to use clocks.

After the Opium War, with the impact of imported clocks and watches, quite a few workshops were converted from making clocks and watches to fixing them, and some had to close down. Only a few workshops or factories which produced in large scale had great potential to develop. In 1906, Sun Tingyuan and Sun Meitang, the father and his son, set up Meihuali Clock Factory in Ningbo, which was moved to Shanghai in 1912. During over ten years, this factory developed into one with more than 300 people, in which machines were used to manufactue several thousands of clocks, such as big clocks, astronomical clocks, pavilion-like clocks, clocks with glass door, clocks used in station and floor clocks. Their products won golden medal in Panama Clock Exposition in 1915. However, the winding mechanisms used in Huameili Clock Factory were from France.

4 The Reasons of Transmission of Clock-Constructing Technology in China

The ways of how the clock technology was transmitted in China were quite special, and there were various reasons.

1. The Superiority of European Clock-constructing Technology

Generally, developed technology was exclusive. Foreign technology would be accepted easily if it could fill a gap, had self-evident functional superiorities, or could complement local technology, otherwise excluded by the local technology, it was difficult to take root and develop. In the end of the Ming China, developed square-pallet chain pump and noria excluded the Archimedean screw which was just introduced into China by Jesuits.

In China, ancient timepieces were clepsydra, gnomon, sundial, sandglass, water-driven clock and so on. Gnomon and sundial were very easy to use, but it couldn't be used when it was cloudy and became dark. Clepsydra was large in size, could work properly and couldn't be moved easily in cold days. Field clepsydra used by folks had lots of errors. Sandglass had obvious shortcomings, too. A water-driven clock was usually one part of astronomic devices. It had water wheel and complicated transmission gear so that it was large and couldn't be moved easily. Matteo Ricci asserted that Chinese clepsydras and water-driven clocks were not so perfect as European timepieces then—"they always make errors and can't tell time accurately" ([3], p. 24). This kind of statement was a little prejudiced but could reflect truth to some degree.

In comparison, European mechanical clocks and watches had complicated, packaged and durable metal devices. They were accurate and had many clear ways to tell time. They were easy to use and had different sizes with elegant shapes and skillful actions. These superiorities made European clocks and watches both practical timepieces and artistic furnishings or interesting toys. Therefore, European clocks and watches were not excluded technologically but welcomed among Chinese people. In 1694, the emperor Kangxi wrote a poem on a fan to praise European clocks—"It is better than any clepsydra without stop day and night; it is compact and tells time euphoniously and accurately" ([5], p. 758).

2. Chinese People's Attitude and Missionaries' Effort

In feudal dynasties the authorities of the highest rank exerted an important influence on the development of technology. Sometime, this influence was decisive. Zhu Yuanzhang, the first emperor of the Ming China, advocated incorruptness and integrity. In 1368, the official who was in charge of the Astronomical Bureau presented an ingenious water clock, in which there were two puppets to sound a bell and a drum to tell time automatically. Considering it useless, Zhu Yuanzhang broke it into pieces. The emperor's disdain for "skill" exerted a negative effect on the invention and improvement of timepieces. At the end of Ming China, the situation was different; the officials were corrupt and their administration was bad. The royalty and their ministers were fond of skillfully made things and they couldn't refuse bribes.

The missionaries realized that their hard work could only lead to results if they had a good relationship with the authorities. When trying to foster a relationship with authorities and emperors, the missionaries found they all liked presents very much, such as clocks, triangular prime and the like. Matteo Ricci said, "These things are necessities in social communication. They are like lube filled in gears of the progress of society." ([3], pp.377-378) If there were some people who were interested in something in high status, there must be people who were also interested in it in low status. Almost all the people who knew clocks and watches in middle and low status dreamed to have this kind of rarity. There was some emperor who hoped prevent folks from using clocks and watches because he "hates their too much skills". However, he failed.

The court was wary of the missionaries' religious activities, but they didn't mind clocks and watches and foreigners with knowledge about clocks and watches were treated in a friendly way. After 1707, the court forbade religious activities; however, the court welcomed the European specialized craftsmen. The foreigners who successively came to work in Beijing as clock-constructing craftsmen and artists tried their best to meet the emperors' requirements. Contrasting with the Western calendar and Catholicism, the transmission of clocks and watches didn't encounter any obstruction from local people. It was plain sailing.

3. The Basis of Technology in China

In the end of Ming China, the Chinese were not unfamiliar with some techniques concerning European clocks. In the aspect of transmission gear and controlling devices, Chinese timepieces adopted gear systems very early. In Song China, complicated transmission gear and skillful, coordinated escapements were adopted in the Water-Powered Armillary Sphere and the Celestial Globe Tower with time-keeping mechanisms constructed by Su Song and Han Gonglian. In the displaying device, there were a dial in a sundial and a floating arrow in a clepsydra. Around the year 1368, Zhan Xiyuan invented a sand-driven clock, on which there was a dial which was divided into twelve hours and over a hundred of quarters. In the aspect of attached puppets, Chinese people had rich imagination and experience. In the aspect of metallic techniques, Chinese coppersmiths and blacksmiths had excellent crafts. The bronze carriages unearthed in Qin Shihuang's mausoleum, skillful gimbals for incense-burners and complex astronomic instruments made during the Tang Dynasty show the craftsmen's capability to make metal mechanical parts. In the aspect of shape, craftsmen were good at expressing domestic esthetics. The traditional technologies mentioned above established a solid base for European technology of clocks and watches to be transmitted to China. This helped Chinese craftsmen imitated European clocks and watches quickly.

The main differences of mechanical clocks between the West and China were power. Generally, Chinese clocks were driven with water wheels and could keep working without escapement although they couldn't be very accurate. However, European clocks depended on stored energy devices (weights or winding mechanisms) as power. Since the energy was controlled-released, rotating speed must be controlled by an escapement. This, perhaps, was one of the reasons why China didn't adopt and improve escapement after the Song Dynasty.

5 Conclusion

The import of technical products was one of the ways to transmit technology in ancient times. Compared with the transmission by means of books, this direct way was not only convenient to show the function and the effect of the imported technology but it was also easy for craftsmen to imitate. Chinese skilled craftsmen began with the simple imitation of the imported products and established their own handicraft industry to make clocks. However, imitation did not imply the mastery of all clock technology involved and the relevant scientific knowledge.

People could imitate the construction of clocks, conjecture their principles and the specific designs, but it was quite difficult for them to deduce quickly their designing ideas, theoretical knowledge and some special techniques involved. Thus, they had to follow others' steps.

The perfectibility of European clocks and watches was based on the successive combination of technological inventions and theoretical knowledge. In modern time, mechanics, mathematical methods and experimentation were adopted in the construction of clocks and watches so that the technologies of clocks and watches were not restricted to accumulation of experience. However, modern mathematics and mechanical methods were not introduced into China together with the clocks and the watches. At the end of Ming China, Wang Zheng and Johannes Terrentius introduced basic mechanical knowledge. When depicting specific western mechanisms, they had some misunderstanding and made some mistakes. Craftsmen were usually unknowledgeable, while well-educated people didn't know technology. Some literati who were interested in technology and crafts did not have the chance to read European monographs on technology and they had no contact with craftsmen in clocks from Europe. Before 1809, a Chinese monograph written by craftsmen of clocks, *Zi Ming Zhong Biao Tu Shuo* (*the Illustrations and Descriptions of Self-sounding Clocks and Watches*), was published. This book reflected the construction experience and the rules used by craftsmen and masters in the handcraft industry of clocks in China then. It was very practical, but it couldn't compete with Western knowledge at the time.

The missionaries' ultimate aim was to spread religious teachings in China and not to help the Chinese develop new technology. Clocks as well as their relevant technology were brought into the Chinese upper classes in ordr to promote their mission. Their activities were limited to churches and the palace. They didn't have much contact with excellent Chinese masters and craftsmen who were absorbed in technology. In the palace of the Qing China, what the emperors cared about was satisfactory toys which were also what the missionaries did with their great efforts. None of them intentionally spread technology of clocks and watches widely in China. Thus, the Clock-Constructing Unit became a workshop controlled by foreigners to make imperial toys. The Clock-constructing Unit did not train a group of Chinese master craftsmen to make clocks and watches. In this situation, the folk handcraft industry of clocks developed by itself with the help of Chinese craftsmen's wisdom. Without the mastery of advanced knowledge, it couldn't develop in time into a powerful industry.

References

[1] Bernard, H.: Tian Zhu Jiao Shi Liu Shi Ji Zai Hua Chuan Jiao Zhi (Recorders of Missionaries in China in the 16th Century). The Commercial Press (1936)

[2] Xu, Z.Z.: Zhong Guo Tian Zhu Jiao Chuan Jiao Shi Gai Lun (A Survey of the History of Catholic Missions in China), pp. 171–172. Shanghai Shishanwan Press (1938)

[3] Ricci, M., Trigault, N.: Li Ma Dou Zhong Guo Zha Ji (Matteo Ricci's Reading Notes about China). Zhonghua Book Company (1983)

[4] Ju, D.Y.: Qing Dai Ye Su Hui Shi Yu Xi Yang Qi Qi (The Jesuits and Western Wonderful Machines in the Qing Dynasty). Palace Museum Journal (1-2) (1989)
[5] Fang, H.: Zhong Xi Jiao Tong Shi (The History of Communication between China and the West). Yuelu Shushe (1987)
[6] Liu, Y.F.: Qing Gong Zuo Zhong Chu (Clock-constructing Unit in the Qing China). Palace Museum Journal (4) (1989)
[7] Zhang, L., Liu, J.T.: Zhong Guo Jiao An Shi (The History of Religious Cases in China), p. 104. Sichuan Social Science Press (1987)
[8] Xu, W.L., Li, W.G.: Tan Qing Dai De Zhong Biao Zhi Zao (On the Manufacture of Clocks in the Qing Dynasty). Cultural Relics (2) (1959)
[9] Wang, Z.: Illustrations and Descriptions of the New-built Machines. Yangzhou (1628)
[10] Song, B.Y.: Guan Yu Wo Guo Zao Zhong Ye De Li Shi Zi Liao (On the Historical Materials about the Clock-making Industry in China). Zhong Biao (Watch & Clock) (2) (1973)
[11] Liu, X.T.: Guang Yang Za Ji (Miscellanea in Guangyang), vol. 3, 2, p. 141, 99. Zhonghua Book Company (1957)
[12] Chen, Z.W.: Ou Zhou Ji Xie Zhong Biao De Chuan Ru He Zhong Guo Jin Dai Zhong Biao Shi Ye De Fa Zhan (The Introduction of European Mechanical Clocks and the Development of Chinese Modern Clock-making Industry). Chinese Historical Materials of Science and Technology (1) (1984)
[13] Song, B.Y.: Qing Dai Mo Nian Nanjing Suzhou Zao Zhong Biao Gong Ye Diao Cha (Investigation of the Clock-making Industry in Nanjing and Suzhou in the end of the Qing Dynasty). Wen Wu (Cultural Relics) (1) (1960)
[14] Chen, Z.W.: Xian Cun Wo Guo Zui Gu Lao De Gu Zhong (The Oldest Clocks Available in China). Zhong Biao (Watch & Clock) (12) (1979)
[15] Chen, K.G.: Qing Dai Suzhou De Zhong Biao Zhi Zao (The Manufacture of Clocks in Suzhou, the Qing Dynasty). Palace Museum Journal (4) (1981)
[16] Shan, Z.N.: Qing Dai Gong Zhong De Guang Dong Zhong Biao (The Clocks made in Guangdong used in the Palace of the Qing Dynasty). Palace Museum Journal (3) (1986)

DMG-Lib – An Open Access Digital Library as an Exploration Tool for Historians and Engineers

Torsten Brix, Rike Brecht, Veit Henkel, and Michael Reeßing

Ilmenau University of Technology

Abstract. End of 2010 the EU started the European project thinkMOTION, whose main goal is the Internet-based providing of information on motion systems for a wide range of users. The content to be collected within the project ranges from historical resources to the latest research results. In doing so, various information sources, such as books, patents, articles, physical models, etc., are digitized, processed and published on the Internet in form of text documents, images, interactive animations, videos, and formalized descriptions of design solutions. From a technical point of view the work in the thinkMOTION project is based on the results of the now established Digital Mechanism and Gear Library (www.dmg-lib.org). The proven DMG-Lib workflow and the proven DMG-Lib tools were adapted to the requirements of a distributed working on European level. The resulting content of thinkMOTION are collected on the DMG-Lib platform and presented in different ways. It is important that the DMG-Lib content will be accessible through the Europeana portal, which already includes as the central European Metaportal millions of digitized objects. Besides a brief introduction to the thinkMOTION project, the following sections focus on key aspects of the used workflow, the rights clarification, the creating of a work environment for library users and the issue of searching/browsing as well as multilingualism.

1 Introduction

A simple and comprehensive access to information, which reflects the current knowledge, experiences, technical skills and historical developments and honours famous scientists and inventors in the field of motion systems, is still not currently given. Main reasons for this situation are the restricted, non-public access and the scattering of many sources, such as books, drawings or physical models, which exist often as non-digitized rarities. Therefore the digitization of the content (Figure 1) and the open access via an online platform are necessary to make it available to a wide range of user groups like scientists, engineers, historians, lecturers, students, pupils and interested laymen.

The technical, organizational and administrative challenges lie not in the digitization and online publication, but in terms of localization, acquisition, rights clarification, indexing, metadata determination in an international environment and in the implementation of a multilingual access via one or several online portals.

T. Koetsier, M. Ceccarelli (Eds.): Explorations in the His. of Machines & Mech., HMMS 15, pp. 579–592.
springerlink.com

Fig. 1. Examples of heterogeneous sources, which must be localized, indexed, legally clarified, digitized, processed and published

1.1 The Projects DMG-Lib and thinkMOTION

In 2004, the DMG-Lib project started the collection, digitization and online presentation of information in the field of mechanism theory. The results of this project are freely accessible via the Internet address www.dmg-lib.org [2]. The German Research Foundation (DFG) supported this work to establish a Center of Excellence for Research Information in the field of motion systems. The core team consisted of contributors from Ilmenau University of Technology, RWTH Aachen University and Dresden University of Technology, but soon the community grew and people from other institutions as well as private persons joined the project. Since 2010, the group of project supporters is international on European level. This was achieved with the help of the project thinkMOTION [9], which is funded by the European Commission. The partners come from:

- Ilmenau University of Technology (IUT, Germany),
- University of Basque Country (UBC, Spain),
- Politehnica University of Timisoara (UPT, Romania),
- RWTH Aachen University (RWTH, Germany),
- French Institute of Advanced Mechanics (IFMA, France) and
- University of Cassino (UCAS, Italy).

In addition, the nonprofit international Association for the Promotion of the Digital Mechanism and Gear Library (DMG-Lib e.V.) and the International Federation for the Promotion of Mechanism and Machine Science (IFToMM [7]) play a key role in the project (see section 1.2). The search for other international partners (other institutions like libraries, museums, research institutions or private people etc.) is one of the ongoing goals of the thinkMOTION consortium.

The work at European level raised new challenges concerning the support of multilingualism, fulfillment of special national requirements (e.g. concerning rights clarification) and the coordination of the work of more than 100 contributing persons. These challenges in combination with the high number of DMG-Lib entries explain the increased demands on the workflow and the tools used in comparison to the

DMG-Lib project. In 2009 ca. 3,000 items were available. At the end of the thinkMOTION project in 2013 more than 60,000 items are expected to be online. Achieving this in a good quality, supporting multilingual access and respecting the rights of use is indeed a big challenge.

1.2 Role of DMG-Lib e. V. and IFToMM

An online publication of digitized information resources always involves the clarification of the intellectual property rights (IPR) rights. The presenting institution or person must have the simple rights of use for these digitized content or the content must be in the public domain. To solve this task, the Association for the Promotion of the Digital Mechanism and Gear Library (DMG-Lib e.V.) was founded. The main task of DMG-Lib e.V. is to receive the simple rights of use from the rights holders of the single items (books, articles, videos etc.) and to provide the content via the web pages on the DMG-Lib webpages [1]. These web pages are indexed and referenced in Europeana [5] (Figure 2).

The DMG-Lib is taking over the role of a juristic person who is responsible for the provision of online content to users and Europeana. New members in this association are always welcome to support the idea of free access to scientific information.

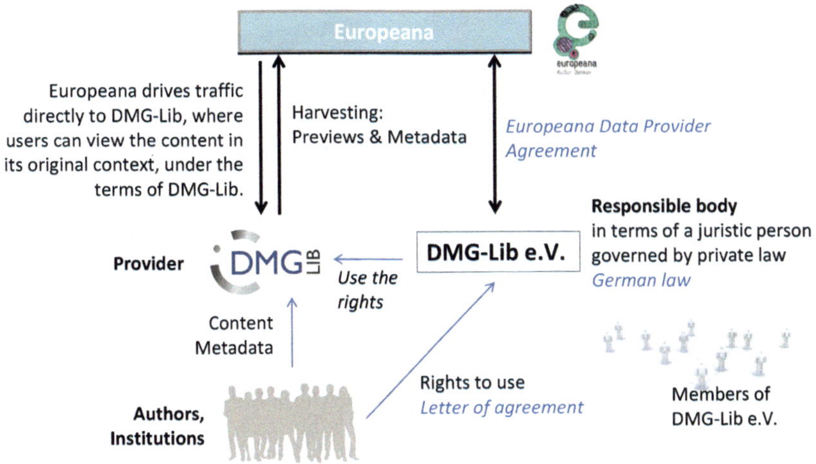

Fig. 2. Role of the DMG-Lib e.V. in the thinkMOTION project

The worldwide active International Federation for the Promotion of Mechanism and Machine Science (IFToMM) is also an important partner in thinkMOTION. The Federation helps to initiate contacts, to bundle activities and to promote the idea of free information access. In addition, there is a specific agreement with the IFToMM, which allows the complete digitization of the IFToMM Archive, which is located in Udine (Italy). The publication of the archive is done using the DMG-Lib platform. The IFToMM Executive Council allows also the publication of all proceedings of the

IFToMM world conferences on the DMG-Lib portal. Furthermore, from mid-2012 the Permanent Commission on Standardization of Terminology will have their work platform for the development of the IFToMM-dictionary in the DMG-Lib.

2 The Workflow of Providing Content for the DMG-Lib Portal within the thinkMOTION Project

The heterogeneous content which should be collected in the thinkMOTION project comprised a brought time range from the cradle of motion systems till this day (Figure 3). The meaning of the old content sources will be often underestimated and reduced at the statement "Only interesting for historian". But it is exactly that old content which contains a rich pool of valuable ideas and solution and which can be very helpful and inspiring for the engineers' work.

The heterogeneity and variety of the content sources have an influence on the necessary steps in the thinkMOTION workflow for the digitizing, processing and making it ready for the presentation in an online portal, like the DMG-Lib [2]. Usually, all the content stemming from the 1980s and before has to be digitized.

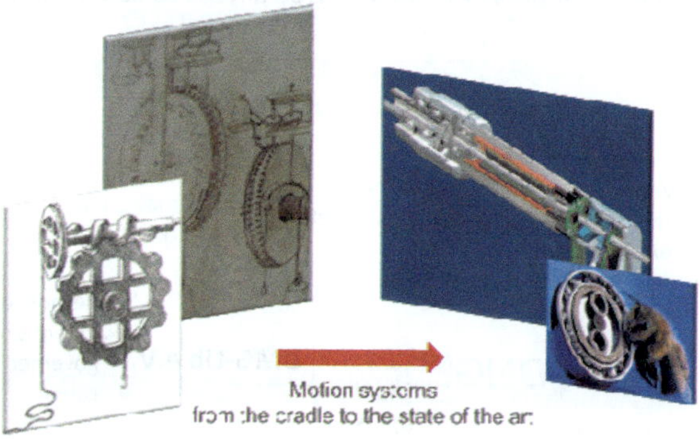

Fig. 3. Similar aspects of motion systems reflected in content sources stemming from different times

Starting with the 1990s, the content was often generated and archived in digital file formats. That means for the thinkMOTION workflow, that the digitization process can be skipped. Nevertheless the content sources must be converted into web compliant derivatives according to their format and size. There are some problems occasionally with reading and converting old file formats.

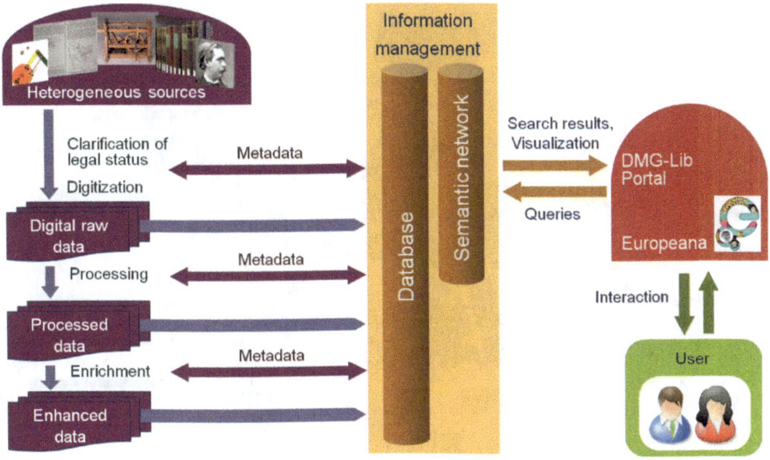

Fig. 4. Overview of the thinkMOTION workflow

Figure 4 gives a simplified overview about the workflow showing the different processing steps, the input and output of each step, such as the *Digital raw data*, the *Processed data* and the *Enhanced data* as well as the role of the database in which the meta data, the raw and the portal ready content data are stored and the DMG-Lib portal, which allows the user to interact with the database.

The workflow starts with the locating and selecting of relevant content sources.

The next step is obtaining the rights of use, which is detailed described in chapter 3.1. Then the content sources have to be provided for the next workflow steps. Libraries, universities and institutions as well as authors from various European countries provide us content sources and support the project by this way substantially.

All content sources which are not yet in a digital format must be digitized. Figure 5 shows some examples of the equipment for the digitalization process which is used in the thinkMOTION project.

The results of the digitization process are digital raw data which are not suitable for the presentation in an online portal. They must be processed in workflow steps which are different depending on the type of content source or media. For paper based documents these necessary steps are: The document or book pages must be cropped, horizontal adjusted, the background must be cleaned and the text has to be recognised (OCR) for the later full-text search. For slides and photos the following steps must be done: The image section with the relevant content must be chosen, the image must be cropped and the image quality has to be improved. With taken image sequences, e.g. from physical models, the same steps analogous to slides and photos have to be done.

Fig. 5. Digitizing equipment used in the thinkMOTION project for books, slides and image sequences of physical models

Additionally, video files or interactive animations must be generated from these image sequences.

In some cases the content is not easy to understand. Such type of content can be enriched with additional information to become clearer. Although content enrichment is very useful, the process of enrichment is often very expensive and can be done only for selected, important content.

An example of enrichment is the overlay of an image of a mechanism model with the results of a simulation tool (Figur 6). This is a useful facility to increase the understandability significantly and to provide an added value for the user of the DMG-Lib portal.

At the end of the thinkMOTION workflow, the content data must be converted into a web compliant data file format and have to be uploaded to the DMG-Lib online portal [2].

The core of the information management of the DMG-Lib portal is a mySQL database which is managed all the digital content, the meta data and processing data. All partners of the thinkMOTION project can access to this database to enter meta data or to upload content.

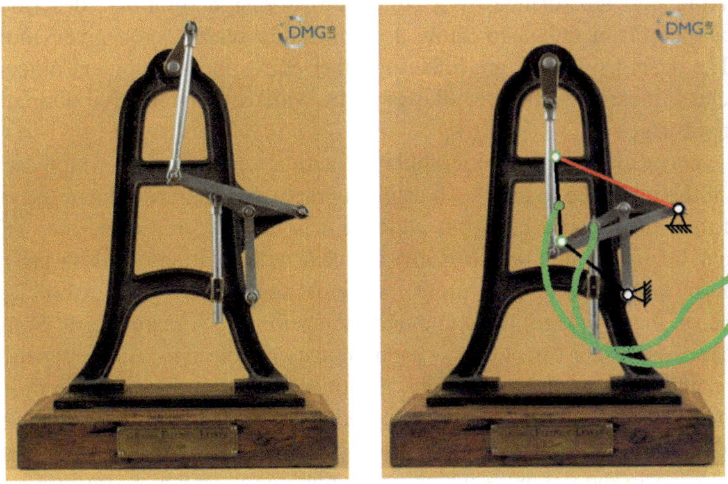

Fig. 6. Overlay of an image of a mechanism model with the results of the simulation tool MASP (Modelling and Analysis of Solution Principles) as an example of enriched content.

The content is distinguishable into four classes (Figure 7). The main class consists of text and image based content sources such as books, articles, photos, animations, videos with their descriptive meta data.

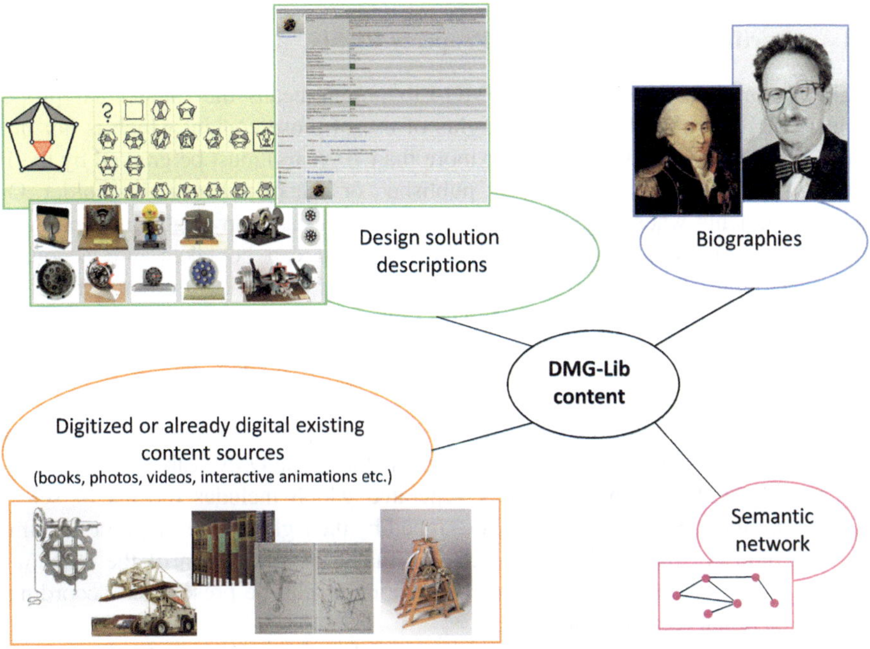

Fig. 7. Classes of content of the DMG-Lib portal

Another big part are formed by a collection of design solutions, described with mechanism specific metadata to allow a comfortable search for a given movement problem. The design solutions are linked to other corresponding content objects from the main class, mentioned before, like pictures, videos or interactive animations as well as text passages.

A third content class are the biographies of author and scientists, which are linked to objects of the main class again such as photographs and publications of the person.

In the future a semantic network, which exists already as a topic map of 6,500 items in four languages and which is cross-linked with relations, will be provide the user a more comfortable and more efficient search over the whole content independent on the language, the name and the spelling of a search term. Sometimes several authors uses their own names synonymously for the same thing, depending on their own point of view, their language, the time were they lived or the opinion of their own teacher's. A simple search can find one search term only, perhaps some more with a little variance in the spelling, but a semantic network can find all references stored in the database which are synonymous or which are in correlation with the search term. The semantic network will forms the fourth class of content.

The content of the database which is accessible free over the DMG-Lib portal will be available also over the Europeana, the portal of the growing European digital library [5].

3 Important Aspects and Functions of DMG-Lib

3.1 Clarification of Intellectual Property Rights

Making content available in the DMG-Lib portal means the intellectual property rights (IPR) must be clarified. The rights of use of those content which is not in the public domain (i.e. author is dead for more than 70 years) must be granted by the IPR holder who can be an author, editor, publisher, or heir of a dead rights holder. Only those documents whose legal situation concerning usage/copy rights is clear can be digitised, made online available in DMG-Lib and can be taken into account as Europeana item. The process of clarifying the rights of use is an important and time consuming step in the thinkMOTION workflow and can be described as follows:

- Each partner identifies and contacts rights holders to get their lists of publications and further contact data.
- The metadata of the rights holders and their publications are entered into the database and can be made online available in the DMG-Lib.
- The Letter of Agreement is generated which includes the list of selected publications. The letter will be signed by the rights holder to grant the simple rights of use to DMG-Lib e.V. To make the generation of the letter more effective different languages and use cases can be preselected according to national requirements.
- The handling of returned letters (archiving of digitized signed letters, activating referenced documents, etc.) is also supported in the database.

Every partner in thinkMOTION consortium has appointed a person in charge for IPR issues. They are responsible to manage the workflow and make the documents online available.

3.2 Multilingualism

The multidisciplinary thinkMOTION consortium of partners in five European countries (Germany, Spain, Italy, Romania, and France) is an excellent base for the proposed multilingual extension of the existing DMG-Lib portal. The digital library portal's translation (English, German, Italian, Spanish, Romanian, and French) is carried out by the different partners to reach as many as possible users in Europe.

To provide specific search functionalities (e.g. queries can be complemented automatically with multilingual synonyms) it is necessary to expand an existing synonym list by additional European languages to a multilingual synonym list for motion systems specific terms. So the users can search in their own language and will get back a result with additional content in other languages. As a further aspect this expanded multilingual synonym list can support the domain specific terminology standardization process in Europe and can deliver a contribution to a better cooperation of the European mechanical engineering industry. Because good translations are usually quite expensive it is necessary to focus on the most important parts. The translation of the DMG-Lib portal is made for:

- static texts used in the web pages like texts for navigation elements (headlines, menus, buttons etc.) and help texts,
- titles of stored items (books, articles, images, videos etc.), and
- metadata fields of items where the set of possible values is small.

This enables users to navigate in a language they understand as well as to find matching content containing texts in other languages. If the non-textual parts of the content (e.g. images, animations, formulas, diagrams) dominate, then the user will probably derive benefit from foreign-language search results too. But even the textual content could be understood based on a raw automatic translation (e.g. provided via web translation services) and the usage of special dictionaries like the IFToMM dictionary [6]. In the future even the workflow for maintaining the IFToMM dictionary shall be based on the translation technology implemented in the DMG-Lib platform [4].

3.3 Applications for Work Environment

DMG-Lib may also serve as the basis for further applications. For instance, the database collects files describing kinematic structures that can be downloaded from DMG-Lib and used with software tools for kinematic modeling and analysis (examples in Figure 8). On the other hand, structures modeled with these tools may be uploaded to form the basis of new solution descriptions in the database.

Another group of applications are research environments that help structuring the information space of motion systems. These environments allow the creation of individual information collections regarding a topic or a project and facilitate communication within research teams. They may also incorporate design and analysis software tools to support the development process. Examples are the experimental myDMG-Lib and the newly started project eKinematix.

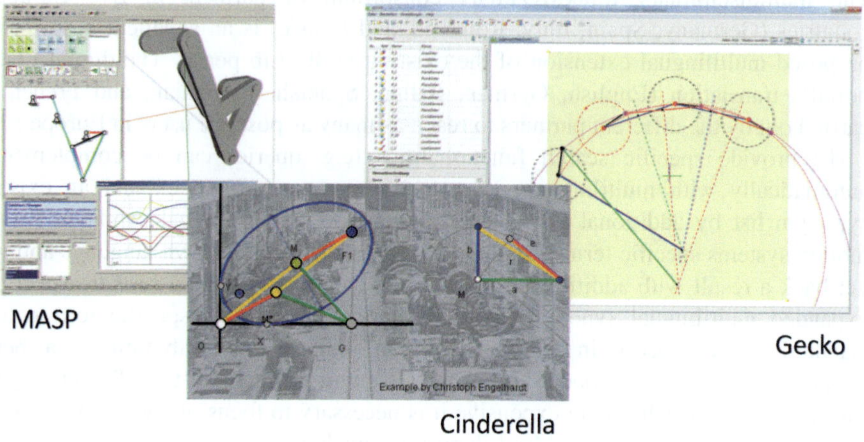

Fig. 8. Examples for software tools that can import solutions downloaded form DMG-Lib (Gecko, MASP) or display them online (Cinderella)

3.4 Searching and Browsing Functions in Heterogeneous Collection

Searching and browsing are the key functions to access all information in the DMG-Lib. For both variants the text-based search is a basic functionality to find all kinds of information from the different sources. DMG-Lib also offers a special mechanism search which allows to search in the three categories: typology of mechanism, guidance function, and transfer function (see tabs in Figure 9).

This allows specific queries for instance by specifying the number of mechanism links, input movement, or the mechanism's dimension (planar, spherical, or spatial). For the representation of search results special options like list or gallery view are available and users can sort result lists by date or refine the results by categories like interactive animations of mechanism descriptions.

Parallel to the mentioned interface other interactive multimedia access applications are developed like the timeline "Mechanisms and machines by the course of time", the "Virtual DMG-Lib Museum", and the hypervideo "Everyday mechanisms". Beside traditional browsing and retrieval methods these applications provide alternative ways to access the knowledge stored in the library especially for users like students or historians.

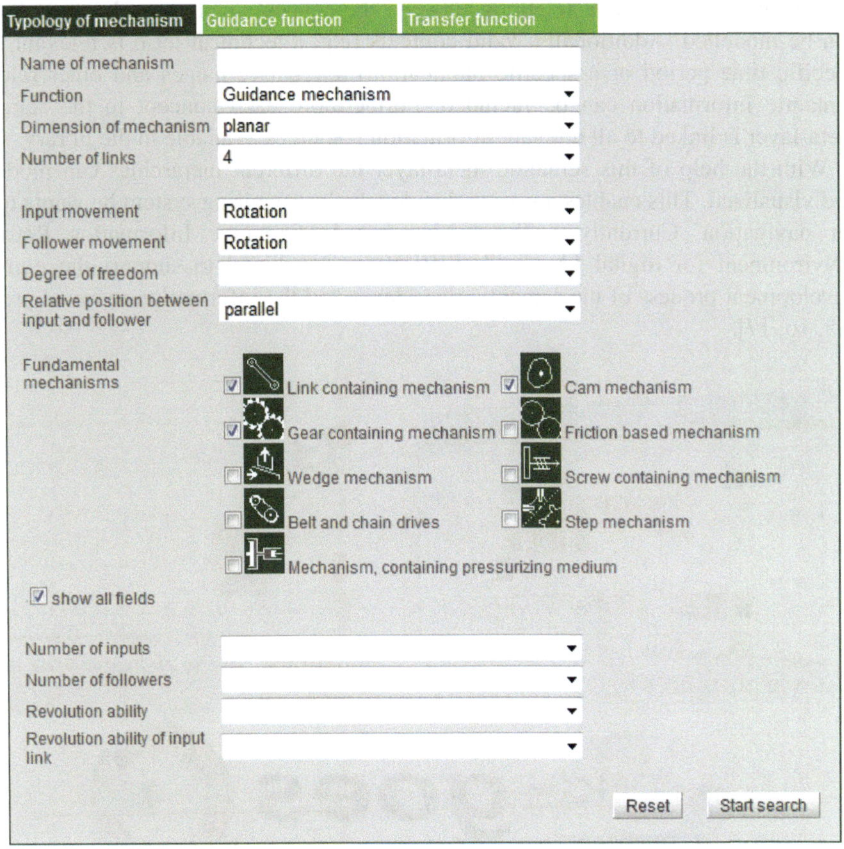

Fig. 9. Example for a text-based search for mechanisms

3.5 Semantic Information Retrieval

A further field of research is the retrieval in heterogeneous information resources using different mechanism and gear hierarchies like the structural system of Reuleaux [11] or other classification systems of well-known publications (e. g. [12]). A visualisation and an efficient navigation over these classifications could help users to get a systematic overview over the huge amount of existing mechanisms and gears. However, the identification and modelling of these classifications and relations between the different technical terms are quite complicated, because for example different opinions of experts and authors, synonyms and foreign language synonyms have to be considered.

To solve this problem semantic web technologies can be used. With the help of Topic Maps, as a special kind of semantic networks, the knowledge of mechanism and gear science can be generalized and explicit modelled in a semantic meta-layer [13, 14]. With the extensive descriptive power of Topic Maps, all relevant concepts (like technical terms) and relations between the concepts of this application domain can be modelled. Additionally, valid contexts (e.g. a technical term is relevant for a specific time period or a specific publication) alternative names and other relevant semantic information can be included. Furthermore each concept in the semantic meta-layer is linked to all relevant information resources available in the library.

With the help of this semantic meta-layer the different hierarchies can modelled and visualized. This enables a user to decide which structuring system he wants to use for navigation. Currently a Topic Map based "Semantic Information Retrieval ENvironment for digital libraries" (SIREN) is developed to support the complex development process of the semantic meta-layer and the information retrieval process [15, 16, 17].

Fig. 10. An essential goal of thinkMOTION: making the content of DMG-Lib available in Europeana

4 Conclusion

The contribution explains that the ambitious goals of thinkMOTION, in particular the provision of at least 60,000 information items, can only be achieved by a combination of qualified staff, a well-structured workflow and optimized tools. The groundwork, done in project DMG-Lib, was vital to establish a workflow and to develop new tools that allow the efficient clarification of rights, procurement, digitization, processing, and enhancement as well as quality assurance and web-based presentation of very heterogeneous sources (books, drawings, photos, demonstration models).

New possibilities for the users arise from the implementation of a multi-lingual access for browsing and searching DMG-Lib. The web-based database supports users who want to contribute information sources by providing means to monitor progress and retrace each work step for each information object anytime.

The thinkMOTION team invites everyone who is interested in supporting the idea of free online access to research and education information. Please help us with locating and providing interesting information sources to be made available for everyone via DMG-Lib and associated internet portals like Europeana (Figure 10).

Acknowledgments. thinkMOTION is funded under the Information and Communication Technologies Policy Support Programme; Area: CIP-ICT-PSP.2009.2.3 – Digital Libraries: European Digital Library Digitising content for Europeana; Funding period: 01.06.2010 – 31.05.2013.

References

[1] Brecht, R., Krömker, H., Kühlewind, A.: Multimedia E-Books for Digital Mechanism and Gear Library. International Journal of Human and Social Sciences 4(3), 193–197 (2009), http://www.waset.org/journals/ijhss/v4/v4-3-25.pdf

[2] Website of the Digitalen Mechanismen- und Getriebebibliothek: http://www.dmg-lib.org

[3] Website of DMG-Lib e.V.: http://www.dmg-lib.org/association

[4] Döring, U., Brix, T., Corves, B.: Suggestion for a More Productive Workflow and Infrastructure of the Permanent Commission on Standardization of Terminology. In: Proceedings of 23rd Working Meeting of the IFToMM Permanent Commission for Standardization of Terminology on MMS, Minsk/Gomel, Belarus, June 21-26, pp. 49–57 (2010), http://www.dmg-lib.org/dmglib/handler?docum=10677009

[5] Website of Europeana: http://www.europeana.eu

[6] IFToMM online dictionaries, Version 2.3 (April 2010), http://www.iftomm.3me.tudelft.nl

[7] IFToMM website: 130.15.85.212

[8] Lucene website: http://lucene.apache.org

[9] thinkMOTION webseite: http://www.thinkmotion.eu

[10] Niemeyer, J.: Methodische Entwicklung von Prinziplösungen bei der Auslegung ungleichmäßig übersetzender Getriebe unter Verwendung eines praxisorientierten interaktiven Wissensspeichers. Shaker Verlag, Aachen (2003)

[11] Reuleaux, F.: Lehrbuch der Kinematik. Vieweg, Braunschweig (1875)

[12] der Technik, K.: Begriffe und Darstellungsmittel der Mechanismentechnik (engl.: Technical terms and representations in mechanism theory). Kammer der Technik, Suhl (1978)

[13] Park, J., Hunting, S.: XML Topic Maps: Creating and using topic maps for the web. Pearson Education Inc., New Jersey (2003)

[14] Pepper, S.: The TAO of Topic Maps (2000), http://www.ontopia.net/topicmaps/materials/tao.Html

[15] Goodall, G.A.: A Time for Digital Libraries (2003), http://www.deregulo.com/facetationpdfs/aTimeForDigitalLibraries.Pdf

[16] Markscheffel, B., Thomas, H., Stelzer, D.: Merlino – a Prototype for semi automated Generation of Occurrences in Topic Maps using Internet Search Engines. In: 3rd European Semantic Web Conference (ESWC 2005), Heraklion, Greece, May 29-June 01 (2005), http://topic-maps.org/lib/exe/fetch.php?cache=cache&media=member%3Aht%3Amerlinoabstract_eswc2005_greece.pdf
[17] Website of Open Archive Initiative: http://www.openarchives.org/

Author Index

9789400741317